Fundamentals of Intracellular Calcium

by Anthony K. Campbell

School of Pharmacy and Pharmaceutical Sciences, Cardiff University
and
Welston Court Science Centre

Registered Offices
John Wiley & Sons, Inc., 111 River Street, Hoboken, NJ 07030, USA
John Wiley & Sons Ltd, The Atrium, Southern Gate, Chichester, West Sussex, PO19 8SQ, UK

Editorial Office
The Atrium, Southern Gate, Chichester, West Sussex, PO19 8SQ, UK

For details of our global editorial offices, customer services, and more information about Wiley products visit us at www.wiley.com.

Wiley also publishes its books in a variety of electronic formats and by print-on-demand. Some content that appears in standard print versions of this book may not be available in other formats.

Library of Congress Cataloging-in-Publication Data

Names: Campbell, Anthony K., author.
Title: Fundamentals of intracellular calcium / by Anthony K. Campbell.
Description: Hoboken, NJ : John Wiley & Sons, 2018. | Includes index. |
   Identifiers: LCCN 2017017745 (print) | LCCN 2017035567 (ebook) | ISBN
   9781118941881 (pdf) | ISBN 9781118942024 (epub) | ISBN 9781118941874 (pbk.)
Subjects: LCSH: Intracellular calcium.
Classification: LCC QP535.C2 (ebook) | LCC QP535.C2 C267 2017 (print) | DDC
   572/.516–dc23
LC record available at https://lccn.loc.gov/2017017745

Cover Design by: Wiley
Cover Images: Courtesy of Anthony K. Campbell

Set in 10/12pt Warnock by Aptara Inc., New Delhi, India

Printed and bound in Singapore by Markono Print Media Pte Ltd

10 9 8 7 6 5 4 3 2 1

Four pioneers who initiated the explosion into investigating intracellular calcium as a universal regulator. (a) Sydney Ringer (1836–1910). © The Royal Society. (b) Otto Loewi (1873–1961). Reproduced with permission from Wellcome Library, London. (c) Lewis Victor Heilbrunn (1892–1959). Reproduced with permission from University of Pennsylvania Archives; (d) Setsuro Ebashi (1922–2006) from Endo, 2006. Reproduced with permission from Nature.

This book is dedicated to my wife Stephanie

Thanks for everything

Contents

About the Author

Professor Anthony K. Campbell MA, PhD, FLS, FLSW

http://www.cardiff.ac.uk/people/view/90830-campbell-anthony
campbellak@cf.ac.uk; tony@theyoungdarwinian.com

 @tonylumi

 Anthony Campbell was born in Bangor, North Wales, in 1945, but grew up in London, attending the City of London School. He obtained an exhibition at Pembroke College, Cambridge, and then a first class degree in Natural Sciences, and a PhD in Biochemistry at Cambridge University. He moved to Cardiff as lecturer in Medical Biochemistry at the then Welsh National School of Medicine in 1970, and then Professor in Medical Biochemistry, followed by Professor in the School of Pharmacy and Pharmaceutical Sciences at Cardiff University. He has studied intracellular calcium as a cell regulator for over 40 years, pioneering the application of Ca^{2+}-activated photoproteins to measure free Ca^{2+} in live animal, plant, bacterial and archaeal cells. He is a world authority on bioluminescence, developing the use of genetically engineered bioluminescence to measure chemical processes in live cells. One of his inventions, using chemiluminescence, is now used in several hundred million clinical tests per year worldwide, was awarded the Queen's Anniversary Prize in 1998, and was selected by the Eureka project of Universities UK in 2006 as one of the top hundred inventions and discoveries from UK Universities in the past 50 years. For the past 15 years his research focus has been lactose and food intolerance, which has led to a new hypothesis on the cause of irritable bowel syndrome, and the mystery illness which afflicted Charles Darwin for 50 years, but was never cured. He is now investigating the relevance of this hypothesis to the current diabetic epidemic, Parkinson's and Alzheimer's disease. He has published 10 books, and over 250 internationally peer-reviewed papers on intracellular calcium, bioluminescence, lactose and food intolerance. Several of his patents have been exploited throughout the world.

Anthony believes passionately in communicating science to the public, and in exciting pupils and students about natural history and cutting edge science. This led him to found the Darwin Centre (www.darwincentre.com) in 1993, now in Pembrokeshire. He also founded the Public Understanding of Science (PUSH) group at Cardiff University in 1994, which organises many events with schools and the public. He has had a laboratory in his house since he was 11 years old. In 1996 he used his patent income to set up Welston Court Science Centre in Pembrokeshire, which is a facility to support the Darwin Centre. He has given regular talks on food intolerance, Darwin, and bioluminescence, at scientific meetings, to schools and the public. He won the Inspire Wales award for Science and Technology in 2011. He is a Fellow of the Linnean Society, and a foreign member of the Royal Society of Sciences in Uppsala, Sweden. In 2013, he was elected a Fellow of the Learned Society of Wales, and to the Council of the

Linnean Society. In 2016 he set up The Young Darwinian, an international journal for school students to publish their projects and scientific experiences (www.theyoungdarwinian.com). He has been a keen musician all his life, as a tenor soloist, conductor and viola player. Now he is developing a project 'DNA sings' to convert light into music. He also makes music in the kitchen, as a keen cook. As a young scientist he was keen as a University and County bridge player, and now, after 40 years, he has started playing again. He has a wife, Stephanie, five amazing children, and four beautiful grandchildren.

Preface

> *I keep six honest serving-men*
> *(They taught me all I know);*
> *Their names are What and Why and When*
> *And How and Where and Who.*
> Rudyard Kipling, *Just So Stories* (1902)

The story of intracellular calcium is a marvellous, inspiring example of how the curiosity of thousands of scientists has led to an understanding of one of the most important regulatory systems in the whole of life – calcium inside cells. Without intracellular calcium we would not have been conceived, born, or remained alive. Curiosity about calcium has catalysed the ingenuity of scientific inventors, who have given us a wide range of molecular, electrophysiological, microscopical, and imaging techniques, which have revolutionised biological and medical research. This has led to breakthroughs in understanding killer diseases, such as heart attacks and cancer, and the development of drugs to treat them. This, quite surprisingly, has produced multi-million dollar markets, with benefits to the world economy and the creation of high-technology jobs. Yet, there has been no Nobel Prize for intracellular Ca^{2+}. Too many people have made seminal contributions and have made major discoveries.

There have been dozens of multi-author books on intracellular calcium published since my first book, *Intracellular Calcium: Its Universal Role as Regulator*, was published by Wiley in 1983 (Figure 1.1), but no single author ones. Multi-author books provide detailed information on highly focussed topics written by world experts. A single-author book offers the opportunity to develop themes within and between chapters. It also allows the author to develop individual creativity, whilst still retaining the consensus view. So, in 2015, a new *Intracellular Calcium* was published, with over 3000 references. *Fundamentals of Intracellular Calcium* is a student edition, focussed on key questions and discoveries, with learning objectives, and recommended reading. The evidence is documented that intracellular calcium is a universal regulator of cell events in all animals, plants, fungi and microbes. *Natural history* is about describing *what goes on* in the Universe. *Natural science* is about understanding *how* the Universe works. This book brings together these two essential approaches, and sings the music of intracellular calcium. As a founder of the renaissance, Albrecht Dürer (1471–1528), wrote 'Be guided by Nature and do not depart from it thinking you can do better yourself. You will be misguided, for truly art is hidden in Nature and he who can draw it out possesses it'.

Everywhere you look, smell, taste, hear, and feel, intracellular calcium is involved. This book is about the biological molecular mechanisms that determine these. But, it focuses first on the real problems that nature has given us. What really matters is not what happens to an artificial tissue culture cell system in the laboratory, but rather how cells in nature work. Thus, throughout I have related Ca^{2+} signalling to the natural physiology and pathology of the cells involved. Two key scientific principles run throughout. First, how intracellular Ca^{2+} acts as a switch, to activate a wide range of cellular events, and how analogue mechanisms can be superimposed

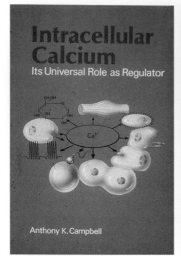

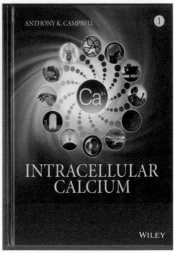

Intracellular Ca²⁺ as
a universal signal
1983

Digital versus analogue signals in
living systems
1994

Intracellular Calcium
2015

Figure 1 Three books by the author. (a) Intracellular Calcium: its universal role as regulator (Campbell, 1983). Front cover reproduced with permission of John Wiley & Sons. (b) Rubicon: the fifth dimension of biology (Campbell, 1994). Front cover reproduced with permission of Gerald Duckworth & Co. Ltd. (c) Intracellular Calcium (Campbell, 2015). Front cover reproduced with permission of John Wiley & Sons.

on this digital signalling process, to alter the timing and strength of the cell event. Secondly, in the tradition of Charles Darwin and Alfred Russel Wallace, the molecular biodiversity of the components of the Ca²⁺ signalling system is highlighted, upon which their BIG idea of evolution by Natural Selection critically depends. These themes are a development of my previous books (Figure 1). *Rubicon: The Fifth Dimension of Biology* provided evidence that life, throughout 4000 million years of evolution, has depended critically on digital events in cells, organisms and ecosystems.

There are 13 chapters. Chapter 1 arouses curiosity about why calcium is special inside cells. Chapter 2 lays down key principles, and identifies how we name things. Chapter 3 shows how the story of intracellular calcium evolved from Ringer's famous experiments on frog heart at the end of the nineteenth century to the present day. Chapter 4 shows how we investigate intracellular Ca²⁺. Chapter 5 explains how Ca²⁺ is regulated inside cells. Chapter 6 describes how cells have exploited the unique chemistry of Ca²⁺. Chapters 7, 8 and 9 deal with the cellular events in animal, microbial and plant cells, which are triggered by intracellular Ca²⁺. Chapters 10 and 11 show how intracellular Ca²⁺ is important in medicine and drug action. Chapter 12 focuses on the fascinating problem of the evolution of Ca²⁺ signalling. The final chapter summarises what we know and what we do not know about intracellular Ca²⁺. I also discuss the importance of intracellular calcium in curricula at school and university, and why it is important for professional scientists to engage with schools and the public. Intracellular calcium is poorly dealt with in schools, and even many university curricula. I believe scientists and engineers must engage with schools, the public, and even politicians. This is crucial for the development of our culture and economy. We need to explain how curiosity has led to the major discoveries and inventions that have revolutionised all of our lives. When I give talks to schools or the public I often start by asking the audience what do they think is the greatest gift evolution has given us? For me, this is *curiosity*. We are the most curious organisms on this planet. I have even been labelled the 'curious Professor'!

Each chapter has a recommended selection of books, reviews and articles. A fuller list is available in the companion web site as a pdf and Endnote library, and *Intracellular Calcium, 2015*. If I have omitted a key paper, or made a mistake, do please email me, and I will try to add these to a web page and any further editions.

There are many people thank. First, my amazing wife, Dr Stephanie Matthews, with whom I have collaborated for over 30 years, and my five children, David, Neil, Georgina, Emma and Lewis, who have been a great inspiration. My mother died before this book was completed, and was a major force in my life. My sister too, Professor Caroline Sewry, who gave me some microscopy pictures for the book. Our dear, late mother, Jennet Campbell gave us our musical genes, and was amazed that she had produced two science Professors. I am grateful to several colleagues, who have given me useful comments about the book, particularly Barry Holland, Paul Luzio, Valerie Morse, Ken Siddle, Tony Trewavas FRS, and Ken Wann. But any errors or omissions are my responsibility. I also thank all those who have worked so hard in Pembrokeshire to make the Darwin Centre such a success there, and the many members of my group over 40 years who helped me investigate intracellular calcium and bioluminescence. Many people at Wiley have worked hard to make this book a success. I thank my editor Jenny Cossham, Sarah Tilley Keegan, and Beth Dufour of RSSP for her vital work on copyright permissions. I have been lucky enough to have my research funded from a wide range of sources. I thank particularly the MRC, BBSRC (formally SRC and AFRC), NERC, The Wellcome Trust, The Arthritis and Rheumatism Council, The Multiple Sclerosis Society, The British Diabetic Association, The Waterloo Foundation and The Royal Society.

Curiosity inspires, discovery reveals.
Bon appétit.

Anthony K. Campbell
November 2016

List of Acronyms

Acronym	Meaning
5-HT	The neurotransmitter 5-hydroxytryptamine, also known as serotonin
ABC	ABC transporters bind MgATP, and form a very large family of membrane proteins in all organisms. They transport nutrients and also get rid of toxic molecules
ABP	Androgen-binding protein, a glycoprotein produced in the testis that binds specifically testosterone, dihydrotestosterone, and 17-β-estradiol
ABRM	Anterior byssal retractor muscle in mussels that keeps them attached to a substrate. It is a catch muscle, like the adductor muscle that keeps bivalves shut, when attacked or when the tide goes out.
ACh	The neurotransmitter acetylcholine that triggers skeletal muscle to contract
ACE	Angiotensin-converting-enzyme inhibitor, a drug used to treat high blood pressure and heart failure
ACTH	Adrenocorticotropic hormone
ADP	Adenosine diphosphate
AIDS	Acquired immune deficiency syndrome caused by human immunodeficiency virus (HIV)
AIF	Apoptosis inducing factor
AKAPS	A-kinase anchor proteins are scaffold proteins that all bind R, the regulatory subunit of protein kinase A (PKA)
ALT	Alanine transaminase
AM ester	Acetoxymethyl ester used to get fluorescent dyes into live cells
AM/FM	Amplitude modulation/frequency modulation
AMP	Adenosine monophosphate
AMPA	α-Amino-3-hydroxy-5-methyl-4-isoxazolepropionic acid, an artificial glutamate analogue
AMPR/AMPAR	α-Amino-3-hydroxy-5-methyl-4-isoxazolepropionic acid receptor, a non-NMDA-type ionotropic transmembrane receptor for glutamate. It mediates fast synaptic transmission in the central nervous system. AMPR is activated by the artificial glutamate analogue AMPA
ASK 1	Apoptosis signal-regulating kinase 1 (mitogen-activated protein kinase kinase kinase 5 – MAP3K5)
AST	Aspartate transaminase
AtCAX	*Arabidopsis* cation exchanger tonoplast-localized, divalent cation/H^+ antiporters in plants

ATF	Activating transcription factor, group of bZIP transcription factors
ATP	Adenosine triphosphate
BAK	Bcl-2 homologous antagonist/killer
BAPTA	1,2-bis(o-Aminophenoxy)ethane-N,N,N',N'-tetraacetic acid, a Ca^{2+} specific chelator, with pK_a for third and fourth H = 5.47 and 6.36. Thus not affected by pH in cells
BAX	Bcl-2-like protein 4. Like BAK, BAX can penetrate the mitochondrial membrane to initiate apoptosis
BAY K8644	A calcium channel agonist used primarily in research, and is an analogue of nifedipine
BCECF	2',7'-bis(2-carboxyethyl)-5-(and-6)-carboxyfluorescein, pH indicator inside live cells. $pK_a = 6.98$; wavelengths, ex 440 nm (isosbestic) and 490 nm, em 535 nm. Measure ratio
BCl2	B-cell lymphoma protein 2, on the outside of mitochondria and important in apoptosis
BFP	Blue fluorescent protein, originally the product of aequorin when triggered by Ca^{2+}, but now used for blue fluorescent protein (EBFP), a mutant of the green fluorescent protein. Ex 383 nm, em 445 nm
BLAST	Basic Local Alignment Search Tool, software that finds regions of amino acid similarity between protein sequences
BRET	Bioluminescent resonance energy transfer, the process whereby aequorin transfers electronic excitation to GFP
C1–9	Complement components 1 to 9
C2 domain	A high affinity Ca^{2+} binding domain created in a protein by a beta-sheet sandwich, e.g. protein kinase C has a C2 domain made from eight β-strands that binds two or three calcium ions. Many C2 proteins interact with membranes
CaCA	Ca^{2+} cation antiporter. The CaCA family exchange Ca^{2+} for another cation, e.g. Na^+, across membranes. Ca^{2+}:H+ Antiporter-2 is CaCA2. CaCA's are found in all organism domains – Bacteria, Eukaryota, Archaea
CACHNA	Voltage-gated Ca^{2+} channel genes: Ca_v 1.1 = S,1. 2 = C, 1.3 = D, 1.4 = F (DHP/HVA); Ca_v 2.2 = A (P/Q – HVA), 2.2 B (N, HVA), 2.3 E (R, IVA); Ca_v 3.1 G, 3.2 H, 3.3 I (T/HVA)
cADPR	cyclic ADP ribose
CALM	Clathrin assembly lymphoid myeloid leukaemia protein is found in endocytic-coated pits, and interacts with clathrin
CALN	Calcineurin (CaN), the Ca^{2+} and calmodulin activated serine/threonine protein phosphatase, also known as protein phosphatase 3, or calcium-dependent serine-threonine phosphatase. It activates the T cells by activating the nuclear transcription factor, which starts in the cytosol (NFATc) by dephosphorylating it
CaM	The high affinity Ca^{2+} binding protein calmodulin, found in all eukaryotic cells, but not bacteria. Humans have three isoforms
CAM	Cell adhesion molecules (CAMs) are proteins located on the cell surface involved in binding with other cells or with the extracellular matrix (ECM)
CaMK.CAMK	Ca^{2+}-calmodulin-activated protein kinase that phosphorylates, from MgATP, serine or threonine residues in proteins
CaLM	Calmodulin-like protein. Humans have six isoforms

CAML/CALMG	Calcium modulating ligand, also known as calcium-modulating cyclophilin ligand, a signalling protein recognized by the TNF receptor TACI
CALP	Calsenilin-like protein (CALP/KChIP), an EF-hand protein that interacts with presenilin 2 and voltage-gated K^+ channel subunit Kv4
cAMP	Cyclic adenosine monophosphate
CAP	Catabolite activator protein, also known as cAMP receptor protein, CRP, a transcriptional activator
CARP	Carbonic anhydrase-related protein
CASQ	Calsequestrin, the major Ca^{2+} binding protein in muscle SR, -1 in fast skeletal, -2 in slow skeletal and heart muscle
CAX	Carbonic anhydrase X protein
CBARA 1	Calcium-binding atopy-related autoantigen (calcium uptake protein 1, mitochondrial) = MICU1, the gatekeeper for MCU for Ca^{2+} uptake by mitochondria
CBL	Casitas B-lineage lymphoma protein. It is E3 ubiquitin-protein ligase involved in cell signalling and protein ubiquitination
CBL	Calcineurin B-like protein. The family are Ca^{2+} sensors that interact with the serine/threonine kinases CBL-interacting protein kinases (CIPKs). They are important in plants
CCCP	Carbonyl cyanide *m*-chlorophenyl hydrazine. It inhibits oxidative phosphorylation in mitochondria, gradually destroying cells
CCD	Charge-coupled device
CCK	Cholecystokinin, previously pancreozymin, a peptide hormone stimulating the exocrine pancreas to release digestive enzymes of fat and protein into the small intestine
cCMP	Cyclic cytosine monophosphate
CD	Circular dichroism, a physical technique to look at changes in the 3D structure of proteins
CDC/cdc	Cell division cycle protein
CDPK	Calcium-dependent protein kinase that phosphorylates serine and threonine residues in proteins, important in plants and protozoans, but not animals, being different from CAMKs
CFA	Colonization factor antigen I subunit D or CFA/I fimbrial subunit D in bacteria
CFA	Cyclopropane-fatty-acyl-phospholipid synthase
CFP	Cyan fluorescent protein = ECFP
CFTR	Cystic fibrosis transmembrane conductance regulator, a Cl^- channel allowing Cl^- to move ions in and out of cells, important for a salt and water balance on epithelia in the lungs or pancreas. CFTR is mutated in cystic fibrosis
cGMP	Cyclic guanosine monophosphate
CGP37157	Inhibits mitochondrial Na^+/Ca^{2+} exchange and SERCA MgATPase, and possibly calcium channels
CHOP	CCAAT-enhancer-binding protein homologous protein. It a transcription factor induced by ER stress and causes apoptosis
CICR	Ca^{2+}-induced Ca^{2+} release. This occurs via the ryanodine receptor 2 in the heart
CIPK	CBL-interacting protein kinases (CIPK). They use Ca^{2+} signals in plant stress
CK	Creatine kinase

CLUSTAL	Computer programs for multiple sequence alignment, widely used to line up amino acid sequences when comparing proteins
CMP	Cytosine monophosphate
CNG	Cyclic nucleotide-gated ion channels, important in vertebrate eyes
COS-1 and -7	Immortalised fibroblast-like cell lines, from monkey kidney tissue, made from CV-1 (simian) in Origin, carrying the SV40 genetic material, hence COS
COX	Cyclooxygenase – prostaglandin-endoperoxide synthase (PTGS), an enzyme responsible for prostanoid synthesis, such as thromboxane and the prostaglandin prostacyclin
CPA	Cyclopiazonic acid
CPVT	Catecholaminergic polymorphic ventricular tachycardia, causing cardiac problems, and death on exercise. It is caused by mutations in RYR2 (autosomal dominant), CASQ2 (autosomal recessive), encoding calsequestrin, TRDN (autosomal recessive), encoding triadin, that interacts with CASQ2 and RyR2 in the SR; and CALM1 (autosomal dominant), encoding calmodulin
CRAC	I_{CRAC} – Ca^{2+} release-activated Ca^{2+} current, or channel, activated by release of Ca^{2+} from the ER as part of SOCE
CRAM1	Calcium release-activated calcium channel protein 1, now called Orai 1
CRAP	Chemiluminescence recovery after photobleaching
CREB	cAMP response element binding protein, a transcription factor regulated by protein kinase A phosphorylation
CRET	Chemiluminescence resonance energy transfer – transfer of electronic excitation by the Förster mechanism
CRP	Cyclic AMP receptor protein
CTP	Cytosine triphosphate
CUP	Cup protein (Oskar ribonucleoprotein complex 147 kDa subunit) plays a crucial role in localization of transcripts in the oocyte and in young embryos of the fruit fly *Drosophila*
CVA	Cerebrovascular accident = a stroke
CYCLOPS	Copy number alterations yielding cancer liabilities owing to partial loss of genes
DACC	Dorsal anterior cingulate cortex in the brain associated with conscious experience
DAG	Diacylglycerol, a product of phosphoinositide hydrolysis by phospholipase C, the other being IP_3
DAPK	Death-associated protein kinase
DCCD	*N,N'*-Dicyclohexylcarbodiimide a chemical used to couple amino acids in artificial peptide synthesis, and can inhibit processes in chloroplasts and mitochondria
DCMU	3-(3,4-Dichlorophenyl)-1,1-dimethylurea, a herbicide that is a specific and potent inhibitor of photosynthesis
DDT	Dichlorodiphenyltrichloroethane, a widely used insecticide
DENV	Dengue virus
DHP	Dihydropyridine
DIDS	4,4'-Diisothiocyano-2,2'-stilbenedisulfonic acid. It inhibits anion exchangers, such as the Cl^-/HCO_3^- exchanger

DIF	Dorsal related immunity factor, a protein in the cytosol during embryo cleavage.
DINS	Drug Identification Numbers
DISI	Discovery, Invention, Scholarship, Impact
DMB	3′,5′-dimethoxybenzoin
DMNPE-4	Tetrasodium-2-[2-[2-[2-[bis(carboxylatomethyl)amino]-1-(4,5-dimethoxy-2-nitrophenyl)ethoxy]ethoxy]ethyl-(carboxylatomethyl)amino]acetate (dimethoxy-*ortho*-nitrophenyl) used to cage Ca^{2+}, EGTA, ATP and luciferin, these being releasable by photolysis
DMSO	Dimethyl sulphoxide
DNA	Deoxyribonucleic acid
DNSpr	Disk large tumour suppressor
DNP	2,4-Dinitrophenol. It is a proton ionophore, destroying pH gradients in mitochondria, and thus blocking ATP synthesis
DOPA	l-3,4-Dihydroxyphenylalanine, used to treat Parkinson's disease
DREAM	Downstream regulatory element antagonist modulator, calsenilin (KChIP3) that binds response elements. It has with EF-hand motifs, a high-affinity Ca^{2+} site, and three low-affinity Ca^{2+} sites
DSC	Differential scanning calorimetry
DSCR1	Down syndrome critical region gene 1, an inhibitor of the Ca^{2+} activated phosphatase calcineurin. Inhibiting the NFAT pathway
DsRed	Red fluorescent protein from the coral *Discosoma* sp., ex 558 nm, em 563 nm
EBFP	Enhanced blue fluorescent protein, mutated from EGFP, ex 383 nm, em 445 nm
EBV	Epstein–Barr virus
EC	Enzyme Commission; this gives four unique numbers for each enzyme
ECFP	Enhanced cyan fluorescent protein, a mutant of GFP, ex 439 nm, em 476 nm
EDG	Endothelial differentiation gene
EDRF	Endothelium-derived relaxing factor. It turned out to be NO, released by endothelium, causing smooth muscle to relax
EDTA	2,2′,2″,2‴-(Ethane-1,2-diyldinitrilo)tetraacetic acid, the first Ca^{2+} chelator used inside cells, replaced by EGTA and BAPTA, since EDTA also binds Mg^{2+}
EF- hand	The first high affinity Ca^{2+} binding domain discovered – a helix-loop-helix, the Ca^{2+} binding loop typically having 12 amino acids, found in calmodulin and troponin C
EGFP	Enhanced green fluorescent protein, S65T, mutated from wild type GFP, ex 488 nm, em 509 nm
EGTA	Ethylene glycol-bis(β-aminoethyl ether)-N,N,N',N'-tetraacetic acid – a high affinity Ca^{2+} chelator, specific over Mg^{2+}
ELC	Essential light chains in myosin
EMCCD	Electron multiplying charge coupled device, aimed at reducing electronic noise
ER	Endoplasmic reticulum, the main Ca^{2+} source in most cells
EYFP	Enhanced yellow fluorescent protein, mutated from EGFP, ex 514 nm, em 527 nm

FACS	Fluorescence activated cell sorter. It enables a mixture of cells to be analysed and sorted on the basis of the light scattering and fluorescent properties of each cell
FAD	Flavin adenine dinucleotide
FBP	Fructose bisphosphate, formally fructose diphosphate (FDP)
FBP(ase)	Fructose bisphosphatase, the enzyme that catalyses the hydrolysis of FBP into fructose 6 phosphate and phosphate in gluconeogenesis
FCCP	Carbonyl cyanide-4-phenylhydrazone, a proton ionophore that disrupts ATP synthesis in mitochondria
FDA	Food and Drug Administration in the USA
FKBP	FK506 binding protein; a protein with prolyl isomerase and chaperone activity, inhibited by calcineurin. FK standing for forskolin used to raise cyclic AMP by activating adenylate cyclase
FLIM	Fluorescence-lifetime imaging microscopy, a technique producing images through differences in the rate of decay of fluorescence. It can be used with confocal microscopy, two-photon excitation microscopy, and multiphoton imaging
FMLP	*N*-Formylmethionyl-leucyl-phenylalanine or N-formyl-met-leu-phe, a potent chemotactic peptide and activator of polymorphonuclear leukocytes (PMN) chemotactic macrophages, based on the N-terminus of bacterial proteins
FMN	Flavin mononucleotide
FRAP	Fluorescence recovery after photobleaching, used to watch the movement of proteins in cells
FRET	Fluorescence resonance energy transfer – the transfer of excited electron energy via a Förster mechanism, without the direct transfer of a photon
G protein	Trimeric GTP binding protein
GABA	Gamma-aminobutyric acid (γ-aminobutyric acid), the main inhibitory neurotransmitter in the mammalian central nervous system
GCAP	Guanylate cyclase-activating proteins, important in eyes
GDP	Guanosine diphosphate
GFP	Green fluorescent protein
GHB	γ-Hydroxybutyric acid, also known as 4-hydroxybutanoic acid, a natural neuro-transmitter and a date rape drug
GPD	FAD-glycerol phosphate dehydrogenase, a mitochondrial enzyme regulated by Ca^{2+}
GFP 78/94	Glucose response proteins, ER Ca^{2+} binding chaperones involved in the ER stress response
GTP	Guanosine triphosphate
HACBP	High affinity calcium binding proteins
HACC	Hyperpolarisation-activated voltage-operated Ca^{2+} channels
HBV	Hepatitis B virus
HC	Heavy chains in myosin
HCN	Hyperpolarising-activated cyclic nucleotide-modulated channels = cyclic nucleotide-gated channels (CNG)
HEK	Human embryonic kidney 293 tissue culture cells, derived from human embryonic kidney cells
HeLa	An immortalised cell type used widely, originating from cervical cancer cells from Henrietta Lacks, who later died of cancer

HEPES	(4-(2-Hydroxyethyl)-1-piperazineethanesulfonic acid) a zwitterionic buffer widely used in culture, pK_a = 3 and 7.5
HIV	Human immunodeficiency virus
HLA	Human leukocyte antigen
HOMER	Homer protein is a nerve protein important in long-term plasticity, named in Brakeman *et al.* (1997) *Nature*, 386, 284–288, known as Vesl and PSD-Zip45
HPV	Human papillomavirus
HTI	High threshold inactivating Ca^{2+} channels
HTN	High threshold non-inactivating Ca^{2+} channels
HU-211	Dexanabinol (also ETS2101[1]), a synthetic cannabinoid
HUGO	Human Genome Organisation, promoting international collaboration for the Human Genome Project
HVA	High voltage activated Ca^{2+} channels
IBD	Inflammatory bowel disease
IBS	Irritable bowel syndrome
ICCD	Intensified charge-coupled device
IDH	NAD-isocitrate dehydrogenase in mitochondria activated by Ca^{2+} through dephosphorylation
IFN	Interferon
Ig	Immunoglobulin A, D, E, G and M
IK	Intermediate conductance potassium channels
IL	Interleukin, e.g. IL-6
INAD	Inactivation no after-potential D
IP_3	Inositol 1,4,5-trisphosphate
IP_4	Inositol 1,3,4,5 tetrakisphosphate
IP_6	Inositol hexaphosphate
IPD	Image photon detector, a photon sensitive imaging device
IR	Infra-red
IRBIT	An inositol 1,4,5-trisphosphate (IP_3) receptor-binding protein, released from the IP_3 receptor upon IP_3 binding to the receptor
IRE1/Ire1	Inositol-requiring enzyme 1, a serine/threonine-protein kinase/endoribonuclease that communicates from the lumen of the ER to the nucleus in the ER stress response
JAK	Janus kinase, a tyrosine kinase transducing cytokine signals via the JAK-STAT pathway, initially named Just Another Kinase
KCNK1	Potassium channel subfamily K member 1, with two pore domains, but may not actually be a channel
KCNK2	Potassium channel subfamily K member 2 (TREK1), opened by anaesthetics, membrane stretching, intracellular acidosis, and heat
KCNK3	Potassium channel subfamily K member 3, a member of the superfamily of potassium channel proteins with two pore-forming P domains, sensitive to pH inside and outside the cell
KCNMA1	The BK calcium-activated potassium channel subunit alpha-1 = the large conductance calcium-activated potassium channel, subfamily M, alpha member 1 ($K_{Ca}1.1$). It is voltage gated with a large conductance of potassium ions (K^+) through the plasma membrane
KDEL	Lys-Arg-Glu-Leu, the retention signal on proteins kept in the ER, and used to keep targeted aequorin and GFP there

LB	Luria–Bertani; a common medium for growing bacteria
LDL	Low density lipoprotein
LED	Light emitting diode
LSD	Lysergic acid diethylamide, a psychedelic drug
LTH	Luteinising hormone, prolactin (PRL), luteotropic hormone or luteotropin, stimulating mammals to produce milk
LTI	Low threshold inactivating Ca^{2+} channels
LVA	Low voltage activated Ca^{2+} channels
M	Membrane spanning domains, numbered in sequence, 1,2,3 etc.
MAP	Microtubule-associated protein
MAPK(K)	Microtubule-associated protein kinase (kinase)
MBA	Marine Biological Association
MCF-7	A breast cancer cell line isolated in 1970 from a 69-year-old Caucasian woman. MCF-7 = Michigan Cancer Foundation-7, in Detroit (now the Barbara Ann Karmanos Cancer Institute), where the cell line was established in 1973 by Soule *et al.*
MCP	Microchannel plate
MCTPs	Multiple C2 domain and transmembrane region proteins
MCU	Mitochondrial calcium uniporter that lets Ca^{2+} into mitochondria when cytosolic free Ca^{2+} rises into the micromolar range
MDMA	3,4-Methylenedioxymethamphetamine – ecstasy
MEK 1	Mitogen-activated kinase 1
MICU 1 and 2	Mitochondrial calcium uniporter that lets Ca^{2+} into mitochondria when cytosolic free Ca^{2+} rises into the micromolar range
MLCK	Myosin light chain kinase, phosphorylates a specific light chain in smooth muscle myosin when activated by Ca^{2+}-calmodulin, causing contraction on actomyosin
MMR	Measles, mumps and Rubella (German measles) vaccine
MPT	Mitochondrial permeability transition, a protein pore formed in the inner membrane under pathological conditions such as stroke
MRSA	Methicillin-resistant *Staphylococcus aureus*
MSH	Melanocyte-stimulating hormone
MSL	Large mechanosensitive ion channels
MTF	Mitosis promoting factor
MTP/MTPP	Mitochondrial transition permeability pore opened by excess Ca^{2+} inside mitochondria
NA	Numerical aperture – the way of quantifying the light trapping ability of a lens in a microscope
NAADP	Nicotinic acid adenine dinucleotide phosphate, an intracellular Ca^{2+} releaser
NAD(P)	Nicotinamide adenine dinucleotide (phosphate)
NAD(P)H	Reduced nicotinamide adenine dinucleotide (phosphate)
NALCN	A voltage-independent, nonselective, non-inactivating channel permeable to Na^+, K^+, and Ca^{2+}, important in insulin beta cells
NBD	4-Chloro-7-nitrobenz-2-oxa-1,3-diazole (4-chloro-7-nitrobenzofurazan) as a chloride, used as a fluorescent label
NCC	Sodium chloride (Na^+-Cl^-) cotransporter, important in the kidney, but that may be regulated by the Ca^{2+} receptor in neurones; also called NCCT, or thiazide-sensitive Na^+-Cl^- cotransporter (TSC)

NC-IUPHAR	International Union of Basic and Clinical Pharmacology Committee on Receptor Nomenclature and Drug Classification
NCKX	The sodium-calcium + potassium exchanger that exploits the K^+ gradient across the plasma membrane, making it more efficient than NCX at extruding Ca^{2+}. NCKX was discovered in the retina, but is important in the brain
NCL	Nucleolin
NCLX	The mitochondrial Na^+/Ca^{2+} exchanger, causing Ca^{2+} to be released by Na^+
NCX	The sodium-calcium exchanger in the plasma membrane that exchanges $3Na^+$ for $2Ca^{2+}$ in or out, depending on Na^+ and Ca^{2+} concentrations.
NDBF	Nitrobenzofuran, a photorelease of caged substances
NFATc	Nuclear factor of activated T-cells
NFκB	Nuclear factor kappa-light-chain-enhancer of activated B cells
NKD	Naked cuticle homolog 1, involved in protein docking, and activating the Wnt signalling pathways
NINAC	Neither inactivation nor after-potential proteins important in fruit fly eyes
NMDA	N-Methyl-d-aspartate, an amino acid derivative, an agonist at the NMDA receptor mimicking glutamate
NMR	Nuclear magnetic resonance
NO	Nitric oxide
NOMPC	No mechanoreceptor potential C, now called the TRPN channel, discovered in fruit flies
NOS	Nitric oxide synthase
NOTCH (DSL)	Transmembrane proteins with repeated extracellular EGF domains and the notch (DSL) domains, involved in inhibition in embryogenesis, from the Notch locus in the fruit fly *Drosophila*
NP-40	Tergitol-type NP-40 (nonyl phenoxypolyethoxylethanol) a detergent used to lyse cells and expose transgenic aequorin to saturating Ca^{2+}
NP-EGTA	Cell-impermeant photolabile EGTA Ca^{2+} chelator, released to bind Ca^{2+} after UV photolysis
NPE	1-(2-Nitrophenyl)ethyl, photoreleasable for caged compounds
NRI	Noradrenaline reuptake inhibitors
NRIP	Nuclear receptor interaction protein, which has an IQ domain that binds Ca^{2+}-calmodulin
NSAID	Non-steroidal anti-inflammatory drug
NSF	N-Ethylmaleimide sensitive fusion protein, crucial for vesicle-membrane fusion
OCR-2	OSM-9 and capsaicin receptor-related, a transient receptor potential channel ((TRPV) vanilloid subfamily) ion channel required for senses, including olfaction, osmosensation, mechanosensation, and chemosensation
Orai	Orai 1, 2 and 3. The calcium ion channel that opens when it binds STIM1, as a result of Ca^{2+} release from then ER
OSM-9	Oncostatin M pleiotropic cytokine in the interleukin 6 group
PCB	Chlorophyll *a/b* light-harvesting protein from *Prochloron didemni*
PCR	Polymerase chain reaction
PDE	Phosphodiesterase

PDP	Pyruvate dehydrogenase phosphatase activated by Ca^{2+} in mitochondria
PDZ	A domain found in many signalling proteins in animals, plants, bacteria and viruses. PDZ, also called Discs large homology repeat (DHR), is derived from the first three proteins discovered in the fruit fly *Drosophila* – PSD-95 (postsynaptic density protein), DLG (the *Drosophila melanogaster* Discs Large protein) and ZO-1 (zonula occludens 1 protein), anchoring membrane proteins to the cytoskeleton
PEP	Phosphoenolpyruvate
PERK	Protein kinase RNA-like endoplasmic reticulum kinase, important in the stress unfolded protein response initiated in the ER
PET	Positron emission tomography
PFK	Phosphofructokinase
PH	Pleckstrin homology domain, about 120 amino acids found in a many signalling proteins or the cytoskeleton. It binds phosphatidylinositol lipids, e.g. phosphatidylinositol (4,5)-bisphosphate, the βγ-subunits of G proteins, and protein kinase C
PHB	Polyhydroxybutyrate
PHB-PP	Polyhydroxybutyrate-polyphosphate, the putative non-protein Ca^{2+} channel in bacteria
PIP_2	Phosphatidylinositol 4,5-bisphosphate or $PtdIns(4,5)P_2$
PIP_3	Phosphatidylinositol (3,4,5)-trisphosphate ($PtdIns(3,4,5)P_3$)
PKA	Protein kinase A, activated by cyclic AMP
PKC	Protein kinase C, activated by diacyl glycerol, formed from PIP_2 by phospholipase C
PKG	Protein kinase G = cGMP-dependent protein kinase
PLA	Phospholipase A
PLC	Phospholipase C
PLD	Phospholipase D
PLDD	Phospholipase delta
PMA	Phorbol 12-myristate 13 acetate
PMCA	Plasma membrane calcium pump
PMCR	Golgi Ca^{2+} pump in yeast, now called SPCA
PMF	Proton motive force or Peter Mitchell force, the pH and potential gradient that drives ATP synthesis in mitochondria
PMR	Plasma membrane related Ca^{2+} pump MgATPase in yeast Golgi
PMT	Photomultiplier tube
PNA	Peptide nucleic acid, for manipulating protein synthesis in live cells
PP	Polyphosphate
PRKC	Genes coding for protein kinase C
PTH	Parathyroid hormone
RACK	Receptor for activated C-kinase that binds active forms of protein kinase C
RDA	Recommended daily amount. For calcium this is 1–2 g, depending on age, gender and physiology (e.g. if pregnant)
RLC	Regulatory light chains in myosin
RNA	Ribonucleic acid; t = transfer; r = ribosomal; m = messenger; si = small inhibitory
RyR	Ryanodine receptor on the SR/ER, of which there are three sub-types. They release Ca^{2+} into the cytosol

SAM	S-adenosyl methionine
SCDF	Alpha sub-units of Ca_v1 channel
SCID	Severe combined immunodeficiency, which led to the discovery of Orai 1, the Ca^{2+} channel in SOCE
SERCA	Sarco/endoplasmic reticulum Ca^{2+} MgATPase pump
SHIM	Second-harmonic imaging microscopy (SHIM). It involves a nonlinear optical effect. Advantages for live cell and tissue imaging include lack of molecular excitation, so no photobleaching or phototoxicity, no labelling needed, as signal is strong, IR light used, 3D images deep into tissues. But it needs a strong laser
SNAP	Synaptosomal-associated protein. For example, SNAP-25 is a part of the trans-SNARE complex needed for the specificity of membrane fusion such as synaptic vesicles to the plasma membranes. It binds the Ca^{2+} target synaptotagmin, and inhibits presynaptic P-, Q- and L-type voltage-gated calcium channels
SNARE	Soluble NSF attachment protein receptor, essential to mediate Ca^{2+} triggered vesicle fusion, e.g. in a nerve terminal
SOCE	Store-operated calcium entry, Ca^{2+} entry via Orai, when Ca^{2+} is released from the ER, triggering punctae from STIM
SOD	Superoxide dismutase
SPCA	Secretory pathway Ca^{2+}-ATPase, pumps Ca^{2+} into the Golgi in animal cells when there is a rise in cytosolic free Ca^{2+}. It can also pump Mn^{2+}, which SERCA cannot do
SSRI	Selective serotonin reuptake inhibitor
SR	Sarcoplasmic reticulum, the main Ca^{2+} store in muscle
STAT	Signal transducer and activator of transcription proteins, activated by membrane receptor-associated Janus kinases (JAKs). These transcription factors mediate immunity, cell proliferation and differentiation, and apoptosis
STEM	Science, Engineering, Technology, Mathematics, used in the UK to highlight these subjects in the education system
STIM	STIM 1 and 2 – Stromal interaction molecule in the ER membrane. It forms punctae when Ca^{2+} is released from the ER, and then binds Orai
SUMO	Small ubiquitin-like modifier proteins, like ubiquitin, attach to and detach from proteins in nuclear-cytosolic transport, transcriptional regulation, apoptosis, protein stability, stress, and the cell cycle
TASK	KCNK3 K^+ channel. Two-pore acid sensitive K channel, activated by the anaesthetics halothane and isoflurane
TBA	Tetrabutylammonium.
TEA	Tetraethylammonium
TES	N-Tris(hydroxymethyl)methyl-2-aminoethanesulfonic acid (2-[[1,3-dihydroxy-2-(hydroxymethyl)propan-2-yl]amino]ethanesulfonic acid) widely used zwitterionic pH buffer, $pK_a = 7.55$
TFEB	Transcription factor EB (TFEB) a master gene for lysosomal biogenesis genes, and expression of genes involves in autophagy
TGN	Trans Golgi network
THC	Tetrahydrocannabinol
TIRF	Total internal reflection fluorescence microscopy, allowing a very thin section (*ca* 200 nm) to be imaged
TM	Transmembrane

TMBR	The neuron-specific tropomyosin isoform
TMEM143A	Another name for the SOCE Ca^{2+} channel Orai
TNF	Tumour necrosis factor
TORC	TORC1 = transducer of regulated CREB activity 1) now called CREB-regulated transcription coactivator 1 (CRTC1)
TPEN	N,N,N',N'-tetrakis(2-pyridinylmethyl)-1,2-ethanediamine, a cell permeant Zn^{2+} chelator
TPI	Triphosphoinositol
TRIM	Tripartite motif proteins, induced by interferons, and help combat infection
Tris	Tris(hydroxymethyl)aminomethane (2-amino-2-(hydroxymethyl) propane-1,3-diol), widely used pH buffer, pK_a = 8.07
TRP	Transient receptor potential ion channels discovered in fruit fly eyes; trp-mutant fruit flies only had a transient response to light, unlike wild-type flies. Many TRPs let Ca^{2+} into cells, but may not be selective. Group 1 = TRPC (canonical), TRPV (vanilloid), TRPM (melastatin), TRPN, and TRPA. Group 2 = TRPP (polycystic) and TRPML (mucolipin)
TRPL	TRP like channel in fruit fly photoreceptors
TRPML1	Transient receptor potential cation channel, mucolipin subfamily, member 1 = mucolipin-1 (MCOLN1), a Ca^{2+} channel in lysosomes
TSH	Thyroid stimulating hormone
TTH-1	Tetra THymosin (four thymosin repeat protein) in the nematode *C. elegans*
TTP	Thymidine triphosphate
TRPM8	Transient receptor potential cation channel subfamily M member 8 (TRPM8) = the cold and menthol receptor 1 (CMR1), Ca^{2+} signals being involved in response to cold
TTX	Tetrodotoxin, a potent blocker of voltage-gated Na^+ channels, produced by bacteria in certain pufferfish, porcupine fish, ocean sunfish, triggerfish, and blue-ringed octopuses, rough-skinned newts, and moon snails
UCH	University College Hospital
UTP	Uridine triphosphate
UV	Ultraviolet (10–400 nm)
VAMP	Vesicle associated membrane proteins – SNARE proteins involved in vesicle fusion
VDAC	Voltage-dependent anion channels, porin type ion channel in the outer membrane of mitochondria
VDCC	Voltage-dependent calcium channels
VGM	Vector geography mapping, used to study the precise conformation of the helix-loop-helix EF-hand
VICC	Voltage-independent cation/Ca^{2+} channels
VSOP	Voltage-sensing domain only protein (voltage-gated proton channels are needed to produce superoxide anion in phagocytes)
YFP	Yellow fluorescent protein, a mutant of GFP
ZAC	Zinc-activated cation channel is a Cys-loop protein in prostate, thyroid, trachea, lung, brain, spinal cord, skeletal muscle, heart, placenta, pancreas, liver, kidney and stomach
γ-GT	Gamma glutamyl transferase, a liver enzyme released after high alcohol intake

About the Companion Web Site

This book is accompanied by a companion web site. Supplementary material relating to this work can be downloaded at:

www.wiley.com/go/campbell/calcium

The website includes:
From Intracellular Calcium (2015)

- Bibliography as a PDF and Endnote file.
- Spreadsheets used for equation and calculations in both books.
- Wav file related to Figure 13.4, 13.5 in Fundamentals.
- Table of contents from *Intracellular Calcium: Its Universal Role as Regulator* by Anthony Campbell, 1983.
- Tables in PowerPoint and PDF from *Intracellular Calcium: Its Universal Role as Regulator* by Anthony Campbell, 1983.
- References as a PDF and Endnote file from *Intracellular Calcium: Its Universal Role as Regulator* by Anthony Campbell, 1983.
- Corrections

In addition from this book – Fundamentals of Intracellular Calcium:

- Recommended reading as a PDF and Endnote file
- Powerpoint slide show of all the figures

Also see www.welstonpress.com for reference updates and further information, and YouTube for demonstrations on chemiluminescence and fluorescence relevant to intracellular calcium. Also see www.theyoungdarwinian.com for reviews of this book, videos of intracellular calcium and other information.

1

Calcium is Special

Ja Kalzium, das ist alles!
Otto Loewi (1959)

Learning Objectives
1. How everyday events in animals, plants and microbes depend on calcium as a molecular switch inside cells.
2. What is special about calcium (Ca^{2+}) that evolution has selected it for this task?
3. How calcium was discovered.
4. The natural history of Ca^{2+}.
5. The requirement of cells for Ca^{2+}.
6. How Ca^{2+} is different from the 28 other elements used in biological processes.
7. The four biological roles of Ca^{2+} – structural, electrical, cofactor and intracellular regulator.
8. The molecular biodiversity of the intracellular calcium signalling system, vital in the process of Natural Selection revealed by Charles Darwin and Alfred Russel Wallace.

This is the amazing story of a simple cation, Ca^{2+}, upon which the whole of life on our planet depends. The bricks and mortar of life depend on the proteins coded for by DNA. But without intracellular calcium, cells, tissues and organisms cannot do anything. Ca^{2+} inside cells is the trigger for life, and its evolution over 4000 million years.

1.1 Calcium and Everyday Events

Who would have thought that the serenity of a Bach chorale, the succulent taste of a coq au vin with a nice glass of Côtes du Rhône, the sensuous smell of a flower meadow in spring, the pleasure we get from seeing a puffin as it flies out of its burrow for some more sand eels, or even the intellectual excitement of a successful experiment all depend on calcium? This is not the calcium in our bones and teeth, that most people think of, but rather tiny puffs of calcium inside the cells that are responsible for all our senses, our movements and the functioning of our brain. The fertilisation of our mother's egg by our father's sperm started our life on a wave of intracellular calcium, and our embryo then developed on calcium signals within the cells as they differentiated into tissues. We were born on a wave of uterine intracellular calcium, as we were thrust out into the world and started to breathe. Throughout our lives we grow, develop

Fundamentals of Intracellular Calcium, First Edition. Anthony K. Campbell.
© 2018 John Wiley & Sons Ltd. Published 2018 by John Wiley & Sons Ltd.
Companion Website: http://www.wiley.com/go/campbell/calcium

and function through intracellular calcium signals within all of our cells. If we are lucky enough to live until 100, we will have generated over 3000 million puffs of calcium within our heart cells to keep them beating. Finally, we will die on a wave of calcium, as it floods into the cells lacking oxygen, when our heart eventually stops beating. Changes in intracellular calcium tell a nerve terminal to fire, a muscle to contract, and a cell to secrete, divide or die. Calcium is a universal switch inside animal, plant and microbial cells from all the three domains – Bacteria, Archaea and Eukaryota. There are two defining principles:

1. Calcium, always as Ca^{2+}, acts as a switch to instruct a cell to cross a threshold, a Rubicon, and do something. But, it is not the energy source.
2. The diversity of molecules that determine when and how a cell fires are a living example of Darwinian variation, upon which Natural Selection depends.

These are dependent on one universal property of cells. All living cells – animal, plant, fungal and microbe – maintain a very low free Ca^{2+} in their cytosol, in the sub-micromolar range. This results in a huge gradient of Ca^{2+} across the outer membrane, 10 000-fold in the case of our own cells. It is this calcium pressure that has allowed evolution to capitalise on the unique chemistry of calcium for it to act as a cellular switch.

1.2   Discovery of Calcium

Humphry Davy (1778–1829; Figure 1.1) was one of the founders of modern chemistry. Working at The Royal Institution in Albemarle Street, off Piccadilly in central London, he came up with an idea that lies at the heart of understanding biological chemistry. He proposed that salts were made up of two parts: one positive, the other negative. To test this hypothesis he connected a liquid salt to a battery, so that this would separate them. In 1808, Davy made a moist mixture of lime and red oxide of mercury, with mercury at the negative electrode (cathode). Removal of the mercury revealed a new metal, 'greyish-white with the lustre of silver', which burnt avidly in air with a brick-red flame. He called this new metal 'calcium', after the Latin for lime – *calx*.

Figure 1.1 Sir Humphry Davy, Bt, FRS, MRIA, FGS (1778–1829). Portrait by Thomas Phillips (died 1845). *Source*: Courtesy of the Royal Institution.

1.3 A Natural History of Calcium

Calcium was one of the earliest elements to form in the Universe. It can be formed in stars and supernova from the fusion of smaller elements, such as argon and helium. Calcium is 11th in abundance in the solar system, and in the Earth's crust it is the 5th most common element, and the third most common metal element. Everywhere you look there are examples of calcium precipitates *outside* cells – in rocks, in the cement holding buildings together, in the sea and every time someone smiles at you. But let us examine two scenarios to help us focus on how calcium plays its unique biological role as a universal regulator *inside* all living cells.

1.3.1 Calcium by the Sea

Wales, where I live, has some of the most beautiful coastline in Europe, with estuaries, sandy beaches, cliffs and coastal paths with superb views, with wonderful wildlife, both above and below water. We have fantastic tides. This means we get some of the best shore marine life and rock pools in the world, with its abundance of animal and plant life (Figure 1.2). Calcium is everywhere to be seen, inside and outside of the cells of the organisms in the rock pool.

The hard material surrounding a rock-pool is made of limestone – calcium carbonate, deposited millions of years ago by extinct calcified organisms. Seawater in the pool contains calcium at a concentration some ten times that free in the blood. Within the pool are many examples of calcium outside cells:

- Calcium carbonate ($CaCO_3$) secreted to cover the pink seaweed *Corallina*.
- The skeletal remains of a dead fish, made of calcium phosphate.
- The 'sheep' of the rocks – limpets, whelks and periwinkles grazing on small algae, as well as mussels and barnacles, with their hard shells made of $CaCO_3$.

Figure 1.2 Calcium by the sea: some organisms easily found on rocks or in rock pools when the tide goes out. Reproduced by permission of Welston Court Science.

The mussels and barnacles under water are open, filtering out planktonic organisms to feed on. But those out of water are firmly shut! They are closed for several hours, until the tide comes back in, because a calcium signal inside their large muscle fibres caused these to contract, keeping their soft insides protected from drying out in the sun. Underneath one of the rocks there is a sea urchin with a cluster of its eggs, ready to develop, because a calcium wave has been triggered by the sperm that fertilised them. A sea anemone clutches your finger, trying to anaesthetise you by injecting its poison from the tiny syringes in its sting cells (nematocysts), triggered to fire by a calcium signal inside each cell. The toxin in its sting works by opening cation channels in your sensory nerves. Also stuck on the underside of the rock is a sea slug, which has synchronised electrical signals, provoked by calcium moving into the cell through special ion channels. There are fish and shrimp darting about, whose movements are all regulated and coordinated by small puffs of calcium in their muscle cells, and at the terminals of their nerves. And then you notice a tiny carpet of what look like small flowers attached to a piece of kelp – a seaweed attached to the side of the pool. These flower-like organisms are in fact animals – the hydroid *Obelia geniculata*, part of the life cycle of a small jellyfish. *Obelia* flashes blue-green light when touched in the dark. This is caused by a chemical reaction within an intracellular protein, obelin, when it binds Ca^{2+}. Its relative, aequorin, from an American jellyfish, *Aequorea*, provided the first universal method for the direct measurement of free Ca^{2+} inside living cells. *Obelia* and *Aequorea* also gave us the green fluorescent protein (GFP), which has had such an impact on cell biology. Even microscopic cyanobacteria that are coating some of the animals and plants in the pool use intracellular Ca^{2+} to regulate their ability to fix nitrogen. Within the limestone sides of the pool there is a fossil ammonite, some 200 million years old, showing us that calcium has been important in life throughout millions of years of evolution. On the overlooking cliff top there are bushes and small trees, shaped by the prevailing wind. These have not been mechanically blown into these streamlined shapes, but rather signalled to grow that way by tiny puffs of intracellular Ca^{2+}, so that the cells on each side of the bush grow at different rates.

1.3.2 Calcium in Your Wake-Up Call

The alarm clock rings. It is 7 a.m. Your eyes open, your heart starts to race a little, and you jump out of bed. After a quick shower, you munch a piece of toast and marmalade, and quickly drink a cup of tea. You rush out of the door to work. You have a lecture on cell signalling, which starts promptly at 9 a.m. As you jump on your bike, a thought crosses your mind. Without little puffs of calcium inside the cells of your brain, heart, leg muscles, pancreas and liver, you wouldn't have been able to wake up and get out of bed, let alone digest your breakfast. And without Ca^{2+} puffs in your parent's gametes you wouldn't even have been conceived!

Yet the timescale over which intracellular Ca^{2+} works varies from milliseconds to hours with each cell type. Intracellular Ca^{2+} triggers a nerve cell to release its neurotransmitter in milliseconds, and the Ca^{2+} signal that induces a heartbeat lasts just 1 s. In contrast, the secretion of insulin and digestive enzymes from the pancreas, together with the stimulation of intermediary metabolism in muscle and liver, after breakfast, last minutes or even hours. The regulation of the cell cycle by Ca^{2+} may take days or even weeks to take effect, through regulation of gene expression. Diseases such as cancer or Alzheimer's, where intracellular Ca^{2+} may be involved, can take months or years to show up.

The intracellular Ca^{2+} signalling system is now a prime target for drug discovery in controlling heart disease, blood pressure, diabetes, arthritis, multiple sclerosis, cancer, diseases of the brain and nervous system, several genetically based diseases, and potentially many infections by bacteria and viruses. Calcium even has a major role in keeping oxygen and nitrogen in the atmosphere at the right level, and in the microorganisms that are involved in controlling global warming.

This then is the fascinating puzzle about calcium. How does one simple cation, Ca^{2+}, do all of this? There is something very special about the chemistry of Ca^{2+} inside living cells. Evolution has selected Ca^{2+} as the universal intracellular switch for an amazing variety of phenomena in animals, plants and microbes. Abundant monovalent cations like Na^+ or K^+, or divalent cations such as Mg^{2+}, Cu^{2+} or Zn^{2+}, just did not have the right chemistry for the task in hand.

1.4 The Elements of Life

During nearly 4000 million years of evolution, living systems have exploited the chemical and electrical properties of 29 elements – 16 metallic and 13 non-metallic (Tables 1.1 and 1.2). Nine metallic elements (sodium, potassium, calcium, magnesium, manganese, cobalt, iron, copper and zinc), and seven non-metals (hydrogen, nitrogen, carbon, oxygen, phosphorous,

Table 1.1 The elements of life – metals.

Metal	Typical biological role
Na^+ and K^+	Osmotic balance Electrical activity across membranes Activation of a few enzymes
Mg^{2+}	ATPMg^{2-} (the substrate for pumps and kinases) Activation of some enzymes Chlorophyll
Ca^{2+}	Hard structures – phosphate, carbonate, sulphate Soft structures – membrane integrity, cell adhesion, granules Electrical excitability across some membranes Cofactor for extracellular proteins and enzymes Wide-ranging intracellular signal
Sr^{2+}	Can be significant in some hard structures such as statocysts
Ba^{2+}	Methanol dehydrogenase in Archaea
V^{3+}	Rare ligand (e.g. in the tunichromes of tunicates)
Cr^{3+}	Rare ligand in some chromophores
Mn^{2+}	Cofactor for several key intracellular enzymes
Co^{2+}	Vitamin B$_{12}$
Ni^{2+}	Methanogenic enzymes in Archaea
Mo^{2+}	Rare catalytic function in some enzymes
Fe^{2+}/Fe^{3+}	Oxygen-carrying pigments Oxido-reduction reactions
Cu^+/Cu^{2+}	Oxygen-carrying pigments Oxido-reduction reactions
Zn^{2+}	DNA-binding proteins in the cell nucleus Insulin granules in some species Catalytic function in some enzymes
Cd^{2+}	Can replace Zn^{2+} in some enzymes when Zn^{2+} is in short supply

Note: Boron is a metalloid, showing both metallic and non-metallic properties. Several other elements can be found in living organisms, for example tungsten, but it is not clear if these other elements have any function.

Table 1.2 The elements of life – non-metals.

Non-metal	Typical biological role
H	H_2O, >70% of the wet weight of all soft tissues H_2 gas produced by bacteria (e.g. in the gut) H^+ in maintaining pH pH gradient in mitochondria and chloroplast energy for ATP Major element of all organic molecules
B	Bacterial quorum sensor and some other rare uses
N	NH_3 and nitrogen fixation Major element in amino acids and proteins Nitric oxide (NO), a novel signalling molecule Chromophore and metal ligand (e.g. haem) Nucleic acid bases (e.g. ligand in double helix) Organics (e.g. fluors, bioluminescence, toxins)
C	*The* major component of all organic molecules Gases – CO_2 by aerobic cells, CH_4 by Archaea
O	Major component of many organic molecules Oxygen metabolite; O_2^-, H_2O_2, OCl^-, OBr^-
F	Found as a strengthener in teeth
P	Internal pH buffer Found as phosphate in DNA and RNA Hard structure in bones and teeth with calcium Regulating proteins through Ser, Thr and Tyr Phosphate in nucleotides (e.g. ATP and GTP) Component of phospholipids in membranes Soluble organics as a regulator (e.g. IP_3)
Si	Hard structures in foraminiferans
S	Reducing agent (e.g. the amino acid Cys) Important catalytic centre in many enzymes H_2S Substitute for O_2 in organisms in thermal vents Oxygen metabolite scavenger (e.g. glutathione)
Cl	Anion in osmotic balance and membrane potential Biocidal agent as OCl^- (e.g. by neutrophils)
Se	Active centre of oxygen scavenger enzymes
Br	Biocidal agent as OBr^- (e.g. by eosinophils)
I	Thyroid hormones T_4 and T_3 Spermicidal agent produced by fertilised eggs

sulphur and chlorine), are essential for virtually all living systems, with metals always working as cations (*positively charged ions*). Interestingly, aluminium, the most common element in the Earth's crust, has no known biological role in any living system. Only toxic effects have been described.

The unique property of Ca^{2+} is to act as an intracellular switch, causing cells, organs and even entire organisms to cross the Rubicon. As a result a biological event occurs. A nerve fires, a muscle contracts, a heart beats, an insulin cell secretes, a luminous jellyfish flashes, an egg divides and differentiates, a plant survives cold shock or wind, a bacterium competes with others in the gut, a cell defends itself from attack or signals itself to die.

1.5 Natural Occurrence of Calcium

1.5.1 Isotopes of Calcium

There are 24 calcium isotopes known. Six occur naturally, ^{40}Ca, ^{42}Ca, ^{43}Ca, ^{44}Ca, ^{46}Ca, and ^{48}Ca, 97% in Nature being ^{40}Ca with 20 protons and 20 neutrons. ^{43}Ca, which has 20 protons and 23 neutrons, is used in nuclear magnetic resonance (NMR) spectroscopy to study solid structures, and can be used to study Ca^{2+} binding to biological molecules such as proteins. Many radioactive isotopes of calcium have been generated. Most have half-lives ($t_{1/2}$) too short for them to occur naturally. However, ^{41}Ca, with a half-life of 102 000 years, is found in the cosmos, produced by neutron activation of ^{40}Ca. The most useful artificial calcium isotope is ^{45}Ca, and has been used to study calcium fluxes across biological membranes and calcium binding to macromolecules.

1.5.2 Geology of Calcium

Calcium is one of the most common elements in the Earth's crust. On a molar basis, aluminium is the most common metal at nearly 38%, after which comes sodium at 14%. Calcium is almost as abundant, at 13.6%. Of the non-metals, oxygen is by far the most abundant at 71.4%, and is the most important biological ligand for Ca^{2+} when it binds to proteins. Over 700 calcium-containing minerals are known. These occur in all three main types of rock: igneous, metamorphic and sedimentary, many forming attractive crystals (Figure 1.3).

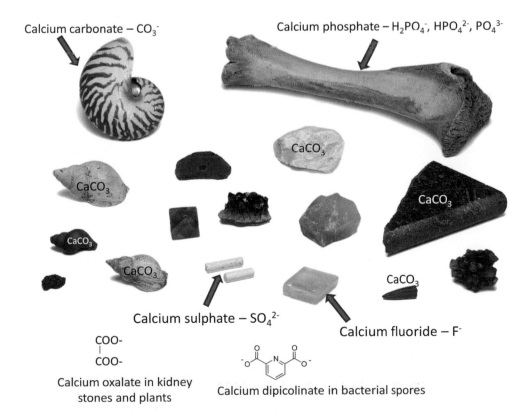

Figure 1.3 Some cations from acids that form precipitates with Ca^{2+}, and some that are found in crystal or non-crystalline form in rocks and in animals. Reproduced by permission of Welston Court Science Centre.

The occurrence of calcified microfossils, cyanophytes and stromatolites, in Precambrian rocks some 3500 million years old shows that calcium played a vital role in evolution from the earliest times. Calcareous protozoa and algae are found throughout the Precambrian era (4570–542 million years ago). Calcareous blue-green algal deposits have been found that are at least 1900 million years old, massive calcified shell deposits of metazoans occurring from the beginning of the Palaeozoic at the start of the Cambrian explosion (542–488 million years ago). Throughout the Palaeozoic (542–251 million years ago), Mesozoic (250–65 million years ago) and Cainozoic (65 million years ago to present) eras, fossil deposits are rich in calcified microscopic foraminifera, molluscs, crustaceans and corals. The origin of vertebrates was around 400 million years ago. Thus, the widespread use of calcium phosphate by living organisms was a much later innovation than the use of calcium carbonate.

1.5.3 Calcium Outside Cells

All animal, plant and microbial cells and tissues contain Ca^{2+}. Some of this is outside cells and some inside, stored in organelles such as the endoplasmic reticulum (ER) and secretory vesicles, and the vacuole in plants and fungi. The total cell Ca^{2+} of an animal cell is typically in the range 1–10 mmol l^{-1} of cell water, though skeletal muscle with large internal Ca^{2+} stores can have as much as 20 mmol Ca^{2+} l^{-1} of cell water. But cells with no internal Ca^{2+} store, such as erythrocytes and bacteria, have only 0.1 mmol Ca^{2+} l^{-1} of cell water. This compares to a total tissue content of K^+, Na^+ and Mg^{2+} of usually more than 100, 20 and 10 mmol l^{-1} of cell water, respectively.

All natural water contains calcium. In the sea the Ca^{2+} concentration is about 10 mM, whereas in fresh rainwater it may only be micromolar. But in hard tap water, such as in London, the concentration can be as high as 1 mM. Ca^{2+} is essential for all animal, plant and several microbial life. Without external Ca^{2+} (Table 1.3) many tissues disintegrate, because Ca^{2+} helps to bind cells together. The weathering of rocks, and the action of bacteria on calcium minerals, causes Ca^{2+} to leach out into fresh water in ponds and rivers. Rivers carry this Ca^{2+} to lakes and the sea.

The total concentration of Ca^{2+} in human plasma is about 2.5 mM. Some 40% of this is bound to albumin, and some to other small and macromolecular ligands. So the free Ca^{2+} to which our cells are exposed is about 1.2 mM. Thus, all tissue and organ culture media contain millimolar levels of Ca^{2+} (Table 1.3). But the highest absolute amount of Ca^{2+} outside cells is found as Ca^{2+} precipitates with anions, in the form of carbonate ($CaCO_3$), phosphate ($Ca_3(PO_4)_2$ or $Ca_{10}(PO_4)_6(OH)_2$), sulphate ($CaSO_4$), oxalate ($Ca(COO)_2\cdot 2H_2O$) or fluoride (CaF_2) (Table 1.4 and Figure 1.3).

Calcium is the fifth most abundant element by weight in the human body. After drying, e.g. after cremation, about one-third of the remaining mass (*ca* 1 kg) is calcium. There are a variety of functions for precipitated calcium minerals in extant organisms. The hard component in the bones and teeth of all vertebrates is apatite, a form of calcium phosphate, $Ca_{10}(PO_4)_6(OH)_2$. Bone also contains other calcium minerals – carbonate, fluoride, hydroxide and citrate. In contrast, the shells of all birds, reptiles, molluscs and arthropods, as well as the hard parts of anthozoans that make up coral reefs, are made of $CaCO_3$. Single calcite crystals are used as a lens by photoreceptors in brittle stars. And the eye of trilobites, which lived some 500 million years ago, may have also used calcite crystals as a lens. Some organisms use 'blackboard chalk' (calcium sulphate, a calcium precipitate), e.g. the small ball (few microns) of $CaSO_4$ inside the balance organ (statocyst) of jellyfish. By rattling around inside its own cage, contact is made with its nerve net, and the jellyfish is able to stay afloat upright. Calcium sulphate is also found in some algae.

Calcium carbonate is the major calcium precipitate in invertebrates, occurring initially around 600 million years ago, leading to the start of the Cambrian era 542 million years ago.

Table 1.3 Concentration of four cations in extracellular fluids.

Extracellular fluid	Total cation concentration (mM)			
	Na⁺	K⁺	Mg²⁺	Ca²⁺
Water				
Sea water	475	10	55	10 (9.3–11.8)
Low-salinity lakes	3.3	1.8	2.7	6.6
Fresh water (Na⁺ + K⁺)	0.1–80		0.02–0.4	0.02–2
Rain water (Na⁺ + K⁺)	0.01		0.004	0.002–0.02
Tap water				0.02–2
Experimental salt solutions				
Locke's saline	156	6		2.5
Krebs–Ringer	148	6	1.3	2.5
Eagle's medium	143	5.4	0.8	1.8
Nematode (*Ascaris*) saline	130–168	3–24	0–16	2–7
Marine invertebrate saline	51.3	12.9	23.6	11.8
Serum and body fluids				
Adult human serum	140	4	1	2.5
Vertebrate serum	87–544	4–12	1–10	1.5–5
Pig intestinal fluid	124	27	6	14
Coelenterate (*Physalia*) gastrovascular fluid	350	33	24	6
Nematode (*Ascaris*) body fluid	129	25	49	6
Molluscan serum				
Marine	475	10–22	55	9–15
Fresh water	16–86	0.4–5	0.1–2.4	1.5–7.8
Land	47–75	2.4–10	1–20	3.3–12.3

A threshold was crossed in evolution some 500 million years ago, the appearance of vertebrates with bones and teeth from calcium phosphate. Other calcium precipitates tend to be pathological, two types of tissue calcification being dystrophic and metastatic. Dystrophic calcification occurs because of dead or damaged tissue, serum calcium usually being in the normal range. In contrast, metastatic calcification arises as a result of disturbances in whole-body calcium metabolism, often involving parathyroid hormone, vitamin D, and the calcium receptor. A further example of pathological calcified deposits is the painful formation of stones (calculi) in the kidney, gall bladder and bladder.

Calcium also binds to the soft outside of all cells. Its removal can cause cells to behave abnormally, affecting electrical excitability and cell–cell adhesion.

1.5.4 Calcium Inside Cells

Most of the Ca²⁺ inside cells is in solution, free or bound to small or macromolecular ligands. But intracellular Ca²⁺ precipitates also occur. In anoxia, calcium phosphate can precipitate in mitochondria. Whereas many plant cells have calcium oxalate as a calcium store inside the

Table 1.4 Examples of calcium precipitates outside and inside cells – biomineralisation.

Organism	Biological structure	Function	Outside or inside cells	Main calcium salt
Unicellular				
Some bacteria	Spore	Survival of DNA without water	Inside	Calcium dipicolinate
Some protozoa	Shell	Protective skeleton	Outside	$CaCO_3$
Some algae	Shell	Protective	Outside	$CaCO_3$
Animal phyla				
Cnidaria (sea corals)	Coral	Protective skeleton	Outside	$CaCO_3$
Molluscs (bivalves and gastropods)	Shell	Protective skeleton	Outside	$CaCO_3$
Arthropods (barnacles)	Shell	Protective skeleton	Outside	$CaCO_3$
Brittle star	'Eyes'	Micro-lens	Inside	$CaCO_3$
Platyhelminths (tapeworms)	Calcareous corpuscles		Outside	$Ca_3(PO_4)_2$
Chordata (vertebrates)	Bones	Skeleton	Outside	$Ca_{10}(PO_4)_6(OH)_2$
Chordata (vertebrates)	Teeth	Breaking up food	Outside	$Ca_{10}(PO_4)_6(OH)_2$
Chordata (vertebrates)	Bird eggs	Protective	Outside	$CaCO_3·H_2O$
Cnidaria (jellyfish)	Statocysts	Balance organ	Outside	$CaSO_4$
Insects	Egg	Protective deterrent	Inside	$Ca(COO)_2·H_2O$
Chordata (vertebrates)	Teeth	Structural strength	Outside	CaF_2
Plants				
Corallina (multicellular algae)	Stem skeleton	Protective skeleton	Outside	$CaCO_3$
Plants	Granules	Ca^{2+} store, pH regulation; support, light gathering	Inside	$Ca(COO)_2·H_2O$

Calcium carbonate and calcite = $CaCO_3$; calcium phosphate = $Ca_3(PO_4)_2$ or $Ca_{10}(PO_4)_6(OH)_2$; calcium sulphate = $CaSO_4$; calcium oxalate = $Ca(COO)_2·H_2O$; calcium fluoride = CaF_2. Calcite crystals ($CaCO_3$) have also been found in fossil trilobites, apparently to form the lens.

vacuole. Calcium oxalate occurs naturally in plants from unicellular to angiosperms and large gymnosperms, where it can be as much as 10% of the dry weight. In some cacti it is as much as 85% of dry weight. Calcium oxalate crystals occur in roots, stem, leaves, fruit and seeds as $CaC_2O_4·H_2O$ (whewellite), $CaC_2O_4·2H_2O$ (weddellite) and $CaC_2O_4·3H_2O$ (trihydrate). Calcium oxalate has one of six functions:

- Calcium regulation.
- Intracellular pH regulation and ion balance.
- Mechanical support.
- Gravity monitor.
- Detoxification.
- Light gathering and reflection.

Calcium oxalate also occurs naturally in fungi, lichens, and in unicellular and multicellular algae. It occurs in some molluscs, such as in the mineral-rich granules in the kidney of

Pecten maximus, where is can be up to 7% of dry weight, and is found in some insects and primitive chordates. In mammals, calcium oxalate precipitates (e.g. as kidney stones) are usually pathological. Calcium dipicolinate is crucial in the formation of spores in bacteria such as *Bacillus*, enabling them to dry out and survive lack of water. This is why anthrax (*Bacillus anthracis*) spores are so stable and can remain dormant in the soil or other material for decades.

The free Ca^{2+} concentration in the cytosol of all cells, animal, plant and microbe, is in the sub-micromolar range. Thus, there is a 1000–10 000 fold gradient of Ca^{2+} across the outer, plasma membrane. In contrast, the free concentration of other major cations is in the millimolar range: K^+ = 120–150 mM, Na^+ = 10–20 mM, Mg^{2+} = 1–2 mM, though in some halophilic Archaea the intracellular K^+ can be as high as 4 M! When a cell is activated the cytosolic free Ca^{2+} rises to between 1 and 10 μM. Damage to the plasma membrane under pathological conditions leads to a level of cytosolic free Ca^{2+} of up to 50–100 μM. The cell can survive this high level for a short while. But unless the cytosolic free Ca^{2+} returns to sub-micromolar levels quickly, the Ca^{2+} is toxic, causing precipitation, irreversible damage to mitochondria, and the cell may die. The free Ca^{2+} concentration in the nucleus is also in the sub-micromolar range in resting cells, rising to micromolar levels when the cell is activated. However, the free Ca^{2+} concentration inside other organelles can be much higher than that in the cytosol. In the ER free Ca^{2+} is around 100–500 μM in resting cells, falling to 1–10 μM when the ER Ca^{2+} is released following cell activation. In the resting cell, mitochondrial free Ca^{2+} is sub-micromolar. But after cell activation it can rise to greater than 10 μM. Similar changes occur in the Golgi. Secretory vesicles, such as those at nerve terminals, in platelets, and in exocrine or endocrine cells, accumulate Ca^{2+} as part of their internal structure. Endosomes initially contain the same concentration of Ca^{2+} as that outside the cell, around 1 mM in mammals and 10 mM if the cell is in sea water. This Ca^{2+} is lost through the endosome pathway, though lysosomal Ca^{2+} may be as high as 500 μM.

Small molecules and macromolecules as potential Ca^{2+} ligands within the cell include inorganic anions, acids, nucleotides, sugars, amino acids, nucleic acids, polysaccharides and proteins. The most important Ca^{2+} ligands inside cells are proteins, with Ca^{2+} affinities in the micromolar range, selective for Ca^{2+} in the presence of millimolar free Mg^{2+}, e.g. troponin C in muscle, and the ubiquitous Ca^{2+}-binding protein calmodulin. Ca^{2+} binding proteins with lower affinities for Ca^{2+} can be physiologically significant if they are in an intracellular compartment where the free Ca^{2+} can be tens to hundreds of micromolar, e.g. calsequestrin in the sarcoplasmic reticulum (SR) of muscle, and calreticulin, in the ER of non-muscle cells. Similarly, proteins in the mitochondria, such as pyruvate dehydrogenase, can also be activated at levels of free Ca^{2+} in the high micromolar range.

1.6 Requirement of Cells for Ca^{2+}

1.6.1 Calcium in External Fluids

Calcium is present in all fluids bathing living cells – animal, plant and microbe. But the concentration of this Ca^{2+} can vary over many orders of magnitude (Table 1.3). The concentration of Ca^{2+} in seawater is around 10 mM. In contrast, in fresh water, Ca^{2+} can be as low as 1–10 μM, though London tap water may have 1 mM Ca^{2+}, and in areas famous for their beers hard water can have a Ca^{2+} concentration as high as 4 mM. The concentration of Ca^{2+} in the body fluids of animals, and in the sap of plants, also varies over several orders of magnitude (Table 1.3). In human blood, the total Ca^{2+} is about 2.5 mM, maintained by hormones such as parathyroid hormone, calcitonin and vitamin D. Over 50% of this Ca^{2+} is bound to albumin and other molecules, leaving a free Ca^{2+} in plasma of about 1.2 mM. The total Ca^{2+} concentration in other animal plasma or haemolymph, the external fluid inside invertebrates, is in the range 1–15 mM.

1.6.2 Requirement of Cell Types for Calcium

Most animals and plants, and their cells in isolation, require Ca^{2+} for normal growth and function. Sydney Ringer (1836–1910), working at University College in London (see Chapter 3), showed that Ca^{2+} is required extracellularly for a frog heart to beat, cell adhesion, and for the normal development of tadpoles and worms. Other workers then showed that external Ca^{2+} was required for the maintenance, morphology, cell adhesion, growth, hormone and drug action, and the function of many animal tissues. Pioneers of organ and tissue culture showed that Ca^{2+} was an essential component of all culture media, together with sodium, potassium and magnesium. Methods for removing calcium from culture media causes changes in cell structure, and a reduction in cell growth. In man and other mammals, even a relatively small drop in plasma free Ca^{2+} from 1.25 to 0.9 mM, e.g. from hyperventilation induced alkalosis or hyperparathyroidism, can cause muscle spasms and tetany as the result of spontaneous firing of nerves and muscle.

Every gardener knows that calcium in the soil can benefit plant growth, hence the use of bone meal. The quality of wines is affected beneficially by calcareous soils. The level for calcium in normal plant growth is 0.1–1 mM, root growth being particularly sensitive to calcium deprivation. Calcium is required for flowering, and development of nodules in leguminous plants infected by the nitrogen fixer *Rhizobia*. Calcium deprivation in plants leads to damage to intracellular organelles, reduction in cell elongation, as well as adverse effects on cell walls and the permeability of cell membranes.

The requirement for calcium in bacteria varies considerably with species. Ca^{2+} is required by marine bacteria, blue-green algae and cyanobacteria, and some other bacteria. But *Escherichia coli* can grow in EGTA, but at a slower rate than with calcium. Interestingly, removal of external Ca^{2+} causes just a 10% increase in generation time. This would lead, within 24 h, to 20 000 times more bacteria of a strain that grows 10% faster than a competitor. Ca^{2+} is also of vital importance in spore-forming bacteria. However, the requirements for Ca^{2+} in other bacteria and in the growth and function of archaeans are not so well documented.

Thus all cells exist in the presence of calcium, containing calcium on the surface and within their cells. Animal, plant and eukaryotic cells need calcium, both for hard extracellular structures and processes inside cells. This is because Ca^{2+} plays a vital role in membrane integrity, catalysis, in the electrical properties of cells and, crucially, as a unique regulator of events inside cells.

1.7 The Four Biological Roles of Calcium

The universal need for Ca^{2+} by cells falls into four distinct functional categories:

1. Structural, in both hard and soft tissues.
2. Electrical across biological membranes.
3. Extracellular cofactor.
4. Intracellular regulator.

1.7.1 Structural

Ca^{2+}, precipitated by various anions, plays a crucial structure role in the shells of invertebrates, bone, teeth, and the statocyst balance organs of jellyfish (Table 1.4). In addition, Ca^{2+} bound to inorganic anions, proteins and phospholipids is vital in maintaining the external and internal soft structures of many animal and plant cells, as well as bacteria and viruses. Ca^{2+} precipitates of phosphate form the hard part of bone and teeth. In contrast, calcium carbonate forms the hard part of coral and the shells of molluscs, as well as the shells of invertebrates, such as foraminiferans and coccolithophores. Ca^{2+} bound to proteins such as adhesins/integrins holds cells in organs together, and, on the outside of cells, Ca^{2+} helps to maintain the stability of the plasma membrane, together with its semipermeable properties. The dissociation of tissues into individual cells in the absence of external Ca^{2+} is due to loss of Ca^{2+} binding to the proteins responsible for tight and gap junctions. Mg^{2+} is the main cation bound to the negatively charged phosphate groups in DNA and RNA. But, Ca^{2+} also binds, for example, in the chromosome in bacteria. Nucleotides such as ATP, GTP, CTP and TTP are all mostly bound to a Mg^{2+}, the form reacting with all kinases and pumps being $ATPMg^{2-}$. However, in secretory vesicles Ca^{2+} bound to nucleotides and other molecules plays a role in maintaining the structural role of the vesicle. It also enables the contents to dissolve very quickly when released from the vesicle after it fuses with the plasma membrane. In contrast, the thyroid hormones T_3 and T_4 are extremely insoluble in water. They have to be stored covalently linked to protein in the thyroid if they are to dissolve quickly in the blood.

1.7.2 Electrical across Biological Membranes

Electrical currents in living systems are carried by ions: K^+, Na^+, Ca^{2+} and Cl^-, and H^+ in the mitochondria. Since Ca^{2+} is a positively charged ion, it moves towards a negative potential. Ca^{2+} changes the electrical potential across a semipermeable membrane, as it moves down a concentration gradient. Many cells have evolved specific Ca^{2+} channels to exploit this electrical activity. All cells maintain an electrical potential difference across their outer membrane, typically 40–100 mV, negative inside, though red blood cells have a membrane potential less than this. The gradient of K^+ ions is responsible for this resting potential of most cells. But excitability of cells is produced through action potentials, which depend on the movement of Na^+ and/or Ca^{2+} across the membrane. The action potential in nerve axons starts as a result of a sudden increase in the permeability to Na^+. This rapidly depolarises the cell, which then repolarises as a result of shutting the Na^+ channels, and/or increasing permeability to K^+. But many other excitable cells have Ca^{2+} channels that are sensitive to the potential across the plasma membrane. For example, the contraction of a barnacle muscle, holding the plates shut when the tide goes out, is provoked by the transmitter glutamate initiating an action potential dependent of Ca^{2+} ions moving into the cell. Our own heart beat depends on the electrical activity of Na^+, Ca^{2+} and K^+. The electrical activity of Ca^{2+} often contributes significantly to a local rise in cytosolic free Ca^{2+}, which is responsible for the cell event (e.g. a heart cell beat). However, the movement of Ca^{2+} through ion channels does not inevitably lead to a global increase in cytosolic free Ca^{2+}. Thus, the electrical role of Ca^{2+} can be considered distinct from its role as an intracellular regulator, even though these two functions interact closely with each other.

1.7.3 Extracellular Cofactor

Ca^{2+} regulates the activity of many extracellular proteins, for example, α-amylase in saliva, C1qrs, the first of complement complex enzymes in blood, trypsin and DNAase I secreted into the intestine by the exocrine pancreas to digest out food, some hormone and transmitter receptors, crystallin in the eye lens, phospholipase A in snake venom, and haemocyanin in mollusc and arthropod haemolymph. This role is one of a cofactor, and is different from the role of Ca^{2+} inside cells as an intracellular regulator, where it is a *change* in free Ca^{2+} inside the cell that triggers a cell event. With proteins outside cells, no change in free Ca^{2+} is associated with their activation. Only by removing Ca^{2+} artificially, using a chelator, can the effect of Ca^{2+} be prevented. Thrombin provokes blood plasma to clot. If the Ca^{2+} is removed from the plasma by addition of citrate, the clot will not form. Ca^{2+} is essential for the prothrombin to thrombin reaction, and for several other reactions in the blood-clotting cascade. But changes in blood Ca^{2+} are not responsible for initiating or propagating the blood-clotting cascade. Similarly, Ca^{2+} is essential for the classical pathway of the complement cascade, via C1q binding to the antibody on a cell. This activates a proteolytic cascade forming ultimately $C5b6789_n$, forming a pore causing the cell to explode, unless this is removed first (see Chapter 10). As with the blood-clotting cascade, Ca^{2+} does not initiate complement activation, nor do changes in blood Ca^{2+} regulate it. The role of Ca^{2+} as a cofactor *outside* cells is distinct from the role of Ca^{2+} as a universal role as regulator *inside* cells.

1.7.4 Intracellular Regulator

Ca^{2+} as a universal intracellular regulator in animal, plant, fungal and microbial cells requires a change in free Ca^{2+}, somewhere inside the cell, that occurs prior to a cell event, and then is responsible for initiating the cell event. This is revealed as the 'cell Ca^{2+} cycle' (Figure 1.4). This 'active' role for Ca^{2+} as an intracellular regulator distinguishes it from its 'passive' role as a cofactor of extracellular proteins. Intracellular Ca^{2+} is a switch, and is not the energy source. This comes from ATP. But ATP does not have an energy rich bond. Rather, the energy for cell events comes from the fact that the MgATP/MgADP + phosphate reaction is maintained well on the side of MgATP. The selection of Ca^{2+} as a universal intracellular regulator during evolution has depended on two critical features:

- The huge gradient of free Ca^{2+} across the outer membrane of all cells.
- The unique chemistry of Ca^{2+}.

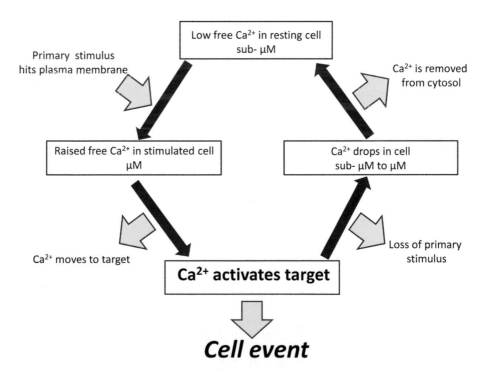

Figure 1.4 The cell Ca^{2+} cycle. Reproduced by permission of Welston Court Science Centre.

All cells have evolved mechanisms to maintain a very low cytosolic free Ca^{2+}, with the consequent large gradient of Ca^{2+} across the plasma membrane, as well as between the inside of organelles and the cytosol. These gradients, typically 10 000-fold across the plasma membrane, generate a Ca^{2+} pressure, which is exploited by physiological and pharmacological stimuli to provoke a cellular event. Then the chemistry of Ca^{2+} makes it ideal as an intracellular regulator. Oxygen ligands provide high-affinity Ca^{2+}-binding sites that enable Ca^{2+} to bind, and come off, proteins quickly at micromolar concentrations of Ca^{2+}, in the presence of millimolar Mg^{2+}. Binding of Mg^{2+} to proteins occurs, but this is not able to produce the necessary structural change to initiate a cellular event. Nor is it possible to generate the necessary gradient of Mg^{2+} across membranes that is possible with Ca^{2+}. Cations such as Zn^{2+}, Mn^{2+}, Fe^{2+}/Fe^{3+} or Cu^{+}/Cu^{2+} have not been selected, because they come off proteins too slowly to allow a bee to buzz or an Olympic athlete to run 100 m in less than 10 s. Furthermore, Ca^{2+} has only one ionisation state, unlike Fe^{2+}/Fe^{3+} or Cu^{+}/Cu^{2+}, which are involved in redox reactions. It is the special electrical and chemical properties of Ca^{2+} that has allowed Natural Selection to select mechanisms that generate the electrochemical gradients and intracellular targets, which trigger such a wide range of biological processes.

In eukaryotes, cell events that are triggered by a rise in intracellular free Ca^{2+} include:

- All forms of muscle contraction.
- Several other types of cell movement, e.g. chemotaxis.
- Secretion through fusion of intracellular vesicles with the plasma membrane.
- Activation of intermediary metabolism – glycogen breakdown in liver and muscle; fat breakdown in adipose tissue; pyruvate oxidation in mitochondria.
- Bioluminescence in luminous animals.
- Sperm maturation, and egg fertilisation.
- Parts of the cell division cycle.
- Defence mechanisms, e.g. the ER stress response and removal of lethal membrane attack complexes of complement.

In plants, events triggered by changes in intracellular Ca^{2+} include:

- Control of transpiration and gas exchange by guard cells in the leaf.
- Formation of the pollen tube, as it seeks out the seed to pollinate.
- Survival against stress, such as cold or heat shock, wind, herbivores, high salt.

In microbes, too, intracellular Ca^{2+} is involved in several important cell events, though the evidence is not as compelling as it is in animal and plant cells:

- Movement by chemotaxis.
- Regulation of gene expression and the cell cycle.
- Heterocyst formation in filamentous cyanobacteria.
- Spore formation.

A rise in intracellular Ca^{2+} is also crucial in several types of cell death, including necrosis and apoptosis (see Chapter 10). Furthermore, changes in intracellular Ca^{2+} play a vital role in the actions of many drugs (see Chapter 11).

The key experiment establishing Ca^{2+} is the initiator of a cell event is measurement of free Ca^{2+} in the live cell, or an intact organ. If Ca^{2+} is the initiator, then a rise in free Ca^{2+} must occur in the cell prior to the cell event. If this is prevented, e.g. by removing external Ca^{2+} or by chelating it inside the cell, then this should stop the Ca^{2+} signal, and thus stop the cell event

(Table 1.5). Then the mechanism responsible for the Ca^{2+} rise, and how this provokes the cell event, can be revealed. This involves:

1. Identifying Ca^{2+} channels and pumps in the plasma membrane.
2. Ca^{2+} release mechanisms from the SR/ER.
3. Regulation of Ca^{2+} signals involving other organelles, such as the mitochondria, lysosomes and Golgi apparatus.
4. Discovering the Ca^{2+}-binding proteins responsible for activating the molecular apparatus inside the cell that is causing the event.

Ingenious inventions to establish intracellular Ca^{2+} as a universal regulator include bioluminescent and fluorescent indicators for measuring free Ca^{2+}, imaging these indicators in live cells and intact tissues, and patch clamping, enabling individual Ca^{2+} channels to be studied in the plasma membrane of live cells. Other crucial inventions include cell biochemical techniques and chromatography, X-ray crystallography and NMR, revealing the three-dimensional structure of proteins, leading to the discovery of a fundamental property of many Ca^{2+} target proteins – the EF-hand or C2 binding site. The DNA revolution has identified thousands of Ca^{2+} signalling proteins in Bacteria, Archaea and Eukaryota. Genetic engineering has given us a wide range of protein indicators for free Ca^{2+} inside live

Table 1.5 Some cellular events triggered or controlled by a rise intracellular Ca^{2+}.

Event	Example of cell type
Movement	All forms of muscle
Secretion	Nerve terminal, exocrine pancreas and endocrine pancreas
Electrical excitation	Heart muscle
Intermediary metabolism in the cytosol and mitochondria	Muscle, liver and adipose tissue
Vision	Photoreceptors
Bioluminescence	Jellyfish
Gene expression	Many cells
ER stress	Cancer cells, yeast and plants
Maturation of sperm	Sperm acrosome reaction
Fertilisation by sperm	Egg
Cell cycle	Many cells
Plant transpiration and gas exchange	Guard cells
Plant defence against cold shock	Roots and leaves
Plant defence against wind	Stem and leaves
Seed germination	Pollen tube
Bacteria	
Chemotaxis	*Escherichia coli*
Gene expression	*E. coli*
Heterocyst formation	Cyanobacteria
Spore formation	*Bacillus*
Cell defence (e.g. vesicular removal of the membrane attack complex of complement)	Many cells
Cell death, necrosis, lysis, apoptosis	Many cells

cells, including the ability to target these to sites within the cell. This technique has also enabled key Ca^{2+} proteins to be altered, or even knocked-out, to test their role in a physiological or pathological process. This cohort of techniques has identified the Ca^{2+} toolbox responsible for intracellular Ca^{2+} as **the** universal regulator in animal, plant, fungal and microbial cells.

1.8   The Puzzle about Ca^{2+} Inside Cells

There are many curious things about intracellular calcium. This can be seen by examining what happens to the *free* Ca^{2+} every second during the beat of our heart. Before any electrical excitation by the action potential has reached the contracting heart cell, the myocyte, the resting free Ca^{2+} in the cytosol is very low, just 50–100 nM. This is more than 10 000 times lower that the free Ca^{2+} in the blood flowing through the coronary arteries of the heart. When the action potential reaches the heart, there is a rapid rise in cytosolic free Ca^{2+}, which falls back to the resting level as the heart beat finishes (Figure 1.5). When you get excited, or run down stairs, the adrenal gland sited over the kidney releases large amounts of adrenaline (epinephrine) into the blood. There will also be release of the neurotransmitter noradrenaline (norepinephrine) from nerves innervating the heart. These adrenergic stimuli cause the cytosolic free Ca^{2+} signal to rise much higher, enabling the myocyte to produce a stronger, faster contraction. The result is your blood pressure goes up and your pulse rate increases.

The puzzles therefore are:

1. How does the heart cell at rest maintain its very low cytosolic free Ca^{2+}?
2. How does it then use the *calcium pressure* resulting from this to cause the cell to contract and the heart to beat?

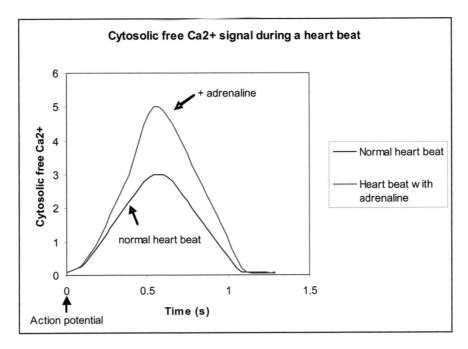

Figure 1.5 Cytosolic Ca^{2+} signals in a heart myocyte, triggered by an action potential. Reproduced by permission of Welston Court Science Centre.

3. How do adrenaline/noradrenaline alter the Ca^{2+} signal, and what is the consequence of this?
4. What goes wrong with the Ca^{2+} signal when the heart malfunctions, such as in an infarction or an arrhythmia?

If the heart cell is to maintain such a large gradient of free Ca^{2+} across its outer membrane there must be efflux mechanisms pumping Ca^{2+} out of the cell to compensate for Ca^{2+} as it leaks in. There must also be Ca^{2+} channels to let Ca^{2+} in during the heart beat, and a store inside to control Ca^{2+} inside the myocyte. And there must be a Ca^{2+} target protein that acts as the switch to activate the contractile apparatus, allowing the cell to contract. Finally, there must be molecular mechanisms activated by adrenaline that can alter the size and timing of the Ca^{2+} transient inside the myocyte. These molecular mechanisms are important drug targets and misbehave in disease. Thus, there are seven key questions we must answer if we are to understand fully how intracellular Ca^{2+} can provoke a cell event (Box 1.1).

Box 1.1 The seven key question.

1. How does Ca^{2+} get into and out of cells, via channels, pumps and exchangers?
2. How is Ca^{2+} regulated inside the cell through internal stores and buffers?
3. What are the Ca^{2+} targets inside cells, and how do these trigger the cell event?
4. How do drugs modify the Ca^{2+} signalling system?
5. What happens to intracellular Ca^{2+} when a cell is injured or dies, and does this explain the symptoms and natural history of disease?
6. What is the role of intracellular Ca^{2+} in cell defence, attack or stress?
7. How did the Ca^{2+} signalling system evolve?

1.9 How Important Intracellular Calcium has been in Science

Web of Science reveals nearly 1.7 million scientific publications involving calcium published between 1900 and 2016, over 500 000 being published in the first decade of the twenty-first century. Over 230 000 publications between 1900 and 2016 relate to intracellular calcium, with more than 64 000 referring specifically to cytosolic calcium (Figure 1.6). This compares with DNA publications being 4.4 million between 1900 and 2016, 1.7 million being published the first decade of the twenty-first century. Over 80% of 'calcium' papers were published between 1980 and 2010. Interestingly, citations for *intracellular* calcium only began in the late 1960s, although one does appear in 1907. By 2010, more than 100 000 had been published, over 80% being published after 1980. The total number of citations for *cytosolic free* calcium is less, nearly 27 000, over 90% being after 1980. There are none found before 1970! Perhaps the most interesting observation from the citation index is when 'Ca^{2+}' was first used as opposed to 'calcium'. Only four citations using Ca^{2+} are to be found before 1950, and none prior to this, with over 450 000 being published after 1970. Ca^{++} was the preferred designation by biologists for calcium until the 1980s. Yet, it is the form of calcium as Ca^{2+} upon which all of its biological properties depend. But the number of citations about calcium may be flattening off! The first cited papers in Web of Science and PubMed are: Ringer and Buxton (1885) and Lorenz and Wehrlin (1900) for Calcium; Noguchi (1907) and Chibnall and Channon (1927) for Intracellular calcium; De Sombre *et al.* (1969) for Cytosolic calcium; Savicevic *et al.* (1959) for Ca^{2+}. But the first true paper on *intracellular* calcium was by Pollack (1928). Cytosolic calcium only begins to emerge as a specific term in the 1970s, the first papers involving the

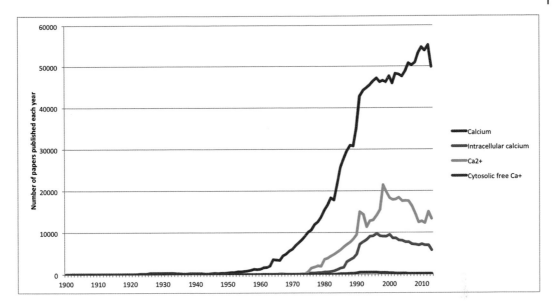

Figure 1.6 Number of calcium citations from 1900 to 2015. Data from PubMed. Reproduced by permission of Welston Court Science Centre.

effect of dantrolene sodium on SR calcium release in muscle, and the action of vasopressin on the kidney. In 1980, Maynard Case at the University of Manchester, United Kingdom had the vision to set up a new international Journal, *Cell Calcium*, first published by Churchill Livingstone, and now Elsevier. Most of the world leaders in intracellular Ca^{2+} have since published in this very useful journal.

1.10   Darwin and Intracellular Ca^{2+}

Natural Selection, revealed to us by Charles Darwin (1809–1882) and Alfred Russel Wallace (1823–1913), is the driving force in evolution. Much of the evidence has been based on lineages of animals and plants found in their 'hard' remains as fossils, and on demonstrations of selective advantages of variations in structures, such as beaks, colour and biochemistry, e.g. pesticide or antibiotic resistance. The variations in the Ca^{2+} signalling toolbox, between different cell types, between individuals of the same species, and between populations of different species, provides a beautiful example of the molecular biodiversity upon which evolution by Natural Selection depends. The fossil record over the past 3800 million years shows that calcium precipitates of $CaCO_3$ have been used throughout the evolution of life on this planet. Evidence from molecular biology, together with the toxic nature of prolonged high Ca^{2+} levels inside cells, argues strongly that primeval cells must have developed Ca^{2+} pumps to keep their free intracellular Ca^{2+} low, setting the scene for the 'calcium pressure' across then plasma membrane to be exploited to act as the source for cell activation. A key chemical property of Ca^{2+} is that it comes on and off proteins fast, in milliseconds. This is essential if fast events, such as nerves firing, or muscles contracting and relaxing quickly, are to occur. Zinc binds avidly to many proteins, but it comes off slowly, so a heart would not be able to beat, and a cheetah would not be able to run and catch an antelope, if evolution had selected Zn^{2+} instead of Ca^{2+} as its universal intracellular regulator.

But Darwin and Wallace's BIG idea of Natural Selection, which Darwin always wrote with initial capital letters, is not only relevant to evolution over millions of years. It works in real-time today, and depends on six things:

1. Species are not constant, they evolve.
2. Variations exist between individuals, within and between species.
3. Hereditary – variations are passed on to offspring.
4. The evolution of a species occurs small change by small change.
5. The struggle for existence – not all offspring survive.
6. Natural Selection results in the best adapted having the best chance of surviving to pass on adaptations.

Natural Selection is a scientific *principle* that applies to the *process* of evolution, and it works in real-time. The beaks of finches in the Galapagos are selected for as particular food supplies become available. Methicillin-resistant *Staphylococcus aureus* (MRSA) is a natural mutation, selected when antibiotics are used. Flu virus survives against our antibody defence system by mutating. Rats and mice are selected for when they become resistant to the pesticide warfarin. Natural Selection is at work in our own bodies, as our brain develops, and when we make anti-bodies. The key feature of a biological process that allows Natural Selection to act is molecular variation. It is molecules that are responsible for shape, colour and behaviour. There are five molecular variations in the Ca^{2+} signalling system (Box 1.2.).

Box 1.2 The five main molecular variations in Ca^{2+} signalling.

1. The type of Ca^{2+} signal, how big it is, how long it lasts and where in the cell it occurs.
2. The amino acid sequences, together with the gene sequences, of the components that produce the Ca^{2+} signal, and those that are its targets.
3. The biochemical and electrical characteristics of these components (e.g. binding constants, kinetics, affinities, specificities, conductance).
4. The level of expression of the components.
5. The number of cells expressing particular components.

These variations occur both within an individual, within a species, and between species. Examples can be found in Ca^{2+} channels, Ca^{2+} pumps and Ca^{2+}-binding proteins, with subtly different biochemical and kinetic properties. In the plasma membrane of different cells there are:

- Four Ca^{2+}-MgATPases.
- Two Na^+/Ca^{2+} exchangers.
- Several Ca^{2+}/H^+ exchangers.

Similarly, there are three types of Ca^{2+} pumps in the ER (SERCA1, 2 and 3), and three types of each receptor (IP$_3$ and ryanodine) in the ER that cause Ca^{2+} to be released into the cytosol. There is a plethora of Ca^{2+} channels in the plasma membrane opened by voltage, ligands outside and inside cells, and events inside the ER. This molecular diversity is compounded by variations within individual cells, even of the same type, as a result of variations in expression and location, caused by differences in transcription, alternative splicing of mRNA, translation, covalent modification, and protein degradation. The combination of these variations that has the best selective advantage gives the individual organism the better chance to pass on its DNA to the next generation. A cheetah can only catch an antelope if its cytosolic free Ca^{2+} signals oscillate fast enough to cause its muscles to contract a high speed.

1.11   The Scene Set

This book aims to excite curiosity about Nature, with its wide range of biological processes, and then attempts to satisfy this by showing how and why a particular process in Nature has evolved to use intracellular calcium as its chemical switch. There are ten crucial questions (Box 1.3).

Box 1.3 Ten crucial questions.

1. What is the evidence that intracellular calcium is the key regulator of a particular phenomenon or process?
2. Is the process analogue or digital, and how does calcium determine this?
3. How does the molecular biodiversity of the calcium signal determine whether a particular cell fires, divides, defends itself against stress or attack, or dies?
4. Why has Nature chosen a specific component for a Ca^{2+}-dependent process?
5. How high does the free Ca^{2+} have to reach to trigger the cell, and how many Ca^{2+} channels are required?
6. Why has a particular cell selected a specific type of Ca^{2+} signal?
7. What have been the key discoveries about intracellular calcium, how were they made, when, and by whom?
8. What goes wrong with Ca^{2+} in disease, and are there drugs to stop this?
9. What has been the importance of Ca^{2+} signalling in the evolution of life over 3800 million years?
10. What don't we know about intracellular calcium?

1.12   'Ja Kalzium, das ist alles!'

Otto Loewi (1873–1961; see Frontispiece) was born in Frankfurt-am-Main, Germany. He won the Nobel Prize in 1936 with Henry Dale for the discovery of neurotransmitters. Before their work, it was not clear whether the signals from nerve to nerve, or from nerve to muscle, were carried electrically or chemically. Loewi had a brain wave during a dream. But, unfortunately, when he woke up in the morning he had forgotten the details of his great idea! So the next night he put paper and pencil by his bedside, in the hope he would have the same dream. He did! Back in the lab, he took two frog hearts; one with the vagus nerve attached, the other with it detached, and suspended them in Ringer's solution. Electrical stimulation of the vagus nerve in heart number one caused a reduction in heartbeat, but not in heart number two, with no vagus nerve connection. He then took the fluid bathing the normal heart, and added it to the second heart, without the vagus connection. This solution also caused a reduction in its heart beat. Loewi therefore argued that this was definitive proof that there had to be a chemical released by the vagus nerve. This turned out to be acetylcholine. Loewi published this key experiment in 1921. Yet, four years earlier, he had shown that calcium was required for the effect of digitalis on the heart. Loewi was one of the first to realise how important calcium was in the excitability of heart muscle and nerve cells, together with the transmission of this excitability between them. By 1939, it was clear to him there was something special about the biological role of calcium. Hence, his famous quote: '*Ja Kalzium, das ist alles!*'

Perceptive observations of Natural history, in the tradition of naturalists like Charles Darwin and Alfred Russel Wallace, are the starting point for understanding what is special about calcium inside cells. Frederick Gowland Hopkins (1861–1947) became interested in biochemistry as a schoolboy, trying to extract the colour from butterfly wings, and the substance squirted out by bombardier beetles. He was the founder of British biochemistry, discovered tryptophan in 1901, and was awarded the Nobel Prize in 1929, along with Christiaan Eijkman, for the discovery of vitamins. He wrote in his Presidential Lecture to the London Natural History Society in 1936:

> *All biologists deserve the coveted name of naturalist. The touchstone of the naturalist is his abiding interest in Nature in all its aspects.*

Curiosity is one of the greatest gifts evolution has given us. As the motto we use for the Darwin Centre I set up in Pembrokeshire in Wales states: *Curiosity inspires, discovery reveals*. There is much to be curious about intracellular calcium.

1.13   Calcium – The Fundamentals

1. Calcium, in the form $^{40}Ca^{2+}$, is a major element in the Universe and the Earth's crust.
2. Ca^{2+} is found inside and outside all cells, and all cells need it – animal, plant, fungal and microbe.
3. Ca^{2+} has four biological roles – structural, electrical, extracellular cofactor, intracellular regulator.
4. Ca^{2+} is the only one of the 29 elements, found in living organisms, which acts as a chemical switch inside cells to trigger movement, secretion, replication, vision, bioluminescence, metabolism, defence, and even death.
5. The key to the universal role of Ca^{2+} as an intracellular regulator is the very low cytosolic free Ca^{2+} in all cells, sub-micromolar, resulting in a Ca^{2+} pressure across the plasma membrane, and across the membrane of intracellular organelles.
6. A small absolute movement of Ca^{2+} into cells, and released from an internal store, causes a large fractional change in cytosolic free Ca^{2+}. This would be impossible with ions such as K^+, Na^+, Mg^{2+} or Cl^-.
7. Ca^{2+} is **the** intracellular signal in birth, life and death.

Recommended Reading

* A must read.

Books

*Campbell, A.K. (1983) Intracellular Calcium: its universal role as regulator. Chichester: John Wiley & Sons Ltd. A wide ranging review of intracellular calcium up to 1983.

*Campbell, A.K. (2015) Intracellular Calcium. Chapter 1, *Setting the scene: What is so special about calcium*? Chichester: John Wiley & Sons Ltd.

Krebs, J. & Michalak, M. (Eds) (2007) Calcium: a matter of life and death. Amsterdam: Elsevier. Comprehensive multi-author book on calcium signalling, but without articles on plants or bacteria.

Reviews

Berridge, M.J. (2012) Calcium signalling remodelling and disease. *Biochem. Soc. Trans.*, **40**, 297–309. Important review on calcium signalling related to disease.

*Carafoli, E. & Krebs, J. (2016) Why calcium? How calcium became the best communicator. *J. Biol. Chem.*, **291**, 20849–20857. Good review on what is special about calcium that makes it a universal regulator.

Carafoli, E. (2005) The symposia on calcium binding proteins and calcium function in health and disease: an historical account, and an appraisal of their role in spreading the calcium message. *Cell Calcium*, **37**, 279–281. Wide ranging papers on calcium signalling.

*Williams, R.J. (2006) The evolution of calcium biochemistry. *Biochim. Biophys. Acta*, **1763**, 1139–1146. Chemical analysis of calcium signalling in evolution.

Articles

*Davy, H. (1808) Electro-chemical researches, on the decomposition of the earths; with observations on the metals obtained from the alkaline earths, and on the amalgam from ammonium. *Philos. Trans. R. Soc. London*, **98**, 333–370. Discovery of calcium.

Pollack, H. (1928) Micrurgical studies in cell physiology. VI. Calcium ions in living protoplasm. *J. Gen. Physiol.*, **11**, 539–545. First demonstration of Ca^{2+} change inside a cell.

Ringer, S. (1883) A further contribution regarding the influence of different constituents of the blood on the contractions of the heart. *J. Physiol.*, **4**, 29–43. First demonstration of the need for Ca^{2+} in a cell event.

2

Intracellular Calcium – Principles

Learning Objectives

- Cell signalling is defined as 'the mechanism that tells a cell what to do'.
- Intracellular Ca^{2+} is the cell signal which tells a nerve to fire, a muscle to contract, an endocrine or exocrine cell to secrete, an egg to be fertilised, a luminous cell to flash, a plant to breath, a bacterium to move, or even a cell to die.
- Disturbances in intracellular Ca^{2+} occur in many diseases, are often involved intimately in the mechanism underpinning the disease, and can be crucial to a cell surviving stress.
- Many pharmaceutical substances, used clinically or experimentally, act directly or indirectly through the Ca^{2+} signalling system.
- The Rubicon principle – intracellular Ca^{2+} acts as a switch. It is digital, other intracellular regulators, such as cyclic AMP, act in an analogue manner to modify the cell response.
- There are three cell types based on whether most of the Ca^{2+} signal comes from outside or inside the cell, or both.
- Ca^{2+} signals can be transient, oscillatory, and localised to a part of the cell, depending on the physiological response of the cell.
- Key experiments identify the specific **Ca^{2+} tool kit** required by each cell type. These require characterisation of how the cell maintains its resting low cytosolic free Ca^{2+}, how the primary stimulus causes Ca^{2+} to rise, how secondary regulators modify this, what the Ca^{2+} target is, how this causes the physiological or pathological cell response, and finally how the cell recovers and returns to rest.
- How things are named, specifically nomenclature relevant to intracellular Ca^{2+}. This includes organisms, genes, proteins, ion channels and pharmaceutical substances.
- How Darwin and Wallace's principle of Natural Selection fits in with intracellular Ca^{2+}, explaining why there is so much molecular biodiversity in the Ca^{2+} signalling system, between both the same and different cell types, and even within one cell.

2.1 Ca^{2+} and the Concept of Cell Signalling

2.1.1 Primary Signals and Intracellular Ca^{2+}

Cell signalling (US spelling = signaling) is essential for cell and organism survival and reproduction. A cell signal instructs a cell to do something – a nerve to fire, a muscle to contract, an endocrine or exocrine cell to secrete, a bioluminescent cell to flash or glow, an egg to be

Fundamentals of Intracellular Calcium, First Edition. Anthony K. Campbell.
© 2018 John Wiley & Sons Ltd. Published 2018 by John Wiley & Sons Ltd.
Companion Website: http://www.wiley.com/go/campbell/calcium

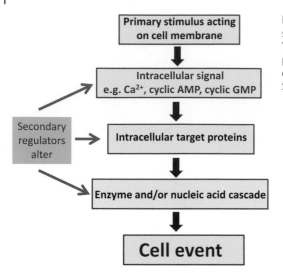

Figure 2.1 Principle of cell signalling. A cell is switched on by a primary external or internal signal. This interacts with a receptor, which activates a protein and enzymatic cascade. This leads to the cell event. Reproduced by permission of Welston Court Science Centre.

fertilised, a cell to defend itself, or even kill itself (Figure2.1). Primary cell signals are physical, chemical or biological, and can be can be generated outside or inside a cell. They include electrical action potentials, neurotransmitters and hormones, bacteria and viruses. The result of the signal is that the cell, and the tissue it is part of, carries out a specialised function, such as movement, secretion, metabolism, growth, division, defence against attack, or death via apoptosis.

Typically, a cell signal interacts with a channel or receptor, which generates an intracellular signal, activating an electrical and/or a chemical cascade, leading to a cell event (Figure 2.1). The event may be digital, when the signal acts as a switch, or analogue, when the cell response is graded, depending on the level of stimulus. Errors in cell signalling are involved in most diseases, either as a cause or consequence of cell damage. Without cell signalling, an organism cannot do anything, or survive attack. Intracellular Ca^{2+} is **the** most important intracellular signal in the whole of life. The Ca^{2+} signal is generated inside a cell, as a result of movement of Ca^{2+} from outside, as well as release from intracellular stores. A Ca^{2+} signal may occur, and act, within milliseconds, e.g. when a nerve or muscle fires, or it may occur over many minutes or even hours, such as in a human egg fertilised by a sperm. The questions we want to answer are:

1. How is the Ca^{2+} signal generated inside the cell?
2. How does this Ca^{2+} signal switch the cell on?
3. How does the cell switch off?

The key to Ca^{2+} as an intracellular signal is the huge Ca^{2+} gradient that every cell maintains across its outer membrane – **the Ca^{2+} pressure**. Also important are the Ca^{2+} gradients maintained between the cytosol and the inside of intracellular organelles – the endoplasmic reticulum (ER), mitochondria, lysosomes, Golgi and vesicles. Measurements of cytosolic free Ca^{2+}, using the Ca^{2+}-activated photoproteins aequorin and obelin, and fluorescent dyes, have established that the cytosolic free Ca^{2+} in all resting cells – animal, plant, fungal and microbe – is sub-micromolar. The cell stimulus causes the cytosolic free Ca^{2+} to rise in the range 1–10 μM, though levels in bacteria can rise up to 20–30 μM. Injury to the cell can cause a rise in cytosolic free Ca^{2+} in the range 10–50 μM. Levels above this mean the cell is dead or dying. With a concentration of free Ca^{2+} of 1.3 mM in human plasma, this means there is at least a 10 000-fold gradient of Ca^{2+} across the outer membrane of every cell in the body. For cells exposed to around 10 mM free Ca^{2+} in seawater, the gradient is even higher. It is this enormous calcium pressure that holds the key to Ca^{2+} switching cells on – to move, to contract, to secrete and so on. Just a small absolute movement of Ca^{2+} into the cytosol will cause a very large fractional rise in cytosolic free Ca^{2+}. For example, when a muscle contracts the cytosolic free Ca^{2+} can rise from less than 0.1 μM to more than 10 μM – a 100-fold rise. Yet, the absolute change osmotically is tiny. A similar fractional change would be impossible for Na^+, K^+, Mg^{2+} or Cl^-!

2.1.2 The Source of Intracellular Ca^{2+}

The source of Ca^{2+} for the intracellular signal is both from outside and intracellular stores, there being much cross talk in Ca^{2+} between organelles. For Ca^{2+} to enter the cell from outside the cell, specific Ca^{2+} channels have to open in the plasma membrane. These can be opened by:

- Receptors.
- Electrically, through changes in membrane potential.
- An intracellular signal acting on proteins exposed to the cytosolic side of the membrane.

There are five main intracellular stores, which can release Ca^{2+} into the cytosol, in animal cells, and another unique to plant and fungal cells:

1. Endoplasmic reticulum, Ca^{2+} being released by inositol trisphosphate, Ca^{2+} itself via ryanodine receptors, sphingosine 1 phosphate, or direct interaction of the ER with the dihydropyridine receptor on the outer membrane.
2. Mitochondria, via changes in energy metabolism or interaction with the endoplasmic reticulum.
3. Lysosomes, Ca^{2+} being released by substances such as NAADP. Key Ca^{2+} channels are mucolipin-1 (TRPML1 = transient receptor potential cation channel, mucolipin subfamily, member 1), and two pore channels through phosphatidylinositol 3,5-bisphosphate (PtdIns(3,5)P2).
4. Intracellular vesicles, Ca^{2+} being released by substances such as cyclic ADP ribose.
5. Golgi, Ca^{2+} being released by various intracellular signals.
6. Vacuole with its tonoplast membrane, unique to plant and fungal cells, Ca^{2+} being released by various intracellular signals.

Imagine you have just had a breakfast of toast and marmalade, and a cup of coffee. Taste buds on your tongue will have been activated, and you will have secreted amylase from salivary glands into your mouth, and digestive juices into your stomach and intestine. Insulin will have been released into your blood, in response to glucose taken up from digested starch from the bread and sucrose in the marmalade. Unfortunately, whilst cutting some bread, you cut your finger. But, after running the cut under the tap, and then pressing tissue paper on to it, the bleeding stops. Amazingly, calcium is involved in every step of this scenario! But only in certain places is Ca^{2+} truly acting as a regulator within cells. For Ca^{2+} to be 'active' as an intracellular regulator, a change in free Ca^{2+} must occur somewhere within the cell, and this must then be the trigger of a cellular event (Table 2.1).

Table 2.1 Examples of cell events triggered by intracellular Ca^{2+}.

Cell response	Cell event	Example
Movement	Contraction	Heart beat, leg movement, smooth muscle in the gut
	Chemotaxis	Neutrophils in infection
Secretion	Substance released	Hormones and neurotransmitters
	Fluid released	Mucus in the gut
	Organelle released	Nematocyst firing in jellyfish
Uptake into the cell	Phagocytosis – particle	Neutrophil engulfing a bacterium
	Endocytosis – membrane	Removal of hormone receptors
	Substrates – fluid	Glucose
	Ions – charged molecule	Ca^{2+} channels
Electrical excitation	Nerve firing	Action potential
	Muscle firing	Action potential
Fertilisation	Egg	Sperm activating the egg
Cell division	Embryo development	Differentiation
Stress	Temperature shock	Yeast
Death	Apoptosis	Lymphocytes after combating an infection
Communication	Bioluminescence	Jellyfish flash
Plants	Wind response	Activation of stress genes
Microbes	Cell replication	Generation time

In the breakfast scenario, Ca^{2+} is acting truly as an 'active' intracellular regulator in many of your cells. First, a rise in free Ca^{2+} occurred in the dendrites of your brain cells, causing them to generate an action potential. When this action potential reached the nerve terminals, it caused a rise in free Ca^{2+} that triggered release of transmitter. This provoked your brain to make decisions. Release of acetylcholine at neuromuscular junctions generated action potentials, which then caused Ca^{2+} signals, provoking your muscles to contract, enabling you to use your knife to cut the bread. TRP (transient receptor potential) channels were activated in your nose and mouth, giving you a smell or taste sensation, as a result of Ca^{2+} entering the taste bud and smell sensory cells. Acetylcholine activated your salivary gland cells inside your cheeks, causing a rise in intracellular Ca^{2+} that stimulated release of amylase. Here, we find also a 'passive' action of Ca^{2+}, as amylase requires Ca^{2+} at its active centre to work effectively as an enzyme. Similarly, when the blood clotted to stop you bleeding to death, after cutting your finger, Ca^{2+} was required by prothrombin to convert it into thrombin, and thus cause the fibrin to coagulate, and form the clot. However, within the clot there were tiny platelets, 1/10 of the size of the red cells, bound together, that secreted substances helping the clot to hold together. This event in the platelets involved a rise in cytosolic free Ca^{2+}. But there were no changes in blood Ca^{2+} to activate amylase, or cause the blood clot to form. Neutrophils and macrophages will have been attracted to the cut, releasing toxic oxygen species and enzymes to kill any invading bacteria, as a result of a rise in cytosolic free Ca^{2+}. Eating food stimulated the release of acetylcholine and cholecystokinin close to exocrine cells in the pancreas, causing a rise in cytosolic free Ca^{2+}. This provoked release of digestive enzymes, such as trypsin and chymotrypsin. Similarly, a rise in Ca^{2+} in the pancreatic endocrine β-cells, caused by a rise in plasma glucose, provoked insulin secretion. The metabolism of your cells around the body then changed, also as a result of rises in cytosolic free Ca^{2+}. Too much coffee might have caused your heart to race, as a result of caffeine acting as secondary regulator of free Ca^{2+} in your heart cells.

The key therefore to all these events was a rise in cytosolic free Ca^{2+}, which then acted as an 'active' regulator. The key experiment to prove this is to measure cytosolic free Ca^{2+} in live cells.

2.2 The Problem

How do cells change their state, and do something (Table 2.1)? How does a nerve fire, and transmit messages in the brain, or to other tissues, such as muscle? How does a muscle contract? How does a heart beat? How does a barnacle shut its lid when the tide goes out? Why do bees buzz? How does a β-cell in the pancreas secrete insulin? How does a sperm fertilise an egg? How do eyes recognise that photons have hit their photoreceptors? How do bacteria move along a nutrient gradient? What causes a jellyfish to flash? What happens to a cardiac myocyte when it suffers a heart attack? And what determines whether any cell survives a stress or dies? These events require:

1. The machinery for the cell event.
2. An energy source.
3. An extracellular signal to set the process off.
4. An intracellular signal to trigger the machinery within the cell.

There has been some confusion about the energy source for a cell event. A rise in Ca^{2+} within the cell is the signal for nerve terminals firing, muscle contraction, and secretion from endocrine and exocrine cells. But, the absolute change in intracellular Ca^{2+} is very small, and thus cannot be the energy source for such phenomena. This is hydrolysis of ATP. But ATP does not drive ion pumps in nerves, muscle contraction or anabolic reactions because it has an energy rich bond. *It does not have such a bond.* MgATP is the energy source for cell events because the cell maintains the MgATP/MgADP + phosphate reaction well on the side of ATP, far from equilibrium. If a cell is at equilibrium it is dead! ATP is always in the form $MgATP^{2-}$, when it reacts inside cells. It is the thermodynamic drive towards equilibrium that is the driving force of all cell events, not any mythical energy in the third phosphate bond of ATP. Unfortunately, the mistaken idea that ATP has an energy-rich bond is still perpetuated in several school and university textbooks, and several web sites.

2.3 Some Key Questions

2.3.1 Question 1 – What is the Primary Stimulus and Secondary Regulator(s)?

All cellular events have to be initiated by a physical, chemical or biological *primary* stimulus, and can be modified in an analogue way by a *secondary* regulator, altering the magnitude and/or timing of the cellular event (Table 2.2). Muscle contraction, secretion, egg fertilisation, death and many other cellular events are all digital, i.e. the cell is switched on by an external signal.

Table 2.2 Some primary stimuli and secondary regulators of cell processes involving intracellular Ca^{2+}.

Primary stimulus	Secondary regulator	Effect of secondary regulator
Action potential on heart	Adrenaline	Increases rate and strength of heart beat
Action potential on heart	Acetylcholine	Decreases rate and strength of heart beat
Action potential on nerve	Adenosine	Reduces amount of transmitter released
Pancreatic β-cell	Adrenaline	Reduces insulin secretion
Bacterial peptides	Adenosine	Reduces superoxide release
Light	Dark	Increased sensitivity of photoreceptors
Prolactin	Dopamine	Reduced release of hormones
Touch	Light	Inhibits bioluminescence of jellyfish

2.3.2 Question 2 – What is the Intracellular Signal?

The primary stimulus, interacting with the outside of the plasma membrane, generates a messenger within the cell. This then activates the appropriate molecular machinery inside the cell that causes the cellular event. Ca^{2+} was the first such intracellular messenger to be identified (see Chapter 3 for historical evidence). The term 'second messenger' was introduced by Earl Sutherland (1915–1974) to describe cyclic AMP. In fact, in many non-excitable cells Ca^{2+} is really a 'third' messenger. For example, histamine or ATP, binding to a cell surface receptor, activates the enzyme phospholipase C (PLC; EC 3.1.4.11). This cleaves the phospholipid, phosphatidyl inositol 4,5-bisphosphate (PIP_2), into diacylglycerol (DAG) and inositol 1,4,5-trisphosphate (IP_3). IP_3 then binds to its receptor on the endoplasmic reticulum (ER), which causes release of Ca^{2+} into the cytosol. As a result Ca^{2+} channels in the plasma membrane open, leading to a large global Ca^{2+} signal, which activates the cell event. DAG activates protein kinase C (EC 2.7.11.13). Thus, IP_3 is the 'second' messenger, extracellular histamine or ATP being the first, the primary stimulus. Ca^{2+} is the 'third' messenger, the intracellular agent that activates synaptotagmin, which provokes the cellular event, secretion. Other intracellular messengers include cyclic GMP, nicotinic acid adenine dinucleotide phosphate (NAADP), cyclic ADP ribose, nitric oxide (NO), IP_4 and other inositol phosphates, and sphingosine-1-phosphate (Figure 2.2). All of these interact with Ca^{2+} signalling.

2.3.3 Question 3 – How do the Secondary Regulators Work?

External *secondary* regulators alter the magnitude and/or timing of the cellular event (Table 2.2). This is the analogue part of cell events triggered by intracellular Ca^{2+}. For example, in the heart, when we run, adrenaline causes a rise in myocyte cyclic AMP. This alters the speed of opening and closing of the Ca^{2+} channels in the plasma membrane, and increases the Ca^{2+} stored inside the sarcoplasmic reticulum (SR). As a result, the cytosolic free Ca^{2+} signal is higher than in the absence of adrenaline, and the heath beat is stronger and faster. In contrast, release of acetylcholine by nerves in the heart reduces the speed and strength of the heart beat.

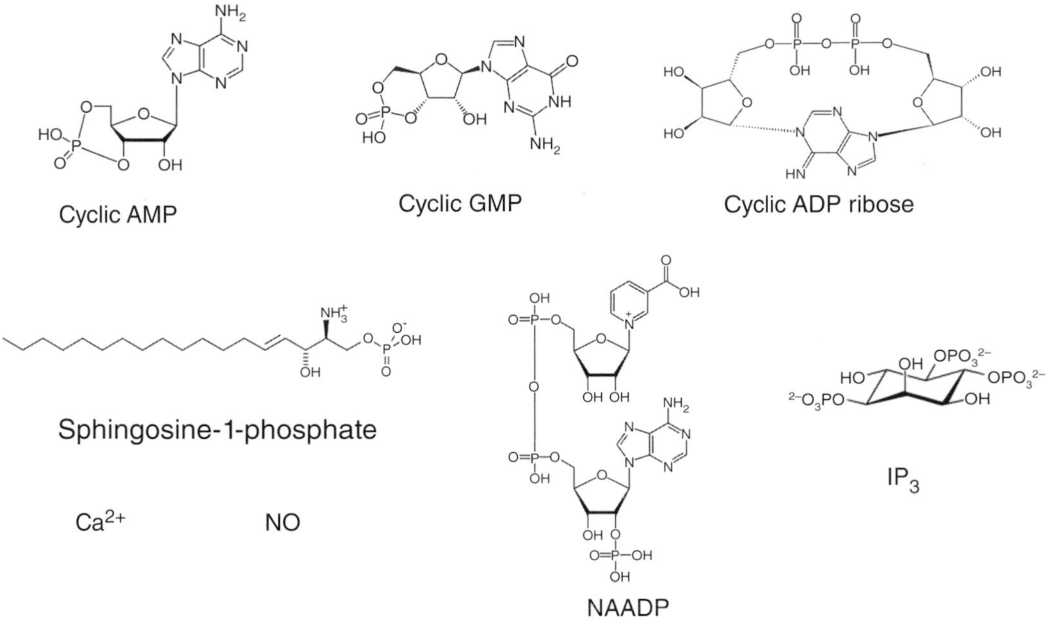

Cyclic AMP

Cyclic GMP

Cyclic ADP ribose

Sphingosine-1-phosphate

Ca^{2+} NO

NAADP

IP_3

Figure 2.2 Some intracellular messengers that interact with Ca^{2+} signalling.

2.4   Types of Intracellular Ca^{2+} Signal

2.4.1   Category of Cell Based on the Ca^{2+} Signal

In virtually all cell events provoked by a rise in intracellular Ca^{2+}, there is movement of Ca^{2+} in and out of organelles inside the cell, as well as movement of Ca^{2+} from outside into the cell, and then out again. Even when the bulk of the Ca^{2+} for the cytosolic free Ca^{2+} signal comes from inside, as a result of release from the ER, there must also be an influx of Ca^{2+} from outside the cell, if the cell event is to be maintained through a continuous rise in cytosolic free Ca^{2+}. Otherwise the cell would be depleted of Ca^{2+}, as a result of it being pumped out of the cell. There are three categories of cells that use a Ca^{2+} signal (Box 2.1).

Box 2.1 Categories of cell based on the source of Ca^{2+} for the intracellular signal.

Type 1: the main source of Ca^{2+} for the global rise in cytosolic free Ca^{2+} is from outside the cell, e.g. neurotransmitter at a nerve terminal, a pancreatic β-cell secreting insulin, a mast cell firing to secrete histamine at the site of an inflammation, and the flash of a luminous jellyfish or hydroid.
Type 2: the main source of Ca^{2+} for the global rise in cytosolic free Ca^{2+} is from inside the cell, e.g. skeletal and heart muscle contraction, and egg fertilisation.
Type 3: the source of Ca^{2+} for the global rise in cytosolic free Ca^{2+} is a sum of Ca^{2+} from both outside and inside the cell, e.g. liver, platelets and the kidney.

2.4.2 Types of Ca²⁺ Signal

Cytosolic free Ca²⁺ signals vary enormously between cell types, the type of signal matching the physiology of the cell event. There are four main types of global intracellular Ca²⁺ signal (Figure 2.3):

- Rapid rise, and then a fall back to the resting level.
- Rise to a plateau, and then fall back to resting level.
- Rise to a peak, and fall to a plateau.
- An oscillation.

Transient cytosolic free Ca²⁺ signals can be a very short spike of milliseconds duration (e.g. in a nerve terminal), or they can last just a few seconds or so (e.g. in the beating heart). In these cases, the cytosolic free Ca²⁺ returns to the basal level after each stimulus. In contrast, when longer rises in cytosolic free Ca²⁺ occur, the Ca²⁺ signal remains at a plateau level, well above that in the resting cell, for many seconds or even minutes. A further interesting feature of the global cellular Ca²⁺ signal is that it can oscillate – first seen in the nerves of the marine sea slug *Aplysia*, where the action potential generates a series of spikes followed by a resting

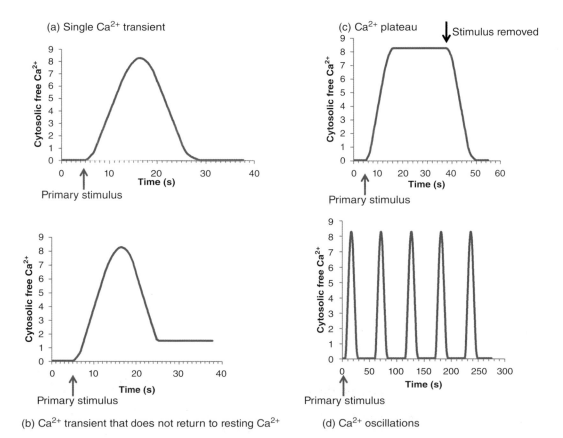

Figure 2.3 Types of cellular Ca²⁺ signals. In (a) the peak in cytosolic free Ca²⁺ in these examples is typical of a many cells, though muscle can twitch in milliseconds. The maximum rise in cytosolic free Ca²⁺ in non-excitable cells is often less than this, typically 1–5 μM. Reproduced by permission of Welston Court Science Centre.

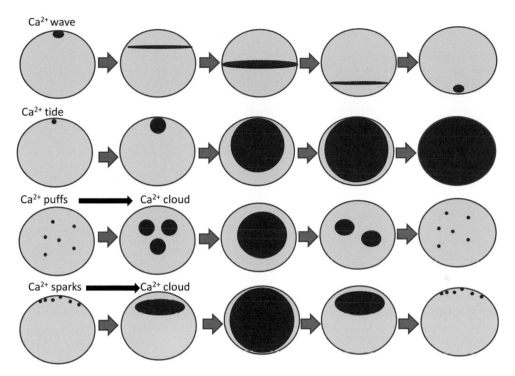

Figure 2.4 Cytosolic free Ca^{2+} signals seen by live cell imaging. Reproduced by permission of Welston Court Science Centre.

period. The action potential opens voltage-gated Ca^{2+} channels in the plasma membrane. Each Ca^{2+} spike leads to a gradual rise in cytosolic free Ca^{2+}. This Ca^{2+} activates K^+ channels, which repolarise the cell, the voltage-gated Ca^{2+} channels close, and the pacemaker cell stops firing. Depolarisation then initiates the firing process to start again. Another type of Ca^{2+} oscillation lasting many minutes was first discovered in hepatocytes by Peter Cobbold, using aequorin to monitor the cytosolic free Ca^{2+} in individual cells. This is a way of maintaining a long-term cell event, such as glycogen breakdown through activation of phosphorylase, without depleting the cell of Ca^{2+}.

Several types of spatial Ca^{2+} rise have been identified (Figure 2.4):

1. Tiny localised clouds – sparklets, sparks, puffs, scintilla and blinks.
2. Large rises in micro-domains.
3. Tides that move to fill up the cell.
4. Waves that move through the cell, and through cell layers.

2.4.3 Ca^{2+} Sparks, Puffs and Sparklets

Small, localised clouds (Box 2.2) of free Ca^{2+} are insufficient in themselves to activate the Ca^{2+} target proteins. A coordinated action of these is required to generate a global cell Ca^{2+} signal, and activate the cell response. Typically a cluster of release proteins, such as the ryanodine or IP_3 receptors on the SR/ER, is required, though channel clusters in the plasma membrane can also give rise to small Ca^{2+} rises.

Box 2.2 Terms used for small, localised Ca²⁺ signals.			
Calcium blinks	Calcium blips	Calcium bursts	Calcium clock
Calcium ember	Calcium glow	Calcium mark	Calcium puff
Calcium quantum	Calcium quark	Calcium scrap	Calcium skrap
Calcium spark	Calcium sparklet	Calcium synapse	Calcium syntilla
Calcium wavelet			
Other terms associated with small local Ca²⁺ changes			
Couplon	Ghost spark	Micro-domain	Sub-surface cistern (SSC)

Ca^{2+} sparks were discovered in cardiac muscle cells, opening ryanodine receptors on the SR. Ca^{2+} puffs were first seen in non-excitable cells as a result of Ca^{2+} release from IP_3 receptors (IP_3R) in the ER. Now Ca^{2+} sparks, puffs and scintilla describe any small release of Ca^{2+} in the cytosol, from RyR, IP_3R, or through plasma membrane Ca^{2+} channels. Sparklets, from opening of Ca^{2+} channels in the plasma membrane, are seen in the heart myocyte, nerve terminals and mechanoreceptors in hair follicle cells. To get the heart cell to contract as part of the beat, they all have to spark simultaneously. Ca^{2+} sparks can open ryanodine receptors on the SR, and Cl^- or K^+ channels in the plasma membrane. Activation of Cl^- channels causes a spontaneous transient inward current (STIC). In contrast, activation of BK^+ channels causes a spontaneous transient outward current (STOC), resulting in local increased negativity in membrane potential. Scintilla are very similar to sparks, being found first in presynaptic nerves in the hypothalamus. In contrast to sparklets and sparks, Ca^{2+} puffs are released as a result of a cluster of IP_3 receptors on the ER. Like Ca^{2+} sparks, accumulation of several Ca^{2+} puffs can generate a global Ca^{2+} signal. But Ca^{2+} blinks show a *loss* of Ca^{2+} in a localised area of the cell.

Ca^{2+} tides and waves occur within cells where the primary stimulus acts at single point on the cell, e.g. when a sperm fertilises an egg. Waves can involve release of Ca^{2+} from the ER by IP_3 and Ca^{2+}-induced Ca^{2+} release, and have been visualised flowing between cells. These intercellular Ca^{2+} waves occur either via gap junctions and IP_3, or via local release of ATP, which stimulates purinergic receptors on the next cell (see Chapter 5). There are three ways by which a Ca^{2+} signal can maintain a cellular event over a time period longer than a few milliseconds:

1. Maintenance of a prolonged rise in cytosolic free Ca^{2+}, e.g. skeletal muscle by repetitive release of acetylcholine from nerves, generating repetitive action potentials travelling down each muscle fibre.
2. Repetitive Ca^{2+} oscillations, e.g. *Aplysia* neurones, hepatocytes.
3. Modification of proteins involved in the cell event, e.g. the muscles of shellfish, such as mussels, keeping their shells firmly shut when the tide goes out.

2.5   Rubicon Principle

2.5.1   Digital Versus Analogue Cell Events

On 10–11 January 49 BC, Julius Caesar (100–40 AD) crossed a small river, the Rubicon, on the east side of Italy, declaring war on the neighbouring province. Now, crossing a Rubicon means that a threshold has been crossed. For example, a nerve fires or it does not, a muscle contracts or it remains relaxed, a cell divides into two or it remains as one, a luminous jellyfish flashes or remains invisible, and a cell dies when stressed or it survives. The cell has switched from one state to another, in a binary fashion. Examples of cells crossing the Rubicon triggered by a rise in cytosolic free Ca^{2+} include:

- Generation of a nerve action potential from summation of dendritic miniature excitatory postsynaptic potentials.
- Release of neurotransmitter at a nerve terminal triggered by the action potential.
- The twitch of a muscle fibre.
- The beat of a heart myocyte, and the concerted beat of these cells to make the heart beat.
- The secretion of insulin from an islet of Langerhans in the endocrine pancreas.
- The production of reactive oxygen metabolites from a neutrophil after an infection, and in rheumatoid arthritis.
- The generation of pain after being stung by a wasp.
- The fertilisation of an egg by a sperm.
- The flash of a luminous jelly-fish, hydroid or sea pansy when touched.
- The movement of the protozoan *Paramecium* away from an object it has hit.
- The decision of a cell to die by apoptosis, or survive a stress.
- The lysis of cells by complement, and the protection mechanism that prevents the cell crossing the lethal threshold.
- The opening and closing of guard cells in a plant leaf.
- The movement of bacteria towards a nutrient, or away from a toxin.

The Rubicon principle can best be seen when examining in detail a typical dose–response curve or time course measured in a cell population (Figure 2.5). For example, the amount of insulin released by pancreatic β-cells is proportional to the number of β-cells switched on. In the cell body of nerves micro-depolarisations have to summate if the nerve is to fire. If a Ca^{2+} signal does not reach its intracellular target, or is too small, the cell remains at rest. In contrast, intracellular messengers such as cyclic AMP, activate or inhibit cellular processes in an analogue manner, e.g. in the liver or heart.

The Rubicon principle is fundamental to understanding how drugs work, and pathological processes including cell injury, stress or death. For example, in cell injury, a rise in cytosolic free Ca^{2+}, produced as result of a pore forming in the plasma membrane, such as the membrane attack complex of complement, activates a protection mechanism, which attempts to remove the potentially lethal complex by vesiculation. If it does not achieve this in time, the cell blows up and lyses. So when haemoglobin release is measured from red cells attacked by complement, or lactate dehydrogenase is released from other damaged cell types and is used to assess cell viability, the process is in fact digital. The amount of haemoglobin or lactate dehydrogenase measured extracellularly reflects the number of cells that have lysed and not a gradual release from all the cells.

Figure 2.5 The Rubicon principle. When a Ca^{2+} signal or a cell end-response is measured in a population of thousands or millions of cells, at half-maximum response, marked by the arrow, half of the cells have been activated, rather than all of the cells activated with half their potential response. Reproduced by permission of Welston Court Science Centre.

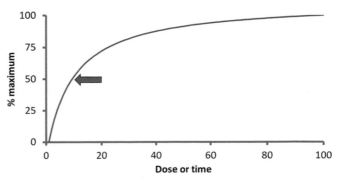

2.5.2 The Ca²⁺ Signaling Toolkit

The components required to produce an intracellular Ca^{2+} signal, and then cause this signal to produce a digital cell event, are the *Ca²⁺ signalling toolkit* (Box 2.3). This includes:

- Proteins that maintain the sub-micromolar cytosolic free Ca^{2+} in the resting cell.
- Channels and enzymes that cause the cytosolic Ca^{2+} signal.
- Targets for the Ca^{2+} signal.
- Ca^{2+} pumps and exchangers that remove Ca^{2+} from the cytosol, so that the cell can return to rest and switch off.

There are thus four ways by which intracellular Ca^{2+} can cause a cell to cross the Rubicon:

1. Generation of a Ca^{2+} signal.
2. The Ca^{2+} signal reaching its target.
3. The concerted action of target molecules.
4. Regulation of one component of an amplification mechanism.

Box 2.3 Major components of the intracellular Ca²⁺ toolkit.

1. *Ca²⁺ pumps and exchangers*, e.g. Ca^{2+}-MgATPases and the Ca^{2+}/Na^+ or /H^+ exchangers.
2. *Channels* in the plasma membrane that open to let Ca^{2+} into the cell.
3. *Enzymes* that can generate intracellular signals, such as IP_3, NAADP and cyclic ADP ribose, which release Ca^{2+} from internal stores.
4. *Organelles* such as the ER/SR, mitochondria, lysosomes, vesicles, and the vacuole in plants and fungi, which store Ca^{2+}, that can be released by IP_3, NAADP, cyclic ADP ribose, sphingosine-1-phosphate and Ca^{2+} itself.
5. *Proteins*, such as Ca^{2+}-MgATPases, in organelles, which can pump Ca^{2+} into them, and channels, such as in MTU mitochondria, and IP_3 or ryanodine receptors in the ER/SR, that can release Ca^{2+} into the cytosol.
6. *Non-protein* ion channels, such as natural ionophores and polyhydroxybutyrate-polyphosphate.
7. *Intracellular targets* for Ca^{2+}, such as troponin C, calmodulin, calcineurin, calpain, and synaptotagmin, which enable the Ca^{2+} signal to trigger the cell response.

2.5.3 Amplification Mechanisms

There are intra- and extra-cellular mechanisms that amplify the initial response of a small amount of primary stimulus. Two important mechanisms are 'cascades' and 'futile' cycles. Amplification cascades multiply at each step. For example, step 1 produces one molecule, step 2 produces 10, step 3 produces 100, and so on. There are several such cascades resulting from a small, local rise in cytosolic free Ca^{2+}, in the heart, many non-excitable cells, and smooth muscle. In contrast, futile cycles involve pathways with a forward and reverse direction. The net flux in a particular direction is the difference in flux between the two (Figure2.6). Examples of intracellular amplification cascades involving cytosolic free Ca^{2+} include:

1. **Heart myocyte** – small, local Ca^{2+} sparks from plasma membrane L-type Ca^{2+} channels activate SR ryanodine receptors. These summate, and propagate, to generate the global cytosolic Ca^{2+} signal which induces the myocyte to contract.
2. **Many non-electrically excitable cells** – Activation of a trimeric G-protein on the plasma membrane activates phospholipase C, which generates IP_3 and diacylglycerol. IP_3 releases

Ca^{2+} from the ER via IP_3 receptors. Loss of ER Ca^{2+} cause STIMs to punctate, move along the ER, and link with Orai 1s in the plasma membrane. These open, and cause the global Ca^{2+} signal.

3. **Egg fertilisation** – sperm releases phospholipase Cζ into the egg. This generates IP_3, which releases Ca^{2+} from an internal store, generating either a propagated Ca^{2+} wave or tide, via Ca^{2+} induced Ca^{2+} release, as the cytosol fills with free Ca^{2+}.

4. **Gap junctions e.g. in smooth muscle** – these connect cells electrically and chemically, allowing ions and molecules some 1200 Da to move between neighbour cells. As a result propagated Ca^{2+} waves move through large sheets of cells.

A cytosolic free Ca^{2+} signal has to be of sufficient size to make a cell fire. Thus, in a heart muscle cell, each voltage-gated Ca^{2+} channel can generate small Ca^{2+} sparks, but only when the action potential opens large numbers of Ca^{2+} channels does the Ca^{2+} concentration close to the SR rise sufficiently for there to be the explosive Ca^{2+} release necessary to provoke the muscle cell to contraction. Large numbers of troponin C molecules must bind Ca^{2+} if a long muscle fibre in an arm, leg or wing is to contract, so that the organ moves. A small local contraction is not enough for a bee to buzz. Another example is the need for a cluster of molecules in the SNARE/SNAP complex, which allows a vesicle to fuse with the plasma membrane, and secrete its contents outside the cell. If insufficient clusters are activated, then a nerve terminal will not secrete enough transmitter to stimulate the next nerve or muscle, and a mast cell will not explode to release its histamine at the site of an inflammation.

There are also longer-term mechanisms where cells cross the Rubicon. Gene expression can be digital or analogue, for example, activation of the nuclear transcription factor κB (NF-κB). Another example is the loss of the enzyme lactase phlorizin hydrolase (EC 3.2.1.62 and 3.2.1.108), which occurs after weaning, in all mammals except white Northern Europeans and few other races. Monkeys do not keep cattle! The amount of lactase in the small intestine is dependent on the number of lactase-expressing cells.

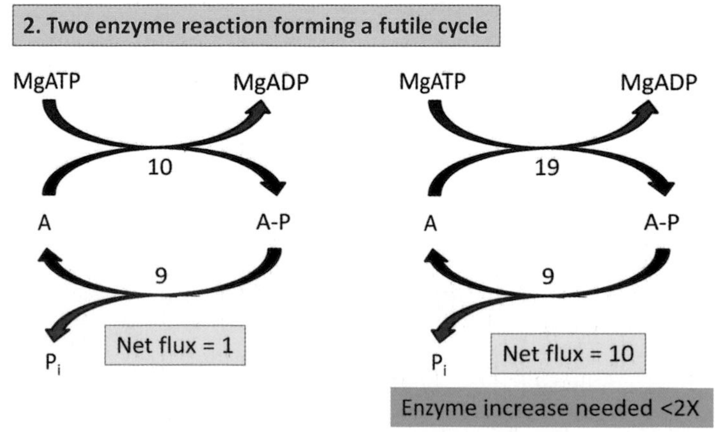

Figure 2.6 Amplification via a futile cycle. The numbers over or under the arrows represent the flux of each enzyme reaction. The figure shows that, with a futile cycle, a much smaller fractional change in enzyme activity is required to produce a large change in net flux of a metabolic pathway. Reproduced by permission of Welston Court Science Centre.

1. Single enzyme reaction with no futile cycle

$$A + MgATP \xrightarrow{1} A\text{-}P + MgADP \qquad A + MgATP \xrightarrow{10} A\text{-}P + MgADP$$

Net flux = 1   Net flux = 10

Enzyme increase needed = 10X

2. Two enzyme reaction forming a futile cycle

MgATP   MgADP   MgATP   MgADP

10   19

A   A-P   A   A-P

9   9

P_i Net flux = 1 P_i Net flux = 10

Enzyme increase needed <2X

2.6 Key Experiments to Answer Key Questions

The key experiments are:

1. Demonstration that increasing Ca^{2+} artificially inside the cell can provoke the cell event, and that preventing this Ca^{2+} rise stops the event.
2. Measurement of the cytosolic free Ca^{2+}, and in organelles, in the live cell activated by the physiological primary stimulus or pathogen, correlating this with the cell event.
3. Identification of the molecular mechanisms causing the Ca^{2+} change, by manipulating proteins in the plasma membrane and inside the cell.
4. Identification of the Ca^{2+} target, by using inhibitors, activators and genetic manipulation, as well as *in vitro* reconstitution of the Ca^{2+}-activated event.
5. How the Ca^{2+} target causes the cell event.
6. Identification of the molecular mechanisms responsible for removing the Ca^{2+}, so that the cell can return to rest.

2.6.1 Raising Cytosolic Free Ca^{2+} Artificially

There are at least six ways that can be used to manipulate intracellular free Ca^{2+}:

1. Removal of extracellular Ca^{2+}, using a Ca^{2+} chelator such as EGTA, stops Ca^{2+} entry and will slowly deplete the cell of Ca^{2+}. Intracellular injection of EGTA, or the use of membrane permeant Ca^{2+} chelators, such as BAPTAAM, will buffer cytosolic free Ca^{2+}.
2. The plasma membrane can be made permeable to Ca^{2+} using ionophores such as A23187 or ionomycin.
3. Ca^{2+} can be released into the cytosol using caged compounds – photo-releasable agents such as nitr-5 and DM-nitophen.
4. Ca^{2+} can be released from internal stores using agents that bind to the IP_3 or ryanodine receptors, e.g. photo-releasable caged IP_3 or caffeine.
5. Ca^{2+} can be released from internal stores, such as the ER, by blocking the SERCA MgATPase that pumps Ca^{2+} from the cytosol into the ER, using thapsigargin or cyclopiazonic acid.
6. Ca^{2+} can be released from mitochondria using blockers of oxidative phosphorylation, such as CN^-.

These methods should all be used in conjunction with an indicator of cytosolic free Ca^{2+} to check that a change has occurred.

2.6.2 Measurement of Free Ca^{2+} in Live Cells

Three widely used methods to measure free Ca^{2+} in live cells are:

- Small molecule fluors, e.g. fura2, fluo3/4.
- Genetically engineered fluorescent proteins, e.g. cameleons, pericams.
- Ca^{2+}-activated photoproteins from luminous jellyfish, e.g. aequorin and obelin.

Ca^{2+}-sensitive microelectrodes have also been developed, but have been superseded by the application of patch clamping to study individual calcium channels in live cells. Free Ca^{2+} changes can be imaged in live cells to examine the heterogeneity of the Ca^{2+} signals on an individual cell basis, and to define the location and type of Ca^{2+} involved (i.e. transient, oscillation, cloud, tide or wave). Protein Ca^{2+} indicators, either fluorescent or bioluminescent, can be engineered so that they target to sites within the cell, such as the ER, mitochondria, nucleus and inner surface of the plasma membrane (see Chapter 4). Endocytosed rhod-4 linked to dextran can be used to measure free Ca^{2+} inside lysosomes, where it is some 0.5 mM. The ability to image Ca^{2+} signals in individual live cells, intact organs and whole organisms has revolutionised the study of intracellular Ca^{2+}.

2.6.3 How the Ca^{2+} Signal is Generated

The Ca^{2+} signal can be generated through a combination of movement through the plasma membrane and release from internal stores. Identifying how much comes from outside the cell, and how much from release by internal stores, requires careful measurement of free Ca^{2+} in the cell, including inside organelles, such as the ER and mitochondria. Types of Ca^{2+} channel in the plasma membrane are identified electrophysiologically, using patch clamping. The receptors for chemical or biological stimuli are identified as ionotropic or metabotropic, using pharmacological agents and biochemical criteria. In non-excitable cells, a link between the plasma membrane and Ca^{2+} release from the ER can be established by measuring IP_3, or by releasing IP_3 by microinjection or photo-releasable caged IP_3.

There are three mechanisms for regulating Ca^{2+} getting into cells:

1. Ca^{2+} channels, opened by depolarisation voltage, hormone or neurotransmitter receptor, intracellular second messenger, or release of Ca^{2+} from the ER (SOCE).
2. Ca^{2+} exchangers – Na^+/Ca^{2+}, H^+/Ca^{2+}, or a $Ca^{2+}/phosphate$ symport.
3. Trapping of external Ca^{2+} in endocytosed vesicles.

The most important releasable store is the ER/SR, Ca^{2+} being released through IP_3 or ryanodine receptors. However, Ca^{2+} can also be released into the cytosol from mitochondria, lysosomes, endosomes, and vesicles, and the vacuole in plants and fungi. Ca^{2+} is removed from the cytosol via pumps and exchangers in the plasma membrane, and the membranes of internal organelles.

2.6.4 Identifying the Ca^{2+} Target

EF-hand Ca^{2+}-binding proteins, such as troponin C in muscle, and calmodulin in non-muscle cells, or C2 Ca^{2+}-binding proteins such as synaptotagmin in secretory cells, are major intracellular Ca^{2+} targets. The role of a particular Ca^{2+}-binding protein can be established biochemically, pharmacologically or by genetic manipulation. Manipulation of gene expression of the Ca^{2+}-binding protein or its target, through over-expression, knock-out or siRNA, tests predicable effects on the cellular event.

2.6.5 How Binding of Ca^{2+} to its Target Causes the Cell Event

This involves a range of biochemical, pharmacological and genetic manipulation techniques, as well as structural studies using X-ray crystallography and nuclear magnetic resonance (NMR) spectroscopy. The key is finding the Ca^{2+} binding protein that binds to the molecular machinery responsible for the cell event. Thus the troponin complex, with troponin C, binds to tropomyosin, which releases its binding to actomyosin in muscle, and synaptotagmin with the SNARE complex in secretory cells.

2.6.6 How the Cell Returns to Rest

To return to rest the Ca^{2+} is pumped out of the cell and back into internal stores. There are three mechanisms for getting Ca^{2+} out of the cell, or for getting Ca^{2+} into intracellular organelles. Since these involve Ca^{2+} moving against an electrochemical gradient, a source of energy is required, either the MgATP/MgADP + phosphate reaction or coupling to another ion:

1. Ca^{2+}-activated MgATPase (i.e. a Ca^{2+} pump) in the plasma membrane, and in the ER or SR, of all eukaryotic cells.
2. Ca^{2+} exchangers: Ca^{2+}/Na^+ exchange in the heart and invertebrate nerves, Ca^{2+}/H^+ exchange in *Neurospora*, intracellular vesicles and mitochondria, Ca^{2+}/phosphate co-entry, a symport in mitochondria.

As the free Ca^{2+} returns towards that in the resting cell, Ca^{2+} drops off the Ca^{2+} target, and the cell event stops. A muscle relaxes, and a nerve terminal stops secreting.

2.7 Nomenclature – How Things are Named

Naming of organisms and molecules in a systematic way is essential if the scientific community is to communicate discoveries and inventions consistently and accurately.

2.7.1 English and Etymology

George Bernard Shaw famously wrote; 'Great Britain and the United States are two countries separated by the same language'! The UK English spelling can be different from the American (Table 2.3). In this book I use spelling and words based on the *Oxford English Dictionary*. This is relevant to using word searches in the electronic version of this book, or when using literature databases such as PubMed (http://www.ncbi.nlm.nih.gov/pubmed) or Web of Knowledge (http://wokinfo .com). It is usually sensible to search using both the UK and US English spelling of key words.

The names of many Ca^{2+} signalling molecules are derived from Latin or Greek. There are also special words. For example, the word 'canonical', derived from the Latin *canonicus* and the Greek κανονικος, meaning 'relating to a rule' is used frequently in biology, and in calcium signalling. To a lay person a 'canon' may be a Christian priest, or a piece of music where the initial theme is repeated in sequence, e.g. in J.S. Bach, and 'Frères Jacques'. But in science

Table 2.3 Some examples of differences between UK English and US English.

UK English	US English	UK English	US English
adrenaline	epinephrine	neurone	neuron
anaemia	anemia	oestrogen	estrogen
characterise	characterize	realise	realize
colour	color	signalling	signaling
centre	center	sulphate	sulfate
haem	heme	targeted	targetted
ionised	ionized	titre	titer
litre	liter	tumour	tumor

and mathematics canonical refers to 'specific rules'. For example, the *standard* genetic code is *canonical*. Thus, canonical Ca^{2+} channels, such as TRP, conform to standard electrical and structural properties, and canonical EF-hand proteins have Ca^{2+}-binding sites that conform to the Ca^{2+} loop originally discovered in parvalbumin between two α-helices, designated E and F, containing 12 amino acids with carboxyls at 1, 3, 5, 7, 9 and 12, with one peptide carbonyl and water always providing the oxygens for Ca^{2+} binding.

2.7.2 Organisms

A glow-worm is not a worm and a firefly in not a fly. Both are actually beetles. Yet the famous New Zealand glow-worm is in fact a dipteran (i.e. a fly). The scientific naming of organisms, whether extant or extinct, follows the binomial system invented by Linnaeus (Figure 3.3a) in the eighteenth century. The name of each animal, plant, fungus or microbe has two parts: the genus and the species. Sometimes a third is added to describe a subspecies. If an individual species is not clear, then the genus is followed by sp. When several organisms within the same genus are involved, then this is followed by spp. The genus and species are written in *italics*, with the genus beginning with a capital letter; sp. and spp. are not in *italics*. But all other parts of the taxonomy (phyla, classes, orders, etc.) are in normal font. So we are *Homo sapiens* in the phylum Chordata. Model systems feature a lot in the story of intracellular Ca^{2+} (see Chapter 3, Section 3.1.2), with both 'trivial' (common) and scientific names. For example, the first cell used to measure cytosolic free Ca^{2+} was a single muscle fibre from a giant barnacle found in the Pacific, off the west coast of the United States. Its scientific name is *Balanus nubilus*. However, taxonomists are continually changing their minds on which group an animal or plant belongs. For example, the jellyfish *Aequorea*, which has given us aequorin, the first really successful indicator for measuring cytosolic free Ca^{2+}, as well as the extraordinary GFP, was named *Medusa aequorea* by Peter Forskål (Figure 3.3b), who discovered it. However, the discoverer of aequorin and GFP, Osamu Shimomura (Figure 3.4a), called it *Aequorea aequorea*. Yet, the distinguished British expert in jellyfish at Plymouth, Frederick Russell called it *Aequorea forskalea*, after its discoverer. Even further confusion was to follow. When the DNA coding for aequorin was cloned, the jellyfish was called *Aequorea victoria*.

2.7.3 Acronyms, Cells and Biochemistry

Everyone is familiar with the acronyms DNA, RNA and ATP. But ones such as ER and SR may be less familiar. ER stands for 'endoplasmic reticulum' and SR for 'sarcoplasmic reticulum', from the Greek *sarx* = flesh, the tubular system inside all eukaryotic muscle cells (skeletal, heart, smooth and invertebrate) which releases Ca^{2+} when the cell is excited. Eukaryotic cells were first named as such because they have a nucleus (Greek εὖ,eu = well or true and κάρυον, karyon, nut or kernel; Latin *nucleus* = nut or kernel). The part of the cell outside the nucleus, but within the outer membrane, the plasma membrane, is the cytoplasm, and includes all the organelles. The free soluble part of this is called the cytosol. Thus, free Ca^{2+} is typically measured in the *cytosol*, unless the Ca^{2+} indicator has been targeted to a specific organelle.

The three-domain system – Archaea, Bacteria and Eukaryota – was introduced in 1990 by Carl Woese, reflecting the three fundamental cell types in evolution, rather than the six-kingdom system of Eubacteria, Archaebacteria, Protista, Fungi, Plantae and Animalia. Bacteria are divided into two groups, based on whether they stain blue with the stain developed by Gram: Gram-positive and Gram-negative. The latter have two outer membranes enclosing a periplasmic space, compared with the one of Gram-positive bacteria, there being similarities and differences in intracellular Ca^{2+} between the two cell types (see Chapter 8). Archaea are a different type of cell. Several species have been found in the human body, including the gut, vagina and mouth. The Ca^{2+} signalling system is only just being unravelled in the Archaea. Viruses and bacteriophages are separate from the three domains, as they are not cells. They can only replicate inside live cells. But, they can stimulate and interact with the Ca^{2+} signals of cells that they bind to or infect.

2.7.4   Genes

A large number of genomes have now been sequenced from dozens of animals, plants and microbes. Each consortia of scientists responsible for giving us these DNA and protein sequences has decided on the way the genes coding for particular proteins are named, resulting in many differences in nomenclature between species. Calmodulin genes are *CALM1, 2 and 3*. Calmodulin-like proteins are CALML1–6. In the fruit fly, *Drosophila*, the gene can have anything from one to four letters. In all organisms, the gene name is *italics*. The protein it codes for is in normal font. But there is a variation between species, in upper case (capitals) versus lower case in both the gene and protein symbols. In the human genome (*Homo sapiens*), genes are in upper case and *italics*, with the proteins also in upper case, but not in italics, and the chicken (*Gallus*) nomenclature is very similar. But in the mouse and rat, the gene, while still in *italics*, only starts with an upper-case letter. In the frog *Xenopus*, both the genes and protein symbols are all in lower case, with only the gene in *italics*. In contrast, in *Escherichia coli*, genes are always in lower-case *italics*, and the protein coded by the gene in normal font with the first letter in upper case. For example, the protein wrongly thought to be Ca^{2+}/H^+ exchanger in *E. coli* is ChaA, coded by the gene *chaA*. In the fruit fly *Drosophila*, genes have a capital first letter, then lower case. Examples of names given to major Ca^{2+} signalling genes are shown in Table 2.4.

Variants from different alleles or tissues are identified by letters or numbers as A, B, C; I, II, III; 1, 2, 3; or Greek letters α, β, γ and so on, though the Human Genome Committee recommends only Latin letters and Arabic numerals. Splice variants have the extension -v1, -v2 and so on, and promoter variants, where transcription starts to make mRNA at different points, being -pr1, -pr2 and so on. Note that the 'v' for variant is written on the line, whereas 'v' for voltage-gated is written as a subscript (e.g. $Ca_v1.1$). Gene deletions are designated by the symbol Δ. But if it is simply a mutant, then a minus (−) sign is used, with a plus (+) sign designating

Table 2.4 Examples of names used for some Ca^{2+} signalling genes and their products in three organisms.

Protein	Human (*Homo sapiens*)		Fruit fly (*Drosophila melanogaster*)		Mouse (*Mus musculus*)	
	Gene symbol	Gene product	Gene	Gene product	Gene	Gene product
Calmodulin 1	*CALM1*	CALM1	*Cam*	CAM	*Calm1*	Calm1
Calpain 1	*CAPN1*	CAPN1	*CalpA*	CalpA	*Capn1*	Capn1
Calcineurin (α subunit)	*PPP3CA*	CALNA	*CalnA*	CalnA	*CalnA*	CalnA
Plasma membrane Ca^{2+} pump	*ATP2B1*	PMCA1	*Pmca1*	Pmca1	*Atp2b1*	PMCA1
ER Ca^{2+} pump	*ATP2A1*	SERCA1	*SercaA1*	SERCA1	*Atp2a1*	SERCA1
Na$^+$/Ca^{2+} exchanger	*NCX*	NCX	*Ncx*	Ncx (Calx)	*Ncx*	Ncx
Voltage-gated Ca^{2+} channel	*CACNA1S*	CACNA1S ($Ca_v1.1$)	*Dmel*	DmcalD	*Cacna1s*	Cacna1s

Calcineurin = protein phosphatase 3.

wild-type. Other variations can be written as superscripts such as *ts*/TS (temperature sensitive) or *cs* (cold sensitive). Knock-outs use the minus sign, '−/−' designating that a gene has been knocked-out in both chromosomes.

In the literature you can find several gene names or protein names for a particular gene and its product. This is because individual workers made up their own names before a consensus from a committee was arrived at. The organisation responsible for the human genome has the acronym HUGO (Human Genome Organisation), the group responsible for its nomenclature being HGNC (HUGO Gene Nomenclature Committee). There can be some confusion regarding the term 'essential' gene. In some cases, mistakenly, the word *essential* is used simply when there is a large decrease in growth rate when the gene in question is deleted. However, in the Keio knock-out collection of *E. coli*, some 300 genes were not found, and were thus designated as *absolutely* essential for the survival and replication of *E. coli*.

2.7.5 Proteins and Protein Sequences

Whilst there are just 20 core amino acids found in proteins, over 100 covalent modifications exist, made after translation of mRNA, several of which are important in Ca^{2+} signalling. For example, the ubiquitous Ca^{2+}-binding protein calmodulin usually has a trimethyl-lysine at position 115, though in the unicellular green alga *Chlamydomonas* and the slime mould *Dictyostelium* lysine 115 is not methylated. Protein sequences are described either in one- or three-letter code (e.g. K or Lys for lysine). Two key amino acids that bind Ca^{2+} are the acidics glutamate (E) and aspartate (D). Mutants are described by the change in amino acid at a particular location. For example, we changed an aspartate to an alanine in one of the Ca^{2+}-binding sites of aequorin to reduce its affinity for Ca^{2+}, thus the mutant is D119A.

Many proteins have shortened names, or ones based on acronyms. For example, the transcription factor NFATc is short for <u>n</u>uclear <u>f</u>actor of <u>a</u>ctivated <u>T</u>-cell, <u>c</u>ytoplasmic. As with other scientific nomenclature, a Protein Naming Utility has been established in an attempt to systematise the naming of new proteins. Proteins directly involved in Ca^{2+} signalling typically start with 'cal-'. Protein variants coded for by separate genes are designated by a letter or number, e.g. CALM1, 2 and 3, and SERCA1, 2 and 3. CALML is for calmodulin-<u>L</u>ike. On the other hand, variants produced as products of the same gene, either by alternative splicing of introns in the mRNA or because of different start sites of the promoter, use the same gene and protein codes, but with 'var' or 'pro' suffixes, or can simply be a, b, c etc. The origins of some of the names of Ca^{2+} binding proteins are:

- Calmodulin = <u>cal</u>cium-<u>mod</u>ulated prote<u>in.</u>
- Calsequestrin = <u>cal</u>cium-<u>seques</u>tering protein.
- Calreticulin = <u>cal</u>cium-binding protein from the <u>reticul</u>um.
- Calpain = <u>cal</u>cium-activated pa<u>pain.</u>
- Calcineurin (protein phosphatase 3 -PPP3CA) = named after its discovery in neurones.
- PMCA = <u>P</u>lasma <u>m</u>embrane <u>Ca</u>$^{2+}$ pump.
- SERCA = <u>S</u>R and <u>ER</u> membrane <u>Ca</u>$^{2+}$-MgATPase.
- NCX = a Ca^{2+} exchanger that exchanges <u>Na</u>$^+$ for <u>C</u>a^{2+}, and vice versa.

2.7.6 Enzymes

Enzymes are classified by an Enzyme Commission (EC) number. This has four parts. The first EC number designates the major class of enzyme: 1 = oxidoreductases; 2 = transferases; 3 = hydrolases; 4 = lyases; 5 = isomerases; 6 = ligases. The second number designates the subgroup, the third a further subgroup, and the fourth is the specific number for the unique reaction catalysed by the particular enzyme.

Calmodulin is not an enzyme, so it does not have an EC number. But, calmodulin-activated kinase II (CaMKII) is EC 2.7.1.17: 2, because it is a 2 = transferase, and 7 = enzymes catalysing that type of phosphate transfer. Similarly, myosin light chain kinase (MLCK) is an enzyme that transfers phosphate, so its number is EC 2.7.11.18. Both these kinases transfer phosphate to either a serine or threonine in the protein target. In contrast, phosphatases, such as the Ca^{2+}-activated calcineurin, are hydrolases, so their first EC number is 3, the next number being 1 for cleavage of phosphate. Thus, calcineurin is EC 3.1.3.16. Ca^{2+}-MgATPases also hydrolyse phosphate off a substrate, in this case not a protein but rather ATP. Calmodulin-activated Ca^{2+}-MgATPase is EC 3.6.1.3. I always call ATPases MgATPase, since MgATP is the real substrate for all enzymes that catalyse reactions involving ATP. The same is true of other nucleotides, GTP, CTP, TTP and UTP, though this is rarely made clear in the literature (e.g. when dealing with G-proteins). In contrast, calpain belongs to the family of Ca^{2+}-dependent, non-lysosomal cysteine proteases, found ubiquitously in mammals and many other organisms. A hydrolase has a 3 as its first EC number, but then a 4 designating it acts on a peptide bond. There are two subtypes, I (μ) and II (m), with very different sensitivities to Ca^{2+}. So the EC numbers are EC 3.4.22.52 and EC 3.4.22.53, respectively.

Enzymes have a systematic name, and often a trivial one as well. Thus, the proper name for the phospholipase that releases IP_3 from PIP_2 is phosphoinositide phospholipase. Its number is EC 3.1.4.11. Its trivial name is phospholipase C (PLC). There are 13 types in mammals alone, classified into six isotypes – β, γ, δ, ε, ζ, η. There are several enzymes that can catalyse the oxidation of coelenterazine to produce light, first discovered in the luminous jellyfish *Aequorea*. The Ca^{2+} activated photoproteins aequorin and obelin, and the coelenterazine luciferase from the sea pansy *Renilla reniformis* have the same EC number – EC 1.13.12.5. This is because they use the same reaction that produces light. This is also the case for organisms, such as decapod shrimp, copepods, arrow worms, squid and fish from eight different phyla. Yet there is very poor sequence similarity between them or with aequorin. None of these other luciferases are directly activated by Ca^{2+}.

2.7.7 Ion Channels

In living organisms, electrical events are usually carried by charged ions – K^+, Na^+, Ca^{2+}, H^+ and Cl^-. Ions move across biological membranes through channels, transporters, exchangers and pumps. Ion channels in biological membranes allow ions to pass through them from one side to another. Some are highly selective for a particular ion, whilst others are non-selective. Ion channels set the resting membrane potential, and determine excitability and action potentials, in, for example, nerves and muscle. Movement of Ca^{2+} through ion channels is a vital feature of Ca^{2+} signalling in most eukaryotic cells, but the role of Ca^{2+} channels in bacteria and Archaea is less well understood. Ion channels occur in all cell types in Eukaryota, Bacteria and Archaea. In eukaryotes, they are found in the plasma membrane, mitochondria, SR/ER, lysosomes, secretary vesicles, and vesicles taken up into the cell. Currents carried by these ions obey the normal laws of electricity, in particular Ohm's law (resistance = 1/conductance = voltage/current) and Faraday's equation (capacitance = charge/potential). The electrical properties of an ion channel are described by:

- Specificity, if any, for a particular ion.
- Conductance (i.e. how much ionic current flows through it at a defined voltage).
- What opens and closes it.
- How its electrical properties can be regulated by interaction with other proteins and regulators.

Patch clamping has shown that ion channels typically obey the Rubicon principle – they are either open or closed, i.e. they behave digitally. Ion channels can be opened by:

- A change in voltage across the membrane.
- Natural or synthetic extra- or intra-cellular ligands interacting directly with the channel.
- Another protein.

Thus, ion channels are described as 'gated', 'sensitive', 'modulated' and 'regulated'. Since ion channels are digital, I prefer to use the terms 'voltage-gated' or 'ligand-gated'. Ion channels also 'inactivate'. There are some 140 related members of human voltage-gated ion channels, making them the largest superfamily of cell signalling proteins in the human genome. There are two types of receptors that can open ion channels:

1. *Ionotropic*: forming the pore, opening ion channels directly.
2. *Metabotropic*: open ion channels indirectly, e.g. linked to a G-protein).

Many ion channels form the actual channel through several subunits, which can be the same protein (homomeric) or different proteins (heteromeric). The International Union of Pharmacology Committee on Receptor Nomenclature and Drug Classification (NC-IUPHAR) has agreed a systematic method for naming voltage-gated or ligand-gated ion channels. However, this means that there is some confusion in the literature, since, prior to this, different names were used by various workers. Ionotropic glutamate receptors were originally named after synthetic agonists that activated them: AMPA, kainate and N-methyl-D-aspartate (NMDA). The NC-IUPHAR has renamed the subunits of these receptors as GluA1–4, GluK1–4 and GluN1–3(A, B). Subunits of the nicotinic, γ-aminobutyric acid (GABA), and glycine receptors are designated by Greek letters (α, β, γ, etc.), with numbers ($\alpha1$, $\alpha2$, etc.) for different alleles. Ligand-gated ion channels are classified into three groups, based on their subunit structure in the membrane:

1. Cys loop receptor (nicotinic acetylcholine) superfamily form pentamers – nicotinic acetylcholine, 5-HT, $GABA_A$, glycine and ZAC receptors.
2. Glutamate receptor family form as tetramers – NMDA, AMPA and kainate receptors.
3. P2X receptor family form as trimers – P2X receptors.

Ion channels are named in two ways:

1. Electrophysiologically, e.g. for Ca^{2+} HVA (L, N, P/Q) or LVA (T, R).
2. Genetically, through sequence similarities.

They are usually named after the ion that they are selective for – potassium, sodium, calcium or chloride. However, TRP channels, discovered as t̲ransient r̲eceptor p̲otentials in the eye of the fruit fly *Drosophila*, are often relatively non-selective, and so are named after the opening mechanism that led to their discovery:

- Group 1 – TRPC (C = canonical), TRPV (V = vanilloid), TRPM (M = melastatin), TRPN (N = no mechanoreceptor potential – NOMPC) and TRPA (A = ankyrin).
- Group 2 – TRPP (P = polycystic) and TRPML (ML = mucolipin).

As with other proteins, the gene is always written in *italics* and the protein is written in normal font. The nomenclature of Ca^{2+} channels uses subscripts identifying an important property of the channel:

- ir = inward rectifying (i.e. the current is stronger in one direction than the other).
- 2P = two pore.

- Ca = Ca^{2+} gated
- v = voltage-gated
- ATP = blocked by intracellular ATP.

The structure of the pore of an ion channel is formed as vestibules, bigger than domains.

There are six broad families of Ca^{2+} channels (L, T, N, P, Q, R), which have two names based on their electrical properties (1) or protein components (2):

1. $Ca_y n_1.n_2$: y is the gating mechanism (e.g. v for voltage or ir for inward rectifying); n_1 is the number of the channel family; n_2 is its individual number within the family.
2. CACNXnY: CACN for <u>CA</u>lcium <u>ChaN</u>nel; X represents the type of subunit, e.g. A for α, the protein component that actually forms the channel, or B for β, the regulator subunit; Ca^{2+} channels can have up to four components (α, β, γ, δ); n represents the number specifying the given subunit; Y identifies the specific location of the channel.

For example, in skeletal muscle $Ca_v1.1$ has the protein CACNA1S, which is the $α_1$ subunit of the voltage-gated Ca^{2+} channel. In contrast, $K_{Ca}1.1$ has KCNMA1, which is the $α_1$ subunit of the Ca^{2+}-activated potassium channel BK. But the human gene coding for $Ca_v1.1$ is *CACHNA1* and the protein is CACHNA1, with CHN being for channel. Other nomenclatures are, however, common in the literature. Ca^{2+} channels were originally designated LVA for 'low voltage activated' or HVA for 'high voltage activated' based in whole-cell electrical recordings in muscle and nerves (see Chapter 5). Then Richard Tsien, brother of late Nobel Laureate Roger, decided on another nomenclature, based on patch clamping data in neurones and muscle. These are L, T, N, P, Q and R:

- L = large and long lasting (HVA).
- T = tiny and transient (LVA).
- N = neither.
- P= Purkinje.
- Q = queer.
- R = residual, the current left after all others are blocked pharmacologically.

All voltage-gated Ca^{2+} channels are tetramers with α, β, γ and δ subunits. The α subunit forms the channel for Ca^{2+}. Thus, another nomenclature uses the type of $α_1$ subunit:

- L-types = $Ca_v1.1$, $Ca_v1.2$, $Ca_v1.3$, $Ca_v1.4$ = S, C, D, F, respectively.
- P/Q = $Ca_v2.1$ = A.
- N = $Ca_v2.2$ = B.
- R = $Ca_v2.3$ = E.
- T = $Ca_v3.1$, 3.2, 3.3 = G, H, I, respectively.

There are three types of Ca^{2+} channel, based on the physiological agent that opens them:

1. Opened by an external physical agent – voltage, light, mechanical, temperature, pH.
2. Opened by an extracellular chemical agent – neurotransmitter, hormone, toxin.
3. Opened by an intracellular chemical – cyclic nucleotide.

Ca^{2+} channels are found in the plasma membrane, in the membranes of internal organelles, such as the ER/SR, mitochondria, lysosomes and secretary vesicles, and in plant organelles, such as chloroplasts, vacuoles and the tonoplast membrane.

2.7.8 Agonists and Antagonists

Agonists are substances that activate cells. Antagonists, sometimes referred to as blockers, are substances that inhibit. In the case of Ca^{2+} signalling, an agonist may activate a cell directly through a change in intracellular Ca^{2+}, or via another signalling pathway that interacts with the Ca^{2+} signalling system. Agonists and antagonists may be natural or synthetic. They can be a protein, peptide or small organic molecule, and may be a hormone, neurotransmitter, neuro-hormone, paracrine, endocrine, drug, toxin or artificial substance.

2.7.9 Chemicals

There is a well-established nomenclature for both inorganic and organic substances, developed by the International Union of Pure and Applied Chemistry (IUPAC). A large number of biologically active organic substances are chiral (i.e. they are 'handed'). In old nomenclature, chirality was designated by (+) or (−), or l/L or d/D, based on what they did to polarised light or what their three-dimensional structure was relative to a standard compound. Now, the convention is to use the R and S nomenclature, though the D/L system still remains in common use in certain areas of biochemistry. The full scientific names of organic substances can be long and daunting. For example, the compound that is oxidised to form the light emitter in the Ca^{2+} indicators aequorin and obelin is called coelenterazine, because it was first discovered in coelenterates, has the full scientific name 6-(4-hydroxyphenyl)-2-[(4-hydroxyphenyl)methyl]-8-(phenylmethyl)-7H-imidazo[3,2-a]pyrazin-3-one (see Chapter 4).

An important issue is how to name substances that have more than one group, such as a phosphate, attached. If two or three of the same group are linked together, then the prefix bi- or tri-, respectively, is used. But if the groups are attached to different atoms, then bis- and tris- should be used. Thus, we have adenosine triphosphate (ATP) and inositol trisphosphate (IP$_3$) (Figure 2.7). Nomenclature often changes with time, as committees establish new naming rules. Thus, a Ca^{2+}-regulated enzyme, 2-oxoglutarate dehydrogenase in the mitochondria was originally called α-ketoglutarate dehydrogenase. A further issue is the naming of synthetic substances produced by the chemical and pharmaceutical industry. These often have a number based on when the substance was made. Thus, a Ca^{2+} ionophore used often to let Ca^{2+} into cells is A23187, but its original designation was in fact A23 187.

Inorganic substances follow the nomenclature based on the elements of which they are composed – sodium, potassium, calcium, chlorine and so on. But, since most of the inorganic substances in living systems are in the form of ions (cations and anions), it is appropriate to designate them as Na^+, K^+, Ca^{2+}, H^+, Cl^- and so on. Ringer and Heilbrunn, two pioneers of calcium signalling, usually wrote calcium as Ca. By the 1960s many researchers used Ca^{++},

Figure 2.7 A triphosphate (ATP) versus a trisphosphate (IP$_3$).

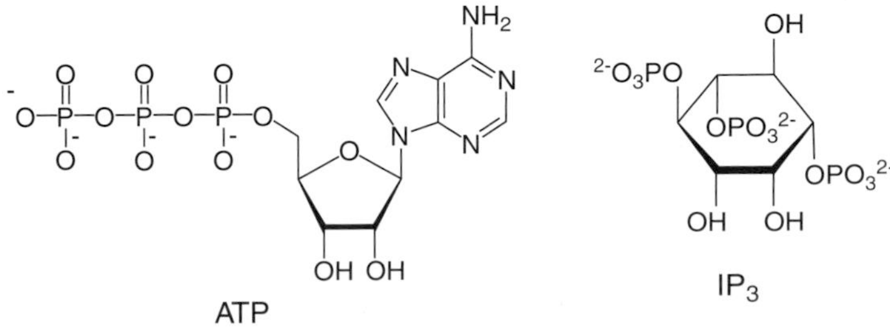

converting to Ca^{2+} by the mid–late 1970s. Now the accepted nomenclature for all ions is 'chemical symbol$^{valency\ charge}$' (with the superscript valency charge usually written as '2+' rather than '++', for example).

2.7.10 Toxins

There are a wide variety of toxins found in nature that interact with the Ca^{2+} signalling system (see Chapter 11). Usually they are named after the organism from which they originate. If there are several toxins isolated from the same organism, then these are distinguished by a Greek letter prefix (e.g. α, μ, ω, δ, κ). A Greek letter may also relate to the structural class and to an ion channel mechanism. Often, a particular group of toxins comes from several, closely related species from the same genus. A particular toxin is identified by a suffix, in upper case. The first letter identifies the species from which the toxin comes. The next one or two letters are Roman numerals, set by the order in which the toxin was identified. The final letter identifies the variant of the toxin from the particular species. Thus, conotoxins, which block various ion channels, are of a group of neurotoxic peptides isolated from the venom of the marine cone snail, genus *Conus*. An example is a conotoxin isolated from *Conus geographus* – ω-conotoxin GIVA or, in shortened form, ω-CTXGIVA (C for *Conus* and TX for toxin; see Chapter 11). This toxin inhibits N-type voltage-dependent calcium channels. In contrast, tetrodotoxin (TTX) made by bacteria in the pufferfish (of the family Tetraodontidae), blocks Na^+ channels.

2.7.11 Drugs

A wide range of pharmaceuticals have been developed to interact directly or indirectly with the Ca^{2+} signalling system, several of which are used clinically (see Chapter 11). They can have trade names, a 'trivial' (generic) scientific name (e.g. an International Non-proprietary Name (INN)), and a systematic scientific name. Two examples of drugs that interact with Ca^{2+} signalling are amlodipine and atenolol. Amlodipine acts directly on the Ca^{2+} signalling system by blocking Ca^{2+} channels of the dihydropyridine class. Amlodipine (generic name) is marketed in various countries under over 30 trade names, such as Amlovasc, Istin and Nelod, and is used as an anti-hypertensive. It works by blocking Ca^{2+} channels in smooth muscle, thereby relaxing smooth muscle in the arterial wall, and hence reducing blood pressure. However, because dihydropyridine receptors are found in many other tissues apart from smooth muscle, amlodipine has several side-effects, including oedema in the feet and gut problems. The drug, scientific name (*RS*)-3-ethyl 5-methyl 2-[(2-aminoethoxy)methyl]-4-(2-chlorophenyl)-6-methyl-1,4-dihydropyridine-3,5-dicarboxylate, is a supplied as a salt of besylate, mesylate or maleate. Atenolol, on the other hand, interacts indirectly with Ca^{2+} signalling, where cyclic AMP can alter Ca^{2+} signals (e.g. in heart muscle). Atenolol is an antagonist of $β_1$-adrenergic receptors, and belongs to the group of drugs known as β-blockers, used to treat cardiovascular disease and hypertension. Atenolol (generic name; trade name: Tenormin) was first used in 1976 as a replacement for the first, widely used, non-selective β-blocker propranolol. Unlike propranolol, atenolol does not cross the blood–brain barrier, and thus avoids side-effects in the central nervous system. The systematic scientific name for atenolol is (*R,S*)-2-{4-[2-hydroxy-3-(propan-2-ylamino)propoxy]phenyl}acetamide, the (*R,S*) showing that it is a racemic mixture of both chiral forms. Chirality can be very important in drug action, such as in several cases of drugs that interact with Ca^{2+} signalling. The *R* and *S* forms can have quite different effects, or only one is active. A similar situation exists in the perfume industry. As with all chemicals, the nomenclature of drugs is set by IUPAC. There are many databases available via the Internet, and graphic packages, such as ChemDraw, for drawing chemical structures. Other useful packages for finding chemical structures and names are ChemBioFinder, ChemINDEX and the Merck Index. WikiPedia has also an amazing list of drug structures.

2.7.12 Ca²⁺ Indicators

The ability to measure and image free Ca^{2+} inside live cells has revolutionised the experimental strategy for elucidating the role of Ca^{2+} signalling in biological processes. There are five essential properties that an indicator must have:

1. It must be able to generate a signal from inside a cell, detectable outside, that changes when there is a change in intracellular free Ca^{2+}.
2. It must be possible to correlate the Ca^{2+} signal with a cellular event.
3. It must be specific for Ca^{2+} at micromolar concentrations, in the presence of millimolar Mg^{2+}.
4. It must be possible to get the indicator into the living cell, without significant damage to the cell.
5. It must be non-toxic and not disturb the Ca^{2+} signal significantly by buffering.

Ideal properties also include:

1. Monitoring free Ca^{2+} in single cells, intact organs and live organisms.
2. Imaging Ca^{2+} signals in single cells, intact organs and live organisms.
3. Targeting the Ca^{2+} indicator to organelles, and specific sites within the cell.
4. Quantification; i.e. it should be possible to relate the indicator signal to absolute concentrations of free Ca^{2+}.
5. Readily available and cheap.

Four techniques satisfy these criteria, and have been widely used to monitor and image changes in free Ca^{2+} in live cells (see Chapter 4):

1. Absorbing dyes.
2. Fluorescent dyes, both small organic and genetically engineered.
3. Bioluminescent Ca^{2+}-activated photoproteins, extracted and genetically engineered.
4. Microelectrodes.

Absorbing and fluorescent Ca^{2+}-sensitive dyes require shining light on them from a lamp or laser, for example arsenazo III and fluo-3/4, respectively. Bioluminescent indicators do not require a light source. Fluorescent and bioluminescent indicators depend on exciting electrons – luminescence. All depend on Ca^{2+} changing one of the physical properties of the indicator. For absorbing indicators this is the extinction coefficient and/or the spectrum. Fluorescent dyes, fluorochromes ('fluors' for short) are characterised by:

- The extinction coefficient.
- Excitation and emission spectra, determining the colour seen by the naked eye.
- Fluorescence quantum yield (Φ_F, i.e. the fraction of excited molecules that emit photons).
- Fluorescence lifetime, usually nanoseconds.

Ca^{2+} binding affects one or more of these. Thus for absorbing and fluorescent dyes binding Ca^{2+} increases or decreases the light detected at a particular wavelength. Bioluminescent indicators depend on Ca^{2+} affecting the rate of the chemical reaction, upon which light emission depends, and thus the rate of light emission. The generic chemical reaction for all bioluminescence is oxidation of a luciferin to an excited oxy-luciferin by oxygen catalysed by a protein luciferase. Aequorin and obelin are photoproteins activated where the luciferin, coelenterazine, and oxygen are tightly bound to the protein. Binding Ca^{2+} triggers the reaction, and blue light is emitted. Engineering peptides proteins to the C-terminus, opens the solvent cage, and allows the apoprotein to turnover like a normal luciferase, the C-terminal P being important. The absolute concentration of free Ca^{2+} in the cell for all Ca^{2+} indicators is calculated by comparing the signal from cells with standard curves of Ca^{2+} affecting these parameters. Ratiometric indicators are useful, as they are independent of the concentration of indicator. For fluorescence these include fura-2, indo-1, cameleons and pericams, and rainbow proteins for bioluminescence (see Chapter 4). Engineering targeting peptides onto the N- or C-terminus of fluorescent or bioluminescent indicators allows free Ca^{2+} to be measured inside organelles. Ca^{2+}-sensitive microelectrodes depend on Ca^{2+} changing the electrode potential and are calibrated by relating its potential to free Ca^{2+} in pure solution.

2.7.13 Units

A wide range of units is found in the scientific literature. *Système international d'unités* (SI for short) units are favoured. Ca^{2+} concentrations are always expressed in molar, typically nanomolar (nM), micromolar (μM) or millimolar (mM), representing 10^{-9}, 10^{-6} and 10^{-3} molar, respectively. A typical cytosolic free Ca^{2+} concentration in a resting cell is 50 nM = 0.05 μM, rising to 1000 nM = 1 μM after stimulation. Units go up or down in factors of 1000: millimoles (mmol), micromoles (μmol), nanomoles (nmol), picomoles (pmol), femtomoles (fmol), attomol (amol), tipomol (tmol) and impossomoles (imol). Imposso = 10^{-24}. You cannot have 1 imol, because of the Avogadro constant, 1.022×10^{23} (i.e. the number of molecules in 1 mole – the molecular weight in grams), but you can have 10 imol.

The affinity of Ca^{2+} for a ligand is an important parameter. Each Ca^{2+}-binding site conforms to the equations:

$$Ca^{2+} + L = CaL \tag{2.1}$$

$$K_a = [CaL]/[Ca][L]; K_a \text{ in } M^{-1} \tag{2.2}$$

$$K_d = \left[Ca^{2+}\right][L]/[CaL]; K_d \text{ in } M, \text{ where } K_d = 1/K_a \tag{2.3}$$

Chemists tend to prefer to express affinities as association constants (Eq. 2.2). However, it is easier to relate the affinity of a ligand to the concentration of Ca^{2+} in the cell, or extracellularly, by using the dissociation constant (Eq. 2.3). The Ca^{2+} ligand is half saturated when the free Ca^{2+} concentration = K_d. Comparison of ion channels uses conductance, in siemens (S), where conductance = 1/resistance = current/voltage. The currents of individual ion channels are so small that these are typically measured in nano- or pico-amperes (nA or pA), and membrane potential in millivolts (mV). Time units for Ca^{2+} signals are typically in seconds or minutes; however, action potentials and channel openings can occur over a millisecond timescale.

2.8 Model Systems

Model systems have played a major role in unravelling the secrets of the intracellular Ca^{2+} signalling system (see Chapter 3). These include whole animals and plants, organ culture, and cell cultures from primary cells (i.e. direct from the organism), as well as cells from cancers or genetic engineering. Model systems enable mechanisms to be worked out, drugs to be tested, and new methodologies established, as well as providing new materials. A model system must be truly relevant to the real situation in the whole organism, and not just a 'simple' system making experiments easy.

2.9 Darwin, Wallace and Intracellular Ca^{2+}

2.9.1 Natural Selection in Real Time

Natural Selection is **the** unifying principle in biology, first revealed by Charles Darwin and Alfred Russel Wallace in 1858 (Figure 2.8). The intracellular calcium signalling system is a beautiful example of how Natural Selection works at the molecular level, in living organisms and throughout evolution. Evolution has chosen Ca^{2+} and not Zn^{2+}, because Ca^{2+} binds to, and comes off, proteins fast. So a muscle can twitch, and a heart can beat. The off rate of Zn^{2+} from proteins would be too slow for Zn^{2+} to work as an acute cell signal. Without molecular mechanisms to regulate intracellular Ca^{2+} the evolution of life over 4000 million years could not have occurred. Natural Selection explains how the best-adapted individuals of a population are selected for over time, pass on this adaptation to their offspring, and eventually lead to the evolution of new species, the poorly adapted following a path to extinction. Natural Selection not only works over periods of millions of years, it works in real time. It explains the beaks of finches on the Galapagos today, as well as MRSA (methicillin-resistant *Staphylococcus aureus*), changes in flu virus, the resistance of mice and rats to warfarin, and mosquitoes to DDT. Natural Selection works all the time in the human body, making sure the best antibodies are made when we have an infection, allowing the right balance of microflora in the gut, and allowing organs such as the brain to develop properly. The Ca^{2+} signalling toolkit is involved in all this. There are five key concepts underpinning Natural Selection as the main mechanism driving the evolution of species (Box 2.4).

Figure 2.8 Charles Robert Darwin (1809–1882) and Alfred Russel Wallace (1823–1913). From a photograph by Maull and Fox circa 1854 as the frontispiece of *Darwin and Modern Science*; Seward, A.C. (1910); and *Alfred Russel Wallace — Letters and Reminiscences*; Marchant (1916).

Charles Darwin
Aged about 50
(1809 – 1882)

Alfred Russel Wallace
Aged 39
(1823-1913)

> **Box 2.4 Key concepts underpinning evolution by Natural Selection.**
>
> 1. Species are not constant. They are continually changing with time – evolving; to use Darwin's word, they 'transform'.
> 2. There are major variations in form, behaviour, cellular and molecular processes, between and within species.
> 3. Small change by small change – these variations and changes within species are often very small.
> 4. There is a 'struggle for existence'.
> 5. Since there are more organisms that are born than can survive, certain small differences give certain individuals, the best adapted, a better chance of producing offspring. This is Natural Selection.

2.9.2 Small Change by Small Change

The consequences of global warming and climate change depend on Darwin's 'small change by small change'. Darwin was the first to realise how coral reefs form, by continuous building up of the $CaCO_3$ containing coral deposited as the coral grows, and then dies. Yet small percentage changes in pH or temperature of seawater will lead to a reef disappearing within a few years, by dissolving its calcium carbonate. Small changes in the body, including Ca^{2+} signalling, are central to understanding how a disease develops over months or years – a timescale far removed from that of the *in vitro* tissue culture system! If every cancer cell divides once a day, within 9 months the tumour would be bigger than the Milky Way. Just a 1% increase in cell division over death, not measurable directly, will give rise to a huge and lethal tumour within a few months. Similarly, rises in cytosolic free Ca^{2+} in *E. coli* speed up the division of genetically engineered bacteria by just 10%, which in the gut would lead to 20 000 times the faster growing cells than the slower ones. Small differences in the conductance of particular Ca^{2+} channels give a selective advantage to cells, and small differences in Ca^{2+} export, and intracellular Ca^{2+} release, mechanisms also have a selective advantage.

Biodiversity is the key to the survival and successful evolution of species, and whole ecosystems. Conventionally there are three types:

1. Diversity of species.
2. Diversity of genetics.
3. Diversity of habitats.

There is, however, a fourth – molecular biodiversity, which is well illustrated by the Ca^{2+} signalling system. This holds the real key to Natural Selection. Molecular biodiversity arises from:

1. Different genes producing sequence-similar proteins, or proteins with very different sequences, but where each has the same molecular function.
2. Different alleles of the same protein, with slightly different sequences, but identical molecular functions.
3. Different protein sequences produced as a result of alternative splicing, or covalent modifications.
4. Different proteins that fulfil the same or similar molecular functions.

This molecular biodiversity subtly affects the biochemical properties of each protein, and is well illustrated in Ca^{2+} signalling:

Type 1: Within an individual, different proteins have the similar or same 'function', e.g.:

- Large variety of channels getting Ca^{2+} into the cell – voltage-gated, receptor-operated, cyclic nucleotide-operated, SOCE.
- Several different mechanisms for getting Ca^{2+} out of the cell – Ca^{2+}-MgATPases (PMCA 1-4), Na^+/Ca^{2+} exchangers (NCX), H^+/Ca^{2+} exchangers.
- Several key Ca^{2+}-binding proteins – calmodulin 1–3, troponin C, synaptotagmin. Three types of IP_3 and ryanodine receptors in many animals.
- Three SR/ER Ca^{2+} pumps (SERCA1–3).
- Alternative splicing of many of these proteins.

Type 2: Between individuals of the same, or closely related, species, as different protein products have similar or the same 'function', e.g.:

- Different colours emitted by luciferases in fireflies and glow-worms.
- Small differences in the sequences of Ca^{2+} signalling proteins, and thus their precise biochemical and electrical properties, between the same or related species.

Type 3: Within or between different phyla, where different genes and their protein products have similar or the same molecular 'function', e.g.:

- Cytosolic Ca^{2+} as the trigger for the emission of the same colour from different chemical reactions in unrelated bioluminescent organisms.
- Small differences in the sequences of Ca^{2+} signalling proteins between different phyla.
- Completely different Ca^{2+} signalling proteins fulfilling the same task in organisms from different taxonomic groups.

Major variations in the timing, magnitude and type of intracellular Ca^{2+} signals occur between different cell types, and within the same cell type, e.g. the level of stimulus required to cause an individual cell to cross the Rubicon and fire. Such molecular biodiversity is dependent on five variations:

1. Amino acid sequences of proteins in the Ca^{2+} signalling toolkit.
2. Biochemical and electrical characteristics, e.g. kinetics, Ca^{2+} affinity, conductance.
3. Covalent modifications.
4. Level of protein expression.
5. Number of cells expressing a particular component.

These variations occur both within a species, and even within an individual, as well as between species. There are a huge number of variants for many key signalling proteins. Further variation occurs in cell types as a result of:

1. Polymorphisms, leading to different allelic forms of the same protein.
2. Splicing variants that produce different mRNAs, and thus protein sequences.
3. Promoter variants that start formation of the mRNA (transcription) at a different point.

There are also variations in the amount of a Ca^{2+} signalling protein through:

1. Different mRNA levels, through variations in rates of transcription, splicing or degradation.
2. Different protein levels, through variations in rates of mRNA translation, covalent modification or degradation.
3. Differences in rates of trafficking to and from its site of action.

These result in subtle biochemical differences including:

1. Affinity for substrates and Ca^{2+}.
2. Catalytic activity of the Ca^{2+} signalling protein.
3. Conductance, ionic specificity, and other electrical properties of Ca^{2+} channels.
4. Effects on their interaction with other proteins.
5. Regulation by natural agonists and antagonists, and covalent modification.
6. Gene structure – such as those differences caused by methylation, and other covalent modifications of DNA.

A further 'hidden' molecular biodiversity is found in 'bad' genes, which give us an inherited disease, or a risk of disease. A large number of polymorphisms have been found in many Ca^{2+} signalling genes (see Chapter 10), which lead to defects in the Ca^{2+} signalling system and organ malfunction. There are several examples of loss of Ca^{2+} signalling genes removed experimentally, or in inherited disease, where the clinical effect is small. This highlights that we should ask: 'What is the 'selective advantage' of a particular gene or protein, or a bad or risk gene? Not what is its 'function'? There is much to be learnt about Darwin–Wallace mechanisms at the molecular biodiversity level and what their real selective advantage is.

2.10   New Knowledge

Our civilisation, culture, economy and health, and that of the planet, depend on the continuous generation of new knowledge. The story of intracellular Ca^{2+}, its hidden mechanisms throughout Nature revealed by thousands of scientists, illustrates the ten guiding principles that are required to generate new knowledge:

1. *Inspiration* – leading to an idea and key questions.
2. *Logical thinking* – giving us an experimental pathway, leading to a key experiment.
3. *Lateral thinking* – thinking 'outside the box', leading to a genuine original idea for discovery.
4. *Invention* – the ability to design a novel technology to answer the key question.
5. *Taking risks* – travelling intellectually into a domain where no one has been before.
6. *Having an open mind, with a positive approach* – discoveries are always made on the basis of positive questions and hypotheses.
7. *Perseverance* – never giving up once you are on the right track.
8. *Hard work* – the pathway is full of intellectual and physical challenges.
9. *Money* – facilities required for new experiments cost money, sometimes millions of pounds; yet with ingenuity, major discoveries and inventions have been made with minimum resources.
10. *CURIOSITY* – the most important driving force of all, and the starting point for generating new knowledge.

One of the great joys of being a human being is our insatiable curiosity. Natural History is the love and curiosity of Nature, and its description in focussed terms, using all our senses – sight, sound, smell, taste and touch. Natural Science, on the other hand, is about mechanism – how the Universe works, from the Big Bang to how Ca^{2+} causes a muscle to fire enabling a cheetah to catch an antelope. The unravelling of the intracellular Ca^{2+} signalling system required a marriage of the skills of the naturalist and the natural scientist, in the tradition of Darwin and Wallace. There were four golden rules:

1. Always being curious.
2. Asking the key question.
3. Designing the key experiment.
4. Always being positive.

2.11   Conclusions

1. Intracellular Ca²⁺ acts as a signal for a wide variety of cell events, but is not the energy source. This comes from the MgATP/MgADP + phosphate reaction, maintained by the cell far from equilibrium on the side of MgATP. ATP does not have an energy rich bond.
2. The *'active'* role of Ca²⁺ as an intracellular signal distinguishes it from its *'passive'* role in enzymes such as amylase, and the blood clotting and complement cascades.
3. The Rubicon principle defines cell events as digital, which are triggered initially by external primary stimuli, and can be regulated in an analogue manner by secondary regulators.
4. Cell events are categorised on the basis of whether the main source of a cytosolic Ca²⁺ signal is extracellular, intracellular or both.
5. Intracellular Ca²⁺ signals occur in a variety of levels, shapes and locations.
6. Scientific naming of the Ca²⁺ signalling toolkit provides a discipline, vital for communicating discoveries about intracellular Ca²⁺.
7. The components of the intracellular Ca²⁺ system, the *toolkit*, exhibit a molecular biodiversity upon which Natural Selection, revealed by Darwin and Wallace, depends.
8. The unravelling of the intracellular Ca²⁺ system is a fine example of the positive DISI model in action in science – Discovery, Invention, Scholarship and Impact, which has revolutionised biomedical science and created several billion dollar markets.

Recommended reading

*A must read

Books

Campbell, A.K. (1994) Rubicon: The fifth dimension of biology. London: Duckworth. Full analysis of digital processes in cells throughout 4000 million years of evolution.
*Campbell, A.K. (2015) Intracellular Calcium. Chapter 2 Intracellular Ca²⁺. Principles and terminology. Chichester: John Wiley & Sons Ltd.
*Haugland, R.P. (1996) Handbook of Fluorescent Probes and Research Chemicals, 6th edn. Eugene, OR: Molecular Probes. Very useful review of all fluorescent indicators, and some bioluminescent ones, with key properties. Dick Haugland set up the company Molecular Probes, now taken over by Invitrogen.

Reviews

*Berridge, M. (2007) Calcium signalling, a spatiotemporal phenomenon. Chapter 19. In: Krebs, J. & Michelak, M. (Eds) Calcium; a Matter of Life and Death. Amsterdam: Elsevier. Good review by Ca²⁺ pioneer with a list of micro-changes in cytosolic free Ca²⁺.
Campbell, A.K. (2003) Save those molecules! Molecular biodiversity and life. *J. Appl. Ecol.*, **40**, 193–203. The principles of biodiversity in molecules, with examples from clinical biochemistry and bioluminescence.
Collingridge, G.L, Olsen, R., Peters, J.A. & Spedding, M. (2009) Ligand gated ion channels. *Neuropharmacology*, **56**, 15. How ion channels are named.
Rudiger, S. (2014) Stochastic models of intracellular calcium signals. *Phys. Rep.- Rev. Sect. Phys. Lett.*, **534**, 39–87. Mathematical models for digital events involving intracellular calcium.
*Tsien, R.Y. (2003) Imagining imaging's future. *Nat. Cell Biol.*, Supplement S, SS16–SS21. Imaging Ca²⁺ in live cells using fluorescence by the late pioneer and Nobel Laureate for GFP.

Papers

*Campbell, A.K. (2012) Darwin shines light on bioluminescence. *Luminescence*, **27**, 447–449. Bioluminescence as a model for one of the key problems in evolution – the origin of a new enzyme.

Darwin, C.R. & Wallace, A.R. (1858, 1 July). On the tendency of species to form varieties; and on the perpetuation of varieties and species by natural means of selection. Communicated by Sir Charles Lyell, FRES, FLS, FRS and J.D. Hooker esq., M.D., F.R.S., F.L.S. *J. Proceed. Linnean Society London. Zool.*, **3**, 45–50. See also http://darwin-online.org.uk/. First public presentation of the principle of Natural Selection.

*Gilkey, J.C., Jaffe, L.F., Ridgway, E.B. & Reynolds, G.T. (1978) A free calcium wave traverses the activating egg of the medaka, *Oryzias latipes. J. Cell Biol.*, **76**, 448–466. First image of a Ca^{2+} wave in a live cell using aequorin.

Woods, N.M., Cuthbertson, K.S.R. & Cobbold, P.H. (1986) Repetitive transient rises in cytoplasmic free calcium in hormone-stimulated hepatocytes. *Nature*, **319**, 600–602. First description of Ca^{2+} oscillations in small cells, using aequorin.

3

A Century plus of Intracellular Ca²⁺

> **Learning Objectives**
> - How the history of intracellular calcium provides insights into scientific discoveries.
> - Asking the right questions.
> - The importance of invention.
> - Three pioneers in the story of intracellular calcium.
> - Key experiments to measure free Ca²⁺ in live cells, beginning in 1967 establishing the sub-micromolar free Ca²⁺ in the cytosol of all cells – animal, plant and microbe.
> - Key discoveries, beginning in the 1960s, of how Ca²⁺ is regulated in the cytosol and organelles, together with the intracellular Ca²⁺ targets, and their special EF hand or C2 Ca²⁺ binding sites.
> - How patch clamping led to the discovery and characterisation of multiple Ca²⁺ channels, upon which the electrical activity of cells depends.
> - How, by the end of the twentieth century, the main features of the intracellular Ca²⁺ toolbox had been revealed.

The calcium ion has an unusual importance in biological phenomena, and the literature concerning its effects extremely voluminous.
Lewis Victor Heilbrunn (1937)
(Quote reproduced with permission from Elsevier.)

3.1 Background

3.1.1 Pathway of Discovery and Invention for Intracellular Ca²⁺

The story of intracellular Ca²⁺ as a universal regulator begins at the end of the nineteenth century with the experiments of Sydney Ringer on frog heart and other systems. This was the first indication that there was something special about calcium, quite distinct from its role in biomineralisation. Ringer was followed by others who showed that removal of external calcium, and addition of pharmacological substances, had damaging effects on a range of animal and plant cells, and organisms. It was not until the 1930s that biologists began to realise that the special feature of calcium they were investigating was inside cells, and not outside. The hunt was then on, during the 1940s, 1950s and 1960s, for how calcium might work inside cells, searching for proteins that could bind Ca²⁺, and systems that could regulate

Fundamentals of Intracellular Calcium, First Edition. Anthony K. Campbell.
© 2018 John Wiley & Sons Ltd. Published 2018 by John Wiley & Sons Ltd.
Companion Website: http://www.wiley.com/go/campbell/calcium

its concentration inside the cell. The major breakthrough at the end of 1960s, and through to the 1980s, was the invention of various indicators to measure directly free Ca^{2+} in live cells. This gave the proof that the special feature of intracellular Ca^{2+} was a signal, to switch on a wide range of cellular processes. The discovery that the free Ca^{2+} inside cells was very low, sub-micromolar, with millimolar free Ca^{2+} outside the cell, led to a search for the molecules that maintained this large gradient, and how the Ca^{2+} pressure could be used to cause a large fractional rise in cytosolic free Ca^{2+}. Measurement of cytosolic free Ca^{2+} also enabled many of the possible protein targets identified in the previous decades to be rejected, as Ca^{2+} only affected them at millimolar concentrations, several hundred times higher than the cytosol free Ca^{2+} ever reached physiologically. This also led to a search for how Ca^{2+} might be a foe, if the cell was overloaded with Ca^{2+}, playing an important role in cell pathology and death. The last 20 years of the twentieth century led to an explosion of experiments, measuring, imaging and manipulating intracellular Ca^{2+}, together with electrophysiology and molecular biology, identifying the complete molecular pathway by which a change intracellular Ca^{2+} leads to cell activation or injury. In the 1970s, X-ray crystallography led to one the most important structural discoveries about how Ca^{2+} bound to proteins with high affinity – the EF-hand. But even as late as 1980, there were still sceptics. How could such an apparently insignificant ion, at such low concentrations, be responsible for such a diverse set of cell responses, from muscle contraction and nerve excitation to egg fertilisation, and the defence of plants against stress. Ca^{2+} was not the energy source for these cell events. Rather it was a chemical switch, releasing a primed up system to act.

3.1.2 Model Systems

The story of intracellular Ca^{2+} has been highly dependent on a wide range of animal, plant and microbial model systems, as well as cells in tissue culture (Table 3.1). These were chosen either because they had a particular end response that was easy to observe and quantify, or because they were easily susceptible to experimental manipulation. A good example of this were the large muscle fibres of the giant barnacle *Balanus nubilus*. These are 'single' cells, at least 1 cm long and several mm in diameter, enabling them to be injected with the Ca^{2+}-activated photoprotein aequorin, to have electrodes inside them, and to be easily attached to a force transducer. As a result cytosolic free Ca^{2+} was measured properly in a live cell for the first time, and correlated with the membrane potential and contractile force of the muscle.

3.2 Why Study the History of Science?

The history of intracellular Ca^{2+} reveals a fascinating story of discovery, invention and scholarship. It reveals how brilliant minds have resolved the puzzle of how one cation does so much, without being the energy source. It also shows how curiosity about calcium led to many surprises, and even established billion dollar markets. Yet, although several calcium pioneers have won the Nobel Prize, there has as yet to be one awarded for calcium itself. So many people have made seminal contributions to the story of intracellular Ca^{2+}, it has been impossible to single out just two or three.

Table 3.1 Examples of model systems that have led to major discoveries in biology and medicine.

Model system	Discovery
Animals	
Frog muscle	Electrical activity of muscle
Starfish larva and *Daphnia*	Phagocytes
Frog heart	Ca^{2+} important in heart beat
Fruit fly (*Drosophila*)	Gene linkage and the foundation of modern genetics
Giant axon of squid (*Loligo*)	Ionic basis (Na^+/K^+) of the action potential in nerves
Sea urchin (*Arbacia punctulata*)	Cyclins in the cell cycle
Star fish oocyte	Ca^{2+} in maturation, cyclins
Jellyfish (*Aequorea, Obelia*)	Aequorin, obelin, GFP
Nematode worm (*Caenorhabditis elegans*)	Development; first genome to be sequenced
Mice (*Mus*)	Production of monoclonal antibodies, gene knock-outs
Plants	
Pea	Inheritance
Tobacco (*Nicotiana*)	Plant development and stress
Weed (*Arabidopsis*)	Plant development and stress
Microbes	
Pneumococcus	DNA as the molecule of inheritance
E. coli	mRNA, gene control
Streptococcus, Staphylococcus	Antibiotic resistance, model for Gram-positive bacteria
Yeast (*Saccharomyces*)	Intermediary metabolism, cell cycle, the role of kinases
Luminous bacteria (Photobacterium, *Vibrio*)	Quorum sensing
Darwin's models	
Giant tortoise, beaks of finches, pigeons, mockingbirds, peacocks, earth worms, orchids, insectivorous plants, climbing plants	Natural Selection

3.3 The Tale of Three Pioneers and What Followed

Three pioneers stand out in the early years – Sydney Ringer, Lewis Heilbrunn and Setsuro Ebashi (see Frontispiece). Ringer and Heilbrunn were physiologists, Ebashi a biochemist (see Box 3.1 below).

3.3.1 Experiments of Sydney Ringer (1836–1910)

The year 1883 was a landmark in the history of calcium. Sydney Ringer, working at University College Hospital (UCH) London, published his famous experiment showing that the frog heart stopped beating if the calcium was removed from the medium. In his first experiments, the isolated beating heart of a frog continued to beat when perfused with a salt solution containing NaCl, KCl, $MgCl_2$ and $CaCl_2$. Removal had no effect. Unfortunately his technician had made up the solutions using London tap water. Ringer found that the tap water contained 38.3 ppm calcium, equivalent in his solution to nearly 1 mM – a concentration we now know is similar to the

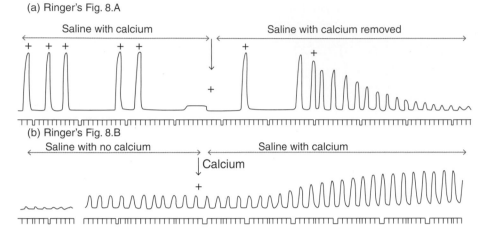

Figure 3.1 Ringer's demonstration that removing calcium stops a heart beat. The preparation was an isolated frog heart with the ventricle connected by a cannula to perfuse the heart with a saline (0.75% NaCl). From Ringer (1883) *J. Physiol.*, **4**, 29–43, Figure VIII.

free concentration of Ca^{2+} in human blood. By removing each constituent one at a time, making up the solutions this time with distilled water, Ringer found that the critical component was calcium, without which the frog heart stopped beating. With saline alone, the heart stopped contracting. But addition of Ca^{2+} allowed the heart to continue beating for 4 h (Figure 3.1). He wrote:

> *I find that calcium, in the form of lime water, or bicarbonate of lime or chloride of calcium, even in minute doses produces the changes in the influence of blood constituents on heart ventricular beat described in my former paper. The heart's contractility cannot be sustained by saline solution nor by saline containing potassium chloride, … but after contractility has ceased, the addition of a lime salt will restore good contractility. … I conclude therefore that a lime salt is necessary for the maintenance of muscular contractility.*

Ringer followed these experiments by investigating the effects of a wide range of salts on frog heart and skeletal muscle, eggs, tadpoles and worms. Calcium was the key electrolyte necessary in all his salines. These experiments led to the now famous 'Ringer' solution. Its exact composition varies all over the world, and was developed by Locke and Hans Krebs, giving us Locke's and Krebs' solutions (Table 3.2). Calcium is very difficult to get rid of completely from solutions, as it leaches off glassware, and even occurs in Analar reagents. To be absolutely sure that your solution has 'no' Ca^{2+}, specially purified NaCl, KCl and $MgCl_2$ (Spec pure) are needed, with water purified on an ion-exchange column, or a chelator such as EGTA must be used.

Ringer spent most of his professional life at University College Hospital (UCH). He was a pioneer of experimental pharmacology, studying the effects of various substances on frog heart and skeletal muscle, tadpoles, worms and eggs. These substances included chloral, opium, atropine, nicotine, alkaloids from the daffodil, arsenic and derivatives, anaesthetics and ergotine. He was also a pioneering biochemist, studying the effects of Ca^{2+} and other substances on egg albumin, casein from milk and blood clotting. His clinical research included a wide range of diseases, such as fever, sneezing, bronchial problems, diabetes, scarlet fever, paralysis and tetanus, but not heart disease. He became a highly respected clinician with a reputation for good diagnosis. Ringer published some 64 scientific papers and one book, a popular book, *A Handbook of Therapeutics*. Ringer retired in 1900, and died in North Yorkshire in 1910.

Table 3.2 Classic salt solutions used to study living tissues *in vitro*.

Salt	Ringer, 1883	Locke, 1901	Krebs–Henseleit, 1932
NaCl	116	154	117
KCl	1.2	5.6	4.7
$MgSO_4$			1.2
$CaCl_2$	1.0	2.1	2.5
$NaHCO_3$	2.7	2.4	24.8
KH_2PO_4			1.2
Glucose		5.6	11.1
Other			5% CO_2
pH	?	?	7.4

Concentrations in mM. See Campbell (2015) for references.

Ringer's experiments prompted a plethora of studies by others, over the following 30 years. All showed that Ca^{2+} was the essential component in physiological salines for animal and plant cells. Without calcium outside, tissues fell apart, and cellular events such as muscle contraction and secretion ceased. But it was an American zoologist and physiologist, Lewis Victor Heilbrunn, at the University of Pennsylvania in Philadelphia, who realised that the unique role of calcium was **inside** cells.

3.3.2 The Vision of Lewis Victor Heilbrunn (1892–1959)

Lewis Victor Heilbrunn was **the** pioneer of *intracellular* Ca^{2+} as a universal regulator. The two editions of his textbook, *An Outline of General Physiology*, are full of references to the role of calcium in cells, with several references to the importance of loss of calcium from internal stores. Heilbrunn began his experimental work at the Marine Biological Laboratory at Woods Hole, Cape Cod, MA. He became aware of a fascinating experiment published in 1928 by Herbert Pollack, a student of Robert Chambers (1881–1957) at Cornell, who was one of the first to develop techniques for microinjecting cells with dyes and other substances. Pollack injected the amoeba, *Amoeba proteus*, with a red dye alizarin sulphonate, which he knew precipitated Ca^{2+}. Immediately, the amoeba rounded up, and stopped moving. The cytoplasm showed fine red granules scattered throughout the cell. When the amoeba tried to move again it put forward a membranous foot, its 'pseudopod'. Pollack wrote:

> *'a shower of red crystals was seen to appear in this area and the pseudopod formation was immediately stopped'.*

Perceptively, he predicted that this precipitate of alizarin sulphonate with calcium had lowered the internal calcium, so the movement stopped. He also argued that the cell must have an internal calcium store – a reserve that enabled the cell to replenish the calcium in the cell matrix. These remarkable proposals by Pollack occurred over 20 years before the sarcoplasmic reticulum (SR) calcium store was first isolated from muscle by Marsh in 1951.

Yet, others did not follow Pollack with similar experiments in other systems. Many assumed, wrongly, that Ca^{2+} could only provoke movement, or muscle contraction, if it was the energy source. And surely this was impossible if the amounts of calcium were so small? The role of ATP as a universal energy source for virtually all biological processes was not yet realised.

And even when it was, the issue was hopelessly confused by one of the greatest misconceptions in biology – the energy-rich bond! Lewis Victor Heilbrunn had the vision to realise that, in contrast, to acting as an energy source, calcium must be a chemical switch. To test this he injected cells with small amounts of Ca^{2+}, and compared this with the effects of injecting other salts, such as KCl, NaCl and $MgCl_2$.

During the 1930s, Heilbrunn carried out experiments with the eggs of marine invertebrates, particularly the sea urchin *Arbacia*, and the ragworm *Nereis*, predicting that release of calcium inside the cell was the trigger for breakdown of the nuclear membrane, following fertilisation by sperm or parthenogenetic stimulation of the *Nereis* egg. Heilbrunn also showed an effect of calcium added to stripped muscle, on what he called muscle protoplasm, the gel-like structure of the inside of cells. Then, he carried out a crucial experiment on muscle with another student, Floyd Wiercinski. Painstakingly making quartz micropipettes, for fear of contamination if Pyrex-type glass was used, they injected various ionic solutions into the skeletal muscle of a frog leg. The response was dramatic! Injection of calcium chloride caused an immediate shortening of the muscle by some 40% or more (Figure 3.2). As Wiercinski wrote to me after the publication of my first book on intracellular calcium, 'It was fantastic to watch. We injected the calcium, and the fibre instantly pulled into a mass.' To their joy, sodium, potassium and magnesium chlorides did not cause the muscle to contract, nor did water alone. Only calcium chloride triggered the fibre to contract. They concluded in their paper published in the *Journal of Cellular and Comparative Physiology*:

> *The calcium ion, in rather high dilution, causes an immediate and pronounced shortening. This effect is not shared by any one of the cations normally present in any quantity in muscle .... The results lend support to the calcium release theory of stimulation and they are opposed to Szent-Györgyi's belief that the potassium ion is primarily responsible for the contraction of muscle.*

However, the Nobel Laureate Archibald Vivian Hill (1886–1977), working at University College in London, argued Heilbrunn was wrong. Hill calculated that the rate of diffusion of calcium from the outer edges of the muscle cell to the contractile machinery, well inside the cell, would be too slow to account for the fast contraction in real muscle. As another of Heilbrunn's students, Marian LeFevre, later remarked, 'In those days, whatever A.V. Hill said was instant law.' It seemed the end of Heilbrunn's vision about calcium. But Hill's overall conclusion was wrong! Muscle has evolved a large internal Ca^{2+} store that lies very close to the fibrils that contract, allowing the calcium to reach its target within milliseconds.

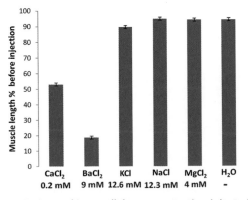

Estimated intracellular concentration injected

Figure 3.2 The first demonstration that injection of Ca^{2+} into muscle caused it to contract. The adductor niagnus muscle from the frog *Rana pipiens* was immersed in Ringer's solution, pH 6, with no Ca^{2+}, to prevent contraction when the fibre was impaled with the micropipette. A single fibre was then injected with approximately 10% of its volume with various salts solutions: $CaCl_2$, $MgCl_2$, KCl, NaCl, $BaCl_2$ or water. Only injection of $CaCl_2$ and $BaCl_2$ caused the muscle to shorten within 5–15 s. Data calculated from Heilbrunn and Wiercinski (1947) *J. Cell Comp. Physiol.*, **29**, 15–32. This paper has over 300 citations – not bad for a paper published in 1947.

In the first edition of his *An Outline of General Physiology* published in 1937, Heilbrunn wrote, 'No other ion exerts such interesting effects on protoplasmic viscosity as the calcium ion.' Herein lay Heilbrunn's problem in getting his universal calcium hypothesis generally accepted. He was obsessed with the idea that changes in the state of the jelly-like protoplasm held the key to cellular events, such as movement and secretion. He even founded a journal, *Protoplasmatologia*, for publications related to this aspect of cell biology. But to many, arguments about the protoplasm sounded like a return to the mysterious 'vital force' that had impeded proper scientific investigation of the chemistry of life in the eighteenth and nineteenth centuries. In fact, the 'protoplasm' turned out to a complex of actin and myosin, with many other proteins. Ca^{2+} binding to gelsolin does convert it from a gel into liquid form.

Heilbrunn was born in Brooklyn, New York, on 24 January 1892, attending Cornell University and obtained a PhD at the University of Chicago in 1914, under the supervision of Frank R. Lillie. After various posts, in 1929 he was appointed to the staff of the Department of Zoology, University of Pennsylvania – an association that continued for 30 years, until his death. He had a deep interest in the arts, particularly in writing and painting, establishing, in 1950, the Ellen Donovan Gallery, in honour of his wife. He became a Trustee of Woods Hole in 1931. Yet a review of work based at this already famous laboratory, published in 1940s, made no mention of Heilbrunn's pioneering calcium work. This must have been very depressing for him. However, he was clearly an inspiration to his students, an enthusiast and highly cultured. As his obituary says,

'His untimely death in an automobile accident on 24 October (1959) snuffed out a creative spirit science can ill afford to lose, but his influence will continue for generations to come.'

3.3.3 Setsuro Ebashi (1922–2006): Pioneer of Intracellular Ca²⁺ in Muscle Contraction

Setsuro Ebashi, working in his home country of Japan for all of his life, discovered two key components of the Ca^{2+} signalling system in muscle during the early 1960s:

- The sarcoplasmic reticulum (SR) as the intracellular Ca^{2+} store, released to provoke contraction.
- Troponin C as the Ca^{2+} target that releases the inhibition of the actomyosin complex by tropomyosin when it binds Ca^{2+}.

Identification of the SR as the internal Ca^{2+} store hit on the head A.V. Hill's objection that the diffusion of Ca^{2+} from the plasma membrane would be too slow for it to provoke rapid twitches of muscle fibres. Then, the discovery of troponin C, the first Ca^{2+}-binding protein with micromolar affinity for Ca^{2+}, opened the door to how Ca^{2+} works inside cells through high affinity Ca^{2+}-binding proteins. This revealed how Ca^{2+} can act a switch, the energy coming from the ATP hydrolysis, as it attempts to reach equilibrium with ADP and phosphate.

Ebashi was an exceptional scientist with an outstanding personality. He was born in Tokyo in the August of 1922. He graduated with a doctorate in Medicine in 1944, from the Faculty of Medicine at the University of Tokyo, obtaining his PhD in 1954, and continuing to work there for most of his professional life. While purifying the enzyme choline-acetylase, he discovered a 'relaxing factor', which in the presence of ATP caused rabbit muscle, permeabilised in glycerin, to relax. He eventually discovered that this 'relaxing factor' was a vesicle with a MgATPase, which could accumulate Ca^{2+} in the presence of MgATP. This was the SR, whose tubular structure was revealed later by electron microscopy.

During the early 1940s, Heilbrunn in the United States and Kamada in Japan had shown that addition of Ca^{2+} to a 'protoplasm' from muscle (i.e. muscle stripped of its outer membrane) caused this to contract. Ebashi followed this in the 1960s by isolating the two key components that transmitted the Ca^{2+} signal to the actomyosin complex, allowing it to contract. Ebashi wrote over 100 scientific papers, many of which were in English, even in his early publications. His first paper, published in *Nature* in 1955, was entitled 'Essential relaxing factor in muscle other than myokinase and creatine phosphokinase'; his last was in 1994 in the *Canadian Journal of Physiology and Pharmacology* and was entitled 'Is phosphorylation the main physiological action of myosin light chain kinase?'

His seminal research attracted, and inspired, many research fellows and associates, who themselves carried out distinguished work, which continued when he moved from Tokyo to Okazaki. Ebashi received many prizes and accolades, but, surprisingly, he was never awarded the Nobel Prize. Like many scientists, Ebashi was highly cultured, had passions outside science, being able to sing unaccompanied (*a cappella*) chorales of J.S. Bach. He also had an interest in stained glass windows in Gothic churches. He died in 2006 aged 83 (Box 3.1).

Box 3.1 Three pioneers and their key discoveries.	
Sydney Ringer (1836–1910)	Ca^{2+} is required for the heart to beat, and many other phenomena in other animals.
Lewis Victor Heilbrunn (1892–1959)	Direct demonstration that a rise in cytosolic free Ca^{2+} triggers muscle contraction.
Setsuro Ebashi (1922–2006)	The sarcoplasmic reticulum Ca^{2+} store and troponin C as the Ca^{2+} target in muscle.

3.4 Ca^{2+} as an Intracellular Regulator

The years that followed the experiments of these three pioneers led to the hypothesis that Ca^{2+} is a universal regulator responsible for a wide range of intracellular events. A rise in cytosolic free Ca^{2+} was shown to be essential for:

- All forms of muscle contraction.
- Many other types of cell movement.
- Nerve excitation.
- Secretion resulting from fusion of internal vesicles with the plasma membrane.
- Metabolic reactions, such as glycogen breakdown and mitochondrial oxidation of pyruvate.
- Gene expression.
- Vision in vertebrates and invertebrates.
- Fertilisation of an egg by sperm.
- Regulation of the cell cycle.
- Some types of bioluminescence.
- Cell defence.

Disturbances in cytosolic Ca^{2+} also were involved in cell injury, reversible cell damage by pore formers and cell death.

The evidence to support the hypothesis that intracellular Ca^{2+} is a universal cell regulator has been accrued through experiments involving physiology, biochemistry, molecular and cell biology, and electrophysiology of individual cells, organs and whole organisms. Manipulation, measurement and imaging of intracellular Ca^{2+}, correlating changes with a cell event, provided the evidence that a change in intracellular Ca^{2+} was the signal. The study of Ca^{2+} fluxes and currents, together the isolation and manipulation of intracellular organelles and proteins, provided the evidence for how changes in intracellular Ca^{2+} occurred, and how these affected intracellular targets to provoke the cell event. The pathway that unravelled Ca^{2+} as an intracellular regulator required many amazing inventions:

- Chemicals that manipulate intracellular Ca^{2+}.
- Pharmacological agents that inhibit specific proteins.
- Radioactive probes to study Ca^{2+} flux across membranes.
- Substances that measure and imaging of free Ca^{2+} in live cells, with advanced microscopy and lasers.
- Patch clamping to characterise Ca^{2+} channels.
- Protein purification to be able to characterise isolated proteins.
- Molecular biology and genetic engineering to isolate genes and manipulate them *in vivo*.
- X-ray crystallography and NMR, enabling the three-dimensional structures of Ca^{2+}-binding proteins and their targets to be determined.

3.4.1 Cell Theory

Although the origin of the cell theory is often ascribed to Schwann (1839), it was Virchow (1821–1902) in 1856 and Sedwick in 1896, followed by Ehrlich, who really established the modern dogma that all life depends on cells. But it was not until the late 1940s and 1950s that the dogma that their reproduction depended on the replication of DNA within them was established. The concept of a cell was fundamental to the development of our understanding of the role of calcium within it. Since the 1970s it has been clear that there are three domains of life – Eukaryota, Bacteria and Archaea – all of which maintain a very low cytosolic free Ca^{2+} in the micromolar range.

3.4.2 Origin of the Use of Ca²⁺

Calcium is in the form Ca^{2+} in all its biological actions and functions. In solution, Ca^{2+} is surrounded loosely by a shell of water molecules. When it passes through a calcium channel these are stripped off, and calcium goes through as Ca^{2+}. The biological actions of Ca^{2+} depend on its ability to complex with anions. These complexes may precipitate, as they do in bones, teeth and shells, or Ca^{2+} may remain in free solution, as it does when Ca^{2+} binds to proteins such as calmodulin. Yet, the form of calcium as Ca^{2+} was not fully realised in the early years of the story of intracellular calcium.

   In 1804, Hisinger and Berzelius stated that neutral salts could be decomposed by electricity – the 'acid' part appearing at one pole, the 'metal' part at the other. In 1887, Svente Arrhenius proposed a new theory to explain the behaviour of electrolyte solutions. For an ionisable solute, there was an equilibrium between undissociated solute molecules and the ions that split off from them. This led to a full ionic theory, developed particularly by Debye and Hückel. Since the valency of calcium had been known to be 2 since the nineteenth century, it is perhaps surprising that there were very few references to Ca^{2+} before the 1950s. Heilbrunn uses Ca, but not Ca^{2+}. Not until the mid-1960s did the use of Ca^{2+} become commonplace in the biological literature, and not until the late 1960s and 1970s was Ca^{2+} used routinely to describe calcium inside cells. The term calcium signalling only became commonplace from the 1990s onwards.

3.4.3 Manipulation of Ca²⁺

The ability to increase or decrease intracellular Ca²⁺ in live cells, and then correlate changes with an effect on cell physiology, has been crucial in providing experimental evidence for Ca²⁺ as an intracellular regulator. This involved:

- Manipulation of extracellular Ca²⁺.
- Injection of Ca²⁺ and Ca²⁺ buffers into live cells.
- Ionophores that carry Ca²⁺ across membranes.

Robert Chambers (1881–1957) pioneered the use of micropipettes to inject substances into live cells, developed by Heilbrunn and his students to inject calcium into live cells. A breakthrough in the 1960s was the use of Ca²⁺ buffers, EDTA and EGTA, with EGTA being selective for Ca²⁺ over Mg²⁺. These enabled the free Ca²⁺ concentration in the cytosol to be set at particular concentrations, but are very sensitive to pH. A major improvement was the development by Roger Tsien of Ca²⁺ buffers such as BAPTA, having pK_as so that they are almost fully ionised at pH 7. Used as an acetoxymethyl ester of BAPTA, this was membrane permeable, and hydrolysed to the free acid Ca²⁺ buffer by intracellular esterases. Ca²⁺ ionophores, such as A23187 or ionomycin, allowed Ca²⁺ to cross biological membranes, and thereby artificially raise free Ca²⁺ in the cytosol, or deplete intracellular stores when used in conjunction with a Ca²⁺ buffer. A further innovation in the 1990s was the use of photosensitive caged compounds that trap Ca²⁺ or a Ca²⁺ buffer, releasing these through a flash of a laser.

3.4.4 Measurement and Location of Free Ca²⁺ in Live Cells

The key to the understanding of Ca²⁺ as an intracellular regulator was the realisation of how low the concentration of cytosolic free Ca²⁺ is. Three techniques were developed during the 1960s and 1970s to establish this:

1. Ca²⁺-activated photoproteins from luminous jellyfish.
2. Indicator dyes, first light absorbing, then fluorescent.
3. Microelectrodes.

In 1925, A.B. MacAllum wrote that a method for measuring Ca²⁺ inside cells was 'a great desideratum'. The first attempt to monitor Ca²⁺ in a live cell was just three years later, by Herbert Pollack – a student of Robert Chambers. But it wasn't until 1967 that free Ca²⁺ transients were measured directly in a live cell, the giant single muscle fibres of the barnacle *Balanus nubilus* (Figure 3.3d), using the Ca²⁺-activated photoprotein aequorin from the luminous jelly fish *Aequorea*, first isolated by Osamu Shimomura (Figure 3.4a), who also discovered GFP. The jellyfish *Aequorea* was first described as luminous by Peter Forskål (1732–1774), a student of Carl Linnaeus in Uppsala, Sweden (Figure 3.3a). He wrote: '*Rasa ligno in pareum adeo in tenebris splendet*'; i.e. 'If you hit it on the head it shines brightly in the dark'. The word 'medusa' was first used by Linnaeus in 1752, alluding to the Medusa in Greek mythology. 'Aequorea' is from the Latin *aequoreus* simply meaning 'of the sea'.

Electrical stimulation led to a flash of light from within the barnacle muscle fibre, followed immediately by a contraction. Injecting the Ca²⁺ chelator EGTA prevented the flash of light from aequorin, and also stopped the muscle contracting. These pioneering experiments were followed in the 1970s by the application of aequorin to measure free Ca²⁺ in the giant axon of the squid, several other invertebrate giant cells, insect salivary gland, an amoeba, an egg from the medaka fish, and single mammalian cells, such as hepatocytes, fibroblasts, heart and adrenal chromaffin cells. I managed to extract a similar protein from its relative *Obelia geniculata*,

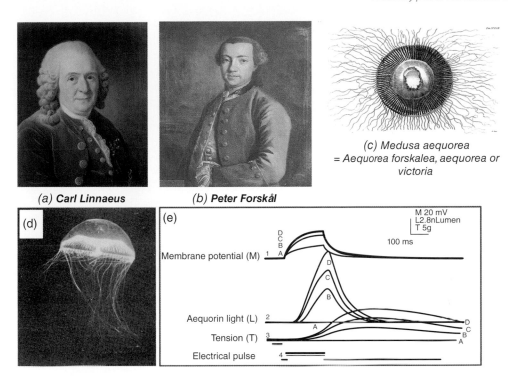

(c) Medusa aequorea
= Aequorea forskalea, aequorea or victoria

(a) **Carl Linnaeus**   (b) **Peter Forskål**

Figure 3.3 The discovery of the jellyfish A*equorea* and the first measurement of cytosolic free Ca²⁺ in a live cell, using aequorin extracted from it. (a) Carl von Linné (Linnaeus) (1707–1778) by Alexander Roslin (1718–1793). Source: Roslin 'Carl von Linné. NMGrh 1053. Photo © Nationalmuseum, Stockholm. (b) Peter Forskål (1732–1763), sometimes spelt Forsskål or Forsskåal, the discoverer of the luminous jellyfish *Aequorea*. In his book, Forskål describes the jellyfish as *Medusa aequorea* (Forskål, 1775), with a drawing. (Forskål, 1776). *Source*: Courtesy of Nils Gyllenadler, Salnecke Castle, Sweden. (c) Jellyfish drawing. Forskål, A. (1776). (d) *Aequorea victoria* from Friday Harbor on the west coast of the United States of America. *Source*: Courtesy of Professor S. Haddock. (e) The first experiment used a single muscle fibre from the giant Pacific barnacle *B. nubilus* injected with purified native aequorin. The fibre was stimulated by increasing single electrical pulses A–D. The aequorin light emission transient started just after the membrane depolarisation and before the rise in muscle tension. As the free Ca²⁺ decreased, so did the muscle tension. Temperature 11–12 °C. Source: Ashley and Ridgway (1970). Reproduced with permission from Wiley.

Figure 3.4 Two Nobel Laureates. (a) Osamu Shimomura (born 1928). Nobel Laureate for Chemistry 2008 and discoverer of the Ca²⁺-activated photoprotein aequorin and the GFP from the luminous jellyfish *A. victoria*. *Source*: Courtesy of Professor Shimomura. (b) Roger Yonchien Tsien (1952–2016). Nobel Laureate for Chemistry 2008 and inventor of the fluorescent indicators that have revolutionised the study of intracellular Ca²⁺. *Source*: Courtesy of R. Tsien.

and to entrap it in erythrocyte ghosts, which were also able to maintain an intracellular free Ca²⁺ at sub-micromolar levels in the presence of mM Ca²⁺ extracellularly. Raising the cytosolic free Ca²⁺, using ionophore A23187 or after attack by complement, raised the free Ca²⁺ to 1–10 μM. By fusing these erythrocyte ghosts with human or rat neutrophils, we entrapped obelin inside a small, nucleated cell, where the free Ca²⁺ was also found to be less than 1 μM, and was increased by the membrane attack complex of complement or the chemotactic peptide f-Met-Leu-Phe (FMLP), which activated the production of reactive oxygen metabolites. All these experiments showed that the cytosolic free Ca²⁺ was submicromolar in the resting cell and rose to around 1–10 μM following cell stimulation.

During the 1970s indicator dyes were used to measure free Ca²⁺ in live cells, first murexide, and then arsenazo III and antipyrylazo III (Figure 3.5). In the large axons of the sea slug *Aplysia californica*, Ca²⁺ oscillations were observed in the micromolar range (Figure 3.6a). But the real breakthrough came from the brilliant work of the late Roger Tsien (Figure 3.4b), originally working with Tim Rink in the Department of Physiology in Cambridge, United Kingdom.

Tsien first synthesised quin-2, whose fluorescence increased when it bound Ca²⁺ in the micromolar range. Quin-2 was based on the Ca²⁺ chelator EGTA, modified so that it was not as sensitive to pH over the range of 6–8 found inside most cells. At physiological pH (i.e. around 7), the Ca²⁺-sensitive dyes were fully ionised with four negative changes, and so did not cross the

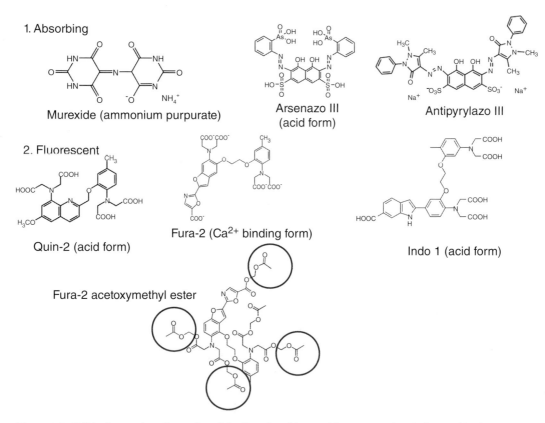

Figure 3.5 Ca²⁺ indicator dyes. Examples of the first absorbing and fluorescent dyes to be used in the measurement of cytosolic free Ca²⁺. The circles show the acetoxymethyl ester groups used to get fluorescent dyes across the outer phospholipid bilayer and into the cytosol, which are then hydrolysed off by a cytosolic esterase. Reproduced by permission of Welston Court Science Centre.

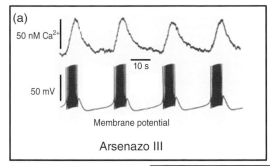

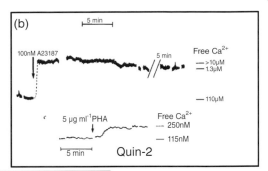

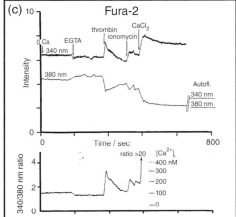

Figure 3.6 Some of the first Ca²⁺ signals detected with indicator dyes. (a) Arsenazo III in neurones of the sea slug *Aplysia*. Source: Gorman and Thomas (1978). Reproduced with permission from Wiley. (b) Quin-2 in lymphocytes from pig mesenteric nodes or mouse thymus. Source: Tsien *et al.* (1982). Reproduced with permission from Nature. (c) Fura-2 in platelets. Fluorescence excitation was at 340 and 380 nm using a beam chopper, with emission at 505 nm. The data show a release of intracellular Ca²⁺ in the absence of extracellular Ca²⁺ (EGTA) induced by thrombin, and then only a small extra rise caused by the ionophore ionomycin, but a huge rise then by addition of Ca²⁺, presumably due to store-operated Ca²⁺ entry (SOCE). The rise virtually saturated the fura-2, hence the dotted arrow. Estimated intracellular fura-2 = 10 μM. Source: Tsien *et al.* (1985). Reproduced with permission from Elsevier.

phospholipid bilayer of biological membranes. Learning from a trick developed by the pharmaceutical industry to get drugs across the gut wall into cells and into the blood, Tsien synthesised quin-2, and the fluorescent dyes that followed, as acetoxymethyl esters (Figure 3.5). These are hydrophobic, and so cross phospholipid bilayers. Once inside the cell, esterases cleaved off the acetoxymethyl group, leaving the now negatively charged free dye trapped within the cell. Quin-2 successfully monitored changes in free Ca²⁺ in lymphocytes and platelets (Figure 3.6b). However, quin-2 had a poor absorbance coefficient and a low fluorescence quantum yield. Tsien therefore designed fura-2 and indo-1. These were not only better fluors than quin-2, but also had the advantage that the fluorescence *excitation* or *emission* spectrum changed when they bound Ca²⁺. They could also be used ratiometrically as a Ca²⁺ indicator. Fura-2 and indo-1 were followed by a host of other fluorescent dyes, such as fura red and fluo-3 and -4, with varying properties for particular applications (see Chapter 4).

A further breakthrough in the 1980s was the development of imaging systems on high-powered microscopes, enabling Ca²⁺ signals to be visualised inside live cells. The first image of a change in free Ca²⁺ in a live cell was in fact after the electrical stimulation on obelin inside the

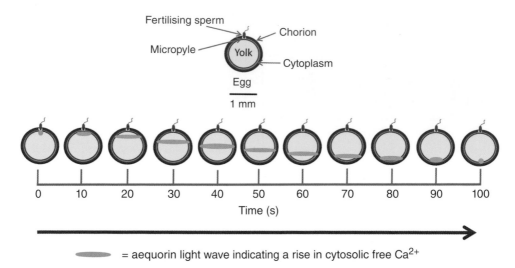

Figure 3.7 Free Ca²⁺ wave in an egg by a sperm. This was from the fresh water medaka fish *Oryzias latipes* injected with purified aequorin. The aequorin light was imaged using a four-stage electromagnetically focussed image intensifier from EMI electron tubes, United Kingdom, pioneered by George Reynolds at the University of Princeton. Data from Gilkey *et al.*

photocytes of *Obelia geniculata*, using a four-stage electromagnetically focussed photomultiplier tube pioneered by George Reynolds at Princeton, United States. This was used to visualise a spectacular Ca²⁺ wave moving down the fertilised egg of a medaka fish (Figure 3.7). Reynolds also set up Loewenstein and coworkers with the same imaging device to visualise aequorin Ca²⁺ signals close to the gap junctions in fly salivary gland. The invention of intensified CCD cameras has enabled aequorin to be imaged now in many other cell types, including live plants. During the 1990s fluorescent imaging systems were developed, with confocal microscopes, enabling Ca²⁺ signals to be imaged in single small cells.

Aequorin was cloned in 1985, the mRNA coding for obelin expressed in live cells in 1988, and GFP cloned in 1992. The cloning of a similar fluor from a coral gave us a red-emitting fluor, DsRed. Genetic engineering led to Ca²⁺ indicators enabling cytosolic free Ca²⁺ to be measured, and imaged, in intact organs and organisms, as well as in defined parts of live cells, such as the ER, nucleus and mitochondria, nucleus, Golgi and plasma membrane for the first time in live cells. GFP folds into a barrel and, in the presence of oxygen, allows three of its amino acids, S65Y66G67 to cyclise, forming a fluor with a very high quantum yield. Forced evolution generated fluorescent proteins emitting blue and yellow light. By linking two different coloured fluors together, cyan fluorescent protein (CFP) and yellow fluorescent protein (YFP), it was possible to engineer a Ca²⁺ sensor. This uses calmodulin, and its target peptide M13 from the muscle MLCK, between them. They are called 'cameleons' (as opposed to the animal 'chameleon'), since they change colour when they bind Ca²⁺. Excitation of CFP resulted in energy transfer, by the Förster mechanism, to YFP. Binding Ca²⁺ increased the efficiency of this transfer and so more yellow light was emitted. Miyawaki has extended this idea by changing the N- and C-termini of GFP around – a process known as circularising, separating the Ca²⁺-binding site from the target peptide, so that the chimera folds on itself in a circular manner. These pericams produce larger colour shifts and are more sensitive to Ca²⁺ in the submicromolar range. Fluorescent and bioluminescent Ca²⁺ indicators led to the discovery during the 1990s of a wide range of types of Ca²⁺ signal including puffs, sparks, sparklets, scintilla, oscillations, tides and waves.

Electrodes that can measure Ca^{2+} concentrations have been available since the 1930s. However, it was not until the 1970s that microelectrodes became available to measure free Ca^{2+} inside live cells. Microelectrodes that form electrical seals for patch clamping have revolutionised the study of a large family of Ca^{2+} channels, opened by voltage, external and internal stimuli, and pharmacological agents.

Indicators for monitoring and imaging free Ca^{2+} in live cells have provided the key evidence for Ca^{2+} as a universal regulator inside animal, plant and bacterial cells. This is a wonderful example of how curiosity has led to a major impact on the mechanisms underlying disease and drug discovery. It has also, quite unexpectedly, created several multi-million dollar markets. For example, the sensitivity of detecting obelin, down to $1-10$ tipomol (10^{-20} to 10^{-21} mol), led me to have the idea of replacing radioactive iodine ^{125}I in immunoassay by a chemical model of its flash. This chemiluminescent technology is now used in several hundred million clinical tests per year world-wide and is used in many microbial tests, including the best test for AIDS, creating a billion dollar market.

3.4.5 Identification of the Components Responsible for Regulating Free Ca²⁺ Inside Cells

The discovery of how Ca^{2+} is regulated inside live cells involved identifying and characterising four cellular components:

1. Ca^{2+} pumps and transporters, discovered first on the SR/ER, and then in the plasma membrane.
2. Ca^{2+} release mechanisms from the SR/ER, in particular Ca^{2+} channels opened by inositol trisphosphate (IP_3) or ryanodine receptors on the SR/ER.
3. Ca^{2+} channels in the plasma membrane, characterised by patch clamping.
4. Ca^{2+} transport across other intracellular organelles, such as the mitochondria, lysosomes and vesicles, which can cause Ca^{2+} signals, and modify or buffer changes in cytosolic free Ca^{2+}.

3.4.6 Discovery of Plasma Membrane Ca²⁺ Pumps and Transporters

Once it was realised how low the cytosolic free Ca^{2+} was, the search was on for the molecular basis of the 10 000-fold Ca^{2+} gradient across the plasma membrane. Internal Ca^{2+} stores can release and remove Ca^{2+}, but they cannot, in the long term, maintain a cytosolic free Ca^{2+} in the submicromolar range with an extracellular Ca^{2+} of millimolar. In the 1950s, the Na^+/K^+-MgATPase was discovered that exchanges three Na^+ out for two K^+ in, to maintain the K^+ inside at about 150 mM and the Na^+ at about $10-20$ mM, with 5 mM K^+ and 140 mM Na^+ outside in the blood. Then, again using erythrocytes, in the 1960s the plasma membrane protein that pumps Ca^{2+} out of all eukaryotic cells was discovered. The pump exchanges one H^+ in for one Ca^{2+} out, and has a calmodulin-binding site on the C-terminal tail that reduces the K_d for Ca^{2+} from $10-20$ to 1 μM. The protein was first cloned and sequenced from rat brain and human teratoma cells, the human protein having 1220 amino acids with a molecular mass of 133 kDa and ten predicted transmembrane domains. There are four isoforms, each with subtly different biochemical properties. This Darwinian molecular biodiversity can be further increased through slice variants.

The second major Ca^{2+} efflux mechanism is a protein that exchanges three Na^+ for one Ca^{2+}, discovered in the late 1960s in invertebrate nerves and muscle. This Na^+/Ca^{2+} exchanger was distinct from the Ca^{2+}-MgATPase and Na^+/K^+ pumps, and was not inhibited by ouabain – a potent inhibitor of the sodium pump. After cloning it was named NCX, with a molecular mass of 120 kDa and ten predicted transmembrane domains. X-ray crystallography and NMR established a domain structure to explain how it works with two Ca^{2+}-binding sites. It is found in most vertebrate and invertebrate cells, and is particularly important in cardiac myocytes and neurones. Alternative splicing gives rise to variants. A second Na^+/Ca^{2+} exchanger, NCKX, discovered in bovine retinal rods, exchanges one Ca^{2+} and one K^+ for four Na^+, in the eyes and photoreceptors of mammals and several invertebrates, including the fruit fly *Drosophila*, extruding Ca^{2+} after the rise caused by Ca^{2+} channels opened by cyclic GMP (see Chapter 7).

3.4.7 Discovery of How Ca^{2+} is Released from the Sarco-/Endo-Plasmic Reticulum

Ca^{2+} release by the 'cortex' inside activated cells was observed in the 1930s, a Ca^{2+} uptake mechanism being discovered in the SR of muscle during the 1960s by Ebashi. But, the question remained whether the ER or mitochondria were the main releasable Ca^{2+} store in non-excitable cells. There were three key discoveries:

- IP_3 as a releaser of Ca^{2+} from the ER via one of three IP_3 receptors.
- The ryanodine receptor on the SR, of which there are three, discovered in muscle, which releases Ca^{2+} from the SR, either by direct coupling to a membrane protein or via Ca^{2+}-induced Ca^{2+} release.
- Store operated Ca^{2+} entry (SOCE), through channels in the plasma membrane, opened as a result of Ca^{2+} release from the ER.

During the 1960s and 1970s, experiments on isolated mitochondria were flawed, because they were typically carried out in Ca^{2+} concentrations in the 100 μM to millimolar range, some 1000–10 000 times the cytosolic free Ca^{2+} in the live cell. Although we now know that mitochondria do take up Ca^{2+} in live cells, it was the discovery of IP_3, as the intracellular messenger that releases Ca^{2+} from the ER, that proved beyond doubt the ER is the major releasable Ca^{2+} store in all non-muscle cells.

Lawaczek in 1928 and Hermann in 1932 predicted that release of Ca^{2+} within the cell was involved in muscle contraction of the heart. But, to overcome criticism that diffusion of Ca^{2+} from the plasma membrane would be too slow to cause a fast muscle twitch, a Ca^{2+} store very close to the intracellular contractile apparatus was required. The breakthrough was the discovery by Marsh in 1951 that a factor isolated from homogenised muscle caused relaxation when added to skinned muscle fibres. This required ATP and Mg^{2+}, and contained vesicles that removed Ca^{2+} from the solution. These vesicles turned out to be fragments of the network of Ca^{2+}-containing tubules first described by Verati in 1902 using a light microscope, and by Bennett and Porter in 1953 using the electron microscope. A major Ca^{2+}-binding protein in the SR is calsequestrin. But oxalate is required to obtain a large uptake of Ca^{2+} in this preparation. The discovery of a Ca^{2+}-activated MgATPase by Ebashi in 1961, and Hasselbach and Makinose in 1963, which makes up over 70% of the protein in the SR membrane, established the molecular basis of Ca^{2+} uptake.

In skeletal muscle, direct communication between the action potential in the T tubules to the SR, via the dihydropyridine (DHP) receptor, provokes the SR to open its Ca^{2+} channels, releasing Ca^{2+}, which then triggers contraction. But a major discovery in the heart was that Ca^{2+} itself was the trigger to release of Ca^{2+} from the SR – calcium-induced calcium release. This was first indicated by experiments on skinned frog skeletal muscle by Ford and Podolsky in 1968, and revealed using permeabilised myocytes by Fabiato and Fabiato in 1975.

Figure 3.8 First demonstration of Ca²⁺ release from the ER by IP₃. IP₃ was added at a concentration of 2 μM to permeabilised rat pancreatic acinar cells in a medium of 110 mM KCl, 6 mM MgCl₂, 5 mM K-pyruvate, 5 mM K₂ATP, 10 mM creatine phosphate, 10 U ml⁻¹ creatine kinase, 25 mM HEPES, pH 7.4. Free Ca²⁺ was measured in the bulk solution by a Ca²⁺-specific electrode. Source: Streb *et al.* (1983). Reproduced with permission from Nature Publishing.

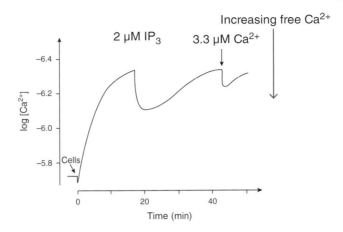

3.4.8 Discovery of IP₃ and its Receptor

The evidence for how Ca²⁺ is released from the ER by IP₃ in non-muscle cells was:

1. Discovery that the plasma membrane lipid phosphatidyl inositol had a rapid turnover when cells were activated by hormones or neurotransmitters by Hokin and Hokin in the 1950s, and Michel and colleagues in the 1970s.
2. Addition of IP₃ to permeabilised cells caused release of Ca²⁺ from the ER, in 1983.
3. Discovery of a G-protein activated phospholipase C (PLC), phosphatidyl inositol phosphatase (PIP₂ase), on the inner surface of the plasma membrane that released IP₃ and diacylglycerol into the cell.
4. Isolation and cloning of the IP₃ receptor on the ER membrane.

The breakthrough came in 1983 when Streb and colleagues added of part of phosphatidyl inositol – inositol 1,4,5-trisphosphate (IP₃) – to permeabilised pancreatic acinar cells in culture in the presence of a Ca²⁺ indicator (an electrode), causing a rapid, large rise in free Ca²⁺ as a result of release from the ER (Figure 3.8). The IP₃ was made from a large batch of red blood cells, by Robin Irvine working with Michael Berridge in Cambridge, UK. Addition of IP₃, with a half maximum of 0.1–3 μM, to a wide range of permeabilised cells showed release Ca²⁺ from the ER. Subcellular fractionation confirmed IP₃ releases Ca²⁺ from a non-mitochondrial Ca²⁺ store – the ER. Cloning was from the mouse cerebellum. X-ray crystallography produced the three-dimensional structure, and atomic force microscopy, with electron microscopy, showed the IP₃ receptor to be huge, 1.2 MDa. Darwinian variation was established by relation of the Ca²⁺ signal to three isoforms, and alternative splice variants in different tissues.

3.4.9 Discovery of the Ryanodine Receptor

Two natural plant alkaloids – ryanodine and caffeine (Figure 3.9) – led to the discovery of the receptors on the SR membrane, which cause Ca²⁺ to be released into the muscle cytosol. Ryanodine is a poisonous alkaloid, originally isolated in 1948 from shrubs in South America of the genus *Ryania*. Natives use crude extracts to poison their arrow heads. During the late 1940s and 1950s, ryanodine was shown to have severe paralytic effects on both skeletal and heart muscle. Caffeine is a natural insecticide, paralyzing and killing insects feeding on the plant. In the 1960s and 1970s, both were found to release Ca²⁺ from the SR by binding the same receptor in the SR membrane. At nanomolar concentrations, ryanodine set the SR Ca²⁺ channel in

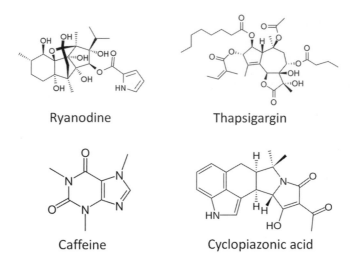

Ryanodine Thapsigargin

Caffeine Cyclopiazonic acid

Figure 3.9 Figure 3.9 Four compounds affecting Ca^{2+} in the ER/SR. Ryanodine from the South American plant *Ryania speciosa* is a ryanodine receptor (RyR) activator at nanomolar concentrations, causing Ca^{2+} release into the cytosol, but is an inhibitor at micromolar concentrations. Caffeine from coffee, tea and other plants is a ryanodine receptor activator. Thapsigargin from the plant *Thapsia garganica*, is a SERCA pump inhibitor. Cyclopiazonic acid (CPA) from *Penicillium cyclopium* and other moulds is a SERCA pump inhibitor. Both inhibitors cause a rise in cytosolic free Ca^{2+}.

its active state, leading to tetanic contraction, and depletion of the SR Ca^{2+} store. But, at concentrations around 100 µM, ryanodine irreversibly blocked channel opening. The isolated protein contained some 5000 amino acids (*ca* 560 kDa) from cloning, and its three-dimensional structure was obtained by electron microscopy and other methods, with localisation by immunomicroscopy. During the 1990s, two other intracellular signals capable of releasing Ca^{2+} from the ER or lysosomes into the cytosol were discovered: cyclic ADP ribose (cADPR) and nicotinic acid adenine dinucleotide phosphate (NAADP).

3.4.10 Discovery of Store-Operated Calcium Entry (SOCE)

A crucial discovery was a remarkable mechanism by which depletion of Ca^{2+} from the ER caused Ca^{2+} channels to open in the plasma membrane, thereby producing a much larger rise in cytosolic free Ca^{2+} than could be obtained by the ER itself. The first clues came in the 1980s, from studies on exocrine cells and hepatocytes. When cells were activated, there was a large Ca^{2+} influx, concomitant with a loss of Ca^{2+} from the ER. Furthermore, the channels in the plasma membrane remained open until the primary stimulus was removed. The loss of Ca^{2+} in the ER could be measured using ER-targeted aequorin, or a cameleon fluorescence resonance energy transfer (FRET) probe. In the absence of extracellular Ca^{2+} there was only a small rise in cytosolic free Ca^{2+} released from the ER, but a large drop in ER free Ca^{2+}. Addition of external Ca^{2+} then resulted in a large rise in cytosolic free Ca^{2+}, hence the name 'Store-operated Calcium Entry' (SOCE). SOCE channels remain open as long as the primary stimulus is present. Initially it was thought that SOCE was simply a means of topping up the ER, after it had lost its Ca^{2+}. It was therefore originally called 'capacitative calcium entry'. It soon became clear that, in many non-excitable cells, SOCE was in fact the main source for the global rise in cytosolic free Ca^{2+}. But the molecular basis of the SOCE channels turned out to be elusive. The predicted Ca^{2+} current associated with SOCE was named the 'inward calcium rectifying activating current'

(I_{CRAC}), and also turned out to be elusive, because this Ca^{2+} current was very small, with a conductance in the femto-siemens range, below that of the electrical noise. However, the amount of Ca^{2+} that has to move to cause a large rise in cytosolic free Ca^{2+} is huge compared with the amount that has to move to depolarise a cell. Thus, there has to be a counter ion. Another puzzle was that in patch clamp studies the Ca^{2+} channels inactivated rapidly, and closed, but in the intact cell SOCE channels remain open as long as the ER has not refilled with Ca^{2+}.

Clear evidence identifying SOCE began from studies in the late 1970s by Putney and colleagues on the parotid gland and rat lacrimal acinar cells. Two toxins – thapsigargin and cyclopiazonic acid (CPA) (Figure 3.9) – caused release of Ca^{2+} from the ER, by blocking the ER Ca^{2+} pump (SERCA), without causing a rise in IP_3, as they do not activate PLC. Blocking of the SERCA pump allows Ca^{2+} to leak out of the ER. Thapsigargin is a sesquiterpene lactone, from the umbelliferous plant *Thapsia garganica* (Figure 3.9), found in the Western Mediterranean, the plant resin having been used for centuries to alleviate rheumatic pain. Pharmacological studies showed that it was a potent releaser of histamine from mast cells, and had effects on lung, heart, neutrophils and platelets. The key experiment was to show, using fura-2 as an indicator of cytosolic free Ca^{2+}, that thapsigargin released Ca^{2+} from the ER by potent, and irreversible, inhibition of the all three SERCA Ca^{2+}-activated MgATPases pumps. It does not affect other P-type ATPases, such as the plasma membrane Ca^{2+} pump, nor Na^+/Ca^{2+} or Ca^{2+}/H^+ exchangers. CPA (Figure 3.9), on the other hand, is a fungal toxin isolated from the mould *Aspergillus*. Mouldy foods, containing CPA, have been known for many years to have toxic effects on humans and animals. CPA, like thapsigargin, blocks SERCA pumps. As with thapsigargin, CPA had no effect on other MgATPases. Primary stimuli, such as histamine and extracellular ATP, stimulated IP_3 production. This releases Ca^{2+} from the ER, causing a decrease in luminal free Ca^{2+}, which triggered SOCE (Figure 3.10).

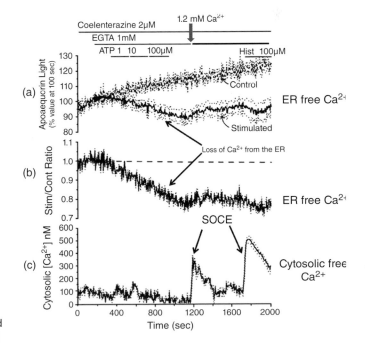

Figure 3.10 First correlation of free Ca²⁺ decrease in the ER with store-operated Ca²⁺ entry. HeLa cells expressing apoaequorin targeted to the ER with a KDEL retention sequence were perfused with 2 μM coelenterazine. With a peptide at the C-terminus apoaequorin acts as a luciferase, producing continuous light emission. In the absence of extracellular Ca²⁺ (EGTA), there was a large decrease in ER free Ca²⁺ (a) and (b) after addition of rising doses of extracellular ATP, but only a small rise in cytosolic free Ca²⁺ (c), from the ER release. Addition of extracellular Ca²⁺ caused a large rise in cytosolic free Ca²⁺, since SOCE channels were open, and just a small increase in ER free Ca²⁺. The initial estimates of the absolute ER free Ca²⁺ published in this paper were too low, due to the incomplete penetration of the luciferin (coelenterazine). Reproduced with permission from Source: Kendall (1996). © Portland Press Ltd.

The next question was: How did the ER signal to the plasma membrane that it had lost Ca^{2+}? It was not caused by a direct effect from the small cytosolic rise in Ca^{2+}, since chelating this using BAPTA did not inhibit SOCE opening. Furthermore, the amount of Ca^{2+} released by the ER was insufficient to fill the cytosol with Ca^{2+}, the Ca^{2+} cloud being restricted close to the ER, and never reaching the plasma membrane. Four hypotheses were put forward:

- A diffusible messenger, called a 'calcium influx factor' (CIF).
- Fusion of vesicles released from the ER with the plasma membrane, thereby either activating or reconstituting a Ca^{2+} channel.
- A 'transient receptor channel' (TRP) was responsible for SOCE.
- Conformational coupling, where an ER protein, sensing luminal free Ca^2 and which traversed the ER membrane, would interact directly with a protein in the plasma membrane.

A form of the last of these turned out to be correct. The breakthrough came during the 2000s, with the discovery of two proteins, one in the ER – STIM1 (STromal Interaction Molecule) – and one in the plasma membrane – Orai1. STIM1 was found first in *Drosophila* S2 cells, and then in HeLa cells. Orai1, originally olf-186F, was found by gene mapping of a family with a rare immunodeficiency disorder called hereditary severe combined immune deficiency syndrome (SCID), infants having a propensity for fungal and viral infections. Also called CRAM1, Orai was chosen as the name after the three keepers of the gates of heaven in Greek mythology – Eunomia for Order or Harmony, Dike for Justice and Eirene for Peace. When Orai1 is closely linked to SOCE it forms the Ca^{2+} entry channel. Two Orais (1 and 2) and three STIMs (1, 2 and 3) have now been identified in animals.

3.4.11 Discovery of Ca^{2+} Channels in the Plasma Membrane

Voltage-gated Ca^{2+} channels, opened by a depolarisation of the membrane potential, were revealed during the 1950s, from studies on crustacean muscle – crab and barnacle, in particular. The giant axon of the squid *Loligo forbesi* had established the ionic basis of the action potential – the transverse electrical signal that travels down a nerve when it fires. 'Pores' in the outer membrane of the axon were predicted, which allowed Na^+ into the cell when the membrane was depolarised, and allowed K^+ to leave the cell in order to repolarise it. These experiments explained why the conduction of an action potential was blocked by removal of external sodium, pioneered by Overton and others in the early 1900s.

During the 1960s and 1970s, the word 'pore' was replaced by the term 'channel', signifying a clearer concept of a mechanism that allowed large numbers of charged ions to move across the lipid bilayer through a specific structure, predicted to be a protein complex. These ion channels opened and closed in a digital fashion, as shown clearly by patch clamping. During the late 1940s and early 1950s, Hodgkin and Huxley had established the 'sodium theory' of the action potential, showing that the initial current that depolarised the membrane of nerves was movement of Na^+ into the cell. This current inactivated, even when the membrane was kept depolarised (e.g. by voltage clamp). Repolarisation of the membrane to a potential of about −90 mV, negative inside, was through movement of K^+ out of the cell. This led several workers to test this in cells other than giant axons of the squid. But, using muscle fibres from the legs of the spider crab, *Maia squinado*, in 1953 Fatt and Katz found that the action potential in these cells was not carried by Na^+. In crab muscle, replacement of Na^+ by choline did not stop the action potential, rather it actually enhanced it. Two other agents, tetraethylammonium (TEA) and tetrabutylammonium (TBA) enhanced the action potential even more, shown previously simply to sustain Na^+ dependent action potentials in squid axons.

Thus, crustacean muscle action potentials are caused by opening of Ca^{2+} channels, not Na^+ channels. Repolarisation occurred again through opening of voltage-gated K^+ channels. External Ca^{2+} could be replaced by either Sr^{2+} or Ba^{2+}. Mg^{2+} was ineffective, whereas Mn^{2+} inhibited. Four key pieces of evidence showed that these Na^+-independent currents were carried by Ca^{2+}:

1. The current was stopped when Ca^{2+} was removed from the external medium, but not when Na^+ was removed.
2. The Ca^{2+} current was not blocked by the pufferfish toxin, TTX – a specific blocker of the Na^+ channel.
3. Measurement of ^{45}Ca showed that sufficient Ca^{2+} had moved to depolarise the membrane potential, the equivalent to 2–6 pmol μF^{-1} of membrane capacitance occurred. This was considerably greater than the 0.5 pmol μF^{-1} of Ca^{2+} needed to depolarise the membrane by 100 mV.
4. Pharmacological substances distinguished Ca^{2+} currents through Ca^{2+} channels from currents through Na^+ or K^+ channels – TTX on Na^+, TEA on K^+, and a verapamil or its analogue D-600 on Ca^{2+}.

Patch clamping of excitable cells, combined with the application of pharmacological agents, led to the discovery of Ca^{2+} channels with different electrophysiological and pharmacological properties. At first named LVA and HVA, then L, T, N, P/Q and R. The first two types were designated as Low and High Voltage Activated (LVA and HVA). The channels were then all designated by a letter: L, T, N, P, Q and R. Certain HVA channels were long lasting (L), and could be blocked by dihydropyridines (DHPs), being found in skeletal, heart and smooth muscle, and neurones, the opening time being enhanced by the agonist BAYK8644. L-type DHP Ca^{2+} channels are very high in skeletal muscle T-tubules, connecting to the ryanodine receptor to release intracellular Ca^{2+} from the SR to provoke contraction. L- and T-type Ca^{2+} channels were distinguished by:

*   The depolarisation voltage that opened them.
*   Current size.
*   How long they stayed open.

Pharmacological agents and natural toxins, such as those from the cone snail and American funnel web spider, then led to the identification of other Ca^{2+} channels. N-type channels, where N = non-L or neuronal, were blocked by the cone snail toxin ω-conotoxin GVIA, whereas P-type channels found in Purkinje cells in the cerebellum were insensitive to both DHP and ω-conotoxin GVIA, but were very sensitive to the ω-agatoxin from the funnel web spider. Another Ca^{2+} channel, exhibiting a more rapid inactivation and lower affinity for the toxin, was designated Q-type. However, P and Q are usually bracketed together as P/Q, and may represent different splice variants. There was still a residual current resistant to DHP and the cone snail and spider toxins, and was therefore designated as R-type (residual).

The DHP Ca^{2+} channel in skeletal muscle was shown to contain five proteins, designated α_1 (170 kDa), α_2 (150 kDa), β (52 kDa), δ (17–15 kDa) and (32 kDa). α_1 bound DHPs, and formed the pore for the Ca^{2+} current. Sequence analysis of the various Ca^{2+} channel types showed that all Ca_v1 were L-type, Ca_v2 were P/Q-, N- and R-type, and Ca_v3 were T-type. All had splice variants, providing another example of Darwinian molecular biodiversity. Many voltage-gated Ca^{2+} channels can be up- or down-regulated by intracellular signalling, such as cyclic AMP protein kinase, and by activation of heterotrimeric G-proteins, first found to inhibit N- and P/Q-type Ca^{2+} channels in neurones.

Thus, several types of voltage-gated Ca^{2+} channel were identified, isolated and sequenced, their ionic selectivity being confirmed by patch clamping. There is now a wide range of pharmacological substances that block Ca^{2+} channels (see Chapter 11).

3.4.12 Discovery of TRP Channels

TRPs are a superfamily of cation channels that let Ca^{2+} or Na^+, or both, into sensory cells responsible for vision, taste, smell, hearing and hot versus cold in invertebrates. They were discovered in the eye of the fruit fly *Drosophila* as 'Transient Receptor Potential' channels, and are found in all animals, and many excitable and non-excitable cells. Their selectivity for Ca^{2+} varies considerably. Some are non-selective for the cation, whereas others can be more than ten-times selective for Ca^{2+} over Na^+. However, with a ratio of Na^+ to Ca^{2+} outside of some 140 : 1, Ca^{2+} will always be competing with Na^+ for entry into the cell. The relevance to Ca^{2+} signalling is two-fold:

- Opening of TRP channels leads directly to a rise in cytosolic free Ca^{2+} from TRP channel opening, or from calcium-induced Ca^{2+} release (CICR).
- Na^+ entry depolarises the plasma membrane, opening voltage-gated Ca^{2+} channels.

TRP channels may be opened by G-protein receptors, or directly by ligand binding, and have a conductance in the tens of pico-siemens. TRPs are distinguished from SOCE because TRP conductances are much higher than the femto-siemens conductances of I_{CRAC}, and TRPs are opened independently of ER Ca^{2+} depletion.

Thus, by the end of the twentieth century five types of Ca^{2+} channel in the plasma membrane were firmly established as crucial to Ca^{2+} as an intracellular signal:

1. Voltage-gated.
2. Store-operated.
3. Receptor-operated.
4. Intracellular messenger-operated.
5. TRPs.

3.4.13 Discovery of G-Proteins

G-proteins bind MgGTP and MgGDP, and have a GTPase activity. There are two types: membrane bound complexes and soluble small G-proteins. Many hormone and neurotransmitter receptors work by being coupled to a membrane bound G-protein complex, whose activation depends on a balance between binding MgGTP and MgGDP. The membrane bound G-proteins were discovered during the 1970s by Martin Rodbell, as a result of a study on adenylate cyclase isolated from the plasma membranes of adipocytes and hepatocytes. These G-protein-coupled receptors were found to be a complex of three components: α, β and γ. Binding frees the α subunit from the $\beta\gamma$ complex, allowing activation of adenylate cyclase. Following the discovery of IP_3, the enzyme phospholipase C in the plasma membrane that produces it was soon found to be activated through G-protein coupled receptors. Thus, G-proteins play a key role in Ca^{2+} signalling.

3.4.14 Ca^{2+} Targets Inside Cells and How They Work

A major breakthrough in Ca^{2+} signalling was the discovery of high-affinity Ca^{2+}-binding proteins, particularly those belonging to the EF-hand and C2 families. The search for effects of Ca^{2+} on intracellular enzymes began in the 1930s and 1940s. But, at that time it was not realised how low the free Ca^{2+} was in the cytosol of live cells. Thus, studies up to the 1960s on the effects of Ca^{2+}, typically mM, on isolated proteins and enzymes were non-physiological. A key experiment was the effect of low concentrations of Ca^{2+} on the contraction of skinned muscle fibres, leading to the discovery by Ebashi of the high-affinity Ca^{2+}-binding protein troponin C, the Ca^{2+} target in all skeletal and heart muscle cells. This led several workers to search for other Ca^{2+}-binding proteins that worked at micromolar concentrations of Ca^{2+}.

An important Ca^{2+}-binding protein in this story was the isolation of a protein by Henrotte in 1952 from carp muscle that had 'albumin-like' solubility properties, and was thus called parvalbumin (Latin *parvus* = small, as it was only 12–14 kDa). Parvalbumin occurs in all fast skeletal muscle and in the brain. It has two high-affinity Ca^{2+} sites with K_d^{Ca} in the micromolar range, and aids movement of Ca^{2+} to and from the SR to the troponin C on the myofibrils. When the 3D structure of parvalbumin was determined by Bob Kretsinger and colleagues in the 1970s, the Ca^{2+}-binding sites turned out to be very interesting. Fifty-two of the 108 residues were found to be in six helices, designated A, B, C, D, E and F (Figure 3.11). A Ca^{2+}-binding site between helices E and F had a structure similar to a 'hand'. This led to the universal concept that a large number of Ca^{2+}-binding proteins inside cells, that work at micromolar concentrations of free Ca^{2+}, have Ca^{2+}-binding sites similar to the EF-hand structure discovered in parvalbumin. This was soon shown for troponin C and calmodulin, and many other Ca^{2+}-binding proteins. When searching a new genome for putative Ca^{2+}-binding proteins, the first thing to do is to search for EF-hand motifs in open reading frames. The EF-hand motif also led to another fundamental discovery about intracellular Ca^{2+}. Ca^{2+}-binding sites that are physiologically relevant usually have seven or eight coordination with oxygen. This enables the protein to select Ca^{2+} at micromolar concentrations in the presence of millimolar free Mg^{2+}. Nitrogen in NH_2 groups can also bind Ca^{2+}. But nitrogen also binds Mg^{2+} well, and so is not selective enough. During the 1990s and early years of the twenty-first century, other high-affinity Ca^{2+}-binding sites were discovered, being formed from a particular arrangement of β-sheets. These included the C2 Ca^{2+}-binding site, for example in synaptotagmin, and the Greek key. Low-affinity Ca^{2+}-binding sites were found in proteins within the SR/ER and extracellularly, formed from clusters of acidic amino acids, and also arrangements of β-sheets.

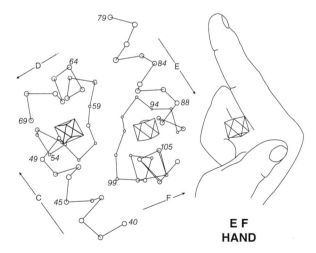

Figure 3.11 The first published EF-hand Ca^{2+}-binding site. The EF-hand is from muscle Ca^{2+}-binding protein (MCBP) from carp muscle. The Ca^{2+}-binding site is octahedral. The numbers refer to the amino acid sequence. The two Ca^{2+}-binding sites with pK_d^{Ca} 6.7 are formed from loops between helices C and D and E and F, respectively. Kretsinger chose the loop from EF helices as the evolutionary prototype of a Ca^{2+}-binding site, since the D helix was slightly distorted due to hydrogen bonding with residues Arg75 and Glu81. Source: Kretsinger and Nelson (1976). Reproduced with permission from Elsevier.

Studies on cyclic AMP phosphodiesterase by Cheung in 1966/67 led to the discovery of calmodulin, which occurs in virtually all animal and plant cells, but not bacteria. Calmodulin has four high-affinity EF-hand Ca^{2+}-binding sites. As well as activating the phosphodiesterase that cleaves cyclic AMP to AMP, calmodulin was found to activate many intracellular proteins at micromolar concentrations of Ca^{2+}, and also to move to the nucleus where it can activate gene expression. Particularly important were calmodulin-activated kinases, proteases, ion channels and transcription factors.

First in the 1940s, and then during the 1960s and 1970s, several metabolic enzymes were discovered that were activated by Ca^{2+} in the micromolar range. The first were succinate dehydrogenase in mitochondria, and the cytosolic kinase that activates phosphorylase in muscle, the liver and adipose tissue. During the 1970s, other enzymes inside the mitochondria were found by Denton and colleagues that could be physiologically activated, or sometimes inhibited, by Ca^{2+}, leading to increased flux of the citric acid/Krebs cycle and more ATP synthesis. All of this was consistent with the concept that cytosolic Ca^{2+} is a switch, but also needs to activate ATP synthesis in an analogue manner, since all of the processes activated by Ca^{2+} require ATP hydrolysis as an 'energy' source.

Other important Ca^{2+}-binding proteins discovered during the 1970s and 1980s were the phosphatase calcineurin, the non-lysosomal cysteine proteinase calpain, and calmodulin-myosin light chain kinase (MLCK). Calmodulin-MLCK has four EF-hand high-affinity Ca^{2+}-binding sites from the calmodulin, and is a key part of the mechanism responsible for contraction of smooth muscle, which, unlike skeletal and heart muscle, does not use troponin C. Also during the 1970s, the Ca^{2+}-binding proteins calsequestrin and calreticulin were discovered inside SR in muscle and in the ER of non-muscle cells, respectively, by MacLennan and colleagues. These have much lower affinity for Ca^{2+} than troponin C, parvalbumin or calmodulin, consistent with the concentration of free Ca^{2+} inside the SR/ER in the tens of micromolar to millimolar range.

Thus, the discovery of Ca^{2+}-binding proteins as the Ca^{2+} targets showed once again how important it was to know the range of cytosolic free Ca^{2+} in live cells. Many proteins have negatively charged amino acids, Glu and Asp, which bind Ca^{2+} at millimolar Ca^{2+} concentrations. Only those that have high-affinity Ca^{2+} sites, which can bind significant amounts of Ca^{2+} at micromolar concentrations in the presence of millimolar Mg^{2+}, turned out to be physiologically relevant. The others were red herrings.

3.4.15 Intracellular Ca^{2+} in Plants

All gardeners know calcium is important for healthy plants. It has been known for more than a century that plants require calcium for growth, for pollen tube formation, flowering, root tip growth and nodule formation. Furthermore, it has been known since the early years of the twentieth century that there was a large calcium store in many plants, inside cells, in the form of calcium oxalate. But, it was not until the 1970s and 1980s that a role for intracellular Ca^{2+} in plant physiology was taken seriously. There were two breakthroughs:

- The discovery of calmodulin in plants, and its ability to regulate intracellular enzymes and gene expression.
- The measurement of free Ca^{2+} in intact plants cells and live plants, using fluorescent dyes and transgenic aequorin.

Measurement of cytosolic free Ca^{2+} soon established that, like animal cells, the free Ca^{2+} in plant cells was in the sub-micromolar range. The development of transgenic plants expressing aequorin in the early 1990s showed that cold shock and mechanical movement triggered cytosolic Ca^{2+} transients, showing for the first time that Ca^{2+} must be an intracellular signal in plants. These free Ca^{2+} changes could be imaged in the whole plant. It was also shown that aequorin could be targeted to plant intracellular organelles, such as chloroplasts, the nucleus, ER and mitochondria. Studies during the 1980s and 1990s identified Ca^{2+} pumps and exchangers, intracellular Ca^{2+} stores, and IP_3-mediated Ca^{2+} release. However, the study of intracellular Ca^{2+} as a regulator in plants generally has lagged behind that in animal cells.

3.4.16   Intracellular Ca²⁺ in Bacteria

Studies in the 1960s showed that extracellular Ca^{2+} was required for the growth and gene expression in some bacteria, but this was far from being universal. Also during the 1960s, it was shown that Ca^{2+} uptake played an essential role in the formation of the spores of Gram-positive bacteria such as *Bacillus*. Ca^{2+} was found to be a critical part of the crystalline structure in the spore. But the potential role of Ca^{2+} as an intracellular signal was not fully recognised until well into the 1990s. The first indication in 1977 that intracellular Ca^{2+} might be a signal for a bacterial event was in chemotaxis. Some flagellate bacteria are attracted to move up the concentration gradient of certain nutrients, but are repelled by toxins. Since the gradient across one bacterium is far too small to be sensed directly, chemotaxis cannot involve a receptor mechanism recognising the gradient. Bacteria move in a straight line, tumble and then set off in another direction. Chemoattractants increase the time between tumbles, and thus the population gradually moves towards higher concentrations. The rate of tumbling was proposed by Adler and colleagues to be inhibited by a rise in cytosolic free Ca^{2+}.

Measurement of free Ca^{2+} in live bacteria was not attempted until the mid-1980s and early 1990s, using fura-2. However, these results appear to be artefacts, as other workers have found it difficult to repeat them. The availability of plasmids containing the gene expressing aequorin enabled the first genuine measurements of free Ca^{2+} in live bacteria to be made. These studies showed for the first time that, like animal and plant cells, bacteria maintain a cytosolic free Ca^{2+} concentration in the micromolar to sub-micromolar range, in the presence of millimolar Ca^{2+} extracellularly. Furthermore, chemoattractants could lower the cytosolic free Ca^{2+} and chemorepellents increase it. During the late 1990s and early to mid-2000s, a range of conditions were discovered that induced Ca^{2+} transients in bacteria, including bacterial metabolic toxins. Cloning, and the availability of genome sequences, led to the identification of three putative Ca^{2+} efflux mechanisms: ChaA, YrbG and PitB. However, these studies failed to correlate the activity of these proteins with measurements of cytosolic free Ca^{2+} in live bacteria. When this was done using a knock-out collection, ChaA, YrbG and PitB were shown not to play a role in Ca^{2+} transport. Calmodulin-like proteins were reported in some bacteria during the 1990s, and, during the 2000s, genomes were searched for EF-hand proteins in bacteria. Ion channels have been found in some bacteria, some claimed to be Ca^{2+} channels, but the only convincing Ca^{2+} channel was non-proteinaceous, identified by Rosetta Reusch, The role of intracellular Ca^{2+} as a signal in bacteria is still not fully established (see Chapter 8).

3.4.17   Pathology of Intracellular Ca²⁺

Calcification of injured or dead tissue has been recognised since the mid-nineteenth century. Furthermore, it was well established in the first half of the twentieth century that there were several clinical conditions associated with the deposition of calcium phosphate precipitates. Heilbrunn recognised that too much calcium inside cells was not a good thing. During the 1960s and 1970s, measurement of total calcium, the use of radioactive ^{45}Ca, and electron microscopy, including X-ray microprobe analysis, showed that there were many examples of tissue and cell damage involving an increase in total intracellular Ca^{2+}. Agents of injury included anoxia, chemical and biological toxins, genetic abnormalities, immune damage, loss of ATP, cancer and vitamin deficiencies.

Once it was realised how large a gradient of Ca^{2+} there was across the plasma membrane of all healthy cells, and that this was maintained by pumps and transporters ultimately dependent on MgATP, it was no surprise to learn that Ca^{2+} flooded into cells when the membrane was damaged, or when the cell lost most its MgATP. What was not clear, however, was whether this increase in intracellular Ca^{2+} was directly involved in either killing the cell, or defending itself against attack. In other words – When is intracellular Ca^{2+} a friend and when is it a foe? During the 1960s it was shown, using subcellular fractionation and electron microscopy, that mitochondria were major targets of cell damage when Ca^{2+} flooded into the cell, resulting in calcium phosphate precipitates, Ca^{2+}-NADH precipitates, and loss of NAD, all leading to loss of ATP synthesis. By the end of the 1970s, it was well established that a persistent rise in intracellular Ca^{2+} in the high micromolar range led to irreversible cell damage, as well as activation of intracellular proteases and nucleases. But, it was also found that Ca^{2+} could activate a protection mechanism against pore formers such as the membrane attack complex of complement. During the 1990s oxidative cell damage by oxygen metabolites was also shown to involve Ca^{2+}.

A breakthrough occurred when a different type of cell death was discovered. It had been known since the late 1960s that blood lymphocytes showed a condensed nucleus after an infection was over. These cells then disappeared. Cell death had also been observed earlier in healthy differentiating tissues and in plants, e.g. along the line at which a leaf falls off in autumn. This turned out to be a signalled cell death, named apoptosis through the pioneering work of Andrew Wyllie and colleagues in Edinburgh, who noticed special dead cells in tumours. It turned out that Ca^{2+} in the mitochondria plays a key role in initiating the pathway to this type of programmed cell death.

Box 3.2   Key historical approaches to discovering intracellular Ca^{2+} as a cell trigger.

1. Manipulation of extra- and intra-cellular Ca^{2+}.
2. Direct measurement of cytosolic free Ca^{2+}.
3. Identification of how Ca^{2+} is released inside, and enters the cell.
4. Identification of the intracellular Ca^{2+} target.
5. Identification of drugs that affect the intracellular Ca^{2+} signalling system.
6. Discovering what goes wrong with Ca^{2+} in disease.

3.5   Conceptual Development of Ca^{2+} as an Intracellular Regulator

The story of intracellular Ca^{2+} is full of creative, unitary hypotheses (Table 3.3) that attempted to bring together a range of apparently disparate facts and mechanisms. Some were naïve, or wrong. However, they all led to new experiments (see Box 3.2), which contributed to the framework we now have about the unique role intracellular Ca^{2+} plays throughout life.

A hypothesis must be testable experimentally. Heilbrunn's initial evidence was based on the effects of removing Ca^{2+} from the medium bathing cells and tissues, on heart and skeletal muscle contraction, contraction, egg fertilisation, drug and anaesthetic action, and many other physiological and pharmacological processes. But, the experiments he used to test his universal hypothesis included the ability of small amounts of injected Ca^{2+} to provoke muscle contraction in skinned fibres, and the release of Ca^{2+} from inside sea urchin eggs.

Table 3.3 Some important concepts establishing Ca^{2+} as a universal intracellular regulator.

Year	Concept or hypothesis	Key scientist(s)
1886	Competitive antagonism or synergy between cations – Na^+, K^+, Ca^{2+}, Mg^{2+}	Sydney Ringer
1928	Change in cytosolic free Ca^{2+} acts as a regulator	Herbert Pollack
1937	Wide-ranging role of intracellular Ca^{2+} as a cell regulator	Lewis Heilbrunn
1940	Ca^{2+} acts on intracellular enzymes	Bailey, Ebashi
1953	Ca^{2+} can move across membranes through specific channels	Fatt and Katz
1947	Ca^{2+} as the mediator of excitation–contraction coupling	Lewis Heilbrunn
1961	Stimulus–secretion analogous to excitation–contraction coupling	Douglas and Rubin
1967	Ca^{2+} acts via activation or inhibition of intracellular enzymes	Bygrave
1970	Inter-relationship between Ca^{2+} and other intracellular messengers such as cyclic AMP	Rasmussen
1974	Ying-yang hypothesis	Goldberg and Haddox
1975	Key role for phosphatidyl inositol for regulating intracellular Ca^{2+}	Michael Berridge
1976	Mono- and bidirectional regulation in cells	Michael Berridge
1979–80	Universal role of Ca^{2+}-binding proteins	Cheung; Ebashi; Means and Dedman
1983	IP_3 and the ER is the major intracellular releasable Ca^{2+} store, and not mitochondria	Michael Berridge
1983	The four biological roles of calcium	Campbell
1983	Intracellular Ca^{2+} and threshold phenomena	Campbell
1992	Intracellular Ca^{2+} can be a friend or foe	Orrenius and Nicotera
	Loss of ER Ca^{2+} triggers a defence stress response	Michelak
1996	Change in free Ca^{2+} can occur in microdomains	Michael Berridge
1997	Frequency versus concentration sensors (AM/FM)	Michael Berridge
2013	Darwin principles of small change and natural selection apply to Ca^{2+} in real time and evolution	This book

An important conceptual breakthrough was the proposal that intracellular Ca^{2+} worked through the activation or inhibition on enzymes inside cells. Although some of the experiments to test this hypothesis were misguided, as they involved the effects of what we now know to be non-physiological concentrations of Ca^{2+}, this hypothesis was supported by the key discovery of high-affinity Ca^{2+}-binding proteins as the targets for rises in cytosolic free Ca^{2+}.

An important step forward was the proposal that intracellular Ca^{2+} acted as a switch, and was not the energy source, of cellular phenomena. Intracellular Ca^{2+} was proposed to interact intimately with other intracellular signals, such as cyclic AMP and cyclic GMP, which were analogue signals, affecting the timing and magnitude of the processes triggered by a rise in intracellular Ca^{2+}. Two unitary hypotheses were put forward to explain the interaction between various intracellular signals:

- Mono/bidirectional pathways.
- Ying-yang.

Ying-yang in Chinese philosophy involves a dualism between opposing forces, between for example cyclic AMP and cyclic GMP. 'Monodirectional' systems were defined as processes where cells were transformed from a non-functional into a functional state. This could involve more than one stimulus, but no inhibiting factors. On the other hand, 'bidirectional' systems involved an interaction between stimulatory and inhibitory factors, which determined the eventual state of the cell. Thus, in a monodirectional system, such as adrenocorticotropic hormone stimulating cortisol secretion from the adrenal cortex, cyclic AMP and cyclic GMP worked together. Whereas in a bidirectional system, such as neutrophil chemotaxis, the two cyclic nucleotides opposed each other. The trouble with these concepts was that they ignored whether the cell response was digital or analogue, and focussed too much on cyclic nucleotides as the primary intracellular stimulus rather than Ca^{2+}.

The unitary hypothesis that cytosolic free Ca^{2+} was very low, in the submicromolar to micromolar range, was based initially on the injection of Ca^{2+} into frog muscle by Heilbrunn and Wiercinski in 1947, and calculations of Ca^{2+} movement in the giant axons of squid by Hodgkin and Keynes in 1957. This hypothesis has been fully vindicated as a result of the ingenious invention of indicators for measuring directly free Ca^{2+} in live cells. This led to the concept that there are transient microdomains of free Ca^{2+} inside the cell, as well as permanent ones close to the plasma membrane.

An important unitary hypothesis for cell pathology involved the mechanisms that determine when a rise in cytosolic free Ca^{2+} is a friend or a foe. This led to the unravelling of the role of cytochrome c release from the mitochondria to trigger cell death by apoptosis, and the defence of cells against heat shock and viruses involving a loss, rather than a rise, in free Ca^{2+} inside the ER.

3.6 Summary

For over a century, beginning with the pioneering experiments of Ringer, massive evidence accumulated that intracellular Ca^{2+} is a universal regulator, triggering a wide range of cellular events in animal, plant and microbial cells. Two pioneers in intracellular Ca^{2+} were Lewis Victor Heilbrunn and Setsuro Ebashi. Heilbrunn, during the 1930s and 1940s, showed that injecting very small amounts of calcium into frog muscle caused it to contract, but sodium, potassium, magnesium or water did not. Ebashi, during the 1960s made two key discoveries, first the uptake mechanism into the ER/SR via a Ca^{2+} MgATPase, and secondly the first high affinity Ca^{2+} binding protein – troponin C in muscle. Key experiments for intracellular Ca^{2+} as a universal regulator were:

1. Indirect evidence, based on the effects of manipulation of extracellular and intracellular Ca^{2+}.
2. Direct measurement of intracellular free Ca^{2+}, using fluorescent dyes and Ca^{2+}-activated photoproteins, with changes being correlated with the timing and magnitude of the event.
3. Identification of the proteins responsible for regulating intracellular Ca^{2+}, and how primary stimuli and secondary regulators act on these to cause Ca^{2+} signals that provoke the cellular event.
4. Identification of the targets for Ca^{2+} inside cells, and how these interact with the proteins and intracellular structures that are responsible for the event.

5. When intracellular Ca²⁺ is friend or foe, and how this explained the role of intracellular Ca²⁺ in cell injury, cell death and disease.
6. The development of drugs that interact with the Ca²⁺ signalling system to treat disease.

These studies distinguished the 'passive' role of Ca²⁺, as a cofactor of some extracellular enzymes, from its 'active' role, as an intracellular signal. There were many key discoveries (Box 3.3).

The unitary hypothesis revealed the six special chemical properties of Ca²⁺ that enable it to be a universal intracellular regulator:

1. Ca²⁺ is doubly charged, so it binds to inorganic anions and $-CO_2^-$ in proteins at low concentrations.
2. Seven or eight coordination with oxygen enables Ca²⁺ to be selective for binding to proteins at micomolar free Ca²⁺ in the presence of millimolar Mg²⁺. Two key sites are the EF-hand and C2.
3. At micromolar concentrations, Ca²⁺ does not bind in large amounts to nucleotides such as ATP, GTP, TTP, CTP and UTP.
4. Ca²⁺, unlike Zn²⁺, comes off proteins quickly when the free Ca²⁺ decreases, allowing fast on/off processes to occur in cells.
5. Ca²⁺ has only one redox state, unlike Fe and Cu.
6. Ca²⁺ at the low free concentrations inside the cell does not contribute to osmolarity directly. Thereby, large fractional changes in free Ca²⁺ can occur without triggering huge water movements. This would be impossible for Na⁺ or K⁺.

Box 3.3 Time scale of key discoveries that established Ca²⁺ as a universal cell trigger.

1. No extracellular Ca²⁺ stops heart beating, and other events – 1883 to 1930s.
2. Injection of small amounts of Ca²⁺ into muscle causes contraction – 1947.
3. Muscle SR vesicles accumulate Ca²⁺, requiring ATP hydrolysis – 1960s.
4. Troponin C, the first high affinity Ca²⁺ binding protein, in muscle – 1960s.
5. Measurement of free Ca²⁺ in live cells establishes sub-micromolar cytosolic free Ca²⁺ in all animal, plant and microbial cells – 1967 to 1990s.
6. Plethora of high affinity Ca²⁺ binding proteins – calmodulin, EF hand, C2 – 1960s to 1980s
7. MgATPase Ca²⁺ pumps and exchangers (Na⁺, H⁺) – 1960s to 1970s
8. Plasma membrane Ca²⁺ channels; voltage-gated, receptor, SOCE – 1980s.
9. IP₃ and ryanodine receptors; how Ca²⁺ is released from internal stores – 1990s.
10. Genome sequencing, genetic engineering, and X-ray crystallography show how Ca²⁺ binding proteins, pumps and channels work – 1990s to 2000s.

Measurement of free Ca²⁺ in live cells, beginning in 1967 with the use of Ca²⁺-activated photoprotein from the bioluminescent jelly fish *Aequorea*, followed, in the late 1970s and early 1980s, by the fluorescent Ca²⁺ indicators of Roger Tsien, revealed a universality of all life – animals, plants, protists, bacteria and Archaea. All living cells maintain a very low free Ca²⁺ concentration in their cytosol, in the micromolar to sub-micromolar range, even in the presence of millimolar Ca²⁺ outside the cell. The consequent Ca²⁺ pressure across the outer membrane of cells has been exploited throughout 4000 million years of evolution to develop intracellular Ca²⁺ as a universal switch.

This history of intracellular calcium is full of key questions, ingenious key experiments and creative hypotheses, providing insights into how scientific discoveries are made. Most important is the beauty and elegance of the phenomenon regulated by Ca^{2+}, and the organisms in which they occur, whether it be a microbe just 1 μm across, invisible to the naked eye, or a blue whale more than 30 m long. By the end of the twentieth century, the key features of the intracellular Ca^{2+} signalling system and its toolbox had been revealed. The story of intracellular calcium is a wonderful example of how curiosity-driven science has led, often quite unexpectedly, to major discoveries of wide importance in biology and medicine, to the development of new diagnostic tests and clinical treatments, and to the development of several billion dollar markets.

Recommended Reading

*A must read

Books

Ashley, C.C. & Campbell, A.K. (Eds) (1979) Detection and Measurement of Free Ca^{2+} in Cells. Amsterdam: Elsevier/North Holland. First collection of papers by pioneers in measuring free Ca^{2+} in cells, based on an EMBO meeting in Oxford, in 1978.

Campbell, A.K. (1983) Intracellular Calcium: Its universal role as regulator, Chichester, John Wiley & Sons Ltd. Contains many historical papers on intracellular calcium. Tables and full reference list, including an Endnote file, are available from the Companion web site.

*Campbell, A.K. (2015) Intracellular Calcium, Chichester: John Wiley & Sons Ltd. Chapter 3. One hundred years plus of intracellular calcium. pp 81–128. Fully referenced history of intracellular calcium.

*Heilbrunn, L.V. (1937/1943). An Outline of General Physiology, 1st and 2nd edns. Philadelphia: Saunders. Contains many historical references to the role of calcium in cell physiology and pathology.

Reviews

Campbell, A.K. (1986) Lewis Victor Heilbrunn. Pioneer of calcium as an intracellular regulator. *Cell Calcium*, 7, 287–296. Review of Heilbrunn's pioneering work on Ca^{2+} signalling.

Carafoli, E. & Klee, C.B. (2008) A tribute to Dr Setsuro Ebashi; memorial issue in honor of Dr. Setsuro Ebashi. *Biochem. Biophys. Res. Commun.*, **369**, 28–276. Several papers reviewing the pioneering contribution of Ebashi to Ca^{2+} signalling.

*Kawasaki, H. & Kretsinger, R.H. (1994) Calcium-binding proteins. 1: EF-hands. *Protein Profile*, **1**, 343–517. Historical review by the discoverer of the EF-hand.

*Miller, D.J. (2004) Sydney Ringer; physiological saline, calcium and the contraction of the heart. *J. Physiol.*, **555**, 585–587. Review of Ringer's pioneering work.

Putney, J.W. (2007) Recent breakthroughs in the molecular mechanism of capacitative calcium entry (with thoughts on how we got here). *Cell Calcium*, **42**, 103–110. Review of store-operated calcium entry by the discoverer.

Research Papers

Ashley, C.C. & Ridgway, E.B. (1970) On the relationship between membrane potential, calcium transient and tension in single barnacle muscle fibres. *J. Physiol.*, **209**, 105–130. Pioneering use of aequorin to measure free Ca^{2+} in a live cell.

Ebashi, S. (1960) Calcium binding and relaxation in actomyosin system. *J. Biochem.*, **48**, 150–151. Discovery of SR Ca^{2+} uptake in muscle.

Ebashi, S. (1961) Calcium binding activity of vesicular relaxing factor. *J. Chir. (Paris)*, **50**, 236–244. Discovery of troponin in muscle.

Edwards, G., Weiant, E.A., Slocombe, A.G. & Roeder, K.D. (1948) The action of ryanodine on the contractile process in striated muscle. *Science*, **108**, 330–332. Discovery of the effect of ryanodine on SR Ca²⁺ release in muscle.

Fatt, P. & Katz, B. (1953) The electrical properties of crustacean muscle fibres. *J. Physiol. (London)*, **120**, 171–204. Discovery of voltage-gated Ca²⁺ channels.

*Heilbrunn, L.V. & Wiercinski, F.J. (1947) The action of various cations on muscle protoplasm. *J. Cell Comp. Physiol.*, **29**, 15–32. First demonstration that small amounts of Ca²⁺ specifically can cause muscle contraction in a live, intact cell.

*Hokin, M.R. & Hokin, L. E. (1953) Enzyme secretion and the incorporation of P-32 into phospholipides of pancreas slices. *J. Biol. Chem.*, **203**, 967–977. First indication that there was something special about inositol phosphates leading to the discovery of IP₃.

Knight, M.R., Campbell, A.K., Smith, S.M. & Trewavas, A.J. (1991a) Recombinant aequorin as a probe for cytosolic free Ca²⁺ in *Escherichia coli*. *FEBS Lett.*, **282**, 405–408. First measurement of free Ca²⁺ in bacteria, using recombinant aequorin.

Knight, M.R., Campbell, A.K., Smith, S.M. & Trewavas, A.J. (1991b) Transgenic plant aequorin reports the effects of touch and cold-shock and elicitors on cytoplasmic calcium. *Nature*, **352**, 524–526. First measurement of free Ca²⁺ in an intact plant, using recombinant aequorin.

*Nilius, B.H.P.L.J.B. & Tsien, R.W. (1985) A novel type of cardiac calcium channel in ventricular cells. *Nature*, **316**, 443–446. Pioneering use of patch clamp to identify a Ca²⁺ channel.

Pollack, H. (1928) Micrurgical studies in cell physiology. VI. Calcium ions in living protoplasm. *J. Gen. Physiol.*, **11**, 539–545. First detection of a rise in cytosolic Ca²⁺ using the dye alizarin, by a student of Chambers.

Prakriya, M., Feske, S., Gwack, Y., Srikanth, S., Rao, A. & Hogan, P.G. (2006) Orai1 is an essential pore subunit of the CRAC channel. *Nature*, **443**, 230–233. Discovery of Orai in the SOCE mechanism.

*Ringer, S. (1883) A further contribution regarding the influence of different constituents of the blood on the contractions of the heart. *J. Physiol.*, **4**, 29–43. Pioneering paper now showing that calcium was essential for the beating heart, correcting his previous paper.

Rink, T.J., Smith, S.W. & Tsien, R.Y. (1982) Cytoplasmic free Ca²⁺ in human platelets: Ca²⁺ thresholds and Ca-independent activation for shape-change and secretion. *FEBS Lett.*, **148**, 21–26. First use of fluorescent Ca²⁺ dyes to measure free Ca²⁺ in live cells by the inventor.

Roos, J., Digregorio, P.J., Yeromin, A.V., Ohlsen, K., Lioudyno, M., Zhang, S.Y., Safrina, O., Kozak, J.A., Wagner, S.L., Cahalan, M.D., Velicelebi, G. & Stauderman, K.A. (2005) Stim1, an essential and conserved component of store-operated Ca²⁺ channel function. *J. Cell Biol.*, **169**, 435–445. Discovery of STIM in the SOCE mechanism.

Shimomura, O.J.F.H. & Saiga, Y. (1962) Extraction, purification and properties of aequorin, a bioluminescent protein from the luminous hydromedusan aequorea. *J. Cell. Comp. Physiol.*, **59**, 223–239. Discovery of aequorin.

Streb, H., Irvine, R.F., Berridge, M.J. & Schulz, I. (1983) Release of Ca²⁺ from a nonmitochondrial intracellular store in pancreas acinar cells by inositol-1,4,5-trisphosphate. *Nature*, **306**, 67–69. Discovery that IP₃ releases Ca²⁺ from the ER.

4

How to Study Intracellular Ca²⁺ as Cell Regulator

Learning Objectives
- The pathway to discovering the role of Ca²⁺ in a cell event.
- What manipulation of extra- and intra-cellular Ca²⁺ tells you.
- The vital importance of measuring free Ca²⁺ in live cells, and whole organisms, as well as within sub-cellular compartments.
- The four methods measuring free Ca²⁺ in live cells, with their advantages and disadvantages.
- How to locate and image free Ca²⁺ in live cells.
- Why, and how to, measure total Ca²⁺ in external fluids and tissues.
- How Ca²⁺ buffers work, and what they are used for.
- The measurement of Ca²⁺ fluxes, and why this is useful.
- How to study ion channels, particularly with patch clamping.
- Discovering how the rise in intracellular Ca²⁺ occurs, and then returns to rest.
- How to identify the intracellular target for Ca²⁺, and how it works.

4.1 Pathway to Discover the Role of Intracellular Ca²⁺ in a Cell Event

Imagine you have just come across a fascinating new phenomenon, or you are looking to investigate the mechanisms underlying a disease such as Alzheimer's or Parkinson's. How would you establish whether Ca²⁺ is the intracellular signal that provokes the cellular event? First, it is necessary to identify and characterise the primary stimuli, and secondary regulators, and whether there are any natural or synthetic pharmacological agents that affect the timing or magnitude of the cell event. Crucially, it is essential to identify whether these agents work in a digital or analogue manner at the individual cell level, following the Rubicon principle (see Section 2.5). There are then eight questions to answer if a role for intracellular Ca²⁺ is to be fully established (Box 4.1).

Fundamentals of Intracellular Calcium, First Edition. Anthony K. Campbell.
© 2018 John Wiley & Sons Ltd. Published 2018 by John Wiley & Sons Ltd.
Companion Website: http://www.wiley.com/go/campbell/calcium

Box 4.1 The eight key questions.

1. Is a rise in intracellular Ca^{2+} necessary to provoke the cell event?
2. If so, where does it come from – intracellular release, extracellularly through the plasma membrane or both?
3. What are the key components in the signalling pathway that causes the rise in intracellular Ca^{2+}, and how do the primary stimuli and secondary regulators affect these?
4. What is happening to Ca^{2+} within intracellular organelles, particularly the ER, mitochondria and nucleus?
5. What is the target for the rise in intracellular Ca^{2+}?
6. How does the target provoke the cell event?
7. What goes wrong in a pathological situation, and can disturbances in intracellular Ca^{2+} explain any of the pathology or cause of a disease?
8. Can we find, or design, pharmacological agents that affect the cell event through Ca^{2+} signalling and that might be useful to treat a disease (or in the case of plants, be used as a herbicide or insecticide)?

There will be two strategies to begin with:

- Whether manipulation of extra- or intra-cellular Ca^{2+} affects the cellular event. Specifically, if removal of extra- or intra-cellular Ca^{2+} stops the event triggered by a primary stimulus.
- Whether the event can be triggered by raising intracellular Ca^{2+} artificially without the primary stimulus.

4.2 Manipulation of Intracellular Ca^{2+}

There are five ways in which to manipulate intracellular Ca^{2+}, to cause an artificial rise, or stop a rise induced by a cell stimulus:

- Removal and manipulation of extracellular Ca^{2+}.
- Addition of Ca^{2+} ionophores that allow Ca^{2+} to cross the phospholipid bilayer in the plasma membrane.
- Removal, or permeabilisation, of the plasma membrane.
- Injection, or photorelease, of Ca^{2+} chelators and buffers into the cell.
- Agents that release Ca^{2+} from internal stores.

Removal of extracellular Ca^{2+} is not as simple as it might at first sight seem. Ca^{2+} is everywhere, in tap, rain and distilled water. Even Analar reagents contain significant calcium, and it leaches off glass. Consequently, a Ca^{2+} chelator must be added to bind contaminating Ca^{2+}, EGTA or BAPTA (Figure 4.1a). If removal of Ca^{2+} prevents the cell event, then a Ca^{2+} dose–response should be carried out (1 μM to 10 mM). Sometimes, a bell-shaped dose–response curve may be observed, e.g. for $Ca^{2+} > 1$–2 mM, the cell response can actually decrease.

Manipulation of intracellular is done by injecting EGTA or BAPTA as Ca^{2+} chelators and Ca^{2+} buffers, into the cell, or loading into the cell using the chelator in its acetoxymethyl (AM) form. Ca^{2+} ionophores (Figure 4.1b) can also be used to allow Ca^{2+} to cross the phospholipid bilayer, both the plasma membrane and internal stores. A further trick is to permeabilise the plasma

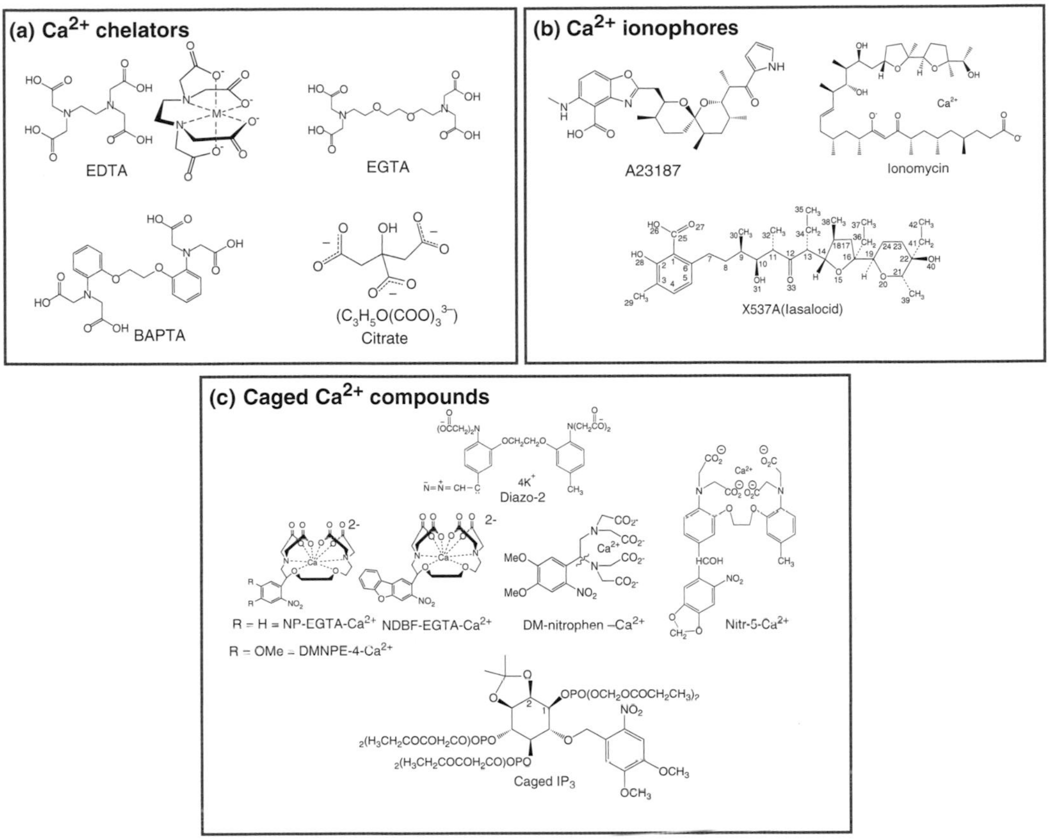

Figure 4.1 Manipulation of extra- and intra-cellular free Ca^{2+}: (a) chelators, (b) ionophores and (c) caged compounds.

membrane by mechanical removal, glycerol, or pore-forming toxins such as alfatoxin or streptolysin. An indicator of cytosolic free Ca^{2+}, such as fluo-3 or fluo-4 (see Figure 4.2b), should be used to check that these manipulations have either stopped any Ca^{2+} transients or caused a rise in cytosolic free Ca^{2+}.

A more sophisticated method for manipulating intracellular Ca^{2+} is to use a 'caged' compound that is photosensitive (Figure 4.1c). These compounds can be synthesised so that they either trap Ca^{2+} or a Ca^{2+} chelator, which can be released inside a cell by a pulse of light from a laser. Caged inositol trisphosphate (IP_3) is also available, so that ER Ca^{2+} can be released at a specific location. Photolysis takes just a few milliseconds. Release of Ca^{2+} from internal stores can also be achieved using thapsigargin or cyclopiazonic acid, which block the Ca^{2+}-MgATPase on the SR/ER.

Useful as these indirect experiments are as initial evidence for intracellular Ca^{2+} as a signal, the key experiment is to measure directly the effect of the primary stimulus on cytosolic free Ca^{2+}. If Ca^{2+} is the intracellular trigger, then the cytosolic free Ca^{2+} must rise before the cellular event is detected. Furthermore, prevention of the rise in cytosolic free Ca^{2+}, by removal of extracellular Ca^{2+} or intracellular chelation using BAPTA, should prevent the cell event. These experiments require an indicator of free Ca^{2+} in the live cell.

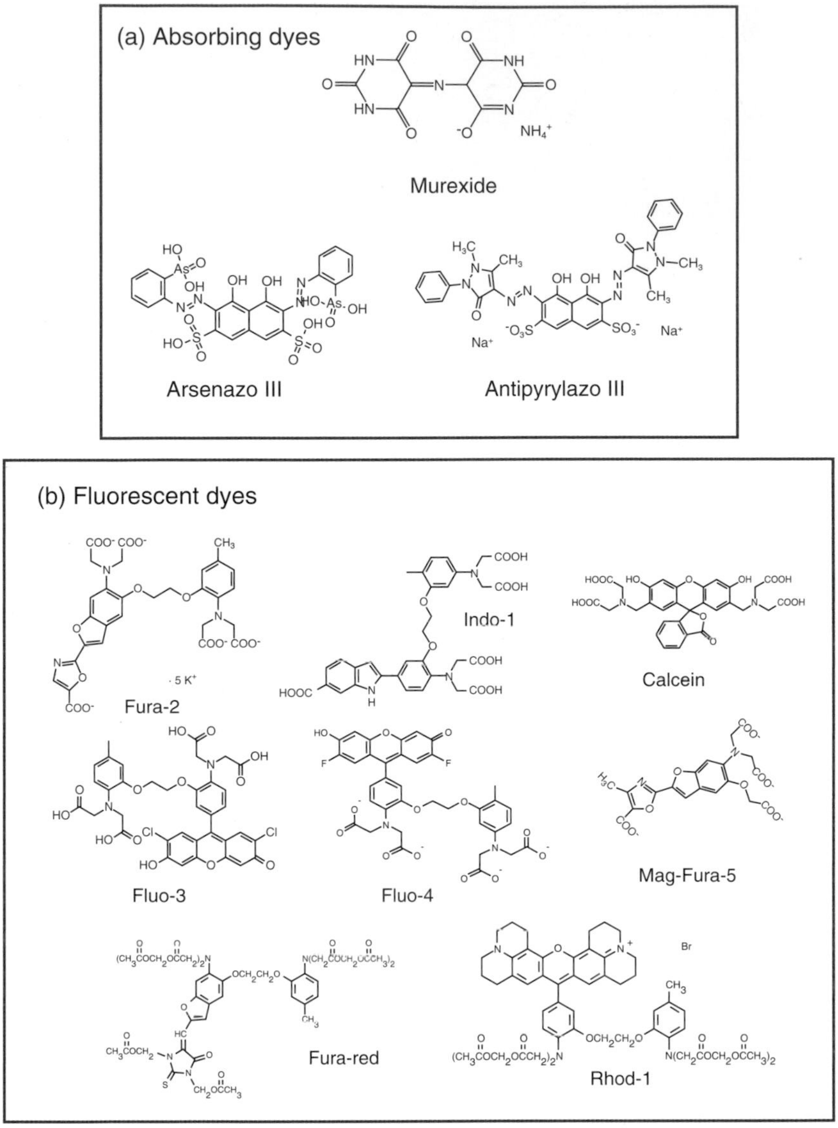

Figure 4.2 Absorbing and fluorescent dyes for measuring intracellular free Ca^{2+}: (a) absorbing and (b) fluorescent.

4.3 Measurement of Intracellular Free Ca^{2+}

Four methods have been well used to measure the concentration of free Ca^{2+} inside cells:

- Absorbing dyes (Figure 4.2a).
- Fluorescent dyes (Figure 4.2b, Table 4.1), and genetically engineered fluorescent proteins (Figure 4.10 below).
- Ca^{2+}-activated photoproteins (Figures 4.6 and 4.7 below).
- Microelectrodes.

Absorbing dyes and microelectrodes had some success in the 1970s, but have been replaced by the fluorescent dyes, the most commonly used method now. The Ca^{2+}-activated

Table 4.1 Properties of organic fluorophores (fluors) measuring free Ca^{2+} in live cells.

Fluor	Excitation (nm)	Emission (nm)	K_d Ca²⁺ (nM)	Fluorescence quantum yield −/+ Ca²⁺	Fluor$_{max}$/Fluor$_{min}$ or R_{max}/R_{min}
Quin-2	350	495	60	0.03/0.14	5−8
Fura-2	363/335	512	145	0.23/0.49	13−25
Indo-1	338	485/405	230	0.38/0.56	20−80
Fluo-3	506	526	325	0.0051/0.18	36−40
Fluo-4	490	516	345	0.14	>100
Rhod-1	556	578	2.3	0.0014/0.021	15
Fura-red	420/480	657	140		5−12
Calcium green-1	490	531	190	0.002/0.5	14
Mag-fura 2 (furaptra)	345	490-500	25000	0.3	6−30

The wavelengths used by different workers vary, and are not necessarily the peak for either excitation or emission. These are chosen on the basis of the exciting light source and good separation. Two wavelengths means it is a ratiometric indicator. The fluorescence quantum yield (Φ_F) represents the fraction of excited molecules that emit a photon. Thus a fluor with a Φ_F of 1 is 100% efficient.

photoproteins aequorin and obelin played a crucial role in the history of intracellular Ca^{2+} as a universal regulator, and still have unique applications, particularly in bacteria and Archaea.

For acute measurement of cytosolic free Ca^{2+} in animal and plant cells in tissue culture, fluorescent indicators are the method of choice, using a spectrofluorimeter for cell populations, or a fluorescent microscope with an imaging camera for studying individual cells. A major problem with all fluors is photobleaching. More than 90% of the fluorescent signal can be lost within minutes under continuous illumination from a bright light source. This can be reduced using imaging in a time-lapse mode, or using resonance excitation from a laser in a confocal microscope. For continuous long-term investigations, 1 h or more, Ca^{2+}-activated photoproteins should be used. For bacteria, the only method at present that works uses Ca^{2+}-activated photoproteins, using a photon-counting chemiluminometer for cell populations, or an imaging camera sensitive to individual photons on a microscope, or attached to a fibre optic, for individual cells. Imaging intracellular free Ca^{2+} in intact organs, such as a whole heart or plant, is best done using genetically engineered Ca^{2+}-activated photoproteins. To monitor free Ca^{2+} inside organelles such as mitochondria, ER, nucleus or Golgi, targeted fluorescent or bioluminescent indicators are used. The cameleons or pericams enable free Ca^{2+} inside organelles to be imaged inside individual cells. But for measurements in cell populations, intact organs or organisms, targeted Ca^{2+}-activated photoproteins are best. If absolute free Ca^{2+} is wanted, then a ratiometric fluorescent Ca^{2+} indicator or Ca^{2+}-activated photoprotein must be used. Fluors get into cells by using the AM ester method. The AM ester often needs a detergent such as Pluronic to increase the permeability of the plasma membrane. Ca^{2+}-activated photoproteins, cameleons or pericams, are transfected into cells from a plasmid.

4.3.1 Absorbing Dyes

Two absorbing dyes have been widely used: arsenazo III and antipyrylazo III (Figure 4.2a). These obey Beer–Lambert's law:

$$\text{Absorbance}\left(A_\lambda\right) = \log_{10}\left(I/I_0\right) = \varepsilon_\lambda cl \tag{4.1}$$

where I_0 is the intensity of the incident light and I is the intensity of the light after it has passed through the indicator; ε_λ (M^{-1} cm^{-1}) is the extinction/absorbance coefficient at wavelength λ; c is the concentration, in this case of the indicator; and l is the path length. The spectrum provides the wavelength where maximum absorbance occurs. Binding of Ca^{2+} affects the spectrum, making arsenazo III change visibly from purple to blue, and antipyrylazo III becomes slightly redder. The ability to measure free Ca^{2+} in cells depends on three parameters:

- The ability to detect the dye, which depends on a large ε_λ.
- The binding constant of the dye for Ca^{2+}.
- The ability of the dye to select Ca^{2+} over other cations such as Mg^{2+}.

Because Ca^{2+} changes the absorbance spectrum, by measuring the ratio of the absorbance at two wavelengths the Ca^{2+} concentration can be estimated independently of dye concentration. Two good wavelengths for arsenazo III are 675 and 685 nm. This allows a discrimination between Ca^{2+} and Mg^{2+} of 4000 : 1, necessary when the free Ca^{2+} is micromolar and Mg^{2+} is millimolar.

4.3.2 Fluorescent Ca²⁺ Indicators

Fluorescent dyes (Figure 4.2b) – fluorophores or 'fluors' for short – absorb light and then re-emit at a longer wavelength. They can be detected at concentrations several orders of magnitude lower than an absorbance indicator. Furthermore, fluors are much more suitable for imaging, and thus locating, Ca^{2+} signals in live cells. They get into cells as acetoxymethyl (AM) esters (e.g. fura-2 AM). These are uncharged, so cannot bind Ca^{2+}. They are hydrophobic, so they diffuse passively through the lipid bilayer of the plasma membrane. Once inside the cell, endogenous esterases hydrolyse off the AM group, leaving the charged fluor trapped inside the cell and able to bind Ca^{2+} (Figure 4.3). Typical intracellular concentrations of the fluorescent dyes are around 10–20 µM. Lower concentrations are difficult to detect over any endogenous autofluorescence, e.g. from NAD(P)H or flavins. Concentrations higher than 50 µM chelate enough Ca^{2+} to distort the Ca^{2+} signal. The dyes can also be injected directly into cells. Thus, by coupling them to dextran, uptake of the dyes into organelles, such as the nucleus and ER, can be prevented. The first fluorescent dye developed by Tsien was *quin-2* derived from a Ca^{2+} chelator, and a fluor, a quinoline.

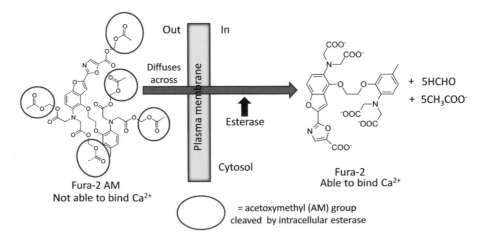

Figure 4.3 Uptake of fluorescent Ca^{2+} dyes as the AM ester. Reproduced by permission of Welston Court Science Centre.

Quin-2 had limited use as it was not a very good fluor, having a relatively low extinction coefficient of less than 5000, and low quantum yield (from 0.03 to 0.14 as the Ca^{2+} increases) – the fraction of excited molecules that emit a photon. Furthermore, the excitation maximum for quin-2 is in the UV at 339 nm, where silica glass lenses are poorly transparent and mirrors poorly reflectant. So Tsien turned to two other compounds – furan and indole, synthesising *fura-2* and *indo-1*. This was followed by exploitation of two other fluors, used widely in microscopy – fluorescein and rhodamine – that have high absorbance coefficients, leading to fluo-3 and -4, and rhod-1 and -2. Fluo-4 has a better quantum yield than fluo-3. These were followed by calcium green, crimson and red, which are particularly useful for fluorescence lifetime measurements and multiphoton imaging. Calcium green is one of the brightest emitters when bound to Ca^{2+}, and has a good dynamic range. Quin-2, fluo-3 and fluo-4 cannot be used ratiometrically, so the intensity of light emitted from a cell depends both on the concentration of the dye and the free Ca^{2+} concentration. In fact, the dye concentration can vary considerably from cell to cell and within different organelles. Fluo-3 and -4 are excited in the visible at 488 nm (Figure 4.4), and therefore do not require a UV laser nor quartz lenses.

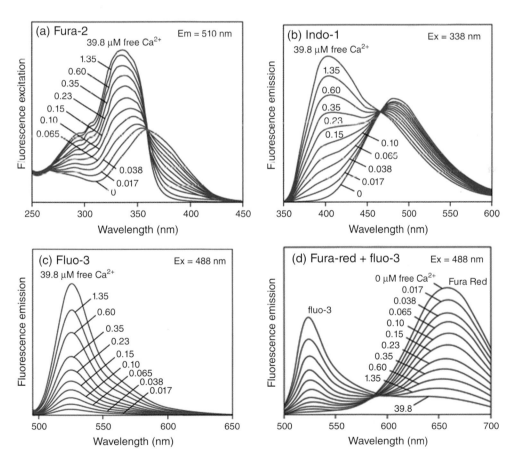

Figure 4.4 Effect of Ca^{2+} on the emission or excitation spectra of fluorescent Ca^{2+} indicators. (a) Excitation spectrum of fura-2, detecting emission at 510 nm. (b) Emission spectrum of indo-1 excited at 338 nm. (c) Emission spectrum of fluo-3 excited at 488 nm. (d) Emission spectrum of fura-red + fluo-3 (10 : 1) excited at 488 nm. Source: Data from Haugland, R.P., 1996. Copyright © 2013 Life Technologies Corporation. Used under permission.

Fura-2 and indo-1 have extinction coefficients of around 33 000 M^{-1} cm^{-1} and quantum yields of around 50% when bound to Ca^{2+}. The two key numbers are:

- The product of the extinction coefficient × the fluorescence quantum yield.
- The equilibrium dissociation constant for Ca^{2+}.

The Ca^{2+} equilibrium dissociation constants for fura-2 and indo-1 are 224 and 250 nM, respectively, when they are half saturated by Ca^{2+} They are thus ideal for monitoring free Ca^{2+} in the range 20 nM to 2 μM. Fura-2 and indo-1 have the further advantage of being ratiometric, fura-2 using two excitation wavelengths and indo-1 two emission wavelengths. This corrects for differences in loading of the dyes in individual cells. But ratiometric indicators complicate the fluorimeter or microscope imaging system, since a beam splitter or spinning wheel with several interference filters is required.

The fluorescent indicator of first choice will be a small organic fluor loaded using its AM ester, chosen from fluo-3, fluo-4, fura-2, indo-1 or fura-red (Figures 4.2, 4.3 and Table 4.1). They are half saturated when the free Ca^{2+} is the same as the dissociation constant. For fura-2 this is 145 nM Ca^{2+} and for fluo-3 it is 390-400 nM Ca^{2+}. So the best range for fura-2 is 10 nM to 2 μM, where it will be more than 90% saturated. The best range for fluo-3 is 40 nM to 5 μM. This is in fact better for most cells, where the free Ca^{2+} can rise to several micromolar. Fluo-4 is a better fluor than fluo-3. Fluo-3 and fluo-4 are also suitable for single-cell analysis using flow cytometry in a FACS. Fura-red is also a ratiometric fluor, Ca^{2+} binding affecting the excitation spectrum, as with fura-2. However, the fluorescence of fura-red decreases when it binds Ca^{2+}. It has the advantage that it is excited by visible light, and neither a UV laser nor quartz lenses, which are expensive, are required. The emission can also be used ratiometrically in conjunction with fluo-3 or calcium green. Rhod-2 has a Ca^{2+} affinity of 1 μM, some four to five times lower than fura-2 or indo-1, whereas mag-fura-2 has a Ca^{2+} affinity of 25 μM. It has thus been used to monitor ER free Ca^{2+} in permeabilised cells. However, since mag-fura-2 also binds Mg^{2+} it is not always clear which cation is being monitored.

Another important characteristic of a Ca^{2+}-binding dye is the rate at which Ca^{2+} binds when the free Ca^{2+} rises, and then dissociates when the free Ca^{2+} drops. For fura-2, the binding rate constant is 6×10^8 M^{-1} s^{-1} and the dissociation rate constant 97 s^{-1}, being insensitive to pH over the range 7–8. This means that the half time of Ca^{2+} dissociation is about 7 ms, fast enough for most cells not to distort the Ca^{2+} signal, although millisecond Ca^{2+} signals will be distorted. It is also vital that the Ca^{2+} dye does not bind Mg^{2+} at the millimolar concentrations free in the cell. The affinity of fura-2 for Mg^{2+} is around 10 mM, and so is hardly affected by Mg^{2+}.

The fluors can leak out of cells, or are pumped out. This may be a particular problem in bacteria, which have active ABC transporters to remove unwanted organic molecules, producing apparent Ca^{2+} signals from fluorescent changes outside the cells. A problem with several fluors is that they bind to intracellular proteins and overload into organelles such as the nucleus and ER. This affects the fluorescence characteristics of the indicator and its affinity for Ca^{2+}, making precise calibration of Ca^{2+} concentrations difficult.

4.3.3 Calibration of Fluors

For single wavelength measurement, when only relative changes in free Ca^{2+} are required, a simple formula provides an estimate:

$$\text{Free Ca}^{2+} \text{ change} = \Delta\text{Ca}^{2+} = \Delta F / F_{\text{rest}} = \left(F_{\text{stim}} - F_{\text{rest}} \right) / F_{\text{rest}}$$

$$(4.2)$$

where F_{rest} is the dye fluorescence in the resting cell and F_{stim} is the dye fluorescence at any time t after addition of the cell stimulus. For absolute free Ca²⁺ the equation changes to:

$$[Ca^{2+}] = K_d \cdot (F_{stim} - F_{min})/(F_{max} - F_{stim}) \tag{4.3}$$

where K_d is the dissociation constant for Ca²⁺, F_{max} is the fluorescence when the dye is saturated by Ca²⁺ and F_{min} is the fluorescence when there is no Ca²⁺ bound to the dye.

Ratiometric dyes can be calibrated in pure solution (Figure 4.5). But, in the cell, at the end of the experiment, the maximum fluorescence is measured when the dye is saturated with Ca²⁺, and then the minimum when all the Ca²⁺ is removed, using a Ca²⁺ chelator such as EGTA. When ratiometric dyes such as fura-2 or indo-1 are used the calibration equation is:

$$[Ca^{2+}] = R_r \cdot K_d \cdot (R_{stim} - R_{min})/(R_{max} - R_{stim}) \tag{4.4}$$

where R is the ratio of fluorescence at the two wavelengths, typically excitation around 350/380 nm for fura-2 or emission around 405/480 nm for indo-1, and R_r is the ratio of fluorescence at the 350/380 nm excitation or 405/480 nm emission, respectively, with the dye in zero or saturating Ca²⁺. However, because there are problems of the precise conditions inside the cell, it is common practice to plot simply the ratio of fluorescence with time for fura-2 and indo-1, and a fluorescence ratio of F_{stim}/F_{rest} for single-wavelength dyes.

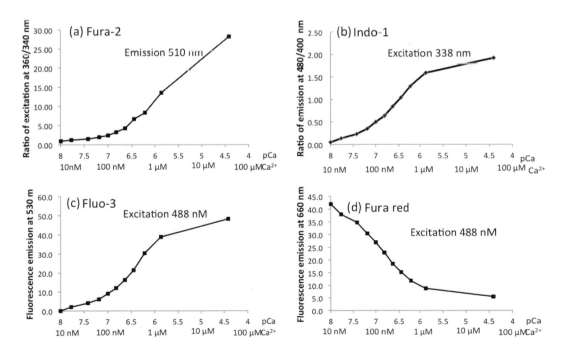

Figure 4.5 Dose–response curves for Ca²⁺ against four fluorescence of Ca²⁺ indicators: (a) fura-2, (b) indo-1, (c) fluo-3 and (d) fura-red. Data from Haugland, R.P., 1996. Copyright © 2013 Life Technologies Corporation. Used under permission.

After basic measurements of intracellular free Ca^{2+}, more advanced microscopical techniques can be used, including fluorescence resonance energy transfer (FRET), multiphoton, second harmonic measurements, fluorescence lifetime imaging microscopy (FLIM), total internal reflection fluorescence (TIRF) and fluorescence recovery after photobleaching (FRAP). The most useful FRET indicator for measuring intracellular free Ca^{2+} uses mutants of GFP – cameleons and pericams (see Figure 4.10 below). A typical donor/acceptor pair is cyan fluorescent protein (CFP) and yellow fluorescent protein (YFP), or a circularised one in a pericam.

Multiphoton fluorescence occurs when two or three photons hit a fluor within the excitation time around a femtosecond (10^{-15} s), then they add up. Thus, fluors such as fura-2, normally excited in the UV at 340 and 380 nm, can be excited in the red at around 700 nm. This has several benefits:

- Photobleaching is much reduced.
- The incident exciting beam has a much sharper focus, as a point, instead of a cone.
- The light penetrates further into intact tissues.

However, this technique requires great skill and very expensive titanium/sapphire lasers, capable of producing pulses of photons lasting a few femtoseconds (10^{-15} s) every picosecond (10^{-12} s).

4.3.4 Ca^{2+}-Activated Photoproteins

The first method to measure cytosolic free Ca^{2+} used the Ca^{2+}-activated photoprotein aequorin, extracted from the luminous jellyfish *Aequorea*, injected into the single-celled muscle fibre of the giant barnacle *Balanus nubilus* (Figure 3.3). A relative of *Aequorea*, the hydroid *Obelia geniculata* provided a similar protein, obelin. Light emission of both aequorin and obelin (Figure 4.6) is triggered by Ca^{2+} binding to three EF hand sites, which allows the oxygen, tightly bound to the apoprotein, to oxidise coelenterazine, also tightly bound, to form coelenteramide in an electronically excited state. This then emits a photon.

Aequorin and obelin have been used extensively to measure and image free Ca^{2+} in live animal, plant and microbial cells. *Aequorea* expresses several aequorins, with slightly different kinetics and Ca^{2+}-binding affinities. Obelin has a faster saturating rate constant than aequorin, but a lower apparent affinity for Ca^{2+}.

The ability to express mRNA coding for photoproteins in live cells, coupled with genetic engineering, has extended enormously the application of the Ca^{2+}-activated photoproteins. By engineering targeting sequences onto these proteins (Figure 4.7), free Ca^{2+} can be measured inside intracellular organelles, including the ER, nucleus, mitochondria, Golgi, the inner surface of the plasma membrane, chloroplasts in plants, and even the periplasm of bacteria. A disadvantage of Ca^{2+}-activated photoproteins is that they only produce a photon once. However, if peptides, such as KDEL, or proteins, are engineered on to the C-terminal proline, this opens up the solvent cage, and converts it into a luciferase that turns over coelenterazine, just like a normal enzyme. The best way to locate a photoprotein is to engineer enhanced GFP (EGFP) on to its N-terminus, so that its green fluorescence can be used to locate it in the live cell (Figure 4.8). If complete energy transfer is to be achieved, then a linker of at least 12 amino acids is required between the aequorin donor and GFP acceptor (Figure 4.12 below)

A problem is the degradation of apoaequorin before it is activated by coelenterazine, or when used in luciferase mode. The half-life of apoaequorin in the cytosol can be as fast as 20 min, and degradation is increased markedly when the free Ca^{2+} is low. This is why it is lost within a few minutes inside the ER when Ca^{2+} is released using, for example, an inhibitor of the SERCA

Aequorin

```
MTSKQYSVKLTSDFDNPRWIGRHKHMFNFLDVNHNGKISLDEMVYKASDI  50
VINNLGATPEQAKRHKDAVEAFFGGAGMKYGVETDWPAYIEGWKKLATDE  100
LEKYAKNEPTLIRIWGDALFDIVDKDQNGAITLDEWKAYTKAAGIIQSSE  150
DCEETFRVCDIDESGQLDVDEMTRQHLGFWYTMDPACEKLYGGAVP      196
```

Obelin

```
MASKYAVKLQTDFDNPKWIKRHKFMFDYLDINGNGQITLDEIVSKASDDI  50
CKNLGATPAQTQRHQDCVEAFFRGCGLEYGKETKFPEFLEGWKNLANADL  100
AKWARNEPTLIREWGDAVFDIFDKDGSGTITLDEWKAYGRISGISPSEED  150
CEKTFQHCDLDNSGELDVDEMTRQHLGFWYTLDPEADGLYGNGVP       195
```

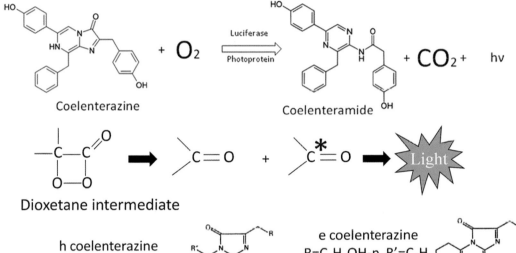

Figure 4.6 The Ca²⁺-activated photoproteins aequorin and obelin, and the bioluminescent reaction. Amino acid sequences of cloned one aequorin and obelin (from *Obelia geniculata*) are 196 and 195, respectively, though the extracted aequorin was 189 amino acids. Calcium binding sites in red.

pump such as cyclopiazonic acid. However, the half-life of apoaequorin can be increased several fold by engineering proteins, such as firefly luciferase or GFP, on to the N-terminus of aequorin. Improving light yields and producing a range of Ca²⁺ affinities can be achieved by activating the apoaequorin using synthetic coelenterazines (Figure 4.6). Although aequorin can be used to quantify Ca²⁺ from 0.1 to 100 μM or more, it decays very quickly at 100 μM. Thus, the photoproteins are consumed rapidly at high Ca²⁺ concentrations, making it impossible to use them to study sites where free Ca²⁺ may be above 10μM for more than a few seconds, such as inside the ER and sarcoplasmic reticulum (SR), Golgi, the inner surface of the plasma membrane, and the mitochondria. By engineering the aspartate at position 119 to an alanine (D119A), a mutant with 20-fold reduction in affinity for Ca²⁺ was produced enabling it to be used to measure Ca²⁺ at high micromolar to millimolar free Ca²⁺, such as the ER and Golgi. Also in luciferase mode aequorin can measure free Ca²⁺ at high micromolar to millimolar concentrations, such as inside the ER.

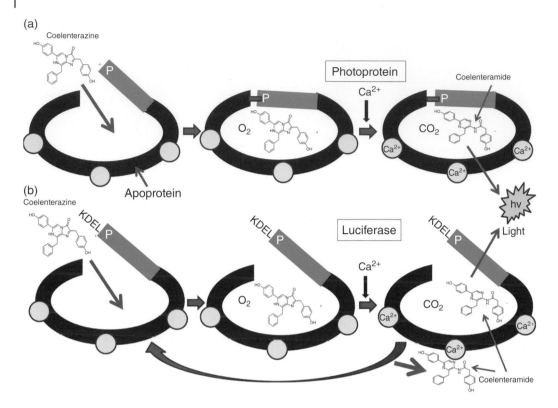

Figure 4.7 Aequorin or obelin as Ca^{2+}-activated photoproteins or luciferases. (a) Coelenterazine enters the solvent cage and is tightly bound with oxygen, the cage being sealed by a crucial C-terminal proline. Binding of Ca^{2+} changes the protein conformation, allowing the oxidation of coelenterazine to coelenteramide in an electronically excited state, which emits a photon with a quantum yield of about 15%. The coelenteramide remains trapped until Ca^{2+} is removed and drops off the protein. (b) Engineering the ER retention signal, or other peptides and proteins, on to the C-terminus of the photoprotein opens up its solvent cage. This allows coelenteramide to diffuse out of the solvent cage, so that new coelenterazine can re-enter and react with oxygen. Thus, the photoprotein is now a luciferase, turning over the substrate coelenterazine continuously.

Although the light emission from Ca^{2+}-activated photoproteins is much less than with fluorescent dyes, aequorin and obelin do have some advantages over fluors:

- Ca^{2+}-activated photoproteins cover a much wider range of free Ca^{2+} (100 nM to 100 μM) than an individual fluor such as fura-2 (10 nM to 2 μM).
- Ca^{2+}-activated photoproteins are used at much lower concentrations than the fluors, and thus are less likely to buffer the Ca^{2+}.
- The photoprotein remains exclusively in the cytosol, unlike the fluors, which overload into the ER and nucleus, causing artefacts.
- Photoproteins do not leak out of cells, and are thus more suitable for studies involving cell injury.
- No illuminating lamp is required when using the photoproteins, so there is no photobleaching or autofluorescence. Thus they are more suitable for measuring free Ca^{2+} in whole organs and organisms, for example live plants. It is difficult to produce a uniform illumination with such specimens, and also intact organisms often exhibit high levels of autofluorescence.
- Bioluminescence is better suited to measurements on cell populations in microtitre plates, whole cover slips, intact organs and organisms.

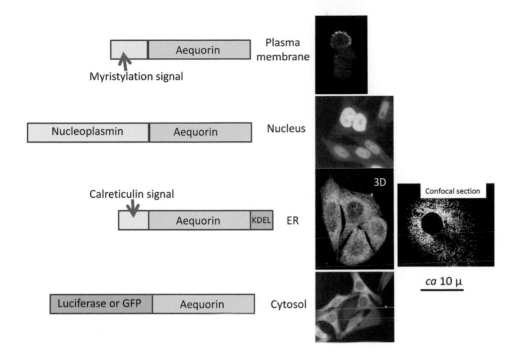

Figure 4.8 Targeted Ca²⁺-activated photoproteins. Aequorin is targeted to the cytosol by coupling it at the N-terminus to GFP or firefly luciferase, so that it is too big to go through the nuclear pore. Targeting to the nucleus is highly efficient using the nuclear protein nucleoplasmin from *Xenopus*, to the ER using the signal from the N-terminus from calreticulin, and to the plasma membrane using a myristylation signal, or a plasma membrane protein such as SNAP, a receptor or an ion channel. The figure shows aequorin, or firefly luciferase (plasma membrane), localised in cell sections using an antibody, except the ER where the aequorin was tagged with GFP. The confocal section shows the apparent vesicular structure of the ER. But the three-dimensional reconstruction shows that the ER is in fact a three-dimensional spider's web. Targeting to mitochondria uses a part of COX, cytochrome *c* oxidase. With thanks to my group: Mike Badminton, Bob Kendall, Angela Ribero Trop, Llewellyn Roderick and Rachel Errington.

- Ca²⁺-activated photoproteins can be genetically engineered to target specifically and efficiently to the mitochondria, ER, Golgi, nucleus, nucleolus and peroxisome.
- Different colours can be produced by using the range of coelenterazines available, and by genetically engineering GFPs onto the N-terminus of the photoprotein.
- Continuous imaging for 24 h or more, at 50 Hz, can be achieved, not only because there is no photobleaching, but because the file size of bioluminescent images can be at least 100 times smaller than with fluorescence.
- Ca²⁺-activated photoproteins are, at present, the only way to measure free Ca²⁺ in live bacteria or Archaea.
- Other intracellular substances and enzymes can be measured simultaneously by genetically engineering a rainbow protein (see Figure 4.11 below).

Ca²⁺-activated photoproteins are expressed in cells using a plasmid engineered to contain the DNA coding for apoaequorin or apoobelin by:

- Transfection of a mammalian cell.
- Microinjection of a giant cell such as an oocyte.
- Using the bacterium *Agrobacterium* or a DNA gun for plants.
- Transformation of a bacterium or archaean.

Expression is via a constitutive promoter such as SV40 or cytomegalovirus, or a controllable promoter. Coelenterazine is then added to form the Ca²⁺-activated photoprotein. Formation of both the photoprotein and GFP require oxygen. But after this, oxygen is only required if the aequorin or obelin is in luciferase mode.

4.3.5 Calibration of Ca²⁺-Activated Photoproteins

Light emission, in counts s⁻¹, from a Ca²⁺-activated photoprotein inside a live cell, depends on how much total photoprotein is present, and how much of this has Ca²⁺ bound to it (Figure 4.9). Therefore, to estimate the absolute free Ca²⁺ in the cell at any one time, it is necessary to measure the rate constant, k (s⁻¹):

$$k = \left(\text{counts s}^{-1}\text{ at a time point}\right)/\left(\text{total counts remaining at that time point}\right) \tag{4.5}$$

This relationship for k is always true, but can only be estimated at the end of the experiment, when the total amount of photoprotein left can be measured. Thus, the total photoprotein at the start, time 0, is the sum of all the counts measured during the experiment + counts remaining at the end. Then the total counts remaining at time t = total counts at time 0 − total counts emitted up to that point. Since each Ca²⁺-photoprotein molecule emits a photon only once, light emission is first order and exhibits an exponential decay:

$$\text{Light emission} = dh\nu/dt = A_0\exp\left(-k_t\right) \tag{4.6}$$

where A_0 is the total light potential at time 0 and k is the rate constant.

At saturating Ca²⁺, the rate constant for aequorin is about 1.4 s⁻¹ ($t_{1/2}$ = 0.69 s) and for obelin is about 4 s⁻¹ ($t_{1/2}$ = 0.18 s). Thus, ½ of a ½ of a ½ of a ½ of a ½ of a ½ (i.e. six half-times) for an exponential decay results in 98% consumption of the photoprotein. For aequorin, this takes 4 s and for obelin 1 s. So, in order to measure the total photoprotein in free solution, 1–10 mM Ca²⁺ is added, and the counts measured for 10 s. When measuring the total active photoprotein inside cells, the membrane is permeabilised by a detergent such as NP40 or by hypotonic solution, and may take several minutes to consume over 90% of the photoprotein.

The reaction of Ca²⁺-activated photoproteins (P) to generate light (Figure 4.6) can be considered in four steps:

Step 1: Ca²⁺ binding, where K_{app} is the apparent affinity for Ca²⁺ binding:

$$P + 3Ca^{2+} \xleftrightarrow{\;K_{app}\;} PCa_3 \tag{4.7}$$

Step 2: The rate-limiting step:

$$PCa_3 \xrightarrow{\;k_{RL}\;} \text{intermediate X} \tag{4.8}$$

The intermediate X is a dioxetane formed from coelenterazine with oxygen.
Step 3: Generation of the electronically excited state of coelenteramide (Y*):

$$X \xrightarrow{\;k_{ex}\;} Y^* \tag{4.9}$$

Step 4: Photon emission ($h\nu$):

$$Y^* \xrightarrow{\;k_{em}\;} Y + h\nu \tag{4.10}$$

Figure 4.9 Aequorin dose–response curves. (a) Obelin chemiluminescence trace after adding Ca²⁺ to saturate the photoprotein. Most of the trace follows a perfect exponential decay, with a rate constant of 4 s⁻¹ and a total light emission of 1.5 × 10⁵ photon counts. The rate constant is independent of the amount of photoprotein present. Inset: plot of rate constant of the exponential decay at different Ca²⁺ concentrations as a percentage of the saturating rate constant. (b) Plot of –log of the ratio of rate constant to saturating rate constant for aequorin and obelin against the Ca²⁺ concentration. The figure also shows the light emission traces at three different Ca²⁺ concentrations (1 μM, 10 μM and 1 mM), 1 mM Ca²⁺ being saturating. This shows how the height of the light emission trace and the slope decrease as the Ca²⁺ concentration decreases; for Ca²⁺ dose–response curve for aequorin using the ratio of L/L_{max}, where L is the peak height of light emission. Reproduced by permission of Welston Court Science Centre.

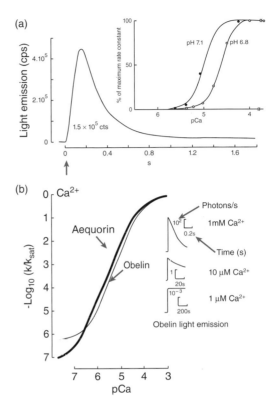

Ca²⁺ binding is fast, and only limited by diffusion, the on rate for Ca²⁺ being about 500 s⁻¹ ($t_{1/2}$ = 1.4 ms). Step 4, the decay of a singlet excited state, is extremely fast, and, like fluorescence, k_{em} is in nanoseconds. Step 3, the formation of the excited coelenteramide, is also fast, and can be measured by saturating the photoprotein with Ca²⁺, and then removing the Ca²⁺ in the solution very rapidly using a stop flow apparatus. This removes all the residual PCa₃, leaving X to react. This produces a rate constant k_{ex} of about 100 s⁻¹ for aequorin and 300 s⁻¹ for obelin. The rate constant of the rate-limiting step, k_{RL}, is easily measured, as it is equivalent to the rate constant when the photoprotein is saturated by millimolar Ca²⁺, when all the aequorin present will form PCa₃ within a few milliseconds. For aequorin this is 1–1.4 s⁻¹ ($t_{1/2}$ 0.5–0.69 s) depending on the variant, and 4 s⁻¹ ($t_{1/2}$ 0.18 s) for obelin.

Thus, when the photoprotein is saturated by Ca²⁺, the rate of photoprotein reacting is:

$$d(PCa_3)/dt = -k_{RL}(PCa_3) = -k_{sat}(PCa_3) \tag{4.11}$$

The minus sign is there because the light emission is decaying, as the coelenterazine on the photoprotein molecules is consumed. Integration of Eq. (4.11) produces an exponential decay:

Photoprotein bound to Ca²⁺ at time $t = (PCa_3)_t = PCa_0 \exp(-k_{sat}t)$ (4.12)

where PCa_0 is the photoprotein at time 0.

As with all chemiluminescence, this is independent of volume:

Light emitted by a photoprotein $= \Phi_{CL} P_R$ (4.13)

where P_R is the number of photoprotein molecules that have reacted and Φ_{CL} is chemiluminescence quantum yield:

$$\Phi_{CL} = \text{number of photons/number of photoproteins reacted} \qquad (4.14)$$

Thus, from Eq. (4.13) :

$$I = \text{light intensity} = \mathrm{d}h\nu/\mathrm{d}t = k_{sat}\Phi_{CL}P_0\exp(-k_{sat}t) \qquad (4.15)$$

$\mathrm{d}h\nu/\mathrm{d}t$ will be measured in counts s^{-1}, but will depend in absolute terms on the efficiency of the chemiluminometer or imaging camera. This efficiency itself depends on three parameters:

- Geometry, i.e. how many of the photons emitted hit the detector.
- Quantum efficiency of the detector.
- Efficiency of the electronics.

The overall efficiency of a good chemiluminometer is never likely to be better than 1% and often is only 0.1% (i.e. only one in 1000 photons emitted from the sample are recorded as counts).

Two models have been proposed for Ca^{2+} binding in order to relate Ca^{2+} to light emission (Figure 4.9). One is based on three Ca^{2+}s binding being at equilibrium with a rate-limiting step leading to light emission, the other assumes the photoprotein exists in two states, with or without Ca^{2+} being bound.

Model 1:

$$P + Ca^{2+} \xleftrightarrow{K_1} PCa + Ca^{2+} \xleftrightarrow{K_2} PCa_2 + Ca^{2+} \xleftrightarrow{K_3} PCa_3 \xrightarrow{k_{sat}} h\nu \qquad (4.16)$$

Where P is the photoprotein, Ks are equilibrium dissociation constants, and k_{sat} is the rate constant when the photoprotein is saturated by Ca^{2+}. When measuring chemiluminescence, such as photoprotein emission, the light emitted is independent of volume (i.e. it is not a concentration term):

$$\mathrm{d}h\nu/\mathrm{d}t = k_{app}\Phi_{CL}P_0\exp(-k_{app}t) = I(\text{light intensity}) \qquad (4.17)$$

where:

$$k_{app} = k_{app}K_1K_2K_3(Ca)^3/\left(1 + K_1Ca + K_1K_2(Ca)^2 + K_1K_2K_3(Ca)^3\right) \qquad (4.18)$$

where Φ_{CL} = chemiluminescence quantum yield.

Model 2

$$T = R + 3Ca = RCa_3 \rightarrow h\nu \qquad (4.19)$$

In this model $k_{app} = k_{sat}[(K_{RCa}Ca)/(K_{RCa} + K_{RCa}K_{TR} + Ca)]^3$, where T = the tight form which does not bind Ca^{2+}, and R is the relaxed form which does bind Ca^{2+}. The main kinetic difference between the two models is that in model-1 Ca^{2+} binding is ordered, whereas in model-2 Ca^{2+} binding is considered to be random.

There is a Ca^{2+} independent light emission for aequorin and obelin, which limits the detection of free Ca^{2+} to around 0.1 μM. The apparent affinity of aequorin for Ca^{2+} of about 6 μM, which like obelin, is affected by other ions, including pH, monovalent cations such as Na^+ and

K⁺, and several naturally occurring divalent cations, such as Mg^{2+}, Zn^{2+} and Mn^{2+}. Thus, it is important to carry out a Ca^{2+} standard curve in an ionic environment that is similar to that inside the cell (i.e. 150 mM K⁺, 10–20 mM Na⁺, 1 mM Mg^{2+}, pH 7 in most eukaryotic cells).

4.3.6 Ca²⁺ Indicators from Engineered GFP

GFP has revolutionised cell biology and biomedical research, enabling organelles, intracellular structures and individual molecules to be lit up, and their movement imaged, in live cells. Although GFP itself is not a Ca^{2+} indicator, chimeras of mutated GFPs have been linked to Ca^{2+} binding domains, providing a family of fluorescent Ca^{2+} indicators, which can be targeted to various locations within live cells. Genetically engineered GFPs, together with other fluorescent proteins, such as the red fluorescent protein DsRed (cherry red) cloned from the coral *Discosoma* sp., have extended enormously the application of fluorescent proteins. It is beautiful example of how curiosity about an apparently obscure phenomenon has led, quite unexpectedly, to major discoveries and inventions, and a Nobel Prize in 2008 for Osamu Shimomura, Roger Tsien and Martin Chalfe. GFP in bioluminescent jellyfish, hydroids, sea pansies and sea pens shifts the colour of the light emitted from blue to green via Förster energy transfer, sharpening the spectrum and shifting the peak emission from about 480 nm to about 510 nm. In the jellyfish *Aequorea* and *Obelia*, the green fluorescence cells are around the umbrella at the base of the tentacles, and in various locations in the hydroid stage. Its location in hydroids provides a very good identification marker. Interestingly, the gonadal location of fluorescence in non-luminous species fits my hypothesis that the selective advantage of these fluors is about temperature. Fluors do not warm up in sunlight, because they re-emit the sun's energy as a photon. In contrast, the light energy absorbed by non-fluorescent organisms or cells is converted into heat. Such small changes will have a selective advantage. GFP-aequorin chimeras produce more light when expressed inside cells, because the half-life of the apoaequorin is increased substantially when linked to another protein, and not because GFP increases the quantum yield. This issue of half-life is often ignored when using these proteins as gene reporters of promoter activation. Bioluminescent proteins are good because they have a short half-life. The half-life of GFP is too long to see genes switching on and off.

Forced evolution is a technique whereby a population of bacteria containing a plasmid are exposed to a mutagen, either a chemical or UV light. This generates mutations in the DNA coding for the proteins coded for by the plasmid. When GFP was exposed to forced evolution, a wide range of mutants with a rainbow of colours was generated. These can be made to have a codon using sequence that is best for a particular organism, be it human, fly, nematode worm, bacteria, archaean, or plant. The fluor in GFP is made by an oxidation and the cyclisation of three amino acids at positions 65, 66, and 67. Gly at position 67 is essential, but the two other amino acids can be mutated without losing the fluorescence. Expression of fluorescent wild-type GFP from DNA turned out to be difficult in many animal cells, only a small percentage of GFP molecules forming the fluor correctly. However, a S65T mutant – enhanced EGFP – was found that formed the fluor much more efficiently, and is now the GFP of choice. Two key mutants are the yellow and blue or cyan fluorescent proteins, although their quantum yields and absorbance coefficient make the proteins less bright than wild-type GFP in *Aequorea*. The fluorescence properties of the GFP mutants vary from wild-type in extinction coefficient, quantum yield, as well as peak excitation and emission wavelengths (Table 4.2). GFP-like proteins have been also found in non-coelenterates, such as copepods. But many other bioluminescent and non-bioluminescent organisms have fluors that are not similar to GFP at all, e.g. bioluminescent dragon fish *Malacosteus*, *Aristostomias* and *Pachystomias*. Some luminous bacteria also exhibit energy transfer.

Table 4.2 Key properties of GFP and engineered GFP Ca²⁺ indicators.

Indicator	Mutation	Extinction coefficient (M⁻¹ cm⁻¹)	Quantum yield	K_d Ca²⁺ (µM)	Excitation wavelength (nm)	Emission wavelength (nm)
GFPs						
Wild-type	None	21000	0.77	NA	395/470	509/509
EGFP	F64L; S65T	55000	0.7	NA	488	507
EBFP	F64L; S65T; Y66H; Y145F	31000	0.2	NA	380	440
ECFP	F64L; S65T; Y66W; N146I; M153T; V163A; N212K	26000	0.4	NA	439	476
EYFP	S54G; S72A; T203/Y	36500	0.63	NA	513/488	527
DsRed	None	23900	0.46	NA	558	583
Ca²⁺ indicators						
Cameleon-2	CFP/YFP	NA	As per FP	4.4	440	535/480
Flash pericam	T203Y	NA	As per FP	0.7	410/488	514

FP = fluorescent protein; NA = not applicable. The association and dissociation rate constants for cameleon-1 were 2.4×10^6 M⁻¹ s⁻¹ and 13 s⁻¹, respectively, giving a half-time of Ca²⁺ dissociation of about 50 ms, fast enough for free Ca²⁺ changes in most cells. Mutations in the calmodulin can produce a family of Ca²⁺ indicators covering the range 10 nM to 10 mM Ca²⁺. Note: the precise wavelengths used with cells often varies slightly from those here (see for example Figure 4.10).

The key to making GFP into a Ca²⁺ indicator was to engineer a Ca²⁺-binding domain on to the GFP so that Ca²⁺ binding affected the fluorescence via energy transfer – fluorescence resonance energy transfer (FRET). FRET is really a misnomer, it being more correct to call it Förster energy transfer, after the pioneer who defined the mathematics. The trick was to engineer calmodulin from the frog *Xenopus*, with a Gly–Gly spacer, to a calmodulin-binding site, M13 – the 26-amino-acid peptide in the myosin light chain kinase (MLCK) from skeletal muscle, in between EBFP or ECFP and either EGFP or EYFP (Figure 4.10). By exciting the BFP or CFP at 370-380 or 339-440 nm, respectively, the excited fluor transferred its excitation to the EGFP or EYFP, which then emitted at around 510 or 535 nm (precise wavelengths 513 and 527 Table 4.2), respectively. These chimeras were named cameleons, after CaM for calmodulin and the amphibian chameleon that can change colour. Binding of Ca²⁺ to calmodulin in the cameleons caused it to wrap around the M13, and thereby increase the energy transfer. BFP and CFP have a peak emission at around 440 and 480 nm (476 precisely), respectively. Thus, an increase in energy transfer is detected by an increase in the ratio of light emitted at 510/440 nm for CFP to GFP, 510/480 nm for BFP to GFP, 535/440 nm for CFP to YFP and 535/480 nm for BFP to YFP. By mutating the Ca²⁺ sites in calmodulin, it was possible to generate a family of FRET Ca²⁺ indicators to measure free Ca²⁺ in the range 10 nM to 10 mM. Sequences were added, to target the cameleons to the ER (MLLSVPLLLGLLGLAAAD) at the N-terminus, and the ER retention signal KDEL at the C-terminus, or a nuclear targeting signal (PKKKRKVEDA). The Kozak consensus sequence was added at the 5′-end of the DNA for optimal translation in mammalian cells.

Although the cameleons were a brilliant invention, and did work inside cells, their fluorescent properties were far from ideal. Differences in the ratio of fluorescence plus or minus Ca²⁺ were at best 2 to 3. A further innovation therefore was to engineer 'circular permutated' (cp) GFPs, based on the original GFP, CFP, YFP and the cameleons. The term 'circular' is rather confusing, as it does not mean that the protein is actually in a circle. Rather, a circularised protein is one where the order of the amino acids in a protein sequence has been changed

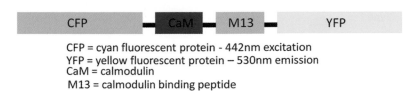

CFP = cyan fluorescent protein - 442nm excitation
YFP = yellow fluorescent protein – 530nm emission
CaM = calmodulin
M13 = calmodulin binding peptide

Figure 4.10 An example of a cameleon. Ca^{2+} binds to calmodulin (CaM), which then binds to the calmodulin-binding site in MLCK (M13). The altered conformation of the whole chimera alters the efficiency of energy transfer between the CFP and the YFP. This is measured by exciting CFP at 422 nm, and measuring the emission at 460 nm (blue) and 530 nm (yellow-green) – the normal emission wavelengths of CFP and YFP, respectively. The ratio of light emission yellow-green/blue is related to the free Ca^{2+}. *Source*: From Miyawaki *et al.* (1997). Reproduced with permission from Nature.

around. For example, a protein designated with three domains as ABC is circularised as CBA. Thus, by making circularised GFPs, which retain the barrel necessary for the fluor to form and be highly fluorescent, it has been possible to make inserted peptides have a greater influence on the fluorophore held within the β barrel solvent cage. This has led to a new family of Ca^{2+} indicators called pericams. Randomised, circular permutation generated three new types of pericam – 'flash', 'inverse' and 'ratiometric', the latter being usable for ratiometric fluorescence monitoring of free Ca^{2+} inside live cells. The pericams were able to monitor and image free Ca^{2+} in the cytosol, nucleus and mitochondria, the latter being difficult with cameleons.

By genetically engineering GFP or other bioluminescent proteins, such as firefly luciferase, on to the N-terminus of aequorin, a family of rainbow proteins were generated (Figure 4.11). These change the emission intensity and/or colour when they react with a substance of interest:

- Firefly luciferase–aequorin chimera enables ATP to be measured by yellow light and Ca^{2+} by blue light.
- Engineering the C-terminus of firefly luciferase generated indicators for kinases and proteases.
- Chimeras of GFP with a bioluminescent protein allowed chemi- or bio-luminescence resonance energy transfer (CRET/BRET) to generate ratiometric indicators for cyclic AMP, other substances, and enzymes such as caspases.

The efficiency of energy transfer can be virtually 100%, with a linker of at least 12 amino acids between GFP and aequorin (Figure 4.12). Rainbow proteins have evolved into Canary proteins, able to detect, identify and measure environmental toxins, including agents of terrorism. To use ratiometric fluorescent or bioluminescent indicators a beam splitter, or spinning disk, is required, so that either the incident or emitted light can be split into two or more wavelengths.

Fluors can emit thousands or even millions of photons per second. But the signal-to-noise ratio depends on autofluorescence. Bioluminescent indicators emit one photon per molecule that reacts, so the signal-to-noise ratio is low, but very sensitive photon detectors and imaging cameras are available. Genetically engineered GFP can bleach within a few minutes under continuous illumination. Interestingly, in the animal, such as *Obelia* hydroids, GFP is much less susceptible to photobleaching. Substances for reducing photobleaching are available, but have not had huge success in fluorescence microscopy. The power of the fluorescent Ca^{2+} indictors has been enhanced considerably by the development of highly sophisticated microscopical imaging systems, such as confocal and multiphoton. Fluorescent indicators are also used in conjunction with fluorescent indicators for pH, Mg^{2+} and other ions, and cyclic AMP.

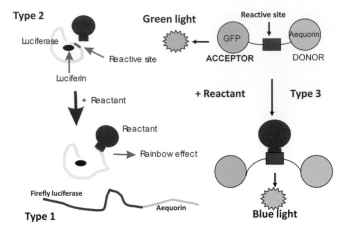

Figure 4.11 Rainbow proteins. Three types of rainbow protein are shown based on the three ways bioluminescent animals produce a rainbow of colours from 400 to 700 nm. Type 1 involves engineering two bioluminescent proteins together. For example, the firefly luciferase–aequorin rainbow protein measures ATP by yellow light and Ca^{2+} by blue light. Type 2 involves engineering a covalent modification site close to the solvent cage where the chemiluminescent reaction occurs (e.g. a protein kinase or protease site). Type 3 uses energy transfer, such as aequorin to GFP or firefly luciferase to biotin–rhodamine complex with a covalent modification site in between, such as caspase, thrombin or enterokinase. Reproduced by permission of Welston Court Science Centre.

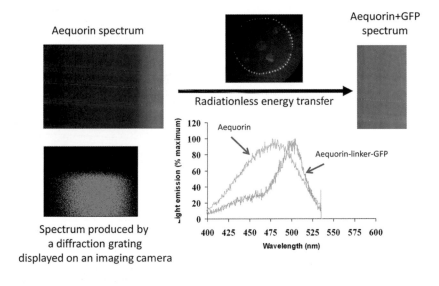

Figure 4.12 Emission spectrum of GFP-linker-aequorin versus aequorin. EGFP was engineered on to the N-terminus of aequorin with a 12-amino-acid linker. The protein was expressed in *E. coli* or eukaryotic cells, extracted and activated with 2 μM coelenterazine. The photoprotein in 200 mM Tris, 0.5 mM EDTA, pH 7.4 was placed in a cuvette in a monochromator housing and triggered by addition of 25 mM $CaCl_2$ (final). The light emission spectrum was spread on to the photocathode of a Photek intensified CCD (ICCD) imaging camera, which was then converted into a trace. The system was calibrated using different coloured light sticks and interference filters with known transmission maxima. Fluorescent photo over the arrow = GFP in an *Obelia* jelly fish. Reproduced by permission of Welston Court Science Centre.

4.3.7 Other Free Ca²⁺ Indicators for Whole Organs and Organisms

Indicators have been developed for intracellular Ca^{2+} in intact organs and organisms, based on NMR, such as fluorine labelled as 5,5'-difluoroBAPTA, and magnetic resonance imaging of Mn^{2+} as a Ca^{2+} analogue, or positron emission tomography (PET). NMR and PET Ca^{2+} probes are difficult to detect, and have yet to find wide use. Both have potential clinical application.

4.4 Detecting and Imaging Photons

There are two detectors for quantifying photons when using Ca^{2+} indicators:

- A photomultiplier tube (PMT).
- A CCD imaging camera.

For bioluminescence detection, it is essential to have a PMT with a very low dark current, with 11–13 dynodes, giving an amplification factor of at least 10^6. Usually the photocathode, the initial photon sensor, is bialkali, and is thus blue sensitive. PMTs with a bialkali photocathode can require some five times as many photons from yellow-emitting firefly luciferase than blue-emitting aequorin to produce the same photon count. Digital processing, as a photon counter, is the most sensitive, using a discriminator to reduce noise, a good chemiluminometer having a background count of less than 10 counts s^{-1}, which can be reduced by cooling. Several proteins, including albumin and insulin, can catalyse coelenterazine chemiluminescence. This can generate a chemical background light emission, reducing the signal-to-noise ratio and thus indicator sensitivity. This is particularly a problem when using *Renilla* or *Gaussia* luciferase as, for example, a gene reporter.

Single-cell analysis using a fluorescent indicator requires either a microscope or a FACS. When using a bioluminescent indicator, single cells can be seen in the microscope, but much better photon capture can be obtained using fibre optics directly optically coupled to a cover slip containing the cells (Figure 4.13a). For fluorescence, an exciting light source is required. Typically, this is a mercury arc, or sometimes xenon, lamp. Higher intensities can be obtained using lasers, though these may have a restricted wavelength range. LEDs are now available. Optimum light capture is essential, being determined by the numerical aperture (NA) of the lens. Three interference filters (Table 4.3) are required for fluorescence:

- For the incident light to be at the optimum wavelength to excite the fluor.
- For the fluorescent light to be measured at the optimum wavelength for emission by the fluor.
- A dichroic filter that allows light of the incident light to pass in one direction to the sample, but does not allow it to pass back to the light detector.

Confocal microscopes enable a live cell to be optically sectioned, each section typically being about a 1 μm. If the light emission per pixel is small, then it may take a second or more to obtain enough photons to obtain a complete image. The software then reconstructs a three-dimensional image of the cell (Figure 4.8). The ER labelled with GFP can be seen moving in a live cell. Lasers may be used as the exciting light source, but have limited wavelengths, so the excitation filter wavelength is chosen as a compromise of the laser wavelength and the absorbance peak of the fluor. The bandwidth of the filters is chosen to balance spectral selection with maximum light passing through. Ca^{2+} noise in a heart cell, and some other signalling events in excitable cells, only takes microseconds. This has led to the development of resonance scanners that can now run as fast as 800 Hz.

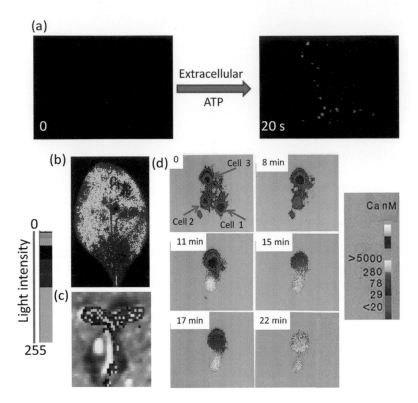

Figure 4.13 Image of free Ca²⁺ signals in live cells. (a) HeLa cells expressing a firefly luciferase–aequorin chimera stimulated by 10 μM extracellular ATP. The light was captured by a bundle of 6 μm optical fibres optically attached to the cells growing on a cover slip. The pseudo-coloured spots represent individual cells. (b) A transgenic tobacco leaf expressing aequorin, showing a Ca²⁺ signal stimulated by cold shock. The pseudo-colour images are over a scale of 0–255 photons per pixel. (c) A transgenic tobacco seedling expressing aequorin, showing a Ca²⁺ signal in the roots and cotyledons. (d) Individual oligodendrocytes loaded with fura-2 and attacked by the membrane attack complex of complement. Cell 1 lyses within 8 min, losing its fluor; cell 2 shows a sustained increase in cytosolic free Ca²⁺ over 22 min; cell 3 shows Ca²⁺ oscillations which start at 15 min. Source: Frith, Hallett and Campbell unpublished (1989). Reproduced by permission of Welston Court Science Centre.

Spatial resolution depends on the number of pixels on the chip. Most CCD cameras on microscopes enable objects 1 μm apart to be resolved. But with specialised microscopes it is now possible to get down to tens of nanometres, beating the apparent limitation of the wavelength of light. CCD cameras also have a pixel depth (i.e. each pixel can accumulate several photons). The speed of the camera is also important. Typical microscope CCD cameras will run at tens of frames s⁻¹. But cameras are now available that will run at hundreds, even a thousand, frames s⁻¹. But a camera with a million pixels being read out at 50 Hz can generate a file of 1 Gbyte within less than 1 min. This makes continuous long-term capture virtually impossible – only solved by using time lapse. Photon counting of bioluminescence reduces the video file size by at least a factor of 100.

When a monochrome camera or PMT is used, the computer-generated image is converted into a colour, to make changes easier to see with the naked eye (see Figure 4.13). This produces a pseudo- or false-colour image, usually using an 8-bit scale of intensities. The intensity range

Table 4.3 Typical interference filter combinations for fluorescent indicators.

Fluor	Excitation filter (nm)	Dichroic filter (nm)	Emission filter (nm)
Dyes			
Fluorescein	450–490	510	515
Fura-2	340/380	400	510
Indo-1	365	440	485/405
Fluo-3	480	505	535
Fura-red	490/440	515	660
Calcium green	480	510	535
Engineered proteins			
EGFP	450–490	510	515
ECFP	395–440	460	470
EYFP	500	515	535
DsRed	545	570	620
Cameleon-2	440	455	535/485
Pericam	415/485	505	535

Bandpass of filters are typically 10 or 20 nm, sometimes 30–50 nm to maximise light transmission. Ratiometric indicators, such as fura-2, indo-1, cameleons and pericams, require a beam splitter or a spinning filter wheel for the excitation or emission to produce or detect at least two wavelengths simultaneously (indicated by the filters separated by a slash '/'). Fura-2 and indo-1 are excited by UV light and require a microscope with quartz lenses. Fluorescein, EGFP, ECFP, cameleons and pericams are excited by blue light, and EYFP by blue-green light. The emission of indo-1 and ECFP is blue, whereas the emission of EGFP, YFP and pericam is green, with the ratiometric cameleon being blue and green. Fura-red and DsRed emit red light.

in each pixel is converted into 0 to 255 (2^8). Then, a look-up table (LUT) converts each number into a colour based on mixing red (R), green (G) and blue (B). Black is R = 0, G = 0, B = 0, white is R = 255, G = 255, B = 255. If GFP is used then the whole scale will be green, or yellow if YFP is used. The software can tell you if two molecules occur in the same place, by adding the colours together, forming a new colour (e.g. red plus yellow = orange). CCD cameras can be cooled for maximum sensitivity or, in a confocal microscope, a scanning pinhole focussed on to a cooled PMT. The software then reconstructs an image. Light is always precious, so it is important to get as much light into the detector as possible. This means lens selection is important. A high NA is required, usually 0.6 to 1.2. For ×100, the lens will be immersed in oil or water. Refraction index affects the NA. The quality of a microscope image is dependent on three parameters:

- Numerical aperture (NA) of the objective lens.
- Total magnification.
- Resolution.

Typical magnification of an objective lens is ×40, ×60 or ×100 for viewing single cells that are 20 μm across. The ×100 will be immersed in oil, or can be a water immersion lens. If the eyepiece has a magnification of ×20, the total magnification will be 2000. If a high NA lens of

1.4 is used, and the emission of GFP is in the green at 510 nm, the resolution will be about 0.18 μm. The larger the cone of light, the more light is captured and the brighter the image. Thus the wider the lens and the closer the focus, the bigger is the NA. The NA, and thus resolution, can be increased by increasing the refractive index. Another important factor is depth of focus, which depends on:

- NA of the lens.
- Refractive index.
- Wavelength of the light.
- Total magnification.

The sensitivity of a light detector depends critically on the noise it generates. In a CCD, this arises from the electronics as the pixels are read out. The faster this is done, the more noise. Noise can be reduced considerably by cooling, particularly by liquid nitrogen. Intensified CCDs (ICCDs) are best for imaging fast changes in photon emission over seconds (e.g. from genetically engineered aequorin). Focussed light is passed through a glass window to a photocathode, which generates electrons. The electrons then accelerate to a stack of microchannel plates (MCPs), which are 6-μm glass fibres, with the inside etched out and sprayed with an electron-amplifying material. These amplify the electrons by 10^6–10^7, keeping the spatial resolution of the original image. The pulse of electrons from each fibre of the MCPs then hits a phosphor screen, which generates an image. This is then read by a CCD in very close proximity. A good signal-to-noise ratio of an ICCD camera depends on a very low dark electron emission from the photocathode, which can be reduced by cooling to between 0 and −40 °C. In contrast, the electronic noise in a CCD camera comes from the read out and amplification of the electrons from the chip. A good ICCD camera has a noise of just 10 photons s^{-1} over the whole 250 000 pixel array, so each pixel can detect a single photon. ICCD cameras are monochrome. An alternative to an ICCD camera is the image photon detector (IPD). This has been used successfully to image aequorin light emission in live zebrafish eggs and embryos. Instead of using a microscope, a fibre optic can be used, in order to increase the capture of photons when using a bioluminescent indicator. Fibre optics can also be used to measure flash bioluminescent spectra in real-time (Figure 4.12).

With ICCD camera systems only a small number of pixels receive photons during each frame, e.g. in 50 ms. So, it is not necessary to record the whole pixel array. Instead, the *xy* coordinates of each positive pixel are recorded, the pixel depth being 0 or 1. The file is saved, and when required the software reconstructs a visible image. This reduces file size by several orders of magnitude. It is possible to record continuously for at least 24 h and only use a few hundred megabytes of disk space. Such cameras can cover 10^6 orders of magnitude of light intensity from bioluminescence. This is essential when using aequorin, and when imaging live bioluminescent animals, which emit visible light so bright that the light will saturate the camera if at maximum sensitivity.

It is always advisable to use a dark room for fluorescence and bioluminescence, with a small lamp. Because of dark adaptation, the human eye can be a hundred or thousand times more sensitive than in a normal room. Furthermore, there will a reduction in phosphorescence from glass and plastic exposed to ceiling fluorescent lights, a reduction in light leaks, and a reduction in the chance of exposing a sensitive photocathode to high light intensity. There are only two ways to damage a PMT or sensitive ICCD: drop them or expose them to bright light when the voltage is on. If the latter occurs, the dark noise will never return to the original low level.

4.5   Measurement of Total Cell Ca²⁺

Measurement of total Ca^{2+} in solutions and tissues can be necessary for three reasons:

- To know the Ca^{2+} contamination in solids and fluids used in cell experiments.
- To measure what happens to total cell and tissue Ca^{2+} when cells have been exposed to primary stimuli, secondary regulators or pathogens.
- To estimate what has happened to intracellular stores, such as the ER/SR or mitochondria, after tissues have been exposed to primary stimuli or secondary regulators.

The Ca^{2+} content is 0.02 µmol kg^{-1} cell water for erythrocytes, 4mmoles kg^{-1} in muscle and 10 mmol kg^{-1} cell water in cartilage. The Ca^{2+} content of necrotic tissue, such as the heart after an infarction or a liver after exposure to a toxin, is much higher. When the heart is exposed to β-adrenergic stimuli, total cell Ca^{2+} increases as additional Ca^{2+} is pumped into the SR.

The most widely used method for total Ca^{2+} measurement involves excitation of Ca^{2+} in a flame, where it emits brick red light. This is flame emission or atomic absorption spectrophotometry, vapourised calcium atoms being excited either by heat or absorption of light respectively. The wavelength of the emitted light is specific for the element, and the intensity directly proportional to the number of atoms excited in the flame, which, in turn, depends on the concentration of atoms in the sample. Flame emission photometry provides a highly sensitive method for measuring sodium and potassium, but is much less sensitive for calcium and magnesium, by a factor as much as 100. Atomic absorption spectrophotometry is some 100 times more sensitive than flame emission for measuring calcium. The method is highly specific for the element, but anions, such as phosphate, reduce the atomisation, and thus reduce sensitivity of detection for calcium. Interference from these can be eliminated by adding La^{3+}. Atomic absorption can measure Ca^{2+} concentrations down to 10–100 nM. For a 2-ml sample, this is equivalent to a detection limit of 0.2 ng atoms for calcium – that found in about 100 µg of tissue. Micromethods can reduce the detection limit even further. The calcium is extracted by digestion of dried tissue in concentrated nitric acid. Total Ca^{2+} can also be measured in a single cell, a frozen or fixed section, using electron X-ray microprobe microscopy. This is a powerful technique, but is not very sensitive for Ca^{2+}, often requiring the equivalent of hundreds of micromolar or millimolar calcium at the location being studied. The Ca^{2+} can be concentrated by precipitation using oxalate or pyroantimonate.

4.6   Calcium Buffers

To study the effects of Ca^{2+} on cells, isolated organelles, or purified proteins, and to be able to calibrate Ca^{2+} indicators, it is essential to be able to use solutions in which the free Ca^{2+} is known precisely. At concentrations above 0.1 mM, this can be done simply be adding a standard Ca^{2+} solution. But, at concentrations in the cytosolic range (i.e. 10 nM to 10 µM), it is essential to use a Ca^{2+} buffer. A good Ca^{2+} buffer should satisfy seven criteria:

1. The free Ca^{2+} produced should be in the cytosolic range of 10 nM to 10 µM.
2. The ligand should bind Ca^{2+} specifically, in the presence of other cations such as K^+, Na^+ and Mg^{2+} and anions such as Cl^-, which exist at millimolar concentrations inside cells. Ca^{2+} binding should not be interfered with significantly by these other ions.
3. The free Ca^{2+} produced by the buffer should not be altered significantly by pH.
4. The ligand should be non-toxic.

5. The ligand should be colourless, unless it is being used as a Ca^{2+} indicator.
6. The rates of association and dissociation of Ca^{2+} should be fast enough to respond to the rate of change of free Ca^{2+} inside cells (i.e. milliseconds to minutes).
7. It should be possible to get the ligand and buffer into the cytosol of living cells.

The two most commonly used Ca^{2+} buffers are EGTA and BAPTA. Both bind Ca^{2+} with high affinity and, although they do bind some Mg^{2+}, this can easily be taken into account when calculating the free Ca^{2+}. A problem with EGTA is that it is very sensitive to pH around the physiological range, because the pKs of the last two hydrogens are 8 and 10. So at physiological pH EGTA is mostly in the LH_2^{2-} form. BAPTA, on the other hand, has much lower pKs for the last two hydrogens and is more than 90% L^{4-} at pH 7. Another artefact is the possible binding of the chelator to proteins inside cells, which can change the affinity constants and spectra. This is a particular problem with some Ca^{2+} indicators such as fura-2. When making up Ca^{2+} buffers use solid $CaCO_3$, or purchase a calibrated Ca^{2+} solution. Ca^{2+} solutions can be calibrated by titration with a dye such as murexide or pH. A Ca^{2+} chelator is best as a buffer when pK_d^{Ca} = pCa (see Table 4.4), BAPTA being best at cytosolic free Ca^{2+} concentrations. The buffering capacity can be calculated as follows:

$$Ca^{2+} + L^{4-} = CaL^{2-} \tag{4.20}$$

$$pCa = -\log_{10}(\text{free } Ca^{2+} \text{concentration}) = pK_d^{Ca} + \log_{10}(L^{4-}/CaL^{2-}) \tag{4.21}$$

Table 4.4 Equilibrium association ($\log_{10}K_a$) constants of some Ca^{2+} buffers.

Ligand	Citrate	EDTA	EGTA	BAPTA
Tetrabasic acids				
$H^+ + L^{4-}$		10.26	9.46	6.36
$H^+ + LH^{3-}$		6.16	8.85	5.47
$H^+ + LH_2^{3-}$		2.67	2.67	<3–4
$H^+ + LH_3^-$		1.99	2.00	<3–4
$Ca^{2+} + LH^{3-}$		3.51	5.33	
$Ca^{2+} + L^{4-}$		10.70	11.0	
$Mg^{2+} + LH^{3-}$		2.28	3.39	
$Mg^{2+} + L^{4-}$		8.69	5.21	
Tribasic acids				
$H^+ + L^{3-}$	5.49			
$H^+ + LH^{2-}$	4.39			
$H^+ + LH_2^{2-}$	3.08			
$Ca^{2+} + L^{3-}$	3.22			
$Mg^{2+} + L^3$	3.20			

Data represent the means of $\log_{10}$ of the association constant (i.e. $\log_{10}K_a$) M^{-1}. pK_d = $-\log_{10}K_a$. The binding constants for last two H^+ for EGTA and EDTA mean that they are very sensitive to pH around pH 7, but those for BAPTA are such that more than 90% BAPTA is ionised at pH 7. Ca^{2+} binding is not sensitive to changes in pH.

Thus, consider a ligand (L) which at pH 7.4 has a $pK_d = 7$ (i.e. $K_d = 0.1\ \mu M$). If the total ligand concentration (L_{total}) = 10 mM, then for a pCa of 6, 90% of the ligand is saturated by Ca^{2+}:

$$\left(L_{total}\right) = \left(CaL^{2-}\right) + \left(L^{4-}\right) = 10^{-2}\,M \tag{4.22}$$

$$\left(Ca_{total}\right) = \left(CaL^{2-}\right) + \left(Ca^{2+}\right) \tag{4.23}$$

Therefore, from:

$$pCa = pK_d{}^{Ca} + \log_{10}\left(L^{4-}/CaL^{2-}\right) \tag{4.24}$$

$$6 = 7 + \log_{10}\left(L^{4-}/CaL^{2-}\right) \tag{4.25}$$

Therefore:

$$\left(CaL^{2-}\right) = 10\left(L^{4-}\right) = \left(L_{total}\right) - \left(L^{4-}\right) = 0.01 - \left(L^{4-}\right) \tag{4.26}$$

So:

$$\left(CaL^{2-}\right) = 0.00909M, \text{ i.e.}\left(Ca_{total}\right) \sim 9.09\ mM \tag{4.27}$$

If 100 μM Ca^{2+} is added to the solution, from a contamination, for example:

$$New\left(Ca_{total}\right) = 0.00919\ M \approx new\left(CaL^{2-}\right) \tag{4.28}$$

Then the new $(L^1) = 0.00081$ M.

From Eq. (4.27), we can see that the new pCa = $7 + \log_{10}(0.00081/0.00919) = 5.95$ (i.e. 1.1 μM). Thus the Ca^{2+} buffer has succeeded in reducing the change in pCa to 0.1 of a pCa unit (i.e. a 10% change in absolute free Ca^{2+}).

The on and off rates for Ca^{2+} must be fast enough to cope with the rates of Ca^{2+} release under physiological conditions, milliseconds in the case of some nerves and muscles, a clear advantage of BAPTA over EGTA.

4.7 Measurement of Ca²⁺ Fluxes

Measurement of Ca^{2+} influx and efflux provides clear evidence for an active role for Ca^{2+}. Ca^{2+} fluxes also enable channels, pumps and exchangers to be identified and quantified, and the role of intracellular Ca^{2+} pools to be defined. There are five main Ca^{2+} pools within a cell:

1. Ca^{2+} bound to the outer surface (the glycocalyx = 3–5%), either as a membrane or a cell wall.
2. Ca^{2+} bound to the inner surface of the plasma membrane.
3. Ca^{2+} in the cytosol: free, and bound to small molecules, proteins, nucleic acids and phospholipids.
4. Ca^{2+} entrapped within organelles: bound and free.
5. Crystalline deposits of Ca^{2+}, as phosphate, carbonate or oxalate, depending on cell type and conditions.

There are three methods used to measure Ca^{2+} flux in intact cells and in isolated organelles:

1. Radioactive Ca^{2+} movement.
2. Absolute Ca^{2+} changes, with or without a precipitating anion such as oxalate.
3. Ca^{2+} currents.

But, the most common is to measure radioactive Ca^{2+} fluxes. There are two useful radioactive isotopes for this: ^{45}Ca and ^{47}Ca:

$$^{45}Ca \rightarrow {}^{45}Sc + \beta^- \left(t_{1/2} = 165 \text{ days}\right)$$
(4.29)

$$^{47}Ca \rightarrow {}^{47}Sc + \beta^- + \gamma\left(t_{1/2} = 4.7 \text{ days}\right)$$
(4.30)

^{47}Ca being a γ emitter can be measured without the need for scintillation fluid. But ^{45}Ca is the most common Ca^{2+} isotope used, because of its longer half-life. Typically radioactive Ca^{2+} fluxes reveal three distinct compartments. But, the major problem in interpreting radioactive Ca^{2+} flux data is the fact that the specific activity (radioactive Ca^{2+}/total Ca^{2+}) of the Ca^{2+} in each compartment is not known. Worse still, it is virtually impossible to have the same specific activity in each compartment. Yet radioactive Ca^{2+} fluxes have provided useful information about the mechanisms of Ca^{2+} influx and efflux used in whole cells, vesicles or liposomes containing purified putative Ca^{2+} transporters.

4.8 How to Study Ca^{2+} Channels

The electrical properties of Ca^{2+} play a crucial part in Ca^{2+} as an intracellular signal. This involves Ca^{2+} currents through Ca^{2+} channels in the plasma membrane and ER, as well as Ca^{2+} activation of K^+ and Cl^- channels. Microelectrodes that measure free Ca^{2+} have been used both inside and outside of cells. But electrodes, in solutions containing proteins, exhibit hysteresis (i.e. the calibration curve for Ca^{2+} is not the same as that obtained in non-protein-containing solutions).

Patch clamping has taken over for studying Ca^{2+} channels, using electrodes filled either with 100 mM $CaCl_2$ or $BaCl_2$. Ba^{2+} currents can be three or more times larger than Ca^{2+}, and are therefore easier to measure. Ba^{2+} also prevents artefacts from K^+ channels. The currents are tiny – nano- to pico-amperes (10^{-9} to 10^{-12} A), and require special electronics to prevent electrical noise hiding the Ca^{2+} current. A current represented as downwards is by convention a cation moving into the cell or an anion moving out. Conversely, an upwards trace represents a cation moving out of the cell or an anion moving in. This is whole patch clamp. It is then possible to pull off the piece of membrane attached to the electrode, giving an inside-out patch, enabling substances to be added to the medium, which influence the ion channel being studied. Using a reference electrode, the effect of voltage, an agonist or antagonist, on the opening and closing of the Ca^{2+} channel can be measured, as well as pressure by applying suction to the electrode, if it is mechanosensitive. A key property of voltage-gated Ca^{2+} channels is the voltage at which they open, and the frequency of opening. The conductance in siemens is then measured by measuring a voltage–current relationship over a range that obeys Ohm's law (conductance, $G = I/V$), and whether the channel is inward or outward rectifying is assessed. For example, a channel that is inward rectifying allows current to flow preferentially inwards.

The ionic selectivity is assessed by altering the ionic composition of the medium, or in the electrode. Some Ca^{2+} channels, such as $Ca_v1.2$ in heart muscle, are highly selective for Ca^{2+} over other cations, such as K^+ and Na^+. However, others, such as TRP (transient receptor channel) channels, can allow several different cations to pass through them, as well as anions such as Cl^-. The driving force for a current carried by an ion is the electrochemical potential, made up of the electrical force driving the ion towards its equilibrium potential, and the concentration gradient across the membrane. A range of natural and synthetic compounds are available to investigate the type of Ca^{2+} channel involved in a cell event (see Chapter 11). An important application of the patch clamp technique is in the study of the effects of specific mutations on the electrical properties of an ion channel. In a typical patch there may be anything up to ten channels for a particular ion. When Ca^{2+} is in the pipette it is often isotonic with normal saline, with a Ca^{2+} concentration of 110 mM. Thus, the question arises: Are these putative Ca^{2+} currents really physiological? Particular characteristics measured for a Ca^{2+} or any ion channel are:

1. Ionic specificity. Some Ca^{2+} channels are very specific, others allow K^+ and Na^+, and even Cl^-, to move through them.
2. Conductance, where the exact pipette and external ion concentrations must be defined. Typical conductances for Ca^{2+} channels vary from pico- to nano-siemens, but some can be as small as femto-siemens.
3. Agent(s) that open the channels, and at what threshold or concentration – voltage, external or internal substances.
4. Inactivation, and by what mechanism.
5. Whether the channel is inward or outward rectifying, allowing more ions to flow in or out, respectively.
6. Pharmacology, particularly Ca^{2+} channel blockers of established specificity for a particular type of Ca^{2+} channel.
7. Intracellular regulation, e.g. by Ca^{2+}, Ca^{2+}-calmodulin or phosphorylation.

Typically, ion channels are digital, i.e. they are either open or closed, agents such as neurotransmitters or Ca^{2+} antagonists affecting the probability of opening, and thus the frequency. The total current through a membrane is thus directly proportional to the number of channels open at a particular time, which in an excitable cell may be 1000 or more. But many ion channels also exhibit sub-conductance states, where the channel appears to be only partially open. This occurs when the channel conductance depends on a number of portals being open into the main pore.

4.9 Discovering How the Rise in Cytosolic Free Ca²⁺ Occurs and then Returns to Rest

In virtually all Ca^{2+} signalling events, Ca^{2+} will be released from internal stores, and will also move into the cell across the plasma membrane. The relative importance of these can be determined by measuring cytosolic free Ca^{2+}, and free Ca^{2+} within organelles using targeted Ca^{2+} indicators, in live cells exposed to primary stimuli, and any secondary regulators, in the presence or absence of extracellular Ca^{2+}. Imaging the Ca^{2+} signal in individual cells will show how the Ca^{2+} signal is generated. Replacing extracellular NaCl by KCl will depolarise the cell membrane, and thus cause voltage-gated Ca^{2+} channels to open. These can be characterised using pharmacological agents that block specific Ca^{2+} channels (see Chapter 11). In non-excitable cells, a key role for store-operated Ca^{2+} entry (SOCE) can be shown. First, by showing that the bulk of the cytosolic free Ca^{2+} signal depends on the presence of extracellular Ca^{2+}. Secondly, showing release of Ca^{2+} from the ER, by the SERCA pump inhibitors thapsigargin or cyclopiazonic acid, causes a large Ca^{2+} influx. The role of IP_3 or ryanodine receptors in generating the Ca^{2+} signal can be assessed by using inhibitors: Li^+, heparin or pharmacological agents to block

IP_3 receptors, or dantrolene sodium to block ryanodine receptors. The presence of a particular receptor type can be confirmed by the polymerase chain reaction (PCR), and its location confirmed microscopically by using specific antibodies. Ruthenium Red can be used to inhibit Ca^{2+} fluxes in mitochondria.

4.10   Discovering the Intracellular Ca^{2+} Target and How it Works

Intracellular Ca^{2+} targets in many types of cell events in eukaryotic cells include:

- Troponin C in skeletal and heart muscle.
- Myosin light chain kinase (MLCK) in smooth muscle via calmodulin.
- The SNARE complex with synaptotagmin in vesicular secretion.
- Calmodulin in many types of cell activation.
- EF-hand or C2 type Ca^{2+}-binding sites in channels and proteins involved in the event.

The role of these can be confirmed by using inhibitors, by manipulation of gene expression by overexpressing a component, or knocking out expression using siRNA. Trifluoperazine (TFP) inhibits Ca^{2+}-calmodulin-activated events, such as cyclic AMP phosphodiesterase, by binding to the hydrophobic region between the two dumbbell EF dual Ca^{2+}-binding domains, thereby preventing the calmodulin binding to its target protein. Inhibitors are also available for particular Ca^{2+}-activated kinases, phosphatases and proteases, such as calmodulin-activated kinases, calcineurin and calpain (see Chapter 11). Ca^{2+} target proteins can be knocked out in live mice, as well as by using siRNA in cell systems. But, compensating mechanisms can remove the specificity of this approach.

Discovering how Ca^{2+} works requires identification of the Ca^{2+}-binding sites in the target, and how these affect the protein when Ca^{2+} binds. This requires a combination of biochemical, molecular biological and structural techniques. The affinity of Ca^{2+} binding can be determined by using ^{45}Ca, and the Ca^{2+} site identified from its amino acid sequence, if the Ca^{2+} site is a conventional EF-hand or C2 site. Three-dimensional structures from X-ray crystallography are obtained from crystals with and without Ca^{2+}. Protein–protein interactions can be discovered, and characterised, using gel electrophoresis or yeast two-hybrid technology. Interactions in live cells can be investigated by using FRET or FLIM (fluorescence lifetime imaging microscopy), using engineering fluorescent markers such as different GFPs. The role of a particular amino acid can be determined by observing the effect of mutation on the properties of the protein. For example, a change from an Asp or Glu to Ala in a predicted Ca^{2+}-binding site will result either in loss in Ca^{2+} sensitivity or reduction in Ca^{2+} affinity. Protein location and movement can be imaged in live cells by engineering the protein with GFP labels.

4.11   Other Ions

K^+, Na^+, Mg^{2+}, H^+ and Cl^-, as well as organic ions such as $ATPMg^{2-}$ and glutamate, can regulate physiological processes inside cells that involve intracellular Ca^{2+}. Most inorganic monovalent ions are in the free form, the cytosolic free K^+ concentration in most bacterial cells being around 150 mM, and the Na^+ concentration 5–20 mM. Mg^{2+} binds to nucleotides, e.g. ATP to form $ATPMg^{2-}$ – the form that is used by ion pumps. Mg^{2+} also binds to, and activates, many enzymes, and can compete for Ca^{2+}-binding sites. Fluorescent Mg^{2+} indicators, such as mag-fura, are now available (Figure 4.2b), but these also bind Ca^{2+} at high concentrations. Thus, with a typical total cell Mg^{2+} of 10–15 mmol kg^{-1} cell water (i.e. 10–15 mM), the cytosolic free Mg^{2+}

is expected to be around 1–2 mM, the rest being bound to nucleotides (concentration around 5–10 mM), nucleic acids, phosphate and some other organic ligands. Surprisingly, relatively little Mg^{2+} appears to be bound to the cell membrane, perhaps less than 0.1%. The concentration of Mg^{2+} bathing cells ranges from 0.5–1 mM in mammalian serum to 24 mM in sea water. Thus, in mammalian cells there is very little Mg^{2+} gradient across the plasma membrane, and even in marine invertebrates there is only a four- to five-fold Mg^{2+} gradient, in contrast to the 10 000- to 100 000-fold gradient of Ca^{2+}. Furthermore, although there are transporters that get Mg^{2+} into cells, and hormones that may affect the concentration of Mg^{2+} inside cells, there is little evidence for a widespread occurrence of voltage-gated or metabolite-gated Mg^{2+} ion channels. However, Mg^{2+} has been shown to go through some ion channels, e.g. a TRP Mg^{2+} channel opens during the cell division cycle. Mg^{2+} is pulled into the cell by the negative membrane potential, so that it can double for the cell to divide. But, neither Mg^{2+} nor monovalent cations are suitable to be an intracellular switch, in the same way as Ca^{2+}.

Box 4.2   The pathway to discovering the role of intracellular Ca²⁺ in a cell event.

1. Define the properties of the cell event:
   a. Digital versus analogue
   b. Time scale
   c. Primary stimuli and secondary regulators
   d. Drugs that modify it
   e. Proteins involved.
2. Establish the effects of manipulating intracellular Ca²⁺:
   a. Removing extracellular Ca²⁺
   b. Effect of Ca²⁺ ionophores
   c. Effect of caged Ca²⁺ and other caged compounds, e.g. IP₃.
3. Measurement of cytosolic free Ca²⁺:
   a. The effect of the primary stimuli on cytosolic free Ca²⁺, and show that any rise occurs before the cell event is detectable
   b. Show that stopping a rise in cytosolic free Ca²⁺ stops the cell event
   c. Establish whether any secondary regulators work via a change in cytosolic free Ca²⁺.
4. Identify what the major source of Ca²⁺ is – external/ER/another organelle:
   a. Use caged compounds and inhibitors to establish the role of the ER, and IP₃ versus ryanodine receptors
   b. Use Ca²⁺ channel blockers on membrane Ca²⁺ channels.
5. Identify the Ca²⁺ target:
   a. Inhibitors of specific Ca²⁺ binding proteins
   b. siRNA and other gene manipulation techniques to establish the role of specific genes.
6. Identify how Ca²⁺ is lost from the cytosol after the cell stimulus is removed.

4.12   Summary

To elucidate whether Ca^{2+} is an intracellular signal, and what the pathway is that makes it works inside a cell, requires a multidisciplinary approach, involving biochemistry, cell and molecular biology, electrophysiology, genetics and structural biology. This has been highly successful in establishing intracellular Ca^{2+} as a universal cellular switch in a wide range of phenomena in animals and plants, and unicellular eukaryotes. But, the strategy used in Eukaryota has not been so successful in Bacteria or Archaea. In eukaryotic cells intracellular Ca^{2+} has a function. It triggers nerve terminals to fire, muscles to contract, and many endocrine and exocrine cells to secrete. But in Bacteria and Archaea, it may be more appropriate to say that Ca^{2+} signalling has a 'selective advantage', rather that a 'function'. Thus, knocking-out nerve terminal secretion or a heart beat would be lethal to an animal, but in a bacterium or archaean stopping Ca^{2+} signalling is not necessarily lethal. The experimental approach to test for selective advantage requires a new experimental strategy, where small change by small change leads to big changes in the cell population over time, and when competing with other microbes.

The pathway to discovering the role of Ca^{2+} in a physiological or pathological process involves at least six steps (see also Box 4.2):

1. Effects of manipulating extracellular and intracellular Ca^{2+}.
2. Measuring the effect of stimuli, pathogens or pharmacological agents on free Ca^{2+} in the cytosol and organelles.
3. Discovering how a rise in intracellular Ca^{2+} occurs via plasma membrane channels and release from internal stores.
4. Discovering how the intracellular Ca^{2+} returns to resting levels, via plasma membrane and organelle pumps and transporters.
5. Identifying the intracellular Ca^{2+} target protein and how it works.
6. Discovering how this system can be damaged in disease.

These are required to obtain a complete understanding of how, and why, evolution has chosen the unique properties of Ca^{2+} to trigger a particular cellular event.

Recommended Reading

*A must

Books

Campbell, A.K. (1983) Intracellular Calcium: Its universal role as regulator (monographs in molecular biophysics and biochemistry). Chichester: John Wiley & Sons Ltd. Chapter 2 The investigation of intracellular Ca^{2+} as a regulator. Full references and Tables available from Companion web site www.wiley.com/go/campbell/calcium.

Campbell, A.K. (1988) Chemiluminescence: Principles and applications in biology and medicine. Ellis Horwood series in biomedicine. Chichester/Weinheim: Horwood/Verlag Chemie. Wide ranging reviews on the applications of chemi- and bio-luminescence.

*Campbell, A.K. (2015) Intracellular Calcium. Chichester: John Wiley & Sons Ltd. Chapter 4, How to study intracellular Ca^{2+} as a cell regulator. Fully referenced, references and other information available from Companion web site www.wiley.com/go/campbell/calcium.

Haugland, R.P. (1996) Handbook of fluorescent probes and research chemicals, 6th edn. Eugene, OR: Molecular Probes. Comprehensive review of fluorescent probes by the founder of Molecular Probes, now Invitrogen.

Reviews

*Blinks, J.R. (1989) Use of calcium-regulated photoproteins as intracellular Ca^{2+} indicators. *Methods Enzymol.*, **172**, 164–203. Classic review of how to use and quantify Ca^{2+}-activated photoproteins in live cells.

Campbell, A.K. (June 2003) Rainbow makers. *Chem. Br.*, 30–33. Genetically engineered bioluminescent proteins that change colour. Pdf available from the author.

*Granatiero, V., Patron, M., Tosatto, A., Merli, G. & Rizzuto, R. (2014) The use of aequorin and its variants for Ca²⁺ measurements. *Cold Spring Harbor Protocol*, **2014**, 9–16. The application of mutants of aequorin in cells.

*Miyawaki, A.M.H.N.T. & Sawano, A. (2003) Development of genetically encoded fluorescent indicators for calcium. *Methods Enzymol.*, **360**, 202–225. Review of cameleon and pericam genetically engineered GFPs for Ca²⁺ in live cells.

Scarpa, A.B., Brinley, F.J., Tiffert, T. & Dubyak, G.R. (1978) Metallochromic indicators of ionized calcium. *Ann. NY Acad. Sci.*, **307**, 86–112. Review of absorbing Ca²⁺ indicators such as arzenazo III for live cells.

Sala-Newby, G.B., Badminton M., Evans, W., George, C., Jones, H., Kendall, J., Ribeiro, A. and Campbell, A. (2000) Targeted bioluminescent indicators in living cells. *Methods Enzymol.*, **305**, 478–498. Review of how to target bioluminescent proteins to organelles in live cells by the pioneers.

Suzuki, J., Kanemaru, K. & Iino, M. (2016) Genetically encoded fluorescent indicators for organellar calcium imaging. *Biophys. J.*, **111**, 1119–1131. Useful comparison of the properties and benefits of engineered Ca²⁺ indicators.

*Tsien, R.Y. (2003) Imagining imaging's future. *Nat. Cell Biol.*, Supplement S, SS16–SS21. Short review of live cell imaging by a pioneer.

Research Papers

Badminton, M.N., Kendall, J.M., Sala-Newby, G. & Campbell, A.K. (1995). Nucleoplasmin-targeted aequorin provides evidence for a nuclear calcium barrier. *Exp. Cell Res.*, **216**, 236–243. First targeting of aequorin to the nucleus.

Grynkiewicz, G., Poenie, M. & Tsien, R.Y. (1985) A new generation of Ca²⁺ indicators with greatly improved fluorescence properties. *J. Biol. Chem.*, **260**, 3440–3450. Pioneering paper describing fura-2 and indo-1 for the first time.

Hallett, M.B. & Campbell, A.K. (1982) Measurement of changes in cytoplasmic free Ca²⁺ in fused cell hybrids. *Nature*, **295**, 155–158. First measurement of free Ca²⁺ in a small live cell using obelin and fusion with an erythrocyte ghost.

Kendall, J.M., Badminton, M.N., Sala-Newby, G.B., Campbell, A.K. & Rembold, C.M. (1996) Recombinant apoaequorin acting as a pseudo-luciferase reports micromolar changes in the endoplasmic reticulum free Ca²⁺ of intact cells. *Biochem. J.*, **318**, 383–387. How to use aequorin in luciferase mode.

*Nilius, B., Hess, P., Lansman, J.B. & Tsien, R.W. (1985) A novel type of cardiac calcium channel in ventricular cells. *Nature*, **316**, 443–446. Use of patch clamping to discover new voltage-gated Ca²⁺ channels.

*Rizzuto, R.S., Simpson, A.W.M., Brini, M. & Pozzan, T. (1992) Rapid changes of mitochondrial Ca²⁺ revealed by specifically targeted recombinant aequorin. *Nature*, **358**, 325–327. Pioneering targeting of aequorin to mitochondria in live cells.

Tsien, R.Y. (1981) A non-disruptive technique for loading calcium buffers and indicators into cells. *Nature*, **290**, 527–528. First use of AM esters to load buffers into live cells.

5

How Ca²⁺ is Regulated Inside Cells

Learning Objectives
- How the resting cell maintains a 10 000 fold gradient of Ca²⁺ across its outer membrane.
- What the sources are for a rise in cytosolic free Ca²⁺, when the cell is activated by a primary stimulus.
- How these sources are mobilised to cause the cytosolic free Ca²⁺ rise.
- How secondary regulators alter the Ca²⁺ signal.
- How the Ca²⁺ is removed from the cytosol, so that the cell can return to rest after the event is over.
- The types and roles of Ca²⁺ channels, pumps and exchangers.

5.1 Principles

All cells from the three domains of life – Bacteria, Eukaryota and Archaea – maintain a cytosolic free Ca²⁺ in the sub-micromolar range, even when the extracellular Ca²⁺ is millimolar. The Ca²⁺ gradient across the outer membrane of all cells holds the key to intracellular Ca²⁺ as a universal regulator. Just small absolute movements of Ca²⁺ from outside the cell, or from an internal store, will cause a large fractional rise in cytosolic free Ca²⁺, enabling it to switch on Ca²⁺-dependent proteins and enzymes. The key questions are:

1. How does the resting cell maintain the large gradient of Ca²⁺ across its outer membrane?
2. What are the sources of the rise in cytosolic free Ca²⁺, when the cell is activated by a primary stimulus?
3. How are these sources mobilised to cause the cytosolic free Ca²⁺ rise?
4. How do secondary regulators alter the Ca²⁺ signal?
5. How is this Ca²⁺ removed, so that the cell can return to rest after the event is over?

The resting cell is able to maintain a sub-micromolar cytosolic free Ca²⁺ for two reasons. First, biological membranes are poorly permeable to charged molecules, such as ions. Specific transport mechanisms are required to carry charged substances across.

Secondly, there are pumps and transporters that counteract the leak of Ca²⁺ into the cell. Organelles, such as the endoplasmic reticulum (ER), Golgi and mitochondria, play a crucial role in regulating the cytosolic free Ca²⁺ when a stimulus hits the cell. But, these internal stores cannot maintain a sub-micromolar cytosolic free Ca²⁺ without a removal mechanism in the outer, plasma membrane. Without this removal mechanism, Ca²⁺ would gradually move passively into the cell until it reached its equilibrium potential. With a negative membrane potential inside of up to 90mV, this could reach at least a level of molar free Ca²⁺!

Fundamentals of Intracellular Calcium, First Edition. Anthony K. Campbell.
© 2018 John Wiley & Sons Ltd. Published 2018 by John Wiley & Sons Ltd.
Companion Website: http://www.wiley.com/go/campbell/calcium

Box 5.1 How Ca²⁺ can get into or out of cells.

- Leak around a protein in the bilayer that traverses the membrane.
- A neutral complex with a hydrophobic molecule, such as an ionophore.
- A neutral permease, usually called a symport, where Ca^{2+} is bound or coupled with an anion, such as HPO_4^{2-}.
- A pore formed by a protein complex, such as the membrane attack complex of complement, or the bacterial toxins alfatoxin and streptolysin.
- An ion channel.
- An ion exchanger, such as the H^+/Ca^{2+} exchanger found in mitochondria, or the Na^+/Ca^{2+} exchanger found in many excitable and some non-excitable cells.
- A pump, such as a Ca^{2+}-MgATPase.

Since the calcium ion Ca^{2+} is charged and hydrated, it is highly hydrophilic and cannot cross the hydrophobic, lipid bilayer of cells without help. There are seven ways by which Ca^{2+} can cross biological membranes (Box 5.1). But the main mechanism by which Ca^{2+} moves into cells when they are stimulated is through a Ca^{2+} channel. There are five ways in which Ca^{2+} channels can be opened (Figure 5.1):

1. Voltage – a drop in the membrane potential.
2. Binding to a cell surface receptor, typically coupled to a channel or trimeric G-protein.
3. An intracellular signal, such as cyclic AMP or cyclic GMP.

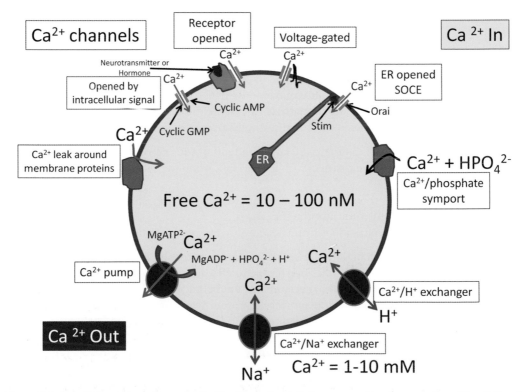

Figure 5.1 How Ca^{2+} gets in and out of cells. The figure shows Ca^{2+} pumps and exchangers for getting Ca^{2+} out of cells through the plasma membrane, and various Ca^{2+} channels opened by voltage, receptors, intracellular messengers and loss of Ca^{2+} from the ER (SOCE) for getting Ca^{2+} into cells. Reproduced by permission of Welston Court Science Centre.

4. Mechanical stretching of the plasma membrane, opening mechanosensitive channels.
5. Release of Ca^{2+} from the ER activating store-operated Ca^{2+} entry (SOCE).

The continuous passive leak of Ca^{2+} in the resting cell is counteracted by pumps and exchangers in the plasma membrane, which use either ATP hydrolysis or the gradient of an ion such as Na^+ to move Ca^{2+} against its electrochemical gradient. These pumps and exchangers are responsible for causing cytosolic Ca^{2+} to return to its sub-micromolar level after the cell event is over. But, the pumps and exchangers of intracellular organelles also play a role, as they supply Ca^{2+} for the cytosolic free Ca^{2+} rise, and can restrict a particular type of Ca^{2+} signal to a wave or oscillation, and at a specific location. There are three types of Ca^{2+} pumps or exchangers in the plasma membrane (Figure 5.2), depending on cell type:

1. Plasma membrane Ca^{2+}-MgATPase (PMCA).
2. $3Na^+/Ca^{2+}$ exchanger (NCX) and $4Na^+/Ca^{2+}$-K^+ exchanger (NCKX).
3. Ca^{2+}/H^+ exchanger.

In addition, after a rise in cytosolic free Ca^{2+}, there are pumps and exchangers in the sarco-endo-plasmic reticulum (SR)/ER, Golgi and mitochondria, which remove some of the Ca^{2+} back into the internal store, or are involved in determining the size, location and type of Ca^{2+} signal:

1. SR/ER Ca^{2+}-MgATPase (SERCA).
2. Mitochondrial Ca^{2+} influx channel.
3. Mitochondrial Na^+ activated Ca^{2+} efflux.
4. Golgi Ca^{2+}-MgATPase (SPCA).

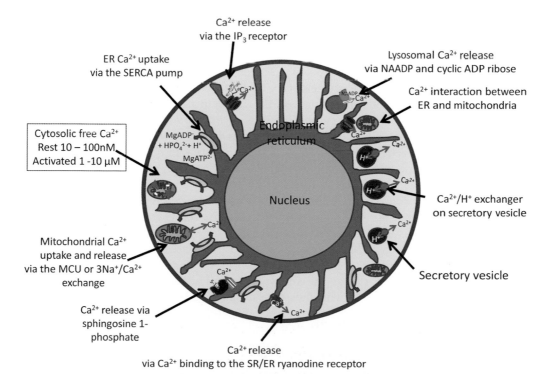

Figure 5.2 How Ca^{2+} is released inside cells. The figure shows the release and uptake of Ca^{2+} by internal stores such as inositol trisphosphate (IP_3) and ryanodine receptors in the ER, sphingosine-activated Ca^{2+} release from the ER, the mitochondrial Ca^{2+} uniporter protein and Na^+-activated Ca^{2+} release from mitochondria, nicotinic acid adenine dinucleotide phosphate (NAADP) and cyclic ADP ribose-activated release of Ca^{2+} from lysosomes. Reproduced by permission of Welston Court Science Centre.

The kinetic properties of the PMCA are well suited to maintaining the low cytosolic free Ca^{2+} in many cells. However, the faster kinetic properties of NCX, with the SERCA pump, make these well suited to restoring the cytosolic free Ca^{2+} to sub-micromolar levels after a heart beat or in a nerve terminal after an action potential. There are also Ca^{2+} transporters on secretory vesicles to get Ca^{2+} into the vesicle, where it plays an important structural role. There are also Ca^{2+} influx and efflux mechanisms allowing Ca^{2+} in and out of other organelles, such as lysosomes. Ca^{2+} will be inevitably trapped inside endosomes after they have been invaginated into the cell, and will contain the same Ca^{2+} influx and efflux mechanisms found in that cell type.

5.2 How Resting Cells Maintain their Ca^{2+} Balance

To maintain the very large electrochemical gradient of Ca^{2+} across the plasma membrane, with a sub-micromolar cytosolic free Ca^{2+}, and mM extracellular free Ca^{2+}, there has therefore to be a mechanism that continuously 'pumps' Ca^{2+} out of the cell. Without it, with a membrane potential of tens of millivolts, negative inside, the cytosolic free Ca^{2+} would eventually rise to hundreds of millimolar or even molar concentrations. Such a rise in free Ca^{2+} cannot be prevented by internal stores, such as the ER or mitochondria. Ca^{2+} has to be pumped out against a large electrochemical gradient, so Ca^{2+} efflux requires energy from either:

1. ATP hydrolysis.
2. An ion gradient, such as Na^+ or H^+.

Cells maintains the MgATP/MgADP + phosphate reaction well away from equilibrium, on the side of MgATP. It is this 'potential' energy that is used by Ca^{2+}-activated MgATPases in the plasma membrane and SR/ER. ATP does not have an energy-rich bond, as is mistakenly described in several textbooks and web sites. Ca^{2+}-activated MgATPase, distinct from the Na^+/K^+ MgATPase, is found in the plasma membrane of most eukaryotic cells, and is reversibly activated by binding Ca^{2+}-calmodulin. In contrast, the Ca^{2+} activated protease calpain irreversibly activates the Ca^{2+} pump by cleaving peptide off the C-terminus, which faces the cytosol. This is important pathologically, when there is a prolonged elevation in cytosolic free Ca^{2+}.

There are four Ca^{2+}-activated MgATPases in human cells – PMCA1–4 (genes *ATPB1–4*). Alternative splicing generates a number of variants, designated by the letters, a, b and so on. Each has slightly different kinetics in terms of Ca^{2+} affinity and turnover number, the balance being selected for according to the needs of a particular cell type – a typical molecular Darwinian selection mechanism. These pumps are P-type MgATPases. P-type ATPases are different from V-ATPases, and the F-type responsible for oxidative ATP synthesis in mitochondria and bacteria. P-type MgATPases have a phosphorylated amino acid as an intermediate in the transport process. A phosphate is transferred from MgATP to a key aspartate residue and Ca^{2+} binds. Part of the protein swivels round allowing Ca^{2+} to move out of the cell and one H^+ is released inside. The phosphate drops off, so that the cycle can begin again. These P-type MgATPases are inhibited by orthovanadate, a phosphate analogue, La^{3+}, which displaces Ca^{2+} from its high-affinity site, and the peptide toxin caloxin A1, which blocks the first extracellular domain. All P-type ATPases alternate between two states, E1 and E2, and a phosphorylation–dephosphorylation cycle. E1-P, phosphorylated on the key Asp residue, has a high affinity for Ca^{2+}, leading to a conformational change to E2-P. This exposes Ca^{2+} to the extracellular side, or luminal in the case of SERCA. This promotes dephosphorylation and the Ca^{2+} drops off, the pump returning to the E1 state. The plasma membrane Ca^{2+} pump exchanges one H^+ for each Ca^{2+}, so it is electrogenic, in contrast to SERCA that is electroneutral, exchanging two H^+ for each Ca^{2+}. The PMCA plasma membrane Ca^{2+} pumps have ten predicted transmembrane domains. The bulk of the protein, including the N- and C-termini, faces the cytosol. The three-dimensional structure is very similar to the SERCA pumps. The K_d of the PMCAs for Ca^{2+} is 10–20 μM, which is decreased by calmodulin to 1 μM. PKA, in the absence of calmodulin, also activates the pump by increasing V_{max} and increasing Ca^{2+} affinity through a decrease in K_m to 1 μM. The 3D structure of a plasma membrane Ca^{2+}-MgATPase calcium pump has been obtained using X-ray crystallography. This showed that there is an essential interaction between membrane phospholipids, and Trp and Arg/Lys residues crucial to the properties the protein.

In contrast to the PMCAs, in excitable cells, such as cardiac myocytes and nerve terminals, the Na^+/Ca^{2+} exchanger (NCX), of which there are three in the human genome, are more important for Ca^{2+} efflux. Another, NCKX, of which there five, in the eye, nerves, smooth muscle, and skin, takes one K^+ with the Ca^{2+} in exchange for four Na^+. NCXs have a lower affinity for Ca^{2+} but are faster than the PMCAs. So they are suited to removing Ca^{2+} quickly from the cytosol after cell activation.

5.3   Electrophysiology of Intracellular Ca²⁺

A crucial feature of Ca^{2+} as an intracellular signal is how it interacts with the electrical properties of the cell. In nerve axons, action potentials depend on Na^+ and K^+ currents. But, in many invertebrate muscles, e.g. crabs and barnacles, action potentials are wholly or partially dependent on Ca^{2+} as a current carrier, rather than Na^+. The driving force for an ion moving through a channel is the electrochemical potential, which is made up of two forces: the membrane potential and the concentration gradient (Eq. 5.1). These come into play when Ca^{2+} channels open, allowing Ca^{2+} to move into the cell down its electrochemical gradient.

The membrane potential of all animal cells, negative inside, is between −10 and −90 mV, being maintained by the passive permeability of the plasma membrane to ions, mainly K^+. When the membrane depolarises, voltage-gated Ca^{2+} channels open, and the cell fires, e.g. nerve terminals. But, when cytosolic Ca^{2+} activates a Ca^{2+}-activated K^+ channel, e.g. BK, the membrane potential becomes more negative (i.e. it hyperpolarises). Thus, there are four ways to depolarise a membrane:

- Opening of Na^+ channels, e.g. squid and mammalian nerve axons, mammalian muscle.
- Closing of K^+ channels, e.g. pancreatic insulin secreting beta cells.
- Opening of Ca^{2+} channels, e.g. invertebrate muscle, mammalian nerve dendrites, nerve terminals.
- Opening of Na^+ channels, followed by opening of Ca^{2+} channels, e.g. mammalian heart.

If the depolarisation is big enough, in excitable cells this generates an **action potential**. An action potential is a short electrical signal, which involves a rapid depolarisation towards zero or positive values, followed by a repolarisation back to high negative values. When measured with an electrode inside the cell it is a spike (Figure 5.3).

Repolarisation occurs via one of three mechanisms:

- Closing of Na^+ channels (inactivation), e.g. mammalian nerve axons.
- Opening of K^+ channels, e.g. squid giant axons and mammalian heart.
- Closing of Ca^{2+} channels, e.g. mammalian heart.

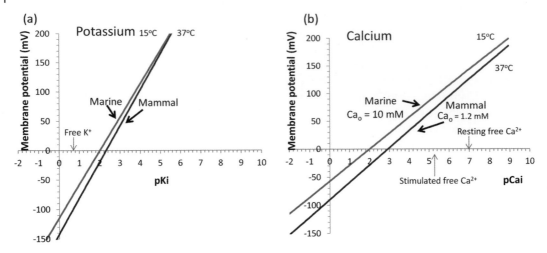

Figure 5.3 Relationship between Ca^{2+} equilibrium potential and cytosolic free Ca^{2+}. The figure shows the membrane potentials that are at equilibrium with cytosolic free concentrations of K^+ (a) or Ca^{2+} (b). A typical cytosolic free K^+ is 150 mM ($pK_i = 0.82$), which is at electrochemical equilibrium at a membrane potential of about −80 to −90 mV, close to the actual membrane potential. In contrast, the cytosolic free Ca^{2+} in resting cells is about 100 nM ($pCa_i = 7$), rising to some 5 µM ($pCa_i = 5.3$) in activated cells such as muscle. The membrane potential would have to be greater than 120 mV, positive rather than negative inside, for this to be at electrochemical equilibrium. This shows the huge Ca^{2+} electrochemical pressure across the plasma membrane of all cells. Reproduced by permission of Welston Court Science Centre.

As Ca^{2+} ions move down their electrochemical gradient, Ca^{2+} attempts to reach its equilibrium (Donnan) potential (E_{Ca}), positive inside. This is predicted by the Nernst equation:

$$Ca^{2+} \text{ equilibrium potential } (E_{Ca}) = (RT/2F)\log_e\left[\left(Ca_o^{2+}\right)/\left(Ca_i^{2+}\right)\right]$$

(5.1)

where Ca_o^{2+} and Ca_i^{2+} are free Ca^{2+} concentrations outside and inside the cell, respectively.

Cytosolic free Ca^{2+} would have to be some 50 to 500 mM for a cell to be at equilibrium with a membrane potential of −50 mV, negative inside.

The number of K^+ ions that have to move to produce a membrane potential of −50 mV is 6.8×10^{-18}, equivalent to 1.6 µM – a tiny fraction of the total intracellular K^+ concentration of 120-150 mM, or extracellular of 5 mM in human blood. But for Ca^{2+}, such a movement, in this case inwards, is equivalent to 0.8 µM, which is significant, compared with the resting cytosolic free Ca^{2+}; though, to produce a change in cytosolic free Ca^{2+} of 1 µM, some 100 µM has to move because of Ca^{2+} buffers in the cytosol. If the Ca^{2+} influx is restricted to only 1% of the cell volume, for example close to the plasma membrane, then the rise would be some 80 µM, enough to trigger a cell event, such as vesicle fusion at a nerve terminal, or the global release of Ca^{2+} from the SR through Ca^{2+}-induced Ca^{2+} release in heart muscle.

But this leaves us with a problem, because, for a channel carrying a current in picoamps and a conductance of 10–100 pS, the number of Ca^{2+} ions moving into the cell with a membrane potential of −90 mV would be some 2.8×10^6 ions s^{-1} (Figure 5.4). For a cell 20 µm in diameter, volume of 4 pl, and 100 Ca^{2+} channels, the change in cytosolic free Ca^{2+} in the whole cell would be about 100 µM s^{-1}! So within less than 1 s the cell would be swamped with Ca^{2+}, and would die! A nerve dendrite, terminal or heart cells may have thousands of Ca^{2+} channels. The problem

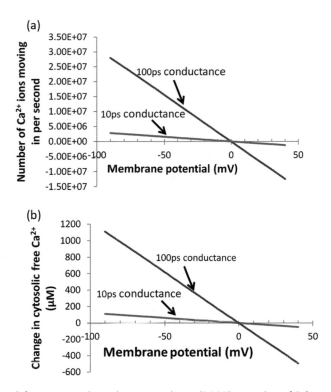

Figure 5.4 How many Ca^{2+} ions move through an open channel? (a) The number of Ca^{2+} ions moving per second through 100 channels with a conductance of 10 or 100 pS. (b) The change in cytosolic free Ca^{2+} per second after Ca^{2+} has moved through 100 channels with a conductance of 10 or 100 pS. Current equivalent $= VG = VG/zF = VG \times N/zF$, mol s^{-1}, so it is possible to calculate how many Ca^{2+} ions would move through a single channel. For a typical channel carrying a current in picoamps and a conductance of 10–100 pS, the number of Ca^{2+} ions moving into the cell with a membrane potential of –90 mV would be 2.8×10^{6} ions s^{-1}. For a cell 20 μm in diameter and a volume of 4 pl, this would mean a change in cytosolic Ca^{2+} in the whole cell of about 1 μM s^{-1}. So if there were 100 Ca^{2+} channels the change would be 100 μM s^{-1}. If the channel had a conductance of 100 pS the numbers would be 10 times this. So within less than 1 s the cell would be swamped with Ca^{2+} and would die! Furthermore, the number of Ca^{2+} channels in a nerve dendrite, terminal or heart cell is likely to be in the thousands, rather than the hundreds. The problem is even worse if the Ca^{2+} change is restricted to a microdomain, as it is in heart muscle, where the Ca^{2+} entering to activate Ca^{2+}-induced Ca^{2+} release from the SR may be restricted to just 1% of the cell volume. Even if the membrane potential dropped to the –20 mV required to open the voltage channels, the change in cytosolic Ca^{2+} would still be some 250 μM. So values for conductance of ion channels measured by patch clamp are too high. Channels with conductance of femto-siemens or less, with currents in the femto-amp range, are very difficult to measure against the electrical noise. Reproduced by permission of Welston Court Science Centre.

is even worse if the Ca^{2+} change is restricted to a micro-domain, as it is in heart muscle. When Ca^{2+} moves into a non-excitable cell (e.g. through SOCE), the total Ca^{2+} that moves in may be as much as 100 μM. Such a movement is equivalent to several volts! This would destroy the cell, so there has to be a counterion.

Ca^{2+} channels are transmembrane proteins, characterised by eight key properties:

1. *Selectivity*: Ca^{2+} relative to Na^+ and K^+.
2. *Conductance*: how good the channel is at conducting Ca^{2+}.
3. *Rectification*: whether the channel allows Ca^{2+} to move in one direction better than another.
4. *Gating*: the mechanism that opens and closes the channel.

5. *Deactivation*: the closing of the channel when the cell is repolarised.
6. *Inactivation*: the spontaneous closing of the channel, e.g. when voltage-dependent.
7. *Pharmacology*: drugs and natural substances that can open or block the channel.
8. *Pathology*: defects in the channel in disease, including inherited channelopathies.

5.4   Primary Stimuli Producing a Cytosolic Ca²⁺ Signal

Three types of extracellular primary stimuli cause Ca^{2+} channels to open in the plasma membrane, or intracellular stores:

1. Physical; electrical, mechanical, light.
2. Chemical; neurotransmitters, hormones, paracrines, intracellular messengers.
3. Biological; microbes, viruses, components of the immune system, toxins.

Physical stimuli open ion channels in the plasma membrane, whereas chemical stimuli work either by direct effects on plasma membrane ion channels or by generating an intracellular messenger, triggering release of Ca^{2+} from an internal store, e.g. the ER. There are four main receptor families for chemical stimuli:

- *Type 1*: channel-linked, known as inotropic, with four or five transmembrane spanning domains, e.g. the nicotinic acetylcholine receptor (Na^+ and K^+) on the endplate of skeletal muscle and inotropic glutamate receptors (Ca^{2+}).
- *Type 2*: G-protein- coupled, known as metabotropic, with seven transmembrane spanning domains, e.g. the muscarinic acetylcholine receptor and catecholamine receptors. Binding of the agonist to the receptor activates an enzyme, e.g. phospholipase C (PLC), phospholipase A2 (PLA2), or adenylate cyclase, via the α subunit of a trimeric G-protein, released from the βγ subunits.
- *Type 3*: kinase-linked receptors, which have an intracellular domain, linked to tyrosine kinase or guanylate cyclase, activated when the agonist binds, e.g. insulin, cytokines, growth factors, and atrial natriuretic peptide.
- *Type 4*: intracellular receptors regulating gene transcription, e.g. steroid and thyroid hormones, vitamin D, and retinoic acid.

All four types of receptor interact with Ca^{2+} signalling, depending on cell type. Type 1 opens Ca^{2+} channels in the plasma membrane (e.g. voltage-gated), leading directly to a rise in cytosolic free Ca^{2+}. Type 2 stimulates production of IP_3 via activation of PLC, releasing Ca^{2+} from the ER. Loss of ER Ca^{2+} provokes opening of store-operated Ca^{2+} channels in the plasma membrane (SOCE). Types 1 and 2 lead directly to cytosolic free Ca^{2+} signals. Types 3 and 4 interact with the Ca^{2+} toolkit. Secondary regulators modify the Ca^{2+} signal, or the threshold of the primary stimulus required to provoke the Ca^{2+} rise.

Ionotropic receptors form an ion channel within their transmembrane portion, and cause a rise in cytosolic free Ca^{2+} in one of two ways:

1. Direct opening of a Ca^{2+} channel, e.g. NMDA glutamate receptor.
2. Generation of an action potential, which leads to opening of voltage-gated Ca^{2+} channels in the plasma membrane, or the release of Ca^{2+} from the SR, e.g. nicotinic acetylcholine receptor at the endplate of skeletal muscle.

In nerve dendrites channel opening causes a rapid rise in cytosolic free Ca^{2+} close to the plasma membrane, as well as depolarisation. These summate to generate a Na^+ dependent action potential that flows down a nerve, opening Ca^{2+} channels in the terminal. The cytosolic receptor tail can be phosphorylated by kinases and dephosphorylated by phosphatases, leading to alterations in its properties. The γ-aminobutyric acid (GABA) ionotropic receptor is a Cl^- channel, making it more difficult for the neurone to generate an action potential.

In contrast, the ionotropic nicotinic acetylcholine receptor forms a non-selective ion channel, responsible for skeletal muscle contraction in all vertebrates. Rapid Na^+ entry depolarises the membrane at the muscle endplate, generating an action potential that flows down the muscle fibre. On reaching the dihydropyridine (DHP) receptors in the T-tubule, the action potential alters its structure, which interacts with the ryanodine receptor on the intracellular SR, causing a massive release of Ca^{2+}. The large rise in cytosolic free Ca^{2+} that ensues triggers the muscle to contract. Rapid degradation of the acetylcholine within the neuromuscular junction leads to dissociation of acetylcholine from the receptor, thereby switching off the ion channel, allowing re-establishment of the membrane potential. On the other hand, most metabotropic receptors involved in Ca^{2+} signalling work by interacting with a G-protein in the plasma membrane, e.g. the metabotropic glutamate receptors mGluR1 and mGluR4–8 bind glutamate in their extracellular domain, activating PLC in the membrane, which generates IP_3, provoking release of ER Ca^{2+} into the cytosol. Thus, there are three ways receptor binding leads to a rise in cytosolic free Ca^{2+}:

1. Direct opening of a Ca^{2+} channel.
2. Opening of an ion channel, which depolarises the membrane, so that voltage-gated Ca^{2+} channels open.
3. Activation of G-proteins coupled to PLC, producing IP_3, which releases Ca^{2+} from the ER, leading to SOCE.

5.5 Plasma Membrane Ca²⁺ Channels

There are seven types of Ca^{2+} channel in the plasma membrane, which can lead to a rise in cytosolic free Ca^{2+} (Box 5.2).

Box 5.2 The seven types of plasma membrane Ca²⁺ channel.

1. Voltage-gated Ca^{2+} channels.
2. TRP (transient receptor potential) channels.
3. Receptor-operated channels.
4. Mechanosensitive channels.
5. Store operated calcium entry (SOCE).
6. Intracellular signal operated.
7. Calcium receptor.

5.5.1 Voltage-Gated Ca²⁺ Channels

Voltage-gated Ca^{2+} channels are opened by a decrease in the membrane potential, towards a more positive value from that in the resting cell. Typical membrane potentials in excitable cells range from -90 to -50 mV, negative inside. On excitation, this depolarises to some $+20$ to $+40$ mV, positive inside. There are two main families of voltage-gated Ca^{2+} channels, HVA and LVA, designated by whether they require a large (**High**) or small (**Low**) depolarisation to open (Figure 5.5, Table 5.1). These are then subdivided on the basis of their electrical properties,

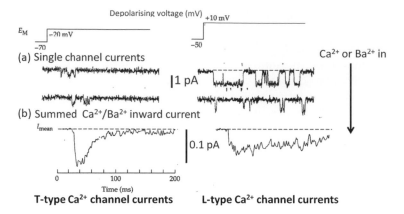

T-type Ca²⁺ channel currents   **L-type Ca²⁺ channel currents**

Figure 5.5 Opening and closing of HVA (L-type) and LVA (T-type) Ca²⁺ channels. The figure shows an example of T- and L-type voltage-gated Ca²⁺ channels. T = tiny/threshold/transient; L = large/long lasting. (a) Currents from a single channel measured by a patch clamp. (b) Net current after a sustained depolarisation (E_m). Source: Nilius *et al.* (1985). Reproduced with permission from Nature.

such as **L**- (Long lasting) and T- (**T**ransient) (Table 5.1). Other voltage-gated Ca²⁺ channels are designated by the letters N, P/Q and R. HVA channels require quite a large depolarisation to open, from −80 to +10 mV (net depolarisation = +90 mV), whereas LVA channels can open with a small depolarisation from −80 to −30 mV (net depolarisation = +50 mV). LVA Ca²⁺ channels inactivate rapidly (i.e. they close even while the cell is still depolarised). In contrast, HVA Ca²⁺ channels can remain open for some time while the cell is depolarised, enabling the two channel types to be distinguished.

Table 5.1 Electrical and biochemical differences between L- and T-type Ca²⁺ channels.

Property	HVA, L-type, slow, persistent	HVA, N-type, fast, inactivating	LVA, T-type, fast, inactivating
Activation threshold (opening) (mV)	−10	−20	−70
Inactivation (closing) (mV)	−60 to −10	−120 to −30	−100 to −60
Decay rate (ms)	Slow; $\tau > 500$	Moderate; $\tau \sim 50$–80	Quite fast; $\tau \sim 20$–50
Deactivation rate	Fast	Slow	Fast
Conductance – single channel (pS)	25	13	8
Kinetics – single channel	Continues to reopen	Long bursts	Short burst, then inactivates
Ba²⁺ versus Ca²⁺	Ba²⁺ > Ca²⁺	Ba²⁺ > Ca²⁺	Ba²⁺ = Ca²⁺
Effect of Cd²⁺	Blocks	Blocks	No effect
Effect of Ni²⁺	Less sensitive than L	Less sensitive than L	Blocks
Block by ω-conotoxin	Weak	Strong	Weak
Effect of DHPs	Nifedipine blocks; Bay K8644 opens	No effect	No effect
Effect of catecholamines	Increases speed of opening and closing	?	No effect

In humans there are two classes of HVA L-type Ca^{2+} channels and three types of LVA T-type Ca^{2+} channel:

1. HVA/L: $Ca_v1.1$, $Ca_v1.2$, $Ca_v1.3$, $Ca_v1.4$.
2. HVA: $Ca_v2.1$ (P/Q), $Ca_v2.2$ (N), $Ca_v2.3$ (R).
3. LVA/T: $Ca_v3.1$, $Ca_v3.2$, $Ca_v3.3$.

N stands for 'neither' or 'neuronal', P/Q are the next letters of the alphabet, with P for 'Purkinje', and R is 'residual'; R remaining after all the other Ca^{2+} channel types are blocked. Other terminologies include LTI = low threshold inactivating, HTI = high threshold inactivating and HTN = high threshold non-inactivating. Eight electrical and pharmacological properties distinguish these channels electrically and biochemically (Table 5.1):

1. Size of depolarisation needed to open the channels.
2. Size of conductance of single channels.
3. Speed of opening and closing.
4. Whether inactivation occurs while the cell is depolarised and is voltage dependent.
5. Ionic selectivity.
6. Sensitivity to blocking cations.
7. Regulation by intracellular signals, such as cytosolic Ca^{2+} and phosphorylation.
8. Sensitivity to pharmacological agents, both natural toxins and drugs (see Chapter 11).

HVA/L-types are particularly important in skeletal muscle, cardiac myocytes, endocrine cells and retina. HVA/P, Q, N and R are important in nerves. LVA/T are important in nerve cell bodies and dendrites, and cardiac myocytes. Thus, many electrically excitable cells have more than one Ca^{2+} channel. Neurones can have all types, in different parts of the cell – dendrites, cell body and terminal. Voltage-gated Ca^{2+} channels are also classified by the sequence similarity of the α_1 subunit (Table 5.2).

Table 5.2 Types of voltage-gated Ca^{2+} channel based on electrophysiology and α_1 subunit.

Voltage type	Classification	α_1 Subunit type	Main cell types, with splice types (a or b type)
$Ca_v1.1$	L	S	Skeletal muscle
$Ca_v1.2$	L	C	Heart a type, smooth muscle b type, nerve (brain), heart, pituitary, adrenal c
$Ca_v1.3$	L	D	Nerve (brain), pancreas, kidney, ovary, cochlea
$Ca_v1.4$	L	F	Retina
$Ca_v2.1$	P/Q	A	Nerve (brain), endocrine (pituitary), cochlea – a and b types
$Ca_v2.2$	N	B	Nerve (brain and nervous system) – a and b types
$Ca_v2.3$	R	E	Nerve (brain), cochlea, retina, heart, endocrine (pituitary) – a type; nerve (brain), cochlea, retina – b type; retina, heart,
$Ca_v3.1$	T	G	Nerve (brain and nervous system) – a type
$Ca_v3.2$	T	H	Nerve (brain), heart, kidney, liver – a type
$Ca_v3.3$	T	I	Nerve (brain) – type

R stands for residual or resistant to blockers of the other channel types. The groupings depend on sequence similarities, revealing three families: α_{1S}, α_{1C}, α_{1D} and α_{1F}; α_{1A}, α_{1B} and α_{1E}; and α_{1G}, α_{1H} and α_{1I}, there being 70% similarity within the same family, but less than 40% between different families. α_{1S} for the initial isoform in skeletal muscle and α_{1C} for the cardiac isoform, but α_{1A} and α_{1B} were discovered afterwards in other tissues. There is extensive alternative splicing, which can alter the voltage dependency (e.g. with $Ca_v1.3$) and make it difficult to distinguish LVA from HVA. The original nomenclature, based on L, T, P/Q, N and R, has been replaced by some authors based on sequence similarity and tissue distribution of the α_1 subunit, a and b types representing alternative splice variants.

Voltage-gated Ca^{2+} channels have a selectivity for Ca^{2+} over Na^+ of about $1000 : 1$, allowing only Ca^{2+} through when the extracellular Na^+ is 500 mM and Ca^{2+} 10 mM in sea water, or Na^+ 140 mM and Ca^{2+} 1.2 mM in human plasma. They are responsible for excitation of nerves in the dendrites, secretion of neurotransmitters at the nerve terminal, muscle contraction, regulation of gene expression in excitable cells, and the release of hormones such as insulin. There are four principle subunits: α_1, β, $\alpha_2\delta$ and γ. The α_1 subunit forms the pore from four similar domains within the one protein (Figure 5.6), with glutamate being the key residue responsible for selectivity for Ca^{2+} over Na^+. Each type of Ca^{2+} channel uses a different α_1 subunit (Table 5.2). These are about 190 kDa in size and have four highly homologous domains, each with six transmembrane domains (1–6), which form the pore (Figure 5.6). Helices 5 and 6 form the inner lining of the pore, whereas helices 1–4 are involved in voltage-gating, helix 4 particularly being the voltage sensor. The rapid inactivation of these Ca^{2+} channels is due either to voltage or Ca^{2+} itself rising close to the cytosolic side of the pore. The cytosol domain can be regulated by Ca^{2+}-calmodulin, through an IQ binding site, and PKA. The β subunit binds to the channel inside the cell, stabilising the shape of the α_1 subunit, playing a role in channel sensitivity and inactivation, and trafficking of the channel to the plasma membrane. The cytosolic side of the voltage-gated Ca^{2+} channel in skeletal muscle, $Ca_v1.1$, interacts directly with the SR ryanodine receptor, releasing Ca^{2+}, which trigger muscle contraction. On the other hand, $Ca_v1.2$ in heart muscle causes a local rise in cytosolic free Ca^{2+}, sufficient to activate Ca^{2+}-induced Ca^{2+} release from the ryanodine receptors in the SR, causing the explosive rise in cytosolic free Ca^{2+} necessary to provoke myocyte contraction.

Na^+ channels open when the cell depolarises below about −30 mV, and only switch off (i.e. deactivate) when the membrane has repolarised. When applied to Ca^{2+} channels, this phenomenon helps us to understand why the heart has evolved an LVA T-type Ca^{2+} channel, the L-type channel in the cardiac cell (Figure 5.6) being activated (i.e. opened) by a voltage pulse of −80 mV. This opens the HVA L-type Ca^{2+} channels, which inactivate rapidly, resulting in a rapid return of the membrane current to zero. A further depolarisation of 60 to +10 mV opens the T-type Ca^{2+} channels, which inactivate slowly compared with the L-type channels. This enables the myocyte to maintain a prolonged action potential, lasting about 1 s – the length of the heartbeat. Thus, Ca^{2+} channels differ markedly in the type and level of agent that opens them, their rate of inactivation and their pharmacology.

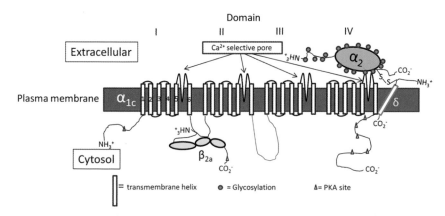

Figure 5.6 Predicted structure of a voltage-gated Ca^{2+} channel. The figure shows the transmembrane helices, protein kinase and glycosylation sites predicted for the cardiac voltage-gated Ca^{2+} channel $Ca_v1.2$ from its protein sequence. The α_1 subunit forms the Ca^{2+} channel from four clusters of six transmembrane helices. The channel can be regulated by it binding to the α_2, β and δ subunits. The channel exhibits Darwinian molecular variation by having several alternative splice variants.

Voltage-gated Ca^{2+} channels have two main advantages, over other voltage-gated channels. First, they are able to maintain an inward current for longer than Na^+ channels, because the inactivation rate is slower. This allows the cell to maintain a depolarisation for longer when voltage-gated Ca^{2+} channels are operating. Secondly, voltage-gated Ca^{2+} channels can cause directly a rise in cytosolic free Ca^{2+}, particularly locally, where it can be sufficient to trigger the cell event, such as heart muscle contraction or secretion from a nerve terminal. A Na^+-dependent action potential, fast for axonal conduction, is too brief to allow enough transmitter to be released, or, on its own, to allow a heart myocyte to beat rather than 'twitch'. Furthermore, modification of voltage-gated Ca^{2+} channels, such as by acetylcholine or phosphorylation induced by catecholamines, enables the cell response to be weakened or strengthened, respectively. Inhibition of Ca^{2+} channels at the nerve terminal results in less transmitter being released. As a result, the time taken for the released transmitter to be degraded or taken back up into the terminal is shortened. Consequently, the next cell is activated for a shorter time.

Voltage-gated Ca^{2+} channels can be blocked, or sometimes activated, by:

1. Lanthanides and transition metals: $La^{3+} > Co^{2+} > Mn^{2+} > Ni^{2+} > Hg^{2+}$.
2. Drugs: DHPs such as nifedipine and nitrendipine, verapamil and D-600, and diltiazem (nano- to micromolar) preferentially block L-type Ca^{2+} channels.
3. Natural toxins such as conotoxin and agatoxin.

Ba^{2+} is frequently used in patch clamp pipettes to study Ca^{2+} channels, because Ba^{2+} currents are larger than Ca^{2+} currents, and Ba^{2+} blocks K^+ channels. Particular Ca^{2+} channels exhibit differing sensitivities to dihydropyridines (DHPs), conotoxin and agatoxin, and the agonist Bay K8644, as well as other drugs and transition metals. For example, only L-type Ca^{2+} channels are blocked by verapamil, D-600 and the DHPs, nifedipine, nitrendipine, and diltiazem. On the other hand, conotoxin reversibly inhibits N-type Ca^{2+} channels. Four subtypes of the Ca_v1 family have been identified, several exhibiting alternative splicing:

1. $Ca_v1.1$ – skeletal muscle.
2. $Ca_v1.2$ – cardiac muscle and in several parts of the brain.
3. $Ca_v1.3$ – brain, neurosecretory cells, pancreatic islets, hair cells in the ear, and cardiac pacemaker cell.
4. $Ca_v1.4$ – eye, a mutation causing one type of congenital night blindness.

There are three sub-types of Ca_v2:

1. $Ca_v2.2$ (ABE) – N type Ca^{2+} in neurones, blocked irreversibly by conotoxin.
2. $Ca_v2.1$ – P/Q type, blocked by both conotoxin and agatoxin, and highly expressed in Purkinje cells and granule cells. Mutations associated with ataxia and cerebella degeneration.
3. $Ca_v2.3$ – R-type central nervous system, particularly cerebellum, Purkinje and granule cells, not blocked by conotoxin.

Finally, Ca_v3 (GA1) interacts with G-proteins, which inhibit Ca^{2+} currents in the presynaptic terminals of some neurones, inhibiting neurosecretion. All types of Ca^{2+} channel can be found in a single neurone, but they are located in exactly the right place for their properties to work, N at the terminal, with P/Q, L and T in the dendrites and cell body. In summary, voltage-gated Ca^{2+} channels exhibit the following properties:

1. They are only found in excitable cells, often more than one channel type being present.
2. Several different ions can compete for the binding site within the channel pore. Removal of divalent cations exposes them to monovalent cations, which can then move through.

3. They can exist in one of three states: deactivated = closed, activated = open, and inactivated = closed.
4. When open, they typically have a conductance of $1–100$ pS for Ca^{2+}.
5. Normally, one Ca^{2+} is always bound in the pore.
6. The sub-units are α_1, α_2, β, γ and δ, the α_1 subunit making the pore.
7. The pore has a filter that selects for Ca^{2+} over other cations, there being a gate that opens and closes at the appropriate voltage.
8. There are at least two main types of Ca^{2+} channel: HVA and LVA, which are themselves divided into subtypes: L and T, and N, P/Q and R, characterised by their voltage dependence, ionic selectivity, rate of inactivation and deactivation, and sensitivity to other cations and pharmacological agents.
9. Ca^{2+} channels can switch modes unexpectedly, and are susceptible to dramatic changes in opening and closing kinetics.
10. Drug receptor sites may differ between channel types and agent.

5.5.2 TRP Channels

TRP channels were discovered in the eye of the fruit fly *Drosophila*. TRP = 'transient receptor potential', the membrane potential change only lasting a short time – transient. TRP channels form a large superfamily of cation channels that let Ca^{2+} or Na^+, or both, into sensory cells responsible for vision, sensing (taste, smell, hearing, pressure and hot versus cold), osmosis, stretch and vibration. They are chemically opened (activated) by marijuana, allicin in garlic, wasabi (allyl isothiocyanate), menthol, peppermint, and capsaicin in hot chilli, depending on channel type. TRPs are found in all animals, in excitable and non-excitable cells. Their selectivity for Ca^{2+} varies considerably. Some are non-selective for the cation (Na^+, Ca^{2+} and Mg^{2+}), whereas others are more than ten times selective for Ca^{2+} over Na^+. However, with a ratio of Na^+ to Ca^{2+} outside of some $140 : 1$, Ca^{2+} will always be competing with Na^+ for entry into the cell. The relevance of TRPs to Ca^{2+} signalling is two-fold:

1. TRP opening causes directly a rise in cytosolic free Ca^{2+}.
2. Na^+ entry via TRP depolarises the plasma membrane, opening voltage-gated Ca^{2+} channels, and regulating the flux of Ca^{2+} entry through the membrane potential.

TRP channels may be opened by G-protein-coupled receptors, or directly by ligand binding. They have a conductance in the tens of pico-siemens. Over 30 TRP channels have now been discovered, with sequence similarities varying from 20% to 60% between them and belonging to seven main families (Table 5.3): TRPC (canonical), TRPM (melastatin), TRPV (vanilloid), TRPA (ankyrin), TRPP (polycystin), TRPML (mucolipin) and TRPN (NOMPC = no mechanoreceptor potential). They are named after how they were first identified, e.g. via an agonist. TRPs are mostly found in the plasma membrane, but at least one is found in the lysosomal membrane, a Ca^{2+} channel – TRPML1. These are mucolipins, mutations causing mucolipidosis.

TRPs have six transmembrane domains, typically forming the pore between domains 5 and 6, and have N- and C-termini facing into the cytosol. They form both homo- and hetero-tetramers, like the K^+ channel. The termini bind other proteins, such as ankyrin, and have phosphorylation and enzymatic sites at the C-termini. Opening of the TRP cation channels depolarises excitable cells, generating an action potential, which then opens voltage-gated Ca^{2+} channels, e.g. at a nerve terminus, stimulating secretion of a neurotransmitter, to tell the animal it has sensed something.

Table 5.3 Examples of TRP channels.

TRP channel	Named after	No. in human family	Ionic selectivity P_{Ca2+}/P_{Na+}	Conductance range (pS)	Cause of gating
TRPC	Canonical	7	0–9	16–66	PKC/DAG
TRPM	Melastatin	8	0–3	16–130	ADP ribose, cold and menthol
TRPV	Vanilloid	6	2.6–>100	35–190	Heat, PKC, low Ca^{2+} and hyperpolarisation
TRPA	Ankyrin	1	0.8–1.4	40–105	Allicin and bradykinin
TRPP	Polycystin		1–5	40–177	Mechanical
TRPML	Mucolipin	3	?	46–84	Raised cytosolic Ca^{2+}
TRPN	NOMPC = no mechanoreceptor potential	1			Mechanical in fruit fly

0 = non-selective, 9 = highly selective for Ca^{2+}.

TRP channels are involved in all six types of taste: sweet, sour, bitter, salty, peppery and umami. For example, TRPV1 can be opened by capsaicin, the 'hot' ingredient in chilli, whereas glutamate in Asian food activates the 'umami' savoury taste receptor on your tongue. Bitter and sweet tastes are triggered by the receptor interacting with G-proteins. Many TRPs are voltage sensitive, and can be opened by hormones such as bradykinin and nerve growth factor. TRPV1 and V2 are also heat sensors, being opened *in vitro* by moderate to high temperatures. Some, such as TRPML1 in lysosomes, may be constitutively active. Menthol, camphor, peppermint and allicin (garlic) are agonists that open TRPM8 channels, and can mimic cold. TRPC2 is at the head of mouse sperm, being involved in the acrosome reaction, necessary for a sperm to successfully fuse with an egg and inject its DNA.

Thus, we need TRPs to enjoy a cordon bleu meal or a curry, or be inspired by a Mendelssohn concerto or a Bach cantata. They are required to smell a rose, or to listen to the dawn chorus. Without TRPs, our hands would soon be scarred, and our mouths full of burn ulcers. Intracellular Ca^{2+} is central to all these.

5.5.3 Receptor-Activated Ca²⁺ Channels

Ca^{2+} signals are generated inside cells by hormones, transmitters, cytokines and other naturally occurring agonists, as a result of binding to a receptor facing the outside of the cell. Natural agonists include acetylcholine, adrenaline and noradrenaline, ATP, GABA glucagon, glutamate, 5-HT (serotonin), histamine, interleukins, parathyroid hormone (PTH), vasopressin, the N-termini of bacterial and mitochondrial proteins, and formylated peptides in coelenterates and other invertebrates. There are four ways by which these agonists can generate a cytosolic Ca^{2+} signal:

1. Opening of a Ca^{2+} channel directly as part of the receptor, e.g. the NMDA glutamate receptor, TRPs.
2. Opening of voltage-gated Ca^{2+} channels as a result of depolarisation, e.g. TRPs, acetylcholine nicotinic receptor.
3. Activation of phospholipase C (PLC) through G-protein coupling, producing IP_3 and opening of SOCE channels, e.g. acetylcholine muscarinic, adrenaline (α), vasopressin, ATP, histamine.
4. Generation of an intracellular second messenger, e.g. cyclic AMP, cyclic GMP, NAADP, cyclic ADP ribose, opening Ca^{2+} channels in the plasma membrane or releasing Ca^{2+} from intracellular organelles.

5.5.4 Mechanosensitive Channels

Many animal and plant cells generate a cytosolic free Ca^{2+} signal in response to touch. This is caused by mechanosensitive channels, linked to receptors that respond to touch, pressure, vibration, proprioception and membrane movement. Although several of these channels allow Ca^{2+} into the cell, many are relatively non-selective for either a cation or anion. They are found in all three domains of life: Eukaryota, Bacteria and Archaea. All animals, plants and microbes have them. In specialised cells in hairs, for example, mechanosensitive channels tell the organism that it has been touched. But the main role in other cells is as a defence against too much stretching or osmotic stress. Several voltage-gated and TRP channels have been identified as mechanosensitive channels in animal cells, allowing Ca^{2+} to enter the cell when opened. Animal cell networks can generate a Ca^{2+} wave that moves through the cells often due to mechanical movement of the cells. The channels are blocked by Gd^{3+} and La^{3+}.

In several bacteria and archaeans the molecular basis of mechanoreceptors has been well defined. The Darwinian selective advantage in bacteria is when the cell wall is attacked, e.g. by antibiotics, lysozyme in tears, or the membrane attack complex of complement. Plants defend themselves against wind, sense objects in the soil, or use supports as they grow, and even sense prey, in the case of insectivorous plants via mechanosensitive channels which allow Ca^{2+} to enter the cell.

5.5.5 Store-Operated Calcium Entry (SOCE)

One to the most remarkable discoveries about the way Ca^{2+} gets into cells was made during the 1980s, whereby release of Ca^{2+} from the ER caused Ca^{2+} channels in the plasma membrane to open, producing a large cytosolic free Ca^{2+} signal. These channels are called 'store-operated calcium entry' channels (SOCE) (Figure 5.7). Initially, it was thought that SOCE was a way of refilling the ER with Ca^{2+} from outside the cell after a cell stimulus had ended. It was thus named 'capacitative calcium entry'. But it is now clear that SOCE is far more significant at causing large cytosolic free Ca^{2+} signals in many cells. There is also an interaction between

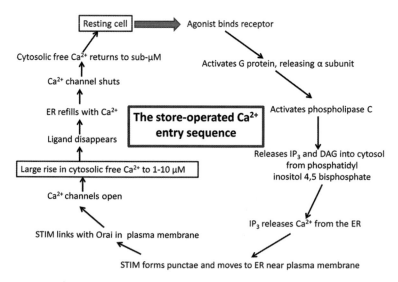

Figure 5.7 The sequence for opening and closing SOCE channels in the plasma membrane. Reproduced by permission of Welston Court Science Centre.

mitochondria and the SOCE channel. Using an elegant mutation analysis, together with the characterisation of the molecule responsible for an immune deficiency disease, two proteins were discovered that are essential for SOCE – STIM (STromal Interaction Molecule) in the ER, and Orai (calcium release-activated calcium channel protein 1 – TMEM142A) in the plasma membrane, named after the three keepers of the gates of heaven in Greek mythology. Orai was discovered through a mutation, causing severe combined immunodeficiency (SCID). There are three in mammals – Orai 1, 2 and 3, with no homology with any other ion channel.

The channel for Ca^{2+} entry is Orai, after binding STIM punctae in the ER, through their cytosolic domains. In the presence of extracellular Ca^{2+} the channel is highly selective for Ca^{2+} over monovalent cations, by some 10 000 : 1, with an extracellular Na^+ of 140 mM and K^+ of 5 mM – the most selective Ca^{2+} channel known. But in the absence of extracellular Ca^{2+}, Orai loses its selectivity and allows Na^+ to move through it. Little or no Ca^{2+} moves when the extracellular Ca^{2+} is micromolar, even if opened by STIM. Each Orai has four transmembrane domains (M1–M4), the ion channel being formed by a hexamer of Orai subunits, with the pore lined with six glutamates, the Ca^{2+} selectivity mechanism, in the centre. The pore is made up of four parts. First, at the extracellular side, six M1 helices from each of the subunits make up the inner ring of the pore, with glutamates acting as the ionic selectivity filter. This is followed by a hydrophobic section, and then a section rich in basic amino acids such as lysine. Finally there is a wider section that extends into the cytosol. The Orai channel is firmly shut until it binds the STIM punctae, which have moved along the ER. A SOCE current can be measured with patch clamp in some cells – I_{CRAC}, highly selective for Ca^{2+}. The key features for opening SOCE are (Figure 5.7):

1. Loss of Ca^{2+} from the ER leads to formation of STIM1 punctae, and then their movement to the part of the ER close to the plasma membrane.
2. Linkage of this STIM1 with Orai1 hexamers in the plasma membrane opens a specific Ca^{2+} channel through Orai.
3. The channel is barely electrogenic, as it has a very small conductance in the femto-siemens range.
4. Ca^{2+} floods into the cell, so long as the channel remains open.
5. The channel switches off when Ca^{2+} is pumped back into the ER, as a result of IP_3 levels returning to those in the resting cell, or experimental removal of a SERCA pump inhibitor such as thapsigargin or cyclopiazonic acid.

5.5.6 Ca²⁺ Receptor

Parathyroid cells, C-cells in the thyroid, and kidney cells have plasma membrane receptors that respond to small changes in extracellular free Ca^{2+}. This enables the body to maintain the total Ca^{2+} concentration in blood within very small limits, approximately 1.9–2.5 mM, with a plasma free Ca^{2+} of 1.1–1.3 mM. This is necessary because, at plasma free Ca^{2+} concentrations below 0.9 mM, nerves start to fire spontaneously, leading to titanic muscle contractions and heart muscle problems. In contrast, at concentrations of free plasma Ca^{2+} above about 1.3 mM, calcium stones can start to form, blocking kidney function and causing damage to other tissues. The parathyroid gland secretes parathyroid hormone (PTH) in response to a small drop in plasma free Ca^{2+}, which then provokes resorption of calcium from bone. The dose–response curve of PTH secretion against extracellular free Ca^{2+} is sharply sigmoid, explaining why secretion is so sensitive to small changes in blood free Ca^{2+}. In contrast, calcitonin (thyrocalcitonin), a 32-amino-acid peptide secreted in humans by thyroid C-cells, opposes PTH, reducing Ca^{2+} in the blood. An increase in plasma free Ca^{2+} provokes secretion of calcitonin, which then inhibits resorption of bone Ca^{2+} and stimulates excretion via the kidney. A decrease in extracellular free Ca^{2+} causes an influx of Ca^{2+} into thyroid C-cells via voltage-sensitive Ca^{2+} channels, resulting in secretion of calcitonin. Vitamin D_3, on the other hand, now known as calcitriol, plays a major role in regulating blood Ca^{2+} through absorption in the gut via a Ca^{2+}-binding protein. The Ca^{2+} receptor belongs to the G-protein C subfamily. As well as activating PLC, leading to an elevation of cytosolic free Ca^{2+}, the Ca^{2+} receptor can also activate PLA2 and phospholipase D, leading to production of arachidonic acid and phosphatidic acid. It also activates several mitogen-activated protein kinases (MAPKs), including ERK1/2, p38 and JNK MAPK, leading to changes in gene expression.

Drugs that interact with the Ca^{2+} receptor are divided into two groups:

1. Calcimimetics, which mimic or potentiate the action of Ca^{2+} on the Ca^{2+} receptor.
2. Calcilytics, which are Ca^{2+} receptor antagonists.

The Ca^{2+} receptor is essential for long-term survival, with a role in many tissues. Changes in expression of the Ca^{2+} receptor play an important role in parathyroid disease, vitamin D deficiency, and bone disorders, such as osteoporosis. Splice variants at exon 5 have been detected in several tissues. The Ca^{2+} receptor must also play a role in invertebrates, particularly those that have shells composed of $CaCO_3$.

5.6 Regulation of Intracellular Ca^{2+} by, and Within, Organelles

All the organelles inside eukaryotic cells can take up and release Ca^{2+}. This plays a role in regulating cytosolic free Ca^{2+}, and in controlling processes activated by Ca^{2+} within the organelle. These processes include SOCE, the stress response inside the ER, enzymes inside the Golgi and mitochondria, and gene expression in the nucleus. The organelles involved are ER/SR, mitochondria, lysosomes, endosomes, secretory vesicles, Golgi, nucleus, acido-calcisomes (calcisomes), and the vacuole in plants, fungi and some other organisms. The ER/SR is the major store of Ca^{2+} in all eukaryotic cells, except plants and fungi, where the vacuole can store large amounts of Ca^{2+}. Mitochondria play a key role in the location and type of cytosolic Ca^{2+} signal that is generated by a primary stimulus. Under some conditions cytosolic free Ca^{2+} may be regulated by uptake or release of Ca^{2+} from lysosomes and secretory vesicles. Ca^{2+} is regulated in the nucleus, where there are transcription factors that can be activated by Ca^{2+}-binding proteins. Whether there is a significant barrier to Ca^{2+} across the nuclear membrane is still controversial. The Golgi needs to regulate its Ca^{2+} because Ca^{2+} can regulate some of its enzymes. Acidocalcisomes (calcisomes) are special organelles that can transport Ca^{2+}, but may also release Ca^{2+} into the cytosol. But, the central controller of intracellular Ca^{2+} is the ER, or SR in muscle. Naturally occurring and synthetic substances can open channels in organelle membranes, and release Ca^{2+} into the cytosol (Figure 5.8).

5.6.1 Endo- and Sarco-Plasmic Reticulum (ER/SR)

The reticulum is a three-dimensional spider's web, wrapped around the nucleus and extending to the plasma membrane, and it is continually moving. The free Ca^{2+} inside the ER/SR, measured by targeted indicators such as aequorin, cameleons or pericams, is much higher than in the cytosol. When full, the free Ca^{2+} in the ER/SR lumen can be hundreds of micromolar, and drop to just a few micromolar if the bulk of the ER/SR Ca^{2+} is released. Two types of channel are

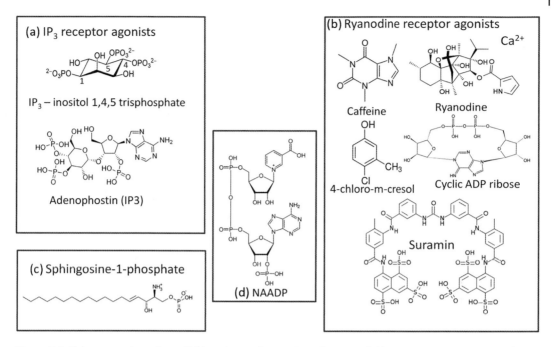

Figure 5.8 Substances that release Ca² from internal stores into the cytosol. (a) IP₃ receptors agonists on the ER. (b) Ryanodine receptor agonists on the ER. (c) Sphingosine-1-phosphate, which can release Ca^{2+} from the ER and possibly other internal stores. (d) NAADP, which can release Ca^{2+} from lysosomes or secretory vesicles.

found in the ER/SR, which release Ca^{2+} into the cytosol when opened by a primary stimulus, such as an action potential or hormone:

- The ryanodine receptor.
- The inositol trisphosphate (IP_3) receptor.

Both of these consist of a complex of four huge subunits, each some 500 kDa, which form the Ca^{2+} pore, with several proteins attached on both the cytosolic and luminal sides of the reticulum. These are able to regulate the channel. Both types of receptor have binding sites for various natural and synthetic ligands, which can open and close the Ca^{2+} channel. Each receptor type has several variants from different genes or alternative splicing, producing a Ca^{2+} release channel suitable for the physiology of the particular cell type. Ryanodine and IP_3 receptor channels have four key properties, there being three sub-types of each in mammals:

1. Ca^{2+} release, as a result of opening the receptor channel by a primary stimulus acting at the plasma membrane.
2. Ca^{2+} release modified by secondary regulators and drugs.
3. Three subtypes, with different properties, are found in particular cells.
4. They can go wrong in disease.

The ER/SR lumen contains several Ca^{2+}-binding proteins, such as calsequestrin, calreticulin, grp78 (BiP) and grp94. These are regulatory proteins, and bind significant amounts of Ca^{2+}. Ca^{2+} is taken up into the ER/SR by the SERCA pump, of which there are three types, the diversity being increased by alternative splicing, each protein having subtly different kinetic properties. In heart myocytes, SERCA2a can be activated by the protein phospholamban, when this is phosphorylated by PKA activated through an increase in cyclic AMP. This increases the Ca^{2+} store, resulting in larger Ca^{2+} transients and a stronger contraction – the basis of the inotropic effect induced by catecholamines. MicroRNA (miRNA) can also regulate SERCAs. Proteins inside the ER require a Ca^{2+} above 50 μM, for protein trafficking to the Golgi, certain enzymatic activities, and the unfolded protein response. Calsequestrin was thought to be the major Ca^{2+} store protein in muscle SR, but its role appears to be more as a regulator, in view of the lack of any dramatic effect in calsequestrin knock-outs. There are two major isoforms, 86% sequence-similar, representing skeletal and cardiac muscles, respectively. Calsequestrin has no transmembrane domains and, unlike its non-muscle equivalent calreticulin, has no KDEL retention sequence. It is anchored inside the SR by two proteins: triadin and junctin, which dock calsequestrin on to the luminal side of the ryanodine receptor. Calsequestrin is highly glycosylated and 30% of its amino acids are acidic, giving it a large Ca^{2+}-binding capacity. Calreticulin is the major ER Ca^{2+}-binding protein in non-muscle cells.

5.6.1.1 Ryanodine Receptor

Two compounds from plants, ryanodine and caffeine, led to the discovery and isolation of the ryanodine receptor (RyR), together with its identification as the channel responsible for release of the Ca^{2+} that triggers muscle contraction. Ryanodine is a poisonous alkaloid, originally isolated from plants in South America of the genus *Ryania*. Caffeine is a bitter, white alkaloid from cocoa. In mammals there are three ryanodine receptor genes, *RyR1–3*, with 70% sequence homology. Each can be modified post-transcriptionally. The ryanodine receptor is a homotetramer. RyR1 is found in muscle, Purkinje fibres and cerebella neurones. RyR2 is heavily expressed in cardiac muscle and is the main receptor in the brain. RyR3 is found at low levels and is more widespread, being found particularly in striated, smooth and cardiac muscle, T-lymphocytes, and parts of the brain associated with memory such as the cortex and hippocampus. In skeletal muscle, the SR ryanodine receptor is opened through direct interaction of the dihydropyridine (DHP) receptor in the T-tubule. In other cells, the ryanodine receptor opens via Ca^{2+}-induced Ca^{2+} release. The ryanodine receptor is activated by low ryanodine (K_d 20 nM), and inhibited by high ryanodine (K_d 10 μM), and is regulated by PKA, protein kinase G and Ca^{2+}-calmodulin protein kinase II (CaMKII). All have sites in the lumen that interact with calsequestrin in muscle, and other ER proteins. Other important proteins that regulate ryanodine receptors include presenilin and pannexin. Ryanodine receptors are also found in non-mammalian vertebrates (reptiles, amphibians, birds and fish) and in invertebrates, such as crustaceans (lobsters), echinoderms (sea urchins), insects (flies) and nematode worms (*Caenorhabditis elegans*).

RyR1 and RyR2 are essential for long-term survival, since mice in which the gene coding for RyR1 was knocked out die at birth with major skeletal muscle abnormalities, and those with the RyR2 gene knocked out die in the embryo with severe heart problems. On the other hand, mice without RyR3 survive and appear to have no major abnormalities in excitation–contraction coupling, but do have defects in learning and in the hippocampus. Experimentally, ryanodine receptors allow a range of divalent and monovalent cations to pass through them. The fact that Ca^{2+} is the major ion released under physiological circumstances is due to the high Ca^{2+} concentration of Ca^{2+} inside the ER/SR compared with other cations. The selectivity filter for Ca^{2+} in ryanodine receptors is the amino acid sequence GIGD, compared with GVGD in IP_3 receptors. The receptor can be opened experimentally by:

- Low concentrations (nM) of ryanodine.
- Direct interaction of the SR with the dihydropyridine (DHP) receptor.
- Ca^{2+} itself – Ca^{2+}-induced Ca^{2+} release.
- Various adenine derivatives, including ATP, ADP, AMP, cyclic AMP, adenine and cyclic ADP ribose.

- Xanthines such as caffeine and pentifylline, 4-chloro-*m*-cresol.
- Sucamin.

But, three agents open ryanodine receptor channels physiologically, and allow Ca^{2+} into the cytosol, depending on cell type:

1. Direct coupling to the DHP receptor in the plasma membrane – skeletal muscle.
2. Ca^{2+}-induced Ca^{2+} release – heart muscle.
3. Cyclic ADP ribose – fertilised eggs.

Naturally occurring and artificial substances which can block, or modify, the ryanodine receptor channel include volatile and local anaesthetics, Ruthenium Red, dantrolene, neomycin, and some peptide toxins, as well as high concentrations of ryanodine.

In skeletal muscle, the dihydropyridine receptor (DHPR), a voltage-gated $Ca_v1.1$ channel, is arranged in ordered arrays of tetrads, directly opposite a 'foot-like' triad arrangement of the ryanodine receptor. The action potential along the T-tubule causes a structural change in the DHPRs, which is transmitted directly to the RyR1 on the SR, opening it. Ca^{2+} release immediately follows. In contrast, in heart muscle, the influx of Ca^{2+} through the DHPR $Ca_v1.2$ channels provokes Ca^{2+}-induced Ca^{2+} release. Ca^{2+}-induced Ca^{2+} release allows Ca^{2+} sparks to combine in an explosive manner, producing a large Ca^{2+} cloud in the cell, which triggers the contraction for the heart to beat.

The ryanodine receptor looks like a mushroom. The 'stalk' spans the membrane, with helices arranged to form the pore. The 'cap' is the cytosolic domain. In the pore there is a sequence GGGIGDE, similar to the selectivity filter in the K^+ channel, TVGYG. Mutations in this motif, and in key Asp and Glu residues, cause large disruptions in Ca^{2+} channel activity. Three proteins on the luminal side of the ER/SR are important in closing the channel at a particular luminal Ca^{2+} concentration, and can determine oscillations in cytosolic free Ca^{2+}:

- Ca^{2+}-binding proteins – calsequestrin in muscle, calreticulin in non-muscle.
- Sorcin.
- Triadin and junctin.

Several proteins outside the ER/SR bind to ryanodine receptors and affect their activity. On the cytosolic side, three are particularly important:

- DHP receptor.
- FKBP, FKBP12 and FKBP12.6.
- Calmodulin and calmodulin-dependent kinase II.

Other interactive proteins include protein kinases and phosphatases, the Ca^{2+}-binding protein S-100, calexcitin, HOMER and snapin. Mutations in the ryanodine receptor cause certain types of heart failure, cardiac arrhythmias, such as catecholaminergic polymorphic ventricular tachycardia (CPVT), and malignant hyperthermia, through defects in the Ca^{2+} signalling mechanism of either cardiac or skeletal muscle, or both (see Chapter 10).

5.6.1.2 Inositol Trisphosphate (IP₃) Receptor

IP_3 is generated by activation of the enzyme phospholipase C (PLC) in the plasma membrane by coupling to the G-protein G_q or by Tyr kinase linked receptors.

IP_3 binds to receptors on the ER, found in all eukaryotic cells, being some of the largest channel complexes known (Figure 5.9a and b). In mammals, there are three types, IP_3R1-3, with 60–80% sequence similarity between them, and formed as hetero- or homo-tetramers. They can be allosterically regulated by Ca^{2+}, giving a bell-shaped dose–response curve;

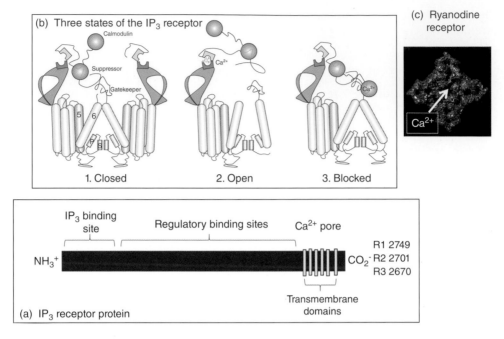

Figure 5.9 Structure of the IP_3 and ryanodine receptors. (a) The domains in the amino acid sequence of the human IP_3 receptor. Reproduced by permission of Welston Court Science Centre. (b) The opening and closing of the IP_3 receptor. Source: Taylor *et al.* (2004). Reproduced with permission from Elsevier. (c) Three-dimensional structure of the ryanodine receptor tetramer. The main portion is the ryanodine receptor tetramer, the arrow indicating the Ca^{2+} pore formed by the tetramer. Courtesy of Dr F. Van Petegem. There are three types of IP_3 and ryanodine receptors in mammals, fish, amphibian, birds and mammals, with their genes on different chromosomes and each with several slice variants. They are both formed from tetramers of large proteins. Both can be regulated by Ca^{2+} and calmodulin, and both can be regulated by proteins in the cytosol and ER lumen, including Ca^{2+}-binding proteins. Both contain several transmembrane domains that combine in the tetramers to form the Ca^{2+} channel, well separated from the agonist site. The IP_3 receptor is around 2700 residues and has the IP_3-binding site near the N-terminus, with six transmembrane domains near the C-terminus some 1600 residues away from IP_3 binding. In between there are several sites that bind regulatory proteins and Ca^{2+}. IP_3 binding enhances the activating Ca^{2+} site and inhibits the inhibitory site. But IP_6 may be an intracellular Ca^{2+} releaser in plants.

300 nM Ca^{2+} activates, increasing the probability of opening, high cytosolic Ca^{2+} inhibits. But Ca^{2+} cannot open the channel without IP_3. This provides a mechanism for regenerative release by neighbouring IP_3 receptors, important in generating a wave or tide of cytosolic free Ca^{2+} moving across the cell. IP_3 receptors can be regulated by the cytosolic enzymes Atk kinase, PKA, CARP, IRBIT, by MgATP, and by proteins in the ER lumen, such as STIM1, calreticulin, and thioredoxin protein ERp44. Mutant IP_3 receptors produce Huntington's or Alzheimer's-like symptoms. IP_3 is generated at the plasma membrane as a result of external stimuli, such as hormones, growth factors, neurotransmitters, neurotrophins, odorants, light and other primary stimuli binding to G-protein-coupled receptors. These receptors activate the enzyme PLC, which catalyses the reaction (Figure 5.10):

$$PIP_2 \rightarrow IP_3 + DAG \tag{5.2}$$

where PIP_2 = phosphatidyl inositol bisphosphate.

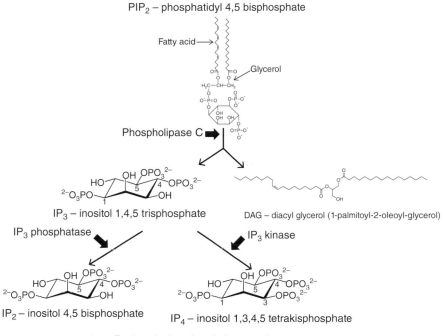

Figure 5.10 Synthesis and loss of IP$_3$. Reproduced by permission of Welston Court Science Centre.

This generates two intracellular signals: inositol trisphosphate (IP$_3$), which releases Ca^{2+} from the ER, and diacylglycerol (DAG), which activates PKC, and provokes its movement to the plasma membrane.

Like the ryanodine receptor, the IP$_3$ receptor is huge, more than 5000 amino acids in size, with four 313-kDa subunits making up the Ca^{2+} channel. Diversity in the IP$_3$ receptor occurs through the three sub-types and alternative splicing. IP$_3$R1 is the most widely distributed. It is found in most tissue types, and all development stages in vertebrates. The importance of IP$_3$R1 was shown in homozygous knock-out or mutated mice. The survival rate ten days after birth was only 20%, the rest dying within 3 weeks. There are two distinct structures, with four-fold symmetry, formed from the four subunits – a windmill-like structure that converts into a square-like structure in the presence of Ca^{2+}, Ca^{2+} acting as an allosteric regulator. The IP$_3$-binding site is a cleft, rich in a cluster of positively charged arginine and lysine residues that bind the three negative phosphates of IP$_3$. The selectivity filter for Ca^{2+} in IP$_3$ receptors is GVGD, similar to the GIGD in ryanodine receptors. A conserved region, distinct from the Ca^{2+} pore and the IP$_3$-binding site, binds HOMER, RACK1, calmodulin, and Ca^{2+}-binding protein 1. IP$_3$ receptors, like ryanodine receptors, are clustered, producing Ca^{2+} 'puffs' prior to a global Ca^{2+} signal.

The IP$_3$ receptor forms a central signalling complex, which acts as a scaffold and controlling structure for signalling events in the cell, and thus has four roles:

1. To generate a large Ca^{2+} cloud or wave for cell activation.
2. To release ER Ca^{2+} into the cytosol when an agonist acts on the plasma membrane to produce IP$_3$, thereby activating the SOCE mechanism.

3. To regulate the type of Ca^{2+} signal via regulation by cytosolic or ER luminal proteins, to produce Ca^{2+} waves, clouds, tides or oscillations.
4. To regulate the activity and location of other proteins.

5.6.2 Mitochondrial Ca^{2+}

Mitochondria play a crucial role in regulating cytosolic free Ca^{2+} signals, via influx and efflux, but are not usually the major source of the Ca^{2+} signal. This is either release from the ER/SR and/or external Ca^{2+} as a result of channels opened in the plasma membrane. Ca^{2+} influx occurs through a mitochondrial calcium uniporter, MCU and MCU1, which is blocked by Ruthenium Red in all cells except yeast. On the other hand, mitochondrial Ca^{2+} efflux occurs through a Na^+/Ca^{2+} exchanger, NCLX, analogous to the Na^+/Ca^{2+} exchanger in the plasma membrane. There is also a H^+/Ca^{2+} exchanger (HCX/Letm1) that works at nM cytosolic free Ca^{2+}, and is bidirectional, working with NCLX when the mitochondrial Ca^{2+} is very high. Many primary stimuli and secondary regulators affect the Ca^{2+} content of mitochondria. Mitochondria are responsible for ATP synthesis by oxidative phosphorylation, which has to be increased when the cell is activated. Mitochondria take up and release Ca^{2+} for one of four reasons:

1. To regulate the cytosolic free Ca^{2+} signal generated by the primary stimulus.
2. To regulate ER Ca^{2+} without global changes in cytosolic Ca^{2+}.
3. To regulate enzymes inside the mitochondrial matrix which are sensitive to Ca^{2+}.
4. To provoke programmed cell death – apoptosis, e.g. by stimulating release of cytochrome *c* to activate the caspase pathway or by regulation of the protein Bcl.

Thus, mitochondria play a crucial role in Ca^{2+} signalling in five ways:

1. Regulation of enzymes inside the mitochondria by Ca^{2+} activation or inhibition to increase substrate supply (NADH) for ATP synthesis.
2. Taking up Ca^{2+} from the cytosol, affecting the size, type and level of the Ca^{2+} signal.
3. Interacting with Ca^{2+} release and uptake from the ER.
4. Releasing Ca^{2+} into the cytosol.
5. Damage to mitochondria by excessive Ca^{2+} uptake, followed by necrosis or activation of apoptosis.

Mitochondria can determine whether a Ca^{2+} signal is a transient or an oscillation, and can restrict it to a micro-domain. Ca^{2+} inside the mitochondria increases ATP synthesis by providing more reducing equivalents for the respiratory chain, through the activation of enzymes involved in pyruvate oxidation and the Krebs cycle. In resting cells, the level of Ca^{2+} in mitochondria is low, typically sub-micromolar, measured with targeted aequorin tagged with part of cytochrome oxidase, COX1. When there is a rise in cytosolic free Ca^{2+}, mitochondria take up Ca^{2+} rapidly, the free Ca^{2+} inside rising to several micromolar. Ca^{2+} is released when the cytosolic free Ca^{2+} returns to sub-micromolar. Too much Ca^{2+} in the mitochondria causes the mitochondrial permeability transition (MPT) pore to open in the mitochondrial membrane, allowing through molecules of less than 1500 Da. This pore plays a role in several pathological conditions, such as stroke and trauma to the brain, neurodegeneration, hepatotoxicity from Reye-type agents, heart necrosis, and various nervous and muscular dystrophies. It may lead to cell death by apoptosis or necrosis. Ca^{2+} overload in the mitochondria also causes Ca^{2+} phosphate to precipitate, causing the cell to become necrotic, as occurs after a heart attack.

The driving force for mitochondrial Ca^{2+} uptake is the electrochemical potential of a pH gradient and membrane potential ($\Delta\Psi$), up to -180 mV negative inside, also the driving force for ATP synthesis. Ca^{2+} uptake into the inner matrix is characterised by:

1. Net Ca^{2+} loss from the medium surrounding mitochondria.
2. Net mitochondrial Ca^{2+} uptake, two H^+ out for one Ca^{2+} in.
3. ^{45}Ca uptake, requiring the respiratory chain, enhanced by ATP or ADP, and blocked by uncouplers such as DNP and FCCP.
4. Increased O_2 uptake by mitochondria by addition of Ca^{2+}.
5. Increase in free Ca^{2+} inside mitochondria.
6. Ca^{2+} uptake blocked by Ruthenium Red (pure form RuRed360).
7. Uptake of Ca^{2+} activates pyruvate oxidation.

Ca^{2+} influx occurs via a 40-kDa protein, mitochondrial uniporter, MCU1 or CBARA1, identified using a database search and experimental evidence. MCU1 is found in all mammalian and other eukaryotic genomes, but not in yeast. Down-regulation using siRNA drastically reduces mitochondrial uptake, and insertion of the protein, made in bacteria, into lipid bilayers produces a Ca^{2+} current which is blocked by Ruthenium Red and La^{3+}. Mutations in the putative pore site inactivate Ca^{2+} uptake in mitochondria, after expression in HeLa cells. This is a 'dominant' inhibition, because the uniporter is a complex of several proteins, with MICU1 (Mitochondrial Calcium Uniporter 1) attached as a Ca^{2+}-dependent regulator. MCU has a low affinity for calcium, 5–10 µM in the cytosol being required for significant uptake into the mitochondria. In humans the complete uniporter consists of MCU, MCU regulatory subunit b (MCUb), essential MCU regulator (EMRE), MICU1, MICU2, and possibly MICU3, MCUI, 2 and 3 binding in the space between the inner and outer membranes. Ca^{2+} passes through the outer membrane via the VDAC. Some mitochondria are closely associated with the endoplasmic reticulum (ER). IP_3 triggers the release of calcium from the ER, creating a microdomain of high Ca^{2+} between the ER and the mitochondria, the conditions for Ca^{2+} uptake by MCU. Mitochondrial Ca^{2+} influx can be activated by naturally occurring compounds, such as polyamines, plant flavonoids and certain oestrogen receptor agonists. Influx is inhibited by Ruthenium Red and RuRed360, nucleotides (ATP > CTP = UTP > GTP), and the thiourea derivative KB-R7943, a Na^+/Ca^{2+} exchange inhibitor. The Ca^{2+} influx channel can be modified through direct binding of other proteins, or covalent modification, e.g. Ca^{2+}-calmodulin, UCP2 and 3, sequence-similar to the chaperone protein uncoupling protein 1 (UCP1), VDAC, kinases and other intracellular signalling systems.

Studies by the mitochondrial Ca^{2+} pioneer Ernesto Carafoli showed that, in excitable cells at least, Na^+, around cytosolic concentrations of 5–20 mM, provokes Ca^{2+} efflux from mitochondria loaded with ^{45}Ca. This Ca^{2+} efflux is inhibited by the diltiazem derivative CGP37157, verapamil, clonazepam and amiloride. There are two mechanisms for mitochondrial Ca^{2+} efflux on the inner membrane:

1. Na^+/Ca^{2+} exchanger, NCLX, with the uptake of three Na^+ per Ca^{2+} out.
2. Na^+/H^+ exchanger, with two H^+ taken up per Ca^{2+} out.

The Na^+/Ca^{2+} exchanger (NCLX) is found particularly in excitable cells, such as muscle and neurones, whereas the Ca^{2+}/H^+ exchanger (HCX) is important in non-excitable cells. NCX has long and short splice variants, and is inhibited by the Na^+/Ca^{2+} exchange inhibitor diltiazem derivative CGP-37157. A knock-out drastically reduces Na^+-dependent Ca^{2+} efflux, and mutants blocked efflux in a dominant-negative manner. Mitochondrial Ca^{2+} efflux is enhanced in cells overexpressing NCLX, and drastically reduced when protein expression is suppressed using siRNA.

Mitochondria can be found localised to particular parts of the cell, specifically to control cytosolic free Ca^{2+} in microdomains. In muscle, there are bands of mitochondria close to the myofibrils, and in sperm, mitochondria are wrapped around the flagellum. In some secretory cells, such as the exocrine pancreas, mitochondria are clustered around the secretory vesicles, where mitochondria can form a Ca^{2+} 'firewall' around the perinuclear, perigranule and sub-plasma membrane microdomains. Mitochondria can prevent a Ca^{2+} signal generated at the apical end of the cell from propagating throughout the cell, thereby allowing the cell to control cytosolic free Ca^{2+} signals independently in different parts of the cell. On the inner surface of the plasma membrane, mitochondria allows sustained Ca^{2+} entry via SOCE, through negative feedback on the mitochondrial Ca^{2+} channel. In HeLa cells, 60% of the mitochondria are located close to the ER, mitochondrial Ca^{2+} acts as a relay, allowing Ca^{2+} to get back into the ER from outside the cell with SOCE open, bypassing the cytosol.

Ca^{2+} is an important regulator of several inner mitochondrial enzymes. These are involved in pyruvate oxidation and the citric acid cycle inside the mitochondria, and are regulated by physiological concentrations of Ca^{2+} in the micromolar range. These enzymes include pyruvate, isocitrate and α-oxoglutarate dehydrogenase phosphatases activated by Ca^{2+}, and pyruvate dehydrogenase kinase inhibited by Ca^{2+}. This increases ATP synthesis. Ca^{2+} also activates the mitochondrial substrate carrier Ca^{2+}-dependent mitochondrial carrier proteins (CaMCs) citrin and aralar, from the Spanish *'araceli hiperlarga'*, meaning 'alter in the sky'. Both are aspartate–glutamate carriers (AGCs), essential for the malate–aspartate–NADH shuttle in and out of the inner mitochondrial matrix. These shuttles are required because NADH, generated by glycolysis, cannot cross the inner mitochondrial membrane directly. The Ca^{2+}-activated shuttle provides a mechanism, independent of direct Ca^{2+} activation of intra-mitochondrial dehydrogenases, for increasing ATP synthesis.

In summary, uptake of Ca^{2+} into mitochondria occurs via a uniporter (MCU), whereas Ca^{2+} release occurs through exchange with either Na^+ (NCLX) or H^+ (HCX) in the cytosol. When the cytosolic free Ca^{2+} rises, mitochondria take up Ca^{2+} in preference to ATP synthesis. This Ca^{2+} uptake depolarises the inner mitochondrial membrane and activates metabolic enzymes that produce NADH for the respiratory chain. Overloading of Ca^{2+} in mitochondria occurs when there is a prolonged, pathological rise in cytosolic free Ca^{2+}, and in oxidative stress. Ca^{2+} uptake and release mechanisms in mitochondria do not resemble the mechanisms found in extant bacteria, arguing against the endosymbiotic theory for the origin of mitochondria in the evolution of eukaryotic cells.

5.6.3   Lysosomal and Endosomal Ca^{2+}

Endocytosis is the pathway in cells trafficking protein vesicles from the plasma membrane to the lysosomes. In between, there is a complex vesicular network of early, mid and late endosomes. Ca^{2+} plays an important role in several parts of the endosome–lysosome trafficking pathway. Both lysosomes and endosomes contain Ca^{2+}, and thus have mechanisms for Ca^{2+} uptake and release. Lysosomes are the garbage disposal system in animal cells. Degradative enzymes inside animal lysosomes include lipases, amylase, proteases and nucleases. Plants and yeast use a similar system called lytic vesicles. Changes in cytosolic free Ca^{2+} can affect the endosome–lysosome pathway and, in some cells, lysosomes release Ca^{2+} into the cytosol, triggered by the intracellular messenger NAADP.

There are three components in the endosome pathway: early, recycling and late. Maturation of the endosome, as it moves towards the lysosomes, involves acidification and loss of Ca^{2+}. Cytosolic free Ca^{2+} may regulate fusion of late endosomes with the lysosomes, and reformation of lysosomes. This requires ATP. The three key ions and ionic movements in lysosomes and endosomes are H^+, Cl^- and Ca^{2+}. The low pH inside lysosomes is responsible for maintaining a high free Ca^{2+} inside, in the millimolar range, compared with the submicromolar Ca^{2+} in the cytosol of resting cells. Measurement of free Ca^{2+} inside lysosomes and endosomes has proved difficult because of the low pH. Four probes have been developed:

1. Fura-2–dextran calibrated in acid conditions.
2. Rhod-dextran.
3. Oregon Green 488 BAPTA-5N.
4. GEM-GECO1.

The latter was genetically engineered using forced evolution, with the CaM binding in myosin light chain kinase, and a single, circularised GFP. This made it pH insensitive, and with the right K_d for the high Ca^{2+} inside endosomes and lysosomes. Thus, the free Ca^{2+} inside early endosomes decreases rapidly, within minutes, to micro-molar levels. The free Ca^{2+} inside lysosomes in intact macrophages is around 0.5 mM, pH inside the lysosome being regulated by the free Ca^{2+}. An increase in lysosomal pH, induced by bafilomycin A1 or ammonium chloride, reduces the free Ca^{2+} inside lysosomes from millimolar levels to micromolar within 10–20 min, reaching 0.1–0.01 μM within 1 h. Re-acidification of the lysosomes returns lysosomal free Ca^{2+} to millimolar levels within minutes. Such measurements led to the discovery that deregulation of lysosomal Ca^{2+} is a key part of Niemann–Pick type C disease, which is a sphingosine storage disease. Endosomal acidification requires both loss of luminal Ca^{2+} and Cl^-. Reformation of lysosomes involves condensation of the luminal contents. This process also requires a Ca^{2+} flux and is blocked by Ca^{2+} chelators loaded into the fused lysosome. Three proteins have been identified as potential Ca^{2+} channels for releasing Ca^{2+} into the cytosol:

1. A TRP channel – TRPL channel found as CUP-5 in *Caenorhabditis elegans*, and TRPML 1/TRPV2 in mammals.
2. Mucolipins, analogues of TRP and designated TRPML channels, TRPML 3 being the most significant and found in the lysosomal membrane.
3. Two-pore channels (TPCs) opened by the second messenger NAADP.

The endosomal pathway is important in manipulating DNA and RNA in live cells, such as the two methods for manipulating mRNA in live cells: siRNA and peptide nucleic acid (PNA). PNAs are non-ionic DNA/RNA constructs, where the ribose has been replaced by uncharged amide links. They are taken up into cells by endocytosis. Ca^{2+} inside lysosomes enables PNAs to be released from endosomes, so that they can interfere with their appropriate mRNA and inhibit protein synthesis. Lysosomal Ca^{2+} also plays a role in apoptosis.

5.6.4 Secretory Vesicle Ca²⁺

Neurotransmitters, hormones such as insulin and adrenaline, histamine, and digestive enzymes are secreted by fusion of internal vesicles with the plasma membrane. All secretory vesicles contain four main components: proteins, divalent cations, nucleotides and small organic molecules. Many contain Ca^{2+} as part of the internal structure for the stored agent. This Ca^{2+} gets into the vesicle via a Ca^{2+}/H^+ exchanger. But in some cells, Ca^{2+} stored within secretory vesicles can be released into the cytosol through IP_3 or NAADP.

5.6.5 Peroxisomal Ca²⁺

Peroxisomes, found in most eukaryotic cells, contain enzymes that get rid of toxic substances via oxidation. In resting cells, the free Ca^{2+} in peroxisomes is higher than in the cytosol, increasing as cytosolic free Ca^{2+} is raised by a primary cell stimulus. Ca^{2+}-calmodulin can activate plant catalase. But it is not clear whether changes in peroxisomal free Ca^{2+} regulate any enzymes *in situ*.

5.6.6 Control of Ca^{2+} by the Golgi

Ca^{2+} and Mn^{2+} have important functions in the Golgi, and are concentrated there mainly by a P-type Ca^{2+}-MgATPase known as Secretory Pathway Calcium MgATPase (SPCA), encoded by the genes *ATP2C1* and *ATP2C2*. These exchange Mn^{2+} or Ca^{2+} for H^+. Protons, on the other hand, are accumulated in the Golgi by V-type ATPases. The Golgi apparatus – the trans-Golgi network (TGN) – consists of a plate-like structure that passes vesicles from one to the other, as the mechanism for producing secretory vesicles, and moving proteins to the plasma membrane. The Golgi adds carbohydrates and lipids to proteins to form glycoproteins and glycolipids. Particularly important are the glycotransferases (GTs) and sulphotransferases (STs). Several, but not all, require Mn^{2+} or Ca^{2+} for maximum activity. Many contain a divalent cation-binding motive, DxD, which binds Mn^{2+} or Ca^{2+} at millimolar concentrations. The Golgi also contains kinases and proteases, several of which are dependent on either Ca^{2+} or Mn^{2+}. For example, casein kinase is activated by millimolar Ca^{2+} or sub-millimolar Mn^{2+} as an important reaction to produce casein phosphate in milk. This binds Ca^{2+} strongly, leading to a net concentration of Ca^{2+} in milk of several mM. Furin, on the other hand, is a protease located in the Golgi, involved in pro-hormone cleavage, and also in processing the enzyme lactase (lactase phlorizin hydrolase) in the small intestine. Like the proteases PC1 and 2, these only require some 5–100 μM Ca^{2+} for maximum activity. Other Ca^{2+}-dependent enzymes in the Golgi include nucleobinding members of the CREC family Cab45, reticulocalbin, ERc45, calcineurin and EF-hand Ca^{2+}-binding proteins. Ca^{2+} in the Golgi is involved in the response to oxidative stress, and anterograde and retrograde vesicle transfer between the Golgi stacks requires Ca^{2+}. All secretory vesicles, processed through the Golgi, contain large amounts of Ca^{2+} and other divalent cations. These help to stabilise the stored contents and make them readily soluble immediately when they fuse with the plasma membrane. The inside of the Golgi is quite acidic, involving Ca^{2+}/H^+, H^+/Na^+ and Na^+/Ca^{2+} exchangers.

Overexpression or depletion of SPCA produce alterations in the cytosolic free Ca^{2+} signals, including changes in Ca^{2+} oscillations. Thus, the Golgi can affect Ca^{2+} signals. But there is no evidence that cytosolic Ca^{2+} signals can be generated by the Golgi, in a way similar to the SR/ER. SPCA1 is high in sperm, where, like yeast, there appears to be little or no SERCA Ca^{2+} pump, and in skin cells, where mutations cause the genetic disorder Hailey–Hailey disease (see Chapter 10). SPCA2, on the other hand, is most abundant in the gastrointestinal tract, the trachea, salivary gland, thyroid, mammary gland and prostate. SPCAs have a higher affinity for Ca^{2+} than SERCA pumps, but their Ca^{2+} turnover is slower.

5.6.7 Nucleus and Ca^{2+}

Changes in free Ca^{2+} in the cytosol and ER play an important role in regulating the expression of particular nuclear genes. Because nucleic acids are negatively charged at physiological pH, both DNA and RNA bind Ca^{2+}. However, the regulation of free Ca^{2+} in the nucleus, and whether there is any barrier to Ca^{2+} between the cytosol and nucleus, is not clear. Calmodulin and transcription factors can be signalled by Ca^{2+} to move to the nucleus. Proteins in the nuclear pore, and others that traverse the ER membrane into the nucleus, can also communicate with the ER in a way dependent on the free Ca^{2+} in the ER lumen, and thereby regulate gene expression. An important example is Ire1P, discovered in yeast, which activates the unfolded protein response.

The nuclear pores, seen using electron microscopy or atomic force microscopy, are large, and would not be expected to act as a barrier to Ca^{2+}. Fluorescent signals, imaged from fura-2 and fluo-3, show that nuclear free Ca^{2+} rises in parallel with that in the cytosol. However, these fluors overload into both the nucleus and ER, the latter affecting the permeability properties of the nuclear pore. Isolated nuclei from plants have a Ca^{2+} barrier under specific conditions, and there appears to be a nuclear membrane Ca^{2+} pump distinct from the SERCA pump in the ER. Furthermore, aequorin targeted to the nucleus has shown that the nuclear free Ca^{2+} does not rise to the same level as that in the cytosol.

5.6.8 Plant Organelles and Ca^{2+}

Ca^{2+} is a key intracellular signal in plants, regulating genes, and the responses of plant cells to hormones, wind and temperature, and light (see Chapter 9), depletion of cytosolic free Ca^{2+} inducing photosynthesis. Plant cells contain two special organelles not found in animal cells: the chloroplasts, and the vacuole entrapped by the tonoplast membrane. Both take up and release Ca^{2+}. The vacuole is a large, membrane-bound organelle found in all plants and fungi. In mature plants the vacuole can make up 30–80% of the total cell volume. The tonoplast is the membrane that surrounds the vacuole. The vacuole has three main functions:

- To get rid of harmful substances.
- To provide turgor pressure to maintain the strength of the plant.
- A Ca^{2+} store.

The tonoplast transports H^+ into the vacuole, making it acid, and can remove toxic heavy metals. It also contains aquaporins to regulate water movement in and out of the vacuole. The tonoplast membrane contains a chloride channel, with a conductance of some 50 pS, which can be activated by Ca^{2+}-calmodulin, acting on the cytosolic side.

Intracellular Ca^{2+} is regulated in, and by, both chloroplasts and the vacuole, the latter via the tonoplast. Both organelles can contain large amounts of Ca^{2+}. Chloroplast total Ca^{2+} can be as high as 4–23 mM. Similarly, the total Ca^{2+} content in vacuoles can be high, stored often as calcium oxalate. Light induces Ca^{2+} uptake in isolated chloroplasts from spinach. The inhibitor DCMU (3-(3,4-dichlorophenyl)-1,1-dimethylurea), a specific and sensitive blocker of the photosynthetic electron transport chain and thus ATP synthesis from light, inhibits this Ca^{2+} uptake. This is consistent with the driving force being the chloroplast membrane potential, negative inside, generated by a Mitchell-type chemiosmotic mechanism. Modification of this membrane potential using lipophilic cations, ionophores + K^+, disrupts Ca^{2+} influx. Using aequorin, targeted either to the cytosol or chloroplasts in two model plant systems, the weed *Arabidopsis* and the tobacco plant *Nicotiana plumbaginifolia*, oscillations in cytosolic free Ca^{2+} have been observed during the normal 24 h light/dark cycle, mimicked by changes in free Ca^{2+} inside the chloroplast soma. Like mitochondria, chloroplasts accumulate a higher free Ca^{2+} than the cytosol, rising from 0.15 μM to a peak of 5–10 μM. Switching off the light causes a large Ca^{2+} influx into the stroma of the chloroplasts, with a lag of about 5 min, reaching a peak by 20–30 min. The magnitude of the Ca^{2+} influx is proportional to the duration of exposure to the light prior to the dark. Ca^{2+} plays a vital role in the thylakoid membrane, which houses a Ca^{2+}/H^+ exchanger. It also regulates several chloroplast enzymes. But chloroplasts also modify cytosolic free Ca^{2+} signals in plants. Tonoplast Ca^{2+} uptake is also mediated by a Ca^{2+}/H^+ exchanger, which in higher plants is voltage sensitive. Specialised plant cells accumulate large amounts of calcium oxalate as a defence against herbivores, but there is no good evidence that this can be mobilised quickly to play a role in acute Ca^{2+} signalling.

5.6.9 Acidocalcisomes

Acidocalcisomes, also known as calcisomes, are membrane-bound vesicles, often just a few hundred nanometres in diameter, which provide a major store of phosphate in cells, and contain quite large amounts of Ca^{2+}. These vesicles contain several other cations, including Na^+, Mg^{2+}, Zn^{2+} and Fe^{2+}, and Ca^{2+}, the main counterion for phosphate. The pH inside is acidic, hence their name. Cells require phosphate for nucleotide, DNA and RNA synthesis, and phosphorylation of proteins. In order to double its DNA a eukaryotic cell requires more phosphate than is available in the total nucleotide pool of ATP, GTP, CTP and TTP. Acidocalcisomes occur in several microbial pathogens, such as *Helicobacter*, *Mycobacterium*, *Shigella* and *Salmonella*, and have also been found in parasites, such as trypanosomes, as well as protists, coccoliths, plants and animals, including humans. They contain large amounts of polyphosphates, and, thus, enzymes, such as polyphosphate kinase and polyphosphate phosphatase, which regulate the number of phosphates in the polyphosphate chain. In bacteria, polyphosphate complexes with PHB, forming a non-proteinaceous Ca^{2+} channel in the membrane.

Acidocalcisomes play a role in development, sporulation, stress and osmoregulation. But it is not clear whether they play a role in regulating cytosolic free Ca^{2+}. In eggs they do not release Ca^{2+} in the presence of IP_3 or ryanodine. The major mechanism for releasing Ca^{2+} involves changes in intravesicular and cytosolic pH and Na^+. These occur after fertilisation. A rise in cytosolic Na^+ or H^+, i.e. acidification outside the acidocalcisomes, causes Ca^{2+} release by exchange across the acidocalcisome membrane involving P- and V-type MgATPases for Ca^{2+}/Na^+, Ca^{2+}/H^+ and Na^+/H^+ exchange. Conversely, an increase in pH (i.e. alkinisation) inside the vesicle causes Ca^{2+} to be released. In yolk-platelets and sea urchin eggs, hydrolysis of the global protein network (GPN)-loop GTPase via cathepsin C increases the osmotic pressure inside the vesicle, causing it to take up water. This leads to the vesicle bursting, thereby releasing its Ca^{2+}. Both acidocalcisomes and yolk-platelets contain bafilomycin A1-sensitive MgATPases, which pump H^+ into the vesicle, making it acidic inside.

5.7 Second Messengers and Regulation of Ca^{2+} Signalling in the Cytosol

Six 'second messengers' have been discovered, which can affect intracellular Ca^{2+}:

1. IP_3.
2. Cyclic ADP ribose.
3. NAADP.
4. Sphingosine 1-phosphate.
5. Cyclic AMP.
6. Cyclic GMP.

Ca^{2+} itself, IP_3, cyclic ADP ribose, NAADP, and sphingosine 1-phosphate can provoke release of Ca^{2+} into the cytosol, depending on cell type and the primary stimulus, thereby raising the cytosolic free Ca^{2+} concentration. Ca^{2+} and IP_3 release Ca^{2+} from the SR/ER. NAADP releases Ca^{2+} from lysosomes. In contrast, cyclic AMP and cyclic GMP alter Ca^{2+} signals by activating kinases that phosphorylate Ca^{2+} channels in the plasma membrane, channels in the SR/ER or other organelles, or regulatory proteins such as phospholamban, the latter regulating the Ca^{2+}-MgATPase in the SR of cardiac muscle.

Three cell systems have been particularly useful in providing key evidence for how cyclic ADP ribose and NAADP are made, how they work, and how they affect intracellular Ca^{2+}:

- Eggs from the sea urchin, *Lytechinus pictus* and *Strongylocentrotus purpurata*.
- Pancreatic acinar cells.
- Pancreatic β-cells.

The first clue that there were second messengers in cells independent of cyclic nucleotides and IP_3 came when it was shown that addition of β-NAD^+ and β-$NADP^+$ to microsomes prepared from sea urchin eggs provoked Ca^{2+} release, independent of IP_3. Two compounds were then isolated: cyclic ADP ribose and NAADP. Both can provoke Ca^{2+} release in a range of invertebrate and vertebrate cells, and plants, distinct, in terms of timing, concentration range and pharmacology, from IP_3-mediated release from the ER. Cyclic ADP ribose is made by cyclising β-NAD^+, catalysed by ADP-ribosyl cyclase, and degraded by a hydrolase, which opens the cyclic ring by breaking the bond between the ribose and the purine of ADP, resulting in formation of adenosine diphosphoribose (ADPR). In contrast, NAADP is formed from β-$NADP^+$ reacting with the ionised form of nicotinic acid, releasing nicotinamide. NAADP is degraded by a 2′-phosphatase, producing phosphate and NAAD. There is also cyclic ADPR-P formed from β-$NADP^+$, in a similar way to cyclic ADP ribose from β-NAD^+.

Cyclic ADP ribose acts by binding to the ryanodine receptor at a different site from ryanodine, opening the Ca^{2+} channel, thereby releasing Ca^{2+} from the ER into the cytosol. The receptor for NAADP is on the membrane of lysosomes or secretory vesicles. An unusual property of the NAADP receptor is the biphasic effect of NAADP. At very low concentrations, in the range 200 pM, NAADP inhibits Ca^{2+} release, whereas at higher concentrations, with a half maximum of 30 nM, it causes Ca^{2+} release into the cytosol. Levels of cyclic ADP ribose and NAADP can reach micromolar concentrations. Two pieces of evidence support cyclic ADP ribose acting on ryanodine receptors and not IP_3 receptors:

1. Selective inhibitors of the ryanodine receptor, such as Mg^{2+}, Ruthenium Red, dantrolene and procaine, block intracellular Ca^{2+} release by cyclic ADP ribose.
2. Heparin, which blocks IP_3-induced Ca^{2+} release, does not block Ca^{2+} release induced by cyclic ADP ribose.

But, it is still not entirely clear why cyclic ADP ribose and NAADP are there, as opposed to IP_3 or Ca^{2+}-induced Ca^{2+} release.

Sphingosine 1-phosphate is unusual as a signalling molecule in that it acts both outside and inside cells. Sphingosine 1-phosphate can generate a cytosolic Ca^{2+} transient, either by binding to a cell surface G-protein-coupled receptor or by direct release of Ca^{2+} from the ER inside the cell, independently of IP_3 or its receptor. A particular puzzle is this release does not appear to activate SOCE, since the Ca^{2+} signal is the same with or without extracellular Ca^{2+}. Sphingosine 1-phosphate may provide a mechanism for releasing Ca^{2+} directly into the nucleus, without generating a global Ca^{2+} signal. Many cells, particularly in the immune system, have G-coupled receptors on their surface that can be activated by sphingosine 1-phosphate. Five such receptors are found in the human genome: S1PR1–5, also known as EDG1, 3, 5, 6 and 8 (where EDG = endothelial differentiation gene). Sphingosine is synthesised from palmitoyl-CoA, and can enter cells, where it is phosphorylated, 'sphing-' coming from the mythical Sphinx. Sphingosine 1-phosphate is involved in the immune system, in cell trafficking, moving lymphocytes from one site to another, in cell proliferation, and plays a role in determining whether a cell enters apoptosis. Sphingosine 1-phosphate may also play a role in neurones, cancer, and multiple sclerosis. Sphingosine 1-phosphate can activate nitric oxide synthase.

Cyclic nucleotides can regulate changes in cytosolic free Ca^{2+}, but are not usually triggers themselves for producing a Ca^{2+} signal. Cyclic nucleotides are 'analogue' intracellular messengers. Their main role in Ca^{2+} signalling is to alter the size, timing and type of cytosolic free Ca^{2+} signal. Cyclic AMP activates PKA, which in the heart phosphorylates Ca^{2+} channels and phospholamban. Phosphorylation of the voltage-gated Ca^{2+} channels speeds them up, and phospholamban activates the SERCA2a pump, increasing the size of the SR Ca^{2+} store. This leads to more Ca^{2+} being released by each action potential. Ca^{2+} also regulates the concentration of cyclic nucleotides. It can inhibit or activate adenylate cyclase, depending on cell type, and activate cyclic nucleotide phosphodiesterases via calmodulin. A further poorly studied messenger is H_2S.

5.8   Pore Formers and Intracellular Ca^{2+}

The membrane attack complex of complement, perforin, and bacterial toxins, such as alfatoxin and streptolysin, insert into the plasma membrane, where they form pores. These pores allows ions and organic substances to leak out of the cell, and Ca^{2+} to enter, which then activates cell responses. Ca^{2+} activates a protection mechanism that removes the pores by vesiculation. If the cell does not do this in time, it lyses. The complement cascade involves a family of proteins in the blood, C1 to C9, and can start via one of three pathways: classical, alternative and other pathway. There are two ways that these pathways can induce Ca^{2+} signals inside cells:

1. Production of the peptides C3a, C3b, and C5a, which, via cytosolic Ca^{2+} signals, act as chemoattractants for phagocytes, induce oxygen metabolite production, and activate mast cells to release histamine.
2. Formation of the membrane attack complex, $C5b6789_n$, whose initial interaction with the plasma membrane allows Ca^{2+} to enter the cell.

The membrane attack complex can cause the cytosolic free Ca^{2+} to rise to greater than 10 µM. This activates Ca^{2+}-dependent processes, such as superoxide production in phagocytes. Multiple C9s, typically nine, form a ring structure, which forms a large pore in the plasma membrane – the next Rubicon. The rise in cytosolic free Ca^{2+} activates a protection mechanism, which tries to remove the potentially lethal C9 complexes by vesiculation. If this is successful, the cell recovers its ATP and survives. If the cell does not succeed in removing the membrane attack complexes in time, it crosses the ultimate Rubicon and lyses. When cell lysis is measured in cell populations it appears to be a graded analogue process. But, the amount of cytosolic enzyme in the extracellular fluid is directly proportional to the **number** of cells that have lysed, not a graded release from all of the cells. Other pore formers also let Ca^{2+} into cells, and will activate a protection mechanism similar to that activated by complement.

Thus pore formers produce rises in cytosolic free Ca^{2+} in the 1–50 µM range, when the cell can recover by removing the potentially lethal complexes. However, if this high level remains for a few minutes or more, mitochondria are irreversibly damaged, proteases and nucleases activated, and the cell then dies through necrosis, apoptosis or lysis.

5.9   Gap Junctions and Connexins

Gap junctions allow electrical and chemical communication between cells. Thus, Ca^{2+} signals can be transmitted between cells. Gap junctions are plaques formed from the protein connexin, of which there are several variants. The junctions can be made of one or more different connexins. They form by a hemi-channel from one cell linking with the hemi-channel of the neighbouring cell. However, hemi-channels can also find their way to the plasma membrane, where, under certain circumstances, they let Ca^{2+} into the cell. By engineering aequorin on to the C-terminus of a connexin, free Ca^{2+} can be measured close to the inner surface of the plasma membrane, where is can be tens of micromolar.

Intercellular gap junctions allow Ca^{2+} and substances up to a 1000 Da or so to move between cells. They also make adjoining cells electrically connected. So depolarisation in one cell can spread electrotonically to the next cell, and action potentials can move across a cell layer. Important examples are the heart, liver, secretory organs, such as the endocrine and exocrine pancreas and salivary gland, smooth muscle, and epithelial tissues. In some organisms, such as hydroids and jellyfish, electrically conducting pathways and nerve conduction occurs via gap junctions. For example, in the luminous hydroid *Obelia geniculata* the Ca^{2+} signal required to trigger bioluminescence in the photocyte involves gap junctions.

In fly salivary gland high cytosolic Ca^{2+} can block gap junctions, and thus stop the movement of large molecules from cell to cell. By imaging Ca^{2+} using aequorin, injected into one cell, it was shown that only when the Ca^{2+} cloud hits the gap junction was electrical communication stopped between cells. Gap junctions also allow micromolar cytosolic Ca^{2+} signals to be transmitted between cells. Thus epithelial cell layers exhibit waves of free Ca^{2+} moving through the cell layer, after one cell is stimulated. There are five possible mechanisms explaining such intercellular Ca^{2+} waves, depending on cell type:

1. Electrotonic movement of action potentials between cells, opening voltage-gated Ca^{2+} channels in neighbouring cells.
2. Ca^{2+} moving directly through gap junctions into the adjoining cell.
3. Movement of IP_3 through the gap junctions, to stimulate release of Ca^{2+} from the ER of the adjoining cell, followed by a Ca^{2+} cascade activating IP_3 receptors.
4. Release of ATP by one cell to activate the purinergic receptors on the adjoining cell, which, via G-protein coupling, generate IP_3 and release Ca^{2+} in this cell, and then also open SOCE.
5. Activation of mechanoreceptors as the result of small cell movements in the cell layer.

Thus, gap junctions play an important role in acute Ca^{2+} signalling, enabling Ca^{2+} signals to be generated across cell layers in whole tissues. Gap junctions also play a role in tissue pathology, switching off when one cell is damaged, preventing substances being lost from a healthy adjoining cell. They also play a role in growth and differentiation.

5.10   Other Ion Channels and Ca²⁺

Several other ion channels interact with Ca^{2+} signalling, particularly K^+ and Cl^- channels, which can be activated by a rise in cytosolic free Ca^{2+}, by direct Ca^{2+} binding or Ca^{2+}-calmodulin. Only three types of K^+ channels are regulated by cytosolic free Ca^{2+}:

1. BK = big conductance, also known as MaxiK, slo-1 or $K_{Ca}1.1$.
2. SK = small conductance, also known as SK1, 2 or 3, or $K_{Ca}2.1$, $K_{Ca}2.2$ or $K_{Ca}2.3$.
3. IK = intermediary conductance, also known as SK4 or $K_{Ca}3.1$.

For example, the four SK channels 1–4 are all activated by Ca^{2+}-calmodulin, and can be phosphorylated by CaMKII and dephosphorylated by phosphatase 2A. Similarly, Cl^- channels in animal and plant cells can be opened as the result of a rise in cytosolic free Ca^{2+}. But potassium and chloride channels in themselves do not cause directly a rise or fall in cytosolic free Ca^{2+}. However, when these ion channels depolarise or repolarise the cell membrane, this will inevitably affect voltage-gated Ca^{2+} channels.

5.11   Summary – How Ca²⁺ is Regulated Inside Cells

The sub-micromolar free Ca^{2+} in all resting animal, plant, protist, bacterial and archaean cells is maintained by plasma membrane Ca^{2+} pumps and exchangers, and has clearly played a crucial role in evolution. Without this low free Ca^{2+}, many proteins and nucleic acids cannot function properly, and Ca^{2+} precipitates with phosphate. Many physical, chemical and biological primary stimuli cause a rise in cytosolic free Ca^{2+} within seconds, or even milliseconds, of interacting with the plasma membrane. These Ca^{2+} signals take different forms depending on the physiology of the cell. Muscle twitches require rapid cytosolic Ca^{2+} transients, whereas heart Ca^{2+} transients last about half a second. In contrast, the liver hepatocyte generates Ca^{2+} oscillations, lasting minutes, which enable the cell to sustain activation of intermediary metabolism without loss of Ca^{2+} from the cell. An egg fertilised by a sperm can generate a wave or tide, which activates processes as it moves down the egg. Both extracellular Ca^{2+} and intracellular Ca^{2+} stores contribute to a rise in cytosolic free Ca^{2+} in all eukaryotic cells. However, the bulk of the Ca^{2+} may come from internal release in some cells (e.g. skeletal muscle), whilst in others the main source of the Ca^{2+} rise is from outside the cell (e.g. in the nerve terminal or a cell with a SOCE mechanism). The mitochondria play a key role in restricting the Ca^{2+} signal to a particular site, when they are located near the ER or plasma membrane. Ca^{2+} also plays an important role in regulating enzymes within organelles, such as the SR/ER, mitochondria, lysosomes, Golgi and acidocalcisomes. Ca^{2+} plays an important structural role in secretory vesicles.

Box 5.3   The four levels of cytosolic free Ca^{2+} in cells.

- Level 1: submicromolar in the resting cell.
- Level 2: micromolar, usually between 1 and 10 µM, in the stimulated cell.
- Level 3: high micromolar, up to about 50 µM, in the injured cell.
- Level 4: millimolar, above 0.1 mM, in the dead cell.

In summary:

1. Cytosolic free Ca^{2+} in all resting cells – Archaea, Bacteria and Eukaryota – is maintained at a sub-micromolar level by Ca^{2+} pumps, requiring MgATP, or exchangers for Na^+ or H^+, in the plasma membrane.
2. There are four levels of cytosolic free Ca^{2+} in cells (Box 5.3), though the amount of Ca^{2+} actually moving may be 10–100 times that of the maximum free Ca^{2+} because of Ca^{2+} buffers and targets.
3. Ca^{2+} is stored in all intracellular organelles, which, under certain circumstances, release Ca^{2+} into the cytosol.
4. All intracellular organelles store some Ca^{2+}, where it regulates enzymes: the ER/SR, mitochondria, Golgi, lysosomes, endosomes, peroxisomes, secretory vesicles, acidocalcisomes and the nucleus.
5. Primary stimuli cause a rise of cytosolic free Ca^{2+} by opening Ca^{2+} channels in the plasma membrane and releasing Ca^{2+} from intracellular organelles, typically resulting in a digital cell event – nerve firing, muscle contracting, endocrine or exocrine secretion, egg fertilisation, plant guard cells opening and closing.
6. Secondary regulators act in an analogue way to modify the size, type and location of the cytosolic Ca^{2+} signal.

7. There are five types of plasma membrane Ca^{2+} channels letting Ca^{2+} into a cell: voltage gated, agonist-receptor, release of Ca^{2+} from the ER (store-operated calcium entry – SOCE), intracellular signal operated, and pore formers.
8. The SR in muscle, and ER in non-muscle cells, is the major source of Ca^{2+} for release into the cytosol, via ryanodine or IP_3 receptors. Lysosomes and secretory vesicles, may release Ca^{2+} under certain circumstances.
9. Intracellular Ca^{2+} signals can be provoked and/or altered by other intracellular signals, such as NAADP, cyclic ADP ribose, cyclic nucleotides, sphingosine 1 phosphate and NO.
10. Mitochondria can restrict the cytosolic free Ca^{2+} signal to microdomains, close to the ER and plasma membrane.
11. Gap junctions allow Ca^{2+} signals to pass between cells, and across cell layers.
12. Other ion channels can interact with cytosolic Ca^{2+}, e.g. K^+ and Cl^-, affecting the membrane potential and the threshold for opening voltage-gated Ca^{2+} channels.
13. The key experiments providing the evidence for all of this have been:
    a. Measurement and imaging of free Ca^{2+} in defined compartments of live cells
    b. Patch clamping characterising Ca^{2+} channels in the plasma membrane.

Recommended Reading

*A must read

Books

*Campbell, A.K. (2015) Intracellular Calcium, Chapter 5. How Ca^{2+} is regulated in cells. Chichester: John Wiley & Sons Ltd. Comprehensive review fully referenced.

Chattopadhyay, N. & Brown, E.M. (2003) Calcium-Sensing Receptor. Boston, MA: Kluwer. Comprehensive reviews on the Ca^{2+} receptor by the discoverer.

Hille, B. (2001) Ion Channels of Excitable Membranes, 3rd edn. Sunderland, MA: Sinauer Associates Inc. Classic text on ion channels by a pioneer.

Krebs, J. & Michalak, M.E. (Eds) (2007) Calcium: A matter of life and death. Amsterdam: Elsevier. Useful multi-author book with several chapters on how Ca^{2+} is regulated in cells.

Reviews

Berridge, M.J. (2016) The inositol trisphosphate/calcium signaling pathway in health and disease. *Physiol. Rev.*, **96**, 1261–1296. Review of how IP_3 is involved in normal cells, and in disease, by the discoverer.

Brini, M., Cali, T., Ottolini, D. & Carafoli, E. (2013) The plasma membrane calcium pump in health and disease. *FEBS J.*, **280**, 5385–5397. How the plasma Ca^{2+} MgATPases are involved in regulating cytosolic free Ca^{2+} and their involvemnent in disease.

Docampo, R. & Moreno, S.N. (2011) Acidocalcisomes. *Cell Calcium*, **50**, 113–119. Review of the vesicular Ca^{2+} store in many cells, and how it can regulate cytosolic free Ca^{2+}.

Dubois, C., Prevarskaya, N. & Vanden Abeele, F. (2016) The calcium-signaling toolkit: Updates needed. *Biochim. Biophys. Acta - Mol. Cell Res.*, **1863**, 1337–1343. The key components of the Ca^{2+} signalling system in cells.

Endo, M. (2006) Calcium ion as a second messenger with special reference to excitation–contraction coupling. *J. Pharmac. Sci.*, **100**, 519–524. Review of how Ca^{2+} is regulated in muscle.

Galione, A. (2015) A primer of NAADP-mediated Ca^{2+} signalling: From sea urchin eggs to mammalian cells. *Cell Calcium*, **58**, 27–47. Evidence for NAADP as a regulator of cytosolic Ca^{2+} by a pioneer.

Hogan, P.G. & Rao, A. (2015) Store-operated calcium entry: Mechanisms and modulation. *Biochem. Biophys. Res. Commun.*, **460**, 40–49. Review of how SOCE works.

*Leybaert, L. & Sanderson, M.J. (2012) Intercellular Ca^{2+} waves: Mechanisms and function. *Physiol. Rev.*, **92**, 1359–1392. Useful review of how waves of Ca^{2+} are generated across cell layers.

Parys, J.B. & De Smedt, H. (2012) Inositol 1,4,5-trisphosphate and its receptors. *Adv. Exp. Med. Biol.*, **740**, 255–279. Useful review of the three types on IP_3 receptor in particular cell types.

*Mammucari, C., Raffaello, A., Reane, D.V. & Rizzuto, R. (2016) Molecular structure and pathophysiological roles of the Mitochondrial Calcium Uniporter. *Biochim. Biophys. Acta - Mol. Cell Res.*, **1863**, 2457–2464. How the key protein in mitochondrial Ca^{2+} uptake works, and what can go wrong.

Munaron, L., Avanzato, D., Moccia, F. & Mancardi, D. (2013) Hydrogen sulfide as a regulator of calcium channels. *Cell Calcium*, **53**, 77-84. How H_2S can act as a cell regulator and regulate intracellular Ca^{2+}.

Pyne, S., Adams, D.R. & Pyne, N.J. (2016) Sphingosine 1-phosphate and sphingosine kinases in health and disease: Recent advances. *Progr. Lipid Res.*, **62**, 93–106. Review of the role of sphingosine 1-phosphate and how it causes Ca^{2+} signals in cells.

Reusch, R.N. (2012) Physiological importance of poly-(R)-3-hydroxybutyrates. *Chem. Biodivers.*, **9**, 2343–2366. A non-protein Ca^{2+} channel, by the discoverer.

*Shuttleworth, T.J. (2012) Orai channels – new insights, new ideas. *J. Physiol.*, **590**, 4155–4156. Brief review of Orai, one of the two key components in SOCE.

*Soboloff, J., Rothberg, B.S., Madesh, M. & Gill, D.L. (2012) STIM proteins: Dynamic calcium signal transducers. *Nat. Rev. Mol. Cell. Biol.*, **13**, 549–65. Review of the ER component of SOCE – STIM.

Van Petegem, F. (2012) Ryanodine receptors: Structure and function. *J. Biol. Chem.*, **287**, 31624–31632. How the structure of the ryanodine receptor explains how it causes a rise in cytosolic free Ca^{2+}.

Research Papers

Calcraft, P.J., Ruas, M., Pan, Z. *et al.* (2009) NAADP mobilizes calcium from acidic organelles through two pore channels. *Nature*, **459**, 596–600. Evidence that NAADP can cause a rise in cytosolic free Ca^{2+} by release from lysosomes or another intracellular compartment.

Fabiato, A. & Fabiato, F. (1975) Contractions induced by a calcium-triggered release of calcium from the sarcoplasmic reticulum of single skinned cardiac cells. *J. Physiol.*, **249**, 469–495. First clear evidence for Ca^{2+}-induced Ca^{2+} release from the SR in heart cells.

Fry, T.E.J.H. & Sanderson, M.J. (2001) Propagation of intercellular calcium waves in C6 glioma cells transfected with connexins 43 or 32. *Microsc. Res. Tech.*, **52**, 289–300. Evidence for Ca^{2+} waves across cell layers.

Gerasimenko, J.V., Sherwood, M., Tepikin, A.V., Petersen, O.H. & Gerasimenko, O.V. (2006) NAADP, cADPR and IP_3 all release Ca^{2+} from the endoplasmic reticulum and an acidic store in the secretory granule area. *J. Cell Sci.*, **119**, 226–238. Evidence for Ca^{2+} release from a vesicular store in pancreatic acini cells.

Jammes, F., Hu, H.C., Villiers, F., Bouten, R. & Kwak, J.M. (2011) Calcium-permeable channels in plant cells. *FEBS J.*, **278**, 4262–4276. Evidence for Ca^{2+} channels in plants.

Mattie, M., Brooker, G. & Spiegel, S. (1994) Sphingosine-1-phosphate, a putative second messenger, mobilizes calcium from internal stores via an inositol trisphosphate-independent pathway. *J. Biol. Chem.*, **269**, 3181–3188. First clear evidence that sphingosine-1-phosphate can cause cytosolic Ca^{2+} signals independent of IP_3.

Norimatsu, Y., Hasegawa, K., Shimuzu, N. & Toyoshima, C. (2017) Protein-phospholipid interplay revealed with crystals of a calcium pump. *Nature*, **545**, 193–198.

Patel, A.K. & Campbell, A.K. (1987) The membrane attack complex of complement induces permeability changes via thresholds in individual cells. *Immunology*, **60**, 135–140. The rubicons in cells following attack by the complement, which starts by a large influx of Ca^{2+}.

*Prakriya, M., Feske, S., Gwack, Y., Srikanth, S., Rao, A. & Hogan, P.G. (2006) Orai1 is an essential pore subunit of the CRAC channel. *Nature*, **443**, 230–233. The discovery of Orai1 as the plasma membrane component in the SOCE and CRAC Ca^{2+} channel.

6

How Ca^{2+} Works Inside Cells

Learning Objectives
1. The special chemical properties of Ca^{2+}, explaining why evolution has selected it as a unique intracellular signal.
2. The types of Ca^{2+} ligand inside and outside of cells.
3. Key thermodynamic properties of Ca^{2+} – binding constants, solvation, activity coefficients, kinetics, diffusion.
4. Equations that predict Ca^{2+} binding on and off a ligand.
5. How Ca^{2+} precipitates form.
6. The difference between high affinity and low affinity Ca^{2+} binding proteins.
7. The range of intracellular high affinity Ca^{2+} affinity binding proteins, and what they do.
8. The key Ca^{2+} binding motifs – EF hand, C2, and acidic clusters.
9. The chemistry of Ca^{2+} interactions with other intracellular signals.
10. The chemistry of Ca^{2+} interacting with other ions.

6.1 Biological Chemistry of Ca^{2+}

6.1.1 The Special Biological Chemistry of Ca^{2+}

Evolution has exploited the unique chemistry of calcium as Ca^{2+} to act as an intracellular trigger of a wide range of cellular events in animal, plant, protist and microbial cells (Box 6.1). Four features make Ca^{2+} special:

1. A 10–100-fold increase in cytosolic free Ca^{2+} occurs without osmotic damage to the cell.
2. Ca^{2+} binds with high affinity, and rapidly, at micromolar concentrations to oxygen ligands in proteins, even in the presence of mM Mg^{2+}.
3. Ca^{2+} comes off ligands rapidly, within milliseconds, when the surrounding free Ca^{2+} returns to sub-micromolar levels.
4. Ca^{2+} has only one valency state, and therefore is not susceptible to oxido-reduction reactions.

Fundamentals of Intracellular Calcium, First Edition. Anthony K. Campbell.
© 2018 John Wiley & Sons Ltd. Published 2018 by John Wiley & Sons Ltd.
Companion Website: http://www.wiley.com/go/campbell/calcium

Box 6.1 Cellular events triggered by a rise in intracellular free Ca^{2+}.

1. Action potentials and other types of electrical activity, through Ca^{2+} currents and regulation of other ion channels.
2. Cell movement, including all forms of muscle contraction, chemotaxis, amoeboid and flagellate movement.
3. Secretion via fusion of intracellular vesicles with the plasma membrane.
4. Activation of intermediary metabolism, such as cytosolic glycogen breakdown and mitochondrial oxidation of pyruvate.
5. Regulation of other intracellular messengers, such as cyclic AMP, cyclic GMP, hydrogen sulphide (H_2S) and nitric oxide (NO).
6. Vision in vertebrates and invertebrates.
7. Bioluminescence.
8. Sperm maturation and egg fertilisation.
9. Cell division and differentiation.
10. Cell defence.
11. Cell death.

Na^+ or K^+, or Cl^-, could not be an alternative to Ca^{2+}, since 10–100-fold changes would have catastrophic osmotic effects. Mg^{2+} could not be used, as it is required at millimolar concentrations inside cells to enable ATP, and other nucleotides, to do their job. Cu^{2+} and Fe^{2+} have more than one oxidation state, and Zn^{2+} comes off ligands slowly, which would prevent a muscle relaxing quickly, or a heart to beat.

The key atom in a ligand that makes Ca^{2+} special is oxygen, giving Ca^{2+} seven or eight coordination, allowing the ligand to bind Ca^{2+} at micromolar concentrations of Ca^{2+} in the presence of millimolar Mg^{2+}. Mg^{2+} usually prefers six coordination. Ca^{2+} and Mg^{2+} can bind nitrogen-containing ligands. But, these are not selective for Ca^{2+}. Ca^{2+} binds significantly to several oxygen containing inorganic and organic anions (Figure 6.1). These form precipitates with Ca^{2+} that provide the hard structures of bones, teeth, shells, balance organs in jellyfish, and calcite crystals in the lenses of starfish and extinct trilobites. Ca^{2+} also forms precipitates with oxalate in plants, and in pathological conditions in animals.

Several oxygen containing organic anions also have significant Ca^{2+}-binding constants, including nucleotides, DNA, RNA, polysaccharides, inosities, such as inositol 1,4,5-trisphosphate (IP_3) and inositol 1,3,4,5-tetrakisphosphate (IP_4), citrate, acidic amino acids, such as aspartate and glutamate, almost fully ionised at pH 7 in the cell cytosol, and acid sugars such as neuraminic acid. However, at the micromolar concentrations of free Ca^{2+}, the fractional binding of Ca^{2+} to these ligands is small. High fractional binding at micromolar free Ca^{2+} requires a cluster of oxygen ligands, charged or uncharged. This occurs in the EF hand loop and C2 domain of high affinity Ca^{2+} binding proteins.

The targets for a rise in cytosolic free Ca^{2+}, or Ca^{2+} within organelles, are binding proteins that use oxygen, typically in aspartate and glutamate, together with oxygen in the peptide chain, to provide high affinity Ca^{2+} binding sites. These allow Ca^{2+} to bind, and come off, the target quickly. Intracellular Ca^{2+} is able to trigger cell events in eight ways:

1. Regulation of action potentials, by opening of Ca^{2+} channels, and the activation by Ca^{2+} of other ion channels, such as K^+ and Cl^-.
2. Generation of a global cytosolic Ca^{2+} signal resulting from loss of Ca^{2+} from the endoplasmic reticulum (SOCE).

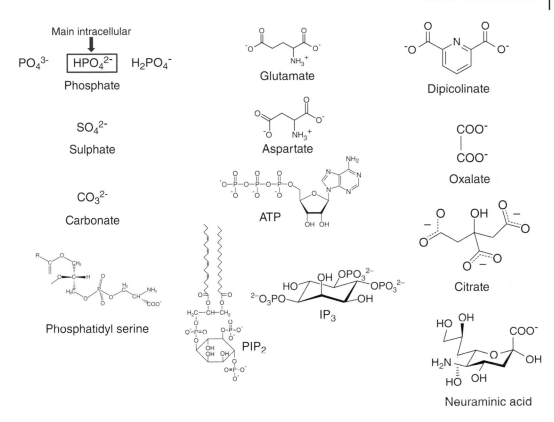

Figure 6.1 Some naturally occurring inorganic and small organic Ca²⁺ ligands; also with octahedral and hexahedral binding.

3. Regulation of the enzymes that control the intracellular and extracellular concentration of other intracellular signals.
4. Activation of intracellular enzymes that cause covalent modification of target proteins, particularly kinases, phosphatases and proteases.
5. Target proteins that cause structural changes in the proteins causing the cell event, such as actin depolymerisation by gelsolin.
6. Deinhibition of protein complexes, such as tropomyosin by troponin, or the release of activators.
7. Activation or inhibition of transcription factors, and binding to response elements, such as the ER stress response or activation of the NFATc (nuclear factor of activated T-cells) pathway.
8. Regulation of RNA processing – splicing, movement out of the nucleus, degradation.

For Ca²⁺ to do these jobs as an intracellular signal, four things are needed:

1. The Ca²⁺ signal must reach its target, and last for the right length of time.
2. The Ca²⁺ target must have the right affinity, so that a high proportion of the target binds Ca²⁺.
3. Ca²⁺ binding must have an effect on the structure of its target.
4. The Ca²⁺ signal must not cause precipitation of inorganic ligands.

6.1.2 Key Chemical Properties of Ca²⁺

Calcium is the fifth most abundant element by weight in the Earth's crust, and the fifth most abundant ion in seawater, after sodium, chloride, magnesium and sulphate. In vertebrates, and shelled invertebrates, calcium is the most abundant element, because of the large amount of calcium in bone, teeth and shells. About 97% of naturally occurring calcium is ^{40}Ca. Calcium has only one ionic state – Ca^{2+}, in which it is always found. Ca^{2+} has the electronic configuration of argon, and thus has no spare electrons to donate. However, Ca^{2+} has empty 3d, 4s and 4p orbitals, which are able to accept electrons from donors, which become coordinating ligands.

6.1.3 Ca²⁺ Ligands

Naturally occurring ligands that bind Ca^{2+} include inorganic, small organic and macromolecular substances. Four parameters characterise interaction between Ca^{2+} and a ligand:

1. Non-metallic coordinating element; nitrogen, oxygen or sulphur, with oxygen the most important biologically.
2. Coordination number; the number of atoms of ions surrounding Ca^{2+} when bound to a ligand. For a polydentate ligand binding Ca^{2+} this is 7–9, typically 8, compared to 5 or 6 for Mg^{2+}.
3. Thermodynamic constants; i.e., the on and off rates for Ca^{2+} binding, which determine the affinity of binding, and how fast it goes on or comes off when free Ca^{2+} changes.
4. Stereochemistry of Ca^{2+} binding, and how this changes the shape of the ligand.

High-affinity Ca^{2+}-binding proteins, with 7 or 8 coordination, are able to select Ca^{2+} over Mg^{2+}, even when the free concentration of Ca^{2+} is 10 000 times less than Mg^{2+}. Typically, in a Ca^{2+}-binding protein involved in intracellular signalling, the oxygen ligands for Ca^{2+} come from acidic acid residues (Asp and Glu), the carbonyl in the peptide chain and water. However, oxygen from inorganic and small organic ligands can come from phosphate, carbonate, sulphate and sugars. The physiological or pathological significance of Ca^{2+} binding to a ligand is judged by relating the binding constant for Ca^{2+} and Mg^{2+} to the free concentrations of Ca^{2+} that exists in the intra- or extra-cellular compartment where the ligand is found. The equation for one Ca^{2+} binding reversibly is:

$$Ca^{2+} + L = CaL \tag{6.1}$$

$$K_d^{Ca} = \left[Ca^{2+}\right][L]/[CaL] \tag{6.2}$$

The total ligand concentration:

$$L_T = [L] + [CaL] \tag{6.3}$$

Thus, the fractional saturation of the ligand by Ca^{2+} is:

$$[CaL]/[L_T] = \left[Ca^{2+}\right]/\left(\left[Ca^{2+}\right] + K_d^{Ca}\right) \tag{6.4}$$

Often binding affinities in reference works are quoted as association constants, K_a, where the units are in M^{-1}. The dissociation constant, K_d, is the reciprocal of K_a:

$$K_d = 1/K_a \tag{6.5}$$

$$pK_d = -\log_{10}\left(K_d\right) \tag{6.6}$$

The benefit of using K_d, with units in Molar (M), is that it is much easier to see at a glance whether binding of a ligand is relevant physiologically. For example, a ligand with a pK_d of 6 (i.e. K_d = 1 μM) will be more than 90% saturated by Ca^{2+} when the free Ca^{2+} is 10 μM, but a ligand with a pK_d of 3 (i.e. K_d = 1 mM), will only be around 1% saturated by Ca^{2+} if the free Ca^{2+} is 10 μM. For Ca^{2+} binding, the on rate is k_1 (M^{-1} s^{-1}):

$$Ca^{2+} + L \xrightarrow{k_1} CaL \tag{6.7}$$

The off rate for Ca^{2+} is k_{-1} (s^{-1}):

$$CaL \xrightarrow{k_{-1}} Ca^{2+} + L \tag{6.8}$$

The on rate k_1 is dependent essentially on the rate of diffusion, and occurs within milliseconds for most Ca^{2+}-binding proteins. But, the off rate, k_{-1}, must also be in the millisecond range, if a muscle is to twitch or a heart myocyte to beat once a second.

The three most significant inorganic ligands for Ca^{2+} are phosphate (HPO_4^{2-}), carbonate (CO_3^{2-}) and sulphate (SO_4^{-}). However, with pK_{Ca} in the range 0.9–6.3, only phosphate binds any significant amount of Ca^{2+} in the cytosol. Phosphoric acid (H_3PO_4) is a tribasic acid and therefore exists in four forms:

$$H_3PO_4 = H^+ + H_2PO_4^- = H^+ + HPO_4^{2-} = H^+ + PO_4^{3-} \tag{6.9}$$

The pK_a for HPO_4^{2-} is 6.8, so this is the major phosphate anion at physiological pH, more than 80% being in this form at pH 7.4. As $Ca_3(PO_4^{3-})_2$ is the least-soluble calcium salt, calcium phosphate precipitates can be induced by making the solution alkaline. However, it is the form of $CaHPO_4$ that is found in many naturally occurring calcium phosphate precipitates such as bone. The pK_{Ca} for HPO_4^{2-} is 2.1, so very little will be bound to Ca^{2+} over the cytosolic range of 0.1–10 μM, particularly with a pK_{Mg} of 2.0 and a free Mg^{2+} of 1–2 mM. However, the pK_{Ca} for PO_4^{3-} is 6.3. So this form of phosphate will bind some Ca^{2+}.

Naturally occurring small organic Ca^{2+} ligands include acids, amino acids, nucleotides, phosphoinositides and sugars (Figure 6.1), the ionised anions being the highest affinity binders. None have a high enough affinity to bind large amounts of Ca^{2+} at the levels of free Ca^{2+} in the cytosol, in the presence of mM Mg^{2+}. But, the anions of dibasic and tribasic acids, such as oxalate and dipicolinate, and ATP^{4-} have higher affinities for divalent cations, with pK_d in the range 3–5 (i.e. 1 mM to 10 μM), and can bind significant amounts of Ca^{2+} at high levels in the cytosol or organelles. Similarly, in the periplasmic space of Gram-negative bacteria, negatively charged oligosaccharides bind significant amounts of Ca^{2+}. Ca^{2+} binding with ATP^{4-} and amine hormones, such as adrenaline and 5-hydroxytryptamine (5-HT), as well as amino acid neurotransmitters such as glutamate, plays an important role in the internal structure of secretory vesicles in nerve terminals, the adrenal gland and other endocrine cells, allowing the substances to dissolve rapidly when the vesicle fuses with the plasma membrane.

Several extracellular enzymes require Ca^{2+} for maximal activity, with pK_d in the millimolar range. In contrast, cytosolic proteins that are regulated by Ca^{2+} signals have pK_d in the micromolar range, e.g. troponin C, calmodulin, the Ca^{2+}-activated phosphatase calcineurin, and the Ca^{2+}-activated protease calpain (Table 6.1). In contrast, pyruvate kinase and phosphofructokinase, with pK_d^{Ca} of 3.7 and 2.7, respectively, are not be regulated by Ca^{2+} under physiological conditions in the cell. But, in the mitochondria, where the free Ca^{2+} can rise to above 10 μM, under physiological conditions, the pK_d^{Ca} of 5.4 for pyruvate dehydrogenase phosphatase is ideal. In addition, in the SR of muscle, where the free Ca^{2+} is hundreds of micromolar, the pK_d^{Ca} of 3.3 (i.e. 0.5 mM) for calsequestrin is appropriate. Extracellular enzymes such as prothrombin also have pK_d^{Ca} in the millimolar range, suitable for a plasma free Ca^{2+} of 1.2 mM. Interestingly, α-amylase in saliva has a much higher affinity (K_d^{Ca} = 1 μM) for Ca^{2+} than might be expected for an extracellular enzyme, because free Ca^{2+} in saliva is much less than in plasma.

Table 6.1 Examples of proteins that bind Ca^{2+} physiologically

Protein	Activates (a) or inhibits (i)	Approximate pK_{Ca}
Cytosolic		
Troponin C	a	6.5
Calmodulin	a	6
Calpain	a	6
Calcineurin	a	6
Plasma membrane Ca^{2+}-MgATPase	a	6
SR/ER Ca^{2+}-MgATPase	a	6.5
Glycerol 3-phosphate dehydrogenase (mitochondrial)	a	7
Phosphatidyl inositol hydrolase	a	6.8
Phosphorylase b kinase	i	6
Inside organelles		
Calsequestrin (SR)	a	3.3
Calreticulin (ER)	a	4
Pyruvate dehydrogenase phosphatase (mitochondria)	a	5.4
Pyruvate carboxylase (mitochondria)	i	3
Endonuclease (nucleus)	a	3
25-Hydroxycholecalciferol 1-hydroxylase	i/a	5/4
Extracellular		
α-Amylase	a	6
Trypsin	a	3.2
Prothrombin (Factor II)	a	3.3
Transglutaminase (Factor XIII)	a	3.6
Acetylcholine receptor (nicotinic)	a	3

6.1.4 Solvation

In solution, ions are surrounded by a sphere of water molecules. The ionic radius of Ca^{2+} is about 100 pm (1 Å). This is similar to La^{3+} and Gd^{3+}, explaining why these are potent inhibitors of many Ca^{2+} channels and Ca^{2+}-binding proteins. Ca^{2+} has a hydrated radius of 412 pm, still small enough to pass through the Orai pore of 600–800 pm. This compares with the smaller Mg^{2+}, which has an ionic radius of 72 pm and a hydrated radius of 428 pm, Na$^+$ with 117 pm and a hydrated radius 358 pm, and K$^+$ with 149 pm and a hydrated radius 331 pm. The similarity between the ionic radii of Ca^{2+} and Na$^+$ explains why several cation channels cannot select between these two ions. When K$^+$ moves through ion channels water is stripped off, K$^+$ binding to glutamate inside the pore providing the Gibbs free energy for this. In contrast, when Ca^{2+} or Na$^+$ pass through an ion channel these may remain hydrated. When bound to protein targets, Ca^{2+} has at most one or two water molecules bound to it. For example, when Ca^{2+} binds to an EF-hand-type protein, one H$_2$O remains as part of the octahedral binding. The ability of Ca^{2+}, with coordination numbers of 7 or 8, to keep water solvation molecules is an important factor determining high-affinity Ca^{2+}-binding sites in proteins over Mg^{2+}.

6.1.5 Activity Coefficients

The equations used so far to describe Ca^{2+} binding have used concentrations. However, strictly, equations should use 'activities':

$$\text{Activity}\,(a) = \gamma m \tag{6.10}$$

where m is the molar concentration (molality) and γ is the activity coefficient.

The behaviour of strong electrolytes deviates from the law of Mass Action based on concentrations in molarities or molalities. The deviation increases as the strength of the electrolyte increases, explained by Debye and Hückel in 1923 through electrical interactions between ions at high concentrations. The more concentrated the solute, the closer are the charged ions to each other, and thus the greater the deviation. It is the activity coefficient that is affected by ionic strength. In Eqn (6.10), for an ideal solution $\gamma = 1$, but for real solutions is less than 1. Thus the true rate of CaL formation is $k_1 a_{Ca} a_L$ and the rate of CaL dissociation is $k_{-1} a_{Ca}$. At equilibrium, the rate of association = the rate of dissociation (i.e. $k_1 a_{Ca} a_L = k_{-1} a_{CaL}$) so:

$$k_1 / k_{-1} = a_{CaL} / a_{Ca} a_L = K_a \tag{6.11}$$

In the cell, where equilibrium is never attained, the rate of change of Ca^{2+} bound to its ligand:

$$da_{CaL} / dt = k_1 a_{Ca} a_L - k_{-1} a_{CaL} \tag{6.12}$$

Such non-steady-state kinetics exist in cells such as muscle, with very fast responses in relation to the on and off rates of Ca^{2+} binding. Activities (a) for Ca^{2+} can be as much as 80% lower than that when molarities or molalities (M) are used. This is rarely taken into account when studying Ca^{2+} binding to intracellular ligands. Changes in dielectric constant also occur close to the membrane during a cellular event. This has dramatic effects on the activity coefficient of a cation such as Ca^{2+}, and would be influenced by the membrane zeta potential.

6.1.6 Fractional Ca²⁺ Binding After a Ca²⁺ Signal

The significance of Ca^{2+} binding to a ligand depends on how much the bound Ca^{2+} changes, when the free Ca^{2+} in the cell rises and falls in response to a stimulus or pathogen. Each binding site will have a K_d^{Ca}. Similarly, an enzyme will have a K_m^{Ca}, the concentration of Ca^{2+} that activates or inhibits the enzyme by 50%. When only one Ca^{2+} binds to an enzyme, the K_d and K_m will be very similar. But when more than one bound Ca^{2+} is required, these can be quite different. An enzyme such as fructose bisphosphatase (FBPase) has a K_d^{Ca} of about 10^{-4} M (i.e. 0.1 mM), and would require some 1 mM to be greater than 90% saturated (Figure 6.2). Over the cytosolic free Ca^{2+} range of 0.1–10 µM, from rest to activated, only 0.1–10% of the enzyme will have Ca^{2+} bound, and thus is unlikely to have any significance inside the cell. However, the individual Ca^{2+}-binding sites of troponin C and calmodulin can be greater than 90% saturated when the cytosolic free Ca^{2+} is micromolar. These proteins, with four Ca^{2+}-binding sites, and an apparent K_d^{Ca} in the micromolar range, are ideally suited to cause a large fractional in enzyme activity when the cytosolic free Ca^{2+} rises from 0.1 to 10 µM (Figure 6.2).

Methods for measuring the affinity of a ligand for Ca^{2+} include binding of radioactive Ca^{2+}, Ca^{2+} indicators, spectrophotometry, nuclear magnetic resonance (NMR), electron spin

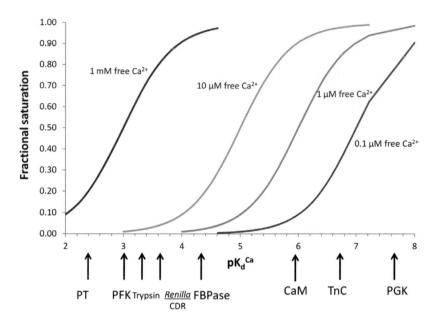

Figure 6.2 Effect of K_d^{Ca} on the fractional saturation of a protein by Ca^{2+}. Only one Ca^{2+}-binding site is considered: $LCa/L_T = [Ca/(Ca + K_d^{Ca})]$, where T indicates the total ligand concentration. TnC = troponin C, PGK = phosphoglycerate kinase, CaM = calmodulin, CDR = Ca^{2+}-dependent regulator in *Renilla*, FBPase = fructose bisphosphatase, PT = prothrombin, PFK = phosphofructose kinase. The graphs show that FBPase, PT, trypsin and PFK require millimolar Ca^{2+} to be at least half-saturated, whereas the individual Ca^{2+}-binding sites of the other proteins would be at least half-saturated at 1 μM cytosolic free Ca^{2+}. Reproduced by permission of Welston Court Science Centre.

resonance and enzyme activity. These depend on measuring the amount of free and bound Ca^{2+} with a known concentration of ligand. The simplest model is a ligand with n Ca^{2+}-binding sites, where binding is random and LCa_n is the only form active with Ca^{2+}:

$$nCa^{2+} + L = LCa_n \tag{6.13}$$

$$\text{Total ligand concentration} = L_T = L_{free} + LCa_n \tag{6.14}$$

$$\text{Fraction of ligand bound to } Ca^{2+} = LCa_n/L_T = \left[Ca/\left(Ca + K_d^{Ca}\right)\right]^n \tag{6.15}$$

where Ca is the free Ca^{2+} after equilibrium has been attained.

A plot of the fractional ligand bound to Ca^{2+} for different numbers of Ca^{2+} sites shows that, as the number of Ca^{2+}-binding sites increase from one to four, the curve becomes steeper (Figure 6.3). Four Ca^{2+} sites are ideal for a Rubicon switch. For a protein with four independent Ca^{2+}-binding sites, each with a K_d^{Ca} of 0.2 μM, the fraction bound to Ca^{2+} changes from <1% to >90% when the free Ca^{2+} rises from 0.1 to 10 μM. Ca^{2+} binding can be cooperative, e.g. in Ca^{2+}-activated photoproteins and calmodulin. The affinity of Ca^{2+} binding sites can also change when the Ca^{2+} binding binds to another protein. For example, Ca^{2+} on troponin C can increase ten-fold when in the troponin complex. The total amount of Ca^{2+} required to activate 90% of 20 μM calmodulin in a cell is nearly 100 μM, at least ten-times the free Ca^{2+}.

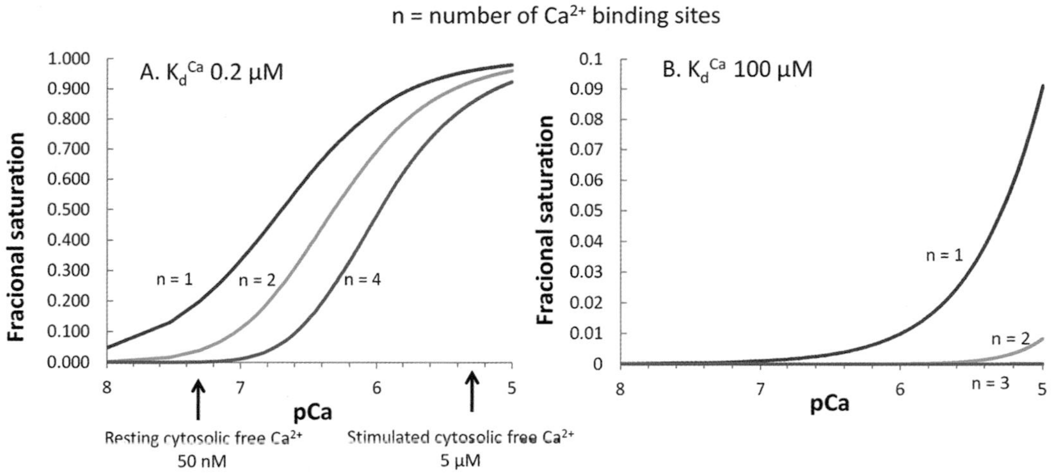

n = number of Ca²⁺ binding sites

Figure 6.3 The fractional saturation of a protein by Ca^{2+}. n = number of Ca^{2+}-binding sites, assumed to be non-cooperative and random. $LCa_n/L_T = [Ca/(Ca + K_d{}^{Ca})]^n$. The $K_d{}^{Ca}$ here was assumed to be (a) 0.2 μM; (b) 100 μM. pCa = $-log_{10}[Ca]$. Reproduced by permission of Welston Court Science Centre.

6.1.7 Kinetics

The speed of an event triggered by Ca^{2+}, and its return to rest, depend on the rate of Ca^{2+} binding, and its dissociation, once the cytosolic free Ca^{2+} has returned to resting levels. For Ca^{2+} binding to one Ca^{2+} site:

$$L + Ca^{2+} \underset{k_{-1}}{\overset{k_1}{\rightleftharpoons}} LCa \underset{k_{-2}}{\overset{k_2}{\rightleftharpoons}} LCa'$$

(6.16)

where LCa is the state of ligand immediately on Ca^{2+} binding and LCa' is the state of ligand after a conformational change induced as the result of Ca^{2+} binding.

The rate of change of LCa:

$$d[LCa]/dt = k_1 [L_t][Ca_t] - k_{-1}[CaL_t]$$

(6.17)

where the suffix t represents the concentration at time t.

The on rate, k_1, for Ca^{2+} binding is determined mainly by the rate of diffusion. In contrast, k_1 for Mg^{2+} is determined mainly by the rate of loss of H_2O, making $k_1{}^{Mg}$ several orders less than $k_1{}^{Ca}$. Thus, Ca^{2+} transients in the millisecond range can affect a Ca^{2+}-binding protein or other ligand, even if it has a relatively high affinity for Mg^{2+}.

6.1.8 Diffusion

The timing of a Ca^{2+} triggered event depends on the rate of diffusion:

$$\text{Flux} = J\left(\text{mol cm}^{-2}\text{s}^{-1}\right) = D(\delta c / \delta x)$$

(6.18)

where D is the diffusion coefficient (cm^{-2} s^{-1}), c is the concentration, x is the distance and $\delta c/\delta x$ is the concentration gradient (mol cm^{-4}).

For Ca^{2+}, D is directly related to its mobility (u):

$$D = (RT/F)(u/|z|) = 0.013u \text{ for } Ca^{2+} \text{ of charge } (z) \, 2 \text{ at } 37^{\circ}C \qquad (6.19)$$

If $u_{Ca} = 4 \times 10^{-4}$ cm^2 s^{-1} V^{-1}, then $D = 5.2 \times 10^{-6}$ cm^2 s^{-1}, which is close to the D_{Ca} in free solution measured as 7×10^{-6} cm^2 s^{-1}. From this it is possible to estimate how long a Ca^{2+} gradient would be expected to last in the cytosol of a live cell. However, in cells, Ca^{2+} diffusion in some ten times less, because of Ca^{2+} uptake by organelles, such as mitochondria and the ER. In live cells Ca^{2+} gradients can exist for seconds, minutes or even longer.

6.1.9 Solubility

Many calcium salts are only sparingly soluble in water. Precipitates of calcium phosphate, carbonate, oxalate and sulphate form the basis of skeletal structures, as well as calcium stores, in many multicellular, and some unicellular, organisms. Microprecipitates of Ca^{2+} salts can form inside organelles such as mitochondria, and occur both intra- and extra-cellularly in pathological conditions, such as kidney stones. The conditions under which a precipitate forms are described by the solubility product K_{so}, such as with a divalent anion like HPO_4^{2-}:

$$Ca^{2+} + X^{2-} = CaX\,(solution) = CaX\,(solid) \qquad (6.20)$$

$$K_{so} = a_{Ca}a_X = \gamma_{Ca}\gamma_X m_{Ca} m_X \qquad (6.21)$$

For a precipitate to form, two conditions must be met. First, the solubility product must be exceeded (Eq. 6.20). Secondly, there has to be a nucleation site upon which the precipitation can start. Without a nucleation site, calcium salts can remain in solution, supersaturated, even if the solubility product is exceeded. This is the case in human plasma. In bone, collagen acts as sites for nucleation, and the precipitation of hydroxyapatite. Furthermore, the apparent solubility product for calcium phosphate in equilibrium with microprecipitates in the cell can be at least an order of magnitude greater than the solubility product measured in pure solutions.

6.2 Ca^{2+}-Binding Proteins

6.2.1 Ca^{2+}-Binding Proteins Inside Cells

The intracellular target when there is a rise in cytosolic free Ca^{2+} is a high affinity Ca^{2+} binding protein. The first Ca^{2+}-binding protein to be discovered in the 1960s was troponin C, the Ca^{2+} target in skeletal and heart muscle, causing them to contract. This was followed by the discovery of calmodulin – the Ca^{2+}-binding protein found in all eukaryotes, but not bacteria or archaeans. Calmodulin binds, and activates, a wide range of soluble and membrane-bound proteins, including kinases, phosphatases, cyclic nucleotide phosphodiesterase, Ca^{2+} pumps and ion channels, including Ca^{2+}, K^+ and Cl^-. There are four ways in which to classify a Ca^{2+}-binding protein:

1. Its action.
2. Its location.
3. Type of Ca^{2+} site in the protein.
4. Ca^{2+} affinity.

The actions of Ca²⁺-binding proteins fall into five categories:

1. Direct effect on the machinery responsible for the cell event.
2. Regulation of the Ca²⁺ signal.
3. Regulation of enzymatic activity and intermediary metabolism.
4. Regulation of gene expression.
5. Regulation within Ca²⁺ stores.

In eukaryotes, key sites for Ca²⁺ action are the cytosol, mitochondria, ER, the nucleus, and the inner surface of the plasma membrane. However, there are also Ca²⁺-binding proteins acting inside the Golgi, lysosomes, secretory vesicles, and specialised organelles. There are four types of Ca²⁺-binding site in Ca²⁺-binding proteins that give them a high affinity for Ca²⁺ in the presence of millimolar Mg²⁺:

1. EF-hand, e.g. parvalbumin, troponin C and calmodulin.
2. C2 binding motif, e.g. synaptotagmin and copines.
3. Multiple Glu and/or Asp residues in clusters, e.g. calsequestrin and calreticulin.
4. Oxygens from acidic amino acids, OH's in Ser and Thr, carbonyls in the peptide chain, and water; brought together by the 3D structure of the protein; e.g., annexins.

Some other Ca²⁺-binding sites have been identified, such as the Greek key, which bind Ca²⁺, but tend to have a lower affinity than EF-hand or C2 motifs. The ER contains several Ca²⁺-binding proteins. Some act as Ca²⁺ stores, and regulate the lumenal domain of the IP_3 or ryanodine receptor. Others, such as BiP (binding immunoglobulin protein; GRP78) and GRP94, regulate gene transcription through proteins that cross from the ER into the nucleus, as part of the ER stress response. The Golgi also has proteins regulated by Ca²⁺. The Ca²⁺ affinity for Ca²⁺-binding proteins depends on location, with K_d^{Ca} ranging from 10^{-4} to 10^{-9} M. Those in the cytosol need a K_d^{Ca} in the micromolar range. In contrast, those in organelles such as the SR or ER have K_d^{Ca2+} in the high micromolar to millimolar range, reflecting differences in free Ca²⁺ between these compartments and the cytosol.

6.2.2 Proteins with the EF-Hand Motif

A breakthrough was the discovery of the EF-hand, the three-dimensional structure of the Ca²⁺-binding site in a large number of Ca²⁺ target proteins. Several, such as parvalbumin, troponin C and calmodulin, are acetylated at the N-terminus, and can have other covalent modifications, such as trimethyl-lysine in calmodulin. At submicromolar free Ca²⁺ the fraction of the EF-hand protein with Ca²⁺ bound is less than a few percent. When the free Ca²⁺ rises to 1−10 μM a high percentage of the protein has Ca²⁺ bound. This is the key to the Ca²⁺ switch.

The EF-hand motif was discovered in 1973 by Kretsinger and colleagues in parvalbumin − a Ca²⁺-binding protein of the muscle of carp. X-ray crystallography showed that the domain binding Ca²⁺ consisted of 29 amino acids, with a 12-amino-acid loop at its centre. This was linked to two helices on either side, designated E (amino acids 1−10) and F (amino acids 19−29) (Figure 6.4); helices A−D were found in other parts of the proteins not directly involved in Ca²⁺ binding. The 12-residue loop was rich in acidic amino acids. The helix−loop−helix structure was originally likened to a cupped hand formed by an outstretched thumb and forefinger. Typically seven or eight coordination gives the loop its selectivity for Ca²⁺ at micromolar concentrations in the presence of millimolar Mg²⁺, with oxygens at positions 1, 3, 5, 7, 9 and 12. Positions 1, 3, 5, 9 and 12 are acidic, either Glu or Asp, with an invariant Glu or Asp giving two oxygens for coordination at position 12 (bidentate). The other two coordination oxygens come from the carbonyl of a peptide bond within the loop, and water. The amino acid in the loop at position

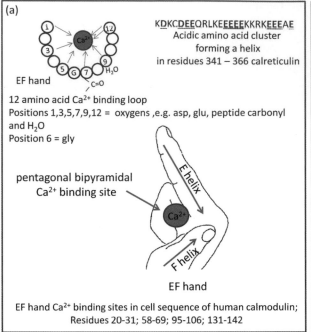

(a)

KDKCDEEQRLKEEEEEKKRKEEEAE
Acidic amino acid cluster
forming a helix
in residues 341 – 366 calreticulin

EF hand

12 amino acid Ca²⁺ binding loop
Positions 1,3,5,7,9,12 = oxygens ,e.g. asp, glu, peptide carbonyl and H₂O
Position 6 = gly

pentagonal bipyramidal
Ca²⁺ binding site

E helix

Ca²⁺

F helix

EF hand

EF hand Ca²⁺ binding sites in cell sequence of human calmodulin;
Residues 20-31; 58-69; 95-106; 131-142

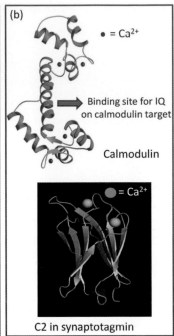

(b)

• = Ca²⁺

Binding site for IQ
on calmodulin target

Calmodulin

⬤ = Ca²⁺

C2 in synaptotagmin

Figure 6.4 Some high-affinity Ca²⁺-binding sites. (a) A classical (canonical) EF-hand and the acidic amino acid cluster in calreticulin, with the acidic residues in red and underlined, see Figure 3.11 for EF hand origin (b) Three-dimensional structures of calmodulin from http://en.wikipedia.org/wiki/File:Calmodulin.png. Manske (2004); C2 in synaptotagmin from http://upload.wikimedia.org/wikipedia/commons/9/93/C2dom .png. Bassophile (2007).

6 is always glycine, and is essential for the necessary conformation of the loop. Amino acids at positions 2, 4, 8, 10 and 11 of the loop are usually hydrophobic, enabling the Ca²⁺-binding loop to sit comfortably within the protein, stabilising the two helices on either side of the loop. Parvalbumin has been claimed to have three EF-hand motifs, but in fact it only binds two Ca²⁺. In contrast, troponin C in skeletal muscle has four Ca²⁺-binding sites, as does calmodulin, giving them a dumbbell-like structure, with two globular EF Ca²⁺-binding domains at each head.

When the free Ca²⁺ concentration rises to 1–10 μM, Ca²⁺ binds to the loop. This causes a conformational change in the orientation of the EF helices. This then affects another site within the protein, exposing a hydrophobic residue, which binds to its receptor protein. For example, Ca²⁺-calmodulin binds to cyclic nucleotide phosphodiesterase, activating it. Troponin C interacts with troponin I and T and tropomyosin in skeletal and heart muscle. Ca²⁺-binding proteins can have anything from one or two to several EF-hands. For example, the protease calpain has five EF-hands. As a result of pairing, many EF-hands exhibit cooperativity.

EF-hand proteins occur in all eukaryotic cells, there being over 60 families. Typical Ca²⁺ binding motifs used for genome searches are D/E-x-D/E-x-D/E-G-x-x-D/E-x-x-E or Dx[DN]x[DN] Gx[ILV][DSTN]xxE. Ca²⁺ proteins in intracellular stores, such as calreticulin and calbindin (D9k), also contain an EF-hand motif. But, calsequestrin in muscle SR has no EF-hands, Ca²⁺ binding being through an acidic amino-acid-rich domain. EF-hand proteins have been found in bacteria, including a calmodulin-like protein calerythrin, though the physiological Ca²⁺ target proteins in *E. coli* remain elusive. Convincing EF-hand proteins have yet to be found in Archaea. Pseudo EF-hand Ca²⁺-binding proteins have also been identified, where the EF-hand loop contains 14 residues, Ca^{2+} binding being mainly from the carbonyls of the peptide bonds within the loop, e.g. the N-termini of S-100-like proteins.

There are dangers in using genome searches and protein sequences alone to identify an EF-hand Ca^{2+}-binding protein. First, the identification of an EF-hand motif does not guarantee that it will bind Ca^{2+} at micromolar levels in the cell. Secondly, even if it does, the primary sequence alone will not tell you whether the putative EF-hand motif is flanked by α-helices, as it is in parvalbumin, troponin C and calmodulin, which are necessary to cause the structural change in the protein for it to trigger the cell event. Thirdly, EF-hands identified from primary sequences do not always fold correctly, and thus do not always bind Ca^{2+}, particularly if they miss the essential glycine at position 6 in the Ca^{2+}-binding loop, as in some of the enzymes in mitochondria.

6.2.2.1 Troponin C

Troponin is a complex of three proteins, C, T and I, discovered by Ebashi and coworkers. The Ca^{2+}-binding sites are in troponin C, which, like calmodulin, is a dumbbell-shaped protein with four EF-hand Ca^{2+}-binding sites (Figure 6.4). Sites I and II are critical in Ca^{2+}-triggered skeletal muscle contraction, and site II in cardiac troponin, though site I does have a role. There are three tissue subtypes: slow skeletal (TNNC1), fast skeletal (TNNC2) and cardiac. The concentration of troponin C in muscle is about 100 μM, which means that some 400 μM Ca^{2+} has to be released from the SR to saturate it completely. Troponin I (I = inhibitory) binds actin in the thin myofilaments, holding the actin–tropomyosin complex in place, preventing contraction until troponin C binds Ca^{2+}. Ca^{2+} binding to troponin C induces a conformational change making troponin I-tropomyosin disengage from myosin, allowing actin and myosin to interact, so the muscle fibre contracts.

6.2.2.2 Calmodulin

Calmodulin is named from CALcium-MODULated proteIN, and exists in all eukaryotic cells – animal, plant, fungal and protist, at a concentration of 1–40 μM. It has been highly conserved during evolution, there being greater than 90% sequence similarity between most calmodulins (Figure 6.5), and also similarity to calmodulin-like proteins. The human genome has three genes coding for calmodulins (CALM1–3). In addition there are six calmodulin-like proteins (CALML1–6). The native protein has two covalent modifications: a trimethyl-lysine at position 115/116, and an acetylated N-terminus. Calmodulin can also be ubiquinated, via a glycyl-lysine isopeptide. The acidic content of Asp + Glu is 35% compared with 9% of basic amino acids Lys + Arg, so calmodulin has a p*I* of 3.9. It has four EF-hand Ca^{2+}-binding sites, each with an affinity around micromolar, giving an apparent affinity for four Ca^{2+}s binding of about 2.4 μM, ideal for a cytosolic free Ca^{2+} change from submicromolar to 1–10 μM by a cell stimulus.

The binding domain of calmodulin is in the spindle of the dumbbell and is known as the 'latch domain' (Figure 6.4). There is a helical domain between the two pairs of Ca^{2+} sites. Binding of Ca^{2+} to the EF-hand loops causes a conformational change, exposing hydrophobic groups, particularly methyls from methionines. These then bind to basic amphiphilic helices in the target protein, which have complementary hydrophobic domains. Calmodulin inhibitors such as trifluoperazine bind to the hydrophobic domain in calmodulin. Three types of Ca^{2+}-calmodulin binding domain have been identified: IQ, and two related domains called 1–14 and 1–5–10, respectively, based on conserved hydrophobic residues that bind to the hydrophobic region in calmodulin exposed when Ca^{2+} binds. The IQ domain is named such after the first two amino acids in its motif, which are usually, but not exclusively, isoleucine (I) and glutamine (Q).

Three sequences in calmodulin are responsible for binding to the α subunit of $Ca_V1.2$, the L-type voltage-gated Ca^{2+} channel in heart myocytes, at amino acids 1609–1628, 1627–1652

(a) Human parvalbumin

```
Ac AMTELLNAED  IKKAIGGAFA  AAESFDHKKF  FQMVGLKKKS  TEDVKKVFHI
   LDKDKSGFIE  EEEGFILKGF  SPDARDLSVK  ETKTLMAAGD  KDGDGKIGAD  EFSTLVSES
```

(b) Troponin C

Human slow skeletal and cardiac

```
Ac MDDIYKAAVE  QLTEEQKNEF  KAAFDIFVLG  AEDGCISTKE  LGKVMRMLGQ
   NPTPEELQEM  IDEVDEDGSG  TVDFDEFLVM  MVRCMKDDSK  GKSEEELSDL
   FRMFDKNADG  YIDLDELKIM  LQATGETITE  DDIEELMKDG  DKNNDGRIDY  DEFLEFMKGV  E
```

Rabbit fast skeletal muscle`

```
Ac DTQQAEARSY  LSEEMIAEFK  AAFDMFDADG  GGDISVKELG  TVMRMLGQTP
   TKEELDAIIE  EVDEDGSGTI  DFEEFLVMMV  RQMKEDAKQK  SEEELAECFR
   IFDRNADGYI  DAEELAEIFR  ASGEHVTDEE  IESLMKDGDK  NNDGRIDFDE  FLKMMEEVQ
```

(c) Calmodulins in cells

```
Human  :      AcADQLTEEQIAEFKEAFSLFDKDGDGTITTKELGTVMRSLGQNPTEAEL
S. pombe:  AcMTTRNLTDEQIAEFREAFSLFDRDQDGNITSNELGVVMRSLGQSPTAAEL
S. cerv.:  AcMSSNLTEEQIAEFKEAFALFDKDNNGSISSSELATVMRSLGLSPSEAEV


Human  :      QDMINEVDADGNGTIDFPEFLTMMARKMKDTDSEEEIREAFRVFDKDGNG
S. pombe:     QDMINEVDADGNGTIDFTEFLTMMARKMKDTDNEEEVREAFKVFDKDGNG
S. cerv.:     NDLMNEIDVDGNHQIEFSEFLALMSRQLKSNDSEQELLEAFKVFDKNGDG


Human  :      YISAAELRHVMTNLGEKLTDEEVDEMIREADIDGDGQVNYEEFVQMMTAK
S. pombe:     YITVEELTHVLTSLGERLSQEEVADMIREADTDGDGVINYEEFSRVISSK
S. cerv.:     LISAAELKHVLTSIGEKLTDAEVDDMLREVSDGSGEI-NIQQFAALLSK
```

Figure 6.5 Amino acid sequences of some high-affinity EF-hand Ca^{2+}-binding proteins. Ca^{2+}-binding sites and loops are in red, bold and underlined. (a) Parvalbumin (108 amino acids, around 12 kDa) with two EF-hand Ca^{2+}-binding sites acts as a slow Ca^{2+} buffer in synaptic plasticity. Domain AB has a deletion of two amino acids in the potential Ca^{2+} loop and so does not bind cations. (b) Slow skeletal and cardiac muscle troponin C (161 amino acids), compared with fast rabbit skeletal muscle troponin C. All have three predicted EF-hand Ca^{2+}-binding sites, but in slow and cardiac muscle only two may be fully active all of the time. (c) Three calmodulins: human (CALM1), and the yeasts *Schizosaccharomyces pombe* and *Saccharomyces cerevisiae* (see P6220, P05933 and P06787 in UniprotKB for further information; www.uniprot.org). Humans have three calmodulins (CALM1–3). The mRNA codes for 149 amino acids, but in the cell the initial methionine is removed and the next alanine acetylated (Ac). The N-terminal methionine in yeast calmodulins is retained, but also blocked with an acetyl group. Human calmodulin also has a trimethyl-lysine at position 115 in the human protein. The 12-amino-acid Ca^{2+}-binding loops are at positions 21–32, 57–68, 94–105, 130–141 for human; 22–33, 58–69, 95–106, 131–142 for *Schizosaccharomyces pombe*; and 21–32, 57–68, 94–105 for *Saccharomyces cerevisiae* (the fourth Ca^{2+} loop in *Saccharomyces cerevisiae* at 131–140 is not complete and probably does not bind Ca^{2+}). Reproduced by permission of Welston Court Science Centre.

and 1665–1685, named A, C and IQ, respectively, Ca^{2+} inactivating the channel. Ca^{2+}-calmodulin regulates several ion channels in animals, plants and protozoa, including voltage- and ligand-gated Ca^{2+}, K^+, Na^+ Cl^- and TRP channels, and can be considered as a subunit of these channels. A rise in cytosolic free Ca^{2+} has two counteracting effects on voltage-gated Ca^{2+} channels, such as $Ca_V1.2$: inactivation and facilitation. The N-terminus of the α subunit is responsible for Ca^{2+}-dependent facilitation, whereas the IQ Ca^{2+}-calmodulin site and EF-hand domain 100 amino acids upstream from the C-terminus are responsible for the rapid Ca^{2+}-dependent inactivation of the channel. The latter includes an essential four-amino-acid sequence, VVTL, in the F-helix of the EF-hand. The two apparently opposing effects have different Ca^{2+} affinities, producing a balance, so that the channel remains open just long enough for a heart to beat every second or so, with a resting period a few 100 ms between.

Calmodulin is an essential gene, since knocking-out of all isoforms is lethal. Ca^{2+} binding activates a wide range of enzymes and other proteins:

1. Enzymes, e.g. kinases, phosphatases, proteases, and cyclic nucleotide phosphodiesterases, the γ subunit of phosphorylase b kinase, MLCK.
2. Structural proteins in microfilaments and microtubules, e.g. mitotic spindle.
3. Pumps and channels in membranes, e.g. Ca^{2+}-MgATPase, BK potassium channel.
4. Promotion of protein translation on the ribosomes, increasing cell cycle progression in yeast, and long-term potentiation in neurones, underlying synaptic plasticity, allowing synapses to alter their strength.
5. Ca^{2+}-calmodulin translocated to the nucleus activates transcription factors and movement of chromosomes during mitosis.

Ca^{2+}-calmodulin acts as either a digital switch or an analogue regulator, depending on the cell event.

6.2.2.3 Calmodulin-Dependent Kinases
A target for Ca^{2+}-calmodulin is a family of kinases (CaMKs) found in all eukaryotic cells. Most are serine/threonine kinases that can phosphorylate many proteins. These include myosin light chain kinase (MLCK), and CaMKI, II and IV, with variations in tissue distribution. CaMKIII, discovered in rat pancreas, has been renamed eEF-2 kinase. Multiple isoforms of CaMKs are coded for by different genes, and produced through alternative splicing. There are two key structural similarities between CaMKs:

1. The catalytic domain at the N-terminus.
2. An autoinhibitory and regulatory domain near the C-terminus.

CaMKs play a role in the cell cycle, particularly in the G_1/S and G_2/mitosis transitions, and in the transition from metaphase to anaphase in mitosis.

6.2.2.4 Calcineurin
Calcineurin is a phosphatase, also called protein phosphatase 3, which cleaves phosphate off serine and threonine in proteins. It consists of two proteins: CNA and CNB. CNA (61 kDa) has the phosphatase enzymatic activity, and also binds calmodulin. CNB (19 kDa) binds Ca^{2+} directly, and regulates CNA activity. In humans there are three catalytic isoforms, coded by genes *PPP3CA–C*, and two genes for the regulatory subunit, *PPP3R1* and *2*. Calcineurin plays a key role in T-lymphocyte activation, and other cells in the immune response, by dephosphorylating the transcription factor cytoplasmic NFAT (NFATc). Calcineurin can be inhibited clinically by cyclosporine, pimecrolimus and tacrolimus, used in transplantation. Calcineurin also plays a role in the action of neurotransmitter receptors in the brain, including *N*-methyl-D-aspartate (NMDA), dopamine and γ-aminobutyric acid (GABA). Genetic engineering of calcineurin in mice induced brain defects analogous to schizophrenia, memory loss, and odd behaviour patterns.

6.2.2.5 Calpains
Calpains are EF-hand cysteine proteases activated by a rise in cytosolic free Ca^{2+}, particularly when Ca^{2+} is restricted to a microdomain. They exhibit large molecular biodiversity within an individual, and between species. Like calcineurin, calpains are heterodimers. Calpains 1 and 2 are linked via a penta-EF-hand motif domain, and also have a C2-like domain. Calpain 3 in skeletal muscle forms a homodimer though its penta-EF-hand domain. Fungi and budding yeast have calpains without penta-EF-hands. Ca^{2+} activation of calpains plays an important role in regulating cell shape, endocytosis and apoptosis.

6.2.2.6 Calsenilin and DREAM

Calsenilin (KChIP3) has EF-hand motifs, modifying the activity of two proteins in neurones and the heart, by binding to the C-terminus of presenilin 2, and the N-terminus of the potassium channel K_v4. Calsenilin also binds to the response elements of the genes coding for pro-dynorphin and c-*fos* genes, decreasing expression. It is also known as 'downstream regulatory element antagonist modulator' (DREAM). DREAM is an EF hand protein with one high-affinity Ca^{2+} site, three low-affinity Ca^{2+} sites, and a distinct Mg^{2+}-binding site. Only Ca^{2+} binding causes a change in the three-dimensional structure of DREAM. A rise in intracellular Ca^{2+} therefore regulates gene expression.

6.2.2.7 Calbindin

Calbindin was discovered as a key protein in vitamin D-dependent Ca^{2+} absorption in the gut of chickens and mammals, but also works in other tissues. The name now refers to a family of different EF-hand Ca^{2+}-binding proteins. For example, calbindin D28k is a 263-residue protein, with six identifiable EF-hand structures, the four loops EF1, 3, 4 and 5 being the active Ca^{2+}-binding sites. In contrast, calbindin D9k is a S-100 protein, with only two EF-hand Ca^{2+} sites, and no homology to calbindin D28k, except in these Ca^{2+}-binding sites. Calbindin D9k is found in the kidney and the uterus. In the intestine, Ca^{2+} is taken up via the TRPV6 channel at the apical surface and binds to calbindin, whose level is regulated by vitamin D. Calbindin carries Ca^{2+} across the enterocyte, where Ca^{2+} is pumped out into the blood by the Ca^{2+}-MgATPase PMCA1. Calretinin is a 28-kDa protein, found in nervous tissue, with 58% homology to calbindin D28k.

6.2.2.8 S-100 Proteins

S-100 proteins are low molecular-weight, high-affinity Ca^{2+}-binding proteins with EF-hand Ca^{2+} sites. S stands for 'soluble' in saturated ammonium sulphate, precipitation in $(NH_4)_2SO_4$ being a classical first step in protein purification. S-100 proteins are classified as A, B, P or Z. They are often homodimers, but can be heterodimers. They are found in a wide range of Eukaryota, but not in Bacteria or Archaea. These Ca^{2+}-binding proteins have classical (canonical) high-affinity Ca^{2+}-binding EF-hand motifs at the C-termini, and pseudo-EF-hand motifs at the N-termini. Binding of Ca^{2+} changes the orientation of one of the helices in the protein, exposing a hydrophobic area in an analogous way when Ca^{2+} binds calmodulin, enabling the S-100 to interact with its target protein. Ca^{2+} binding to S-100 proteins can interact with the SR/ER, and regulate the cytoskeleton and transcription through kinase activation and phosphorylation. Some S-100 proteins are found extracellularly, where they act as cytokines in an inflammatory response.

6.2.2.9 STIM 1 and 2

STIM1 and 2 are proteins that traverse the ER membrane, and have Ca^{2+}-binding sites facing the lumen in the ER. A *decrease* in free Ca^{2+} in the ER induces STIM to form small plaques, punctae, which move so that they can bind Orai in the plasma membrane to activate the SOCE channel. STIM has two EF-hand Ca^{2+} sites that work together.

6.2.2.10 Caldendrin/Calbrain/Calponin and Caldesmon

Caldendrin, also called calbrain or calcium-binding protein 1, is found particularly in neurones in the hippocampus epithalamus, cerebellum, cortex and retina. It has a role in schizophrenia. It has two EF-hand motifs. Calponin and caldesmon are important in smooth muscle. Calponin can be phosphorylated by CaMK, stopping the inhibition of myosin MgATPase in smooth muscle. Caldesmon is a calmodulin binding protein, and also inhibits myosin MgATPase in smooth muscle.

6.2.3 Proteins with the C2 Motif

The C2 motif (Figure 6.4c and 6.6), named from the C2 domain in protein kinase C, is particularly important in proteins associated with membranes. It is formed from beta-pleated sheets. C2 domain proteins bind other proteins, phospholipids and inositol phosphates. Interaction with membranes plays an important role in secretion, membrane trafficking, IP_3 and DAG generation, MgGTPase activation, and protein phosphorylation. Examples of C2 domain proteins are protein kinase C (PKC), synaptotagmins (Figure 6.6), copines, phospholipase C (PLC), rabphilin, Doc2 and Munc13. C2 domains typically have about 130 amino acids, which form an eight-stranded sandwich with two pairs of four β-sheets in opposite directions. In between are two loops, A and B, which face membranes and bind one Ca^{2+} each, though some C2 domains may bind three Ca^{2+} (Figure 6.6). The classic (canonical) seven or eight coordination of Ca^{2+} is made by five oxygens from acidic residues, most often Asp, and two from water. The C2 motif has been found in several hundred proteins from a wide range of eukaryotes, and has been found in some bacterial toxins (e.g. from *Clostridia*). Thus synaptotagmins are membrane-bound proteins, with an N-terminal transmembrane region, a variable linker, and two C-terminal Ca^{2+}-binding C2 domains: C2A and C2B. These proteins are the Ca^{2+} targets that cause nerve synapses to fuse their vesicles containing neurotransmitter with the nerve terminal membrane. Ca^{2+} binding to synaptotagmin activates an electrostatic switch, initiating fusion of the vesicles with the terminal membrane. Ca^{2+} binding to synaptotagmins is necessary for vesicle fusion, and in displacing coplexin from the SNARE complex in the latter stage of exocytosis. It is the C2A domain that controls vesicle fusion.

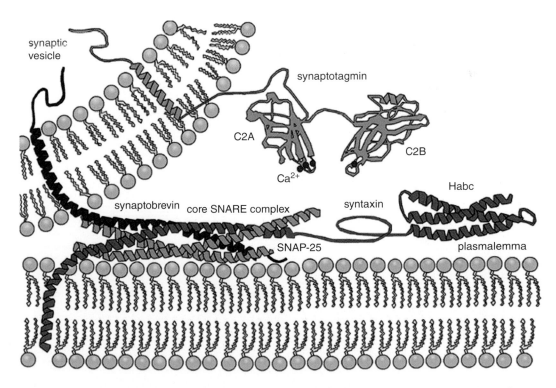

Figure 6.6 Synaptotagmin. Source: Danko Dimchev Georgiev, M.D. [GFDL (www.gnu.org/copyleft/fdl.html) or CC-BY-SA-3.0 (http://creativecommons.org/licenses/by-sa/3.0/)], via Wikimedia Commons.

Copines were discovered in the unicellular protozoan *Paramecium*. They were named after the French 'copine', used to describe a close friend. Copines are widely distributed throughout eukaryotes. They are highly conserved, many organisms exhibiting molecular biodiversity through multiple genes. Copines have two C2 domains near the N-terminus, and an 'A' domain, sometimes called 'I', in the C-terminal domain. The two C2 domains bind Ca^{2+}, and there is also a domain that binds phospholipids, similar to that in PKC, synaptotagmin and PLC, holding proteins attached to the inner surface of the plasma membrane. Thus, copines play a role in cellular events involving membrane interactions, including exocytosis, membrane trafficking, cell division and growth gene transcription, changes in the cytoskeleton, and apoptosis.

Phospholipases (PLs) are enzymes that cleave phospholipids (Figure 6.7). There are four types: A, B, C and D. Two interact with intracellular Ca^{2+}: PLA2 and PLC, both having C2 domains, with PLC also having four tandem EF hand domains. PLA2 plays a role in Ca^{2+} signalling by producing arachidonic acid, being activated by Ca^{2+}. Extracellular PLA2, secreted by the pancreas, and in venoms, also requires Ca^{2+}.

But not all C2 domain proteins bind, or are regulated by Ca^{2+}.

6.2.4 Proteins with a Cluster of Acidic Residues

Two Ca^{2+}-binding proteins, calsequestrin in muscle SR and calreticulin in the ER of non-muscle cells, have a multiple Ca^{2+}-binding site made up of a cluster of acidic amino acid residues (Figure 6.4). Such sites are not as high affinity as EF-hand motifs or C2 domains, but bind significant Ca^{2+} when the free Ca^{2+} is tens or hundreds of micromolar, as it is in the SR/ER. Surprisingly, knock-out mice in embryo survive, questioning their role as essential SR/ER Ca^{2+} stores.

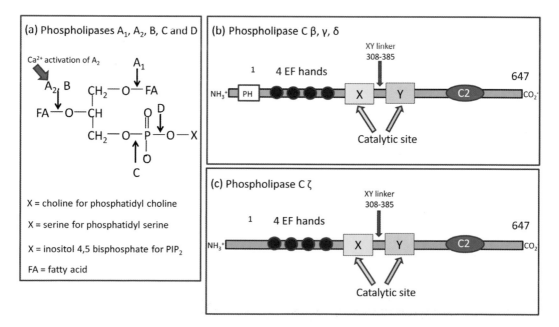

Figure 6.7 The four types of phospholipases and the key PLCs. (a) A_1/A_2, B, C and D. (b) Cβ, γ and δ. (c) Cζ. PLA2 releases arachidonic acid; PLC releases IP_3 and DAG. Reproduced by permission of Welston Court Science Centre.

6.2.4.1 Calsequestrin

Calsequestrin is the major Ca^{2+}-binding protein in the SR of muscle, with a Ca^{2+}-binding capacity of 18–50 per protein molecule. It has three repeating domains, but no EF-hand Ca^{2+} site. Ca^{2+} binding is formed by clusters of acidic amino acid residues Asp and Glu. There are two forms of calsequestrin: CASQ1 in fast skeletal muscle, and CASQ2 in slow skeletal muscle and heart muscle. Its structure is a random coil, the α-helical content being increased by up to 11% after binding Ca^{2+}. Calsequestrin can be phosphorylated by casein kinase II. Calsequestrin is secreted into the intestine, where it reduces free Ca^{2+}, depriving bacteria of Ca^{2+}.

6.2.4.2 Calreticulin

Calreticulin, discovered in muscle SR, is the major Ca^{2+}-binding protein in the ER of all non-muscle cells. Though it is sometimes found in the cytosol and on the outside surface of cells. It has two Ca^{2+}-binding domains: a high-affinity Ca^{2+}-binding site, not of the EF-hand type, in the centre of the protein, and multiple low-affinity sites near the C-terminus, rich in the acidic amino acids Asp and Glu. Calreticulin has had several names – CRP55, calregulin, ER resident protein ERp60, GRP60 and high-affinity calcium-binding protein (HACBP). Calreticulin binds to soluble lectins in the ER lumen. This complex interacts with monoglucosylated proteins in the ER, as well as other proteins. The complex formed inhibits the SERCA pump, when Ca^{2+} is bound, preventing Ca^{2+} overload in the ER. A highly acidic helix binds up to six Ca^{2+}, with an affinity of 590 μM. When a cell is at rest, and the ER is full of Ca^{2+}, the lumenal free Ca^{2+} is several hundred micromolar. At this concentration, the high-affinity Ca^{2+} site on calreticulin is fully occupied, and the lower affinity Ca^{2+} site greater than 50% saturated by Ca^{2+}. After a cell stimulus the lumenal free Ca^{2+} drops to tens of micromolar or less, when the high-affinity Ca^{2+} site may still be occupied, but the low-affinity site will not. This exposes Asn327, allowing it to be glycosylated. There is a KDEL sequence at the C-terminus to retain calreticulin in the ER. Calreticulin is heat stable – fitting for its role as a heat shock protein, acting as a chaperone to help other proteins fold. It can also bind other cations, such as Mg^{2+}, Mn^{2+} and Zn^{2+}, the Zn^{2+}-binding site being in the N terminal domain.

6.2.5 Proteins Forming a Cluster of Oxygen Ligands

Several proteins can form Ca^{2+}-binding sites, with varying affinities, from folding of the peptide chain (Figure 6.4). These have various names, such as type II and the Greek key. There is a large group of proteins which form an 'annexin fold' as a Ca^{2+}-binding site.

6.2.5.1 Annexins

The annexins, often less than 40 kDa in size, originally called calpactins, are a large family of Ca^{2+}-binding proteins found in all eukaryotes, involved intracellularly in endo- and exo-cytosis, changes in cell shape, phagocytosis, membrane trafficking and vesicle organisation, Ca^{2+} channel formation, and cell stress. They are found in a range of tissues. Annexins A 1, 2, and 5 have a role extracellularly, on the cell surface or free in the blood, in coagulation, fibrinolysis and provoking apoptosis. Annexins have two main domains, and can bind several Ca^{2+}:

1. A core domain around the C-terminus, highly conserved between different annexins, with several amino acid repeat sequences (eight in annexin A6), each of which can bind Ca^{2+} with high affinity.
2. A flexible domain at the N-terminus, which varies between annexins.

The Ca^{2+}-binding sites are non-EF-hand, being formed through the three-dimensional structure of the protein. Ca^{2+} binding to the site in the third domain exposes a tryptophan that then interacts with hydrophobic molecules in the membrane or other proteins. The Ca^{2+}-binding sites can be type II, III or AB, type I being the EF-hand motifs not found in annexins. Seven or eight coordination, required for Ca^{2+}-binding to be selective over Mg^{2+}, is satisfied by carbonyls in the peptide chain, carboxyl side-chains and water. In annexin V, all four AB Ca^{2+}-binding sites and three of the predicted B sites are essential for high-affinity binding to phospholipids, the DE site in the first domain contributing a little. The Ca^{2+} affinity of annexins is lower than that of EF-hand Ca^{2+}-binding proteins, because of the larger number of water molecules involved. Annexins bind negative phospholipids such as phosphatidyl serine and PIP_2 in a Ca^{2+}-dependent manner.

6.2.5.2 Calcimedins

Four calcimedins (67, 35, 33 and 30 kDa), each with one high-affinity Ca^{2+}-binding site ($K_d{}^{Ca}$ around 0.4 µM), were originally isolated from muscle. They have Trp residues that can act as hydrophobic binding sites, exposed when the proteins bind Ca^{2+}, and play a role in inflammation. The 67-kDa calcimedin is a member of the calpactin/lipocortin family, and can inhibit PLA2, Ca^{2+}-dependent F-actin binding, and phospholipid binding activity, similar to calpactins (lipocortins).

6.2.5.3 Calelectrin

Calelectrin is a Ca^{2+}-binding protein first isolated from the cholinergically activated electric organs of the ray *Torpedo marmorata*. At sub-micromolar free Ca^{2+}, calelectrin binds to membranes, but at micromolar Ca^{2+} the protein aggregates, affecting exocytosis and the generation of the electric shock. Similar proteins have been found in other organisms, including human and bovine retina. In the electric organ, calelectrin is widely distributed in patches throughout the postsynaptic cell. In the nerve terminal it is associated with the vesicles, but never found outside the cell.

6.2.5.4 Gelsolin

Gelsolin is another protein that has a Ca^{2+} site formed by folding of carbonyls in the peptide chain. This is a ubiquitous protein in eukaryotes, and is responsible for the cytoskeleton converting from a gel into a sol when there is a rise in cytosolic free Ca^{2+}, hence its name. It does this through its interaction with actin, though gelsolin also binds several other proteins, as well as negatively charged phospholipids such as phosphatidyl inositides. Gelsolin is attached to the cytoskeleton through the barbed ends of actin. It is also found extracellularly in the plasma, as the result of different initiation sites and alternative splicing of a single gene. Gelsolin promotes nucleation, pseudopod formation for amoeboid movement, and invagination of particles by phagocytosis. As well as physiological cell movement, gelsolin has also been implicated in apoptosis, cancer and amyloid deposition, important in Alzheimer's disease. The Ca^{2+}-binding sites are type II. The tail of the C-terminus (S/G6) is the Ca^{2+} sensor, acting as a latch dependent on the free Ca^{2+} concentration. When no Ca^{2+} is bound, the tail helix prevents binding to actin. When Ca^{2+} binds, the tail straightens and exposes the actin-binding sites. By gelsolin binding to two actins, the F-actin polymers disintegrate into the G-globular form, and the gel converts into a sol.

6.2.5.5 Calcium Homoeostatic Regulator (CALHM1)

CALHM1 is a transmembrane Ca^{2+}-binding protein in the plasma membrane, and in the ER of neurones, with similarities to the NMDA receptor (NMDAR). CALHM1 can produce a Ca^{2+} current. The protein increases leakage of Ca^{2+} from the ER, and reduces Ca^{2+} uptake into the ER by reducing the affinity of the SERCA pump for Ca^{2+}. This causes a large decrease in the

Ca^{2+} content of the ER, which triggers the ER unfolded protein stress response. This, and the potential to be a regulator of amyloid precursor gene, fit with a role in Alzheimer's disease.

6.2.5.6 Regucalcin

Regucalcin is a non-EF hand Ca^{2+}-binding protein that activates Ca^{2+} transport in the plasma membrane, ER and mitochondria. Hormones cause it to move into the nucleus, where it inhibits Ca^{2+}-activated kinases and phosphatases, and those independent of Ca^{2+}, DNA fragmentation and RNA synthesis. This leads to effects on metabolism, and can affect cell division and survival, as well as being involved in sperm.

6.2.5.7 Other Ca²⁺-Binding Proteins in the SR/ER

Several other medium- to low-affinity Ca^{2+}-binding proteins are found in the SR/ER. These include GRP78 (BiP), GFP94, sorcin and sarcalumenin. GRP78 (BiP) is a Ca^{2+} binding chaperone with MgATPase activity. Loss of Ca^{2+} from the ER exposes a hydrophobic binding site in BiP, and activates an unfolded protein response in the nucleus via a transmembrane protein Ire1P (see Chapter 10). In the heart, sarcalumenin is found in the longitudinal part of the SR, and regulates reuptake of Ca^{2+} by interacting with the luminal side of the SERCA2a pump. This plays a vital part in keeping the heart beating normally when stressed (e.g. in severe exercise). Sarcalumenin may be important in heart failure, when the reuptake of Ca^{2+} after each beat is impaired. Sorcin is a Ca^{2+}-binding protein in the lumen of the heart SR that regulates Ca^{2+} signalling by interaction with the luminal side of the ryanodine receptor.

6.2.5.8 Ca²⁺ Pumps and Exchangers

There are three types of Ca^{2+} pumps and exchangers:

1. Ca^{2+}-MgATPases, e.g. PMCA and SERCA in all eukaryotic cells.
2. Ca^{2+}/Na⁺ exchanger, e.g. NCX and NCKX, or CAX in yeast.
3. Ca^{2+}/H⁺ exchanger, e.g. in secretory vesicles, fungi and plants.

Membrane P-type MgATPases have ten transmembrane α-helices, three of which line the central channel through the lipid bilayer (Figure 6.8). Connected to the helices are three domains that reach into the cytoplasm:

- N: nucleotide-binding domain.
- P: phosphorylation domain.
- A: activator domain.

When a Ca^{2+}-MgATPase is unphosphorylated, two of the helices are disrupted, forming a cavity accessible from the cytosol. This cavity binds two Ca^{2+} ions. MgATP binds to a site that phosphorylates the adjacent P domain. This causes conformational changes, which bring the N and P domains into close proximity. A 90° rotation of the A domain then occurs, causing transmembrane helices 4 and 6 to rearrange. This then releases Ca^{2+} out of the cell in the case of PMCA or, in the case of SERCA, into the lumen of the SR. Ca^{2+}/Na⁺ exchangers are dimers also with ten transmembranes, with C2 domain like structures responsible for Ca^{2+} binding. They play an important role in removing Ca^{2+} in the heart myocyte, nerve terminals, photoreceptors and mitochondria. Some cells, particularly in plants, have a Ca^{2+}/H⁺ antiport exchanging two or three H⁺ for each Ca^{2+}. But, the movement of Ca^{2+} across membranes does not contribute significantly to the membrane potential. Ca^{2+}/Na⁺ and H⁺/Ca^{2+} exchangers are reversible. Depending on the Na⁺ concentration or pH on either side of the membrane Ca^{2+} will move either in or out.

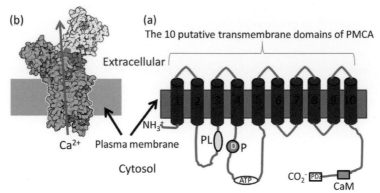

PL = phospholipid sensitive domain   D = key aspartate phosphorylated by ATP for pump
ATP = MgATP binding site   CaM = calmodulin binding site
PDZ = PDZ binding site

Figure 6.8 Structure of a P-type MgATPase Ca^{2+} pump. (a) The ten predicted transmembrane domains of the plasma membrane Ca^{2+} pump (PMCA), based on Ortega *et al.* (2007). Reproduced by permission of Welston Court Science Centre. (b) The three-dimensional structure of the PMCA. PDZ is a domain of 80–90 amino-acids in the signalling proteins of bacteria, yeast, plants, viruses and animals, anchoring membrane proteins to the cytoskeleton, derived from the first three proteins discovered in the fruit fly *Drosophila* – PSD-95 (postsynaptic density protein), DLG (the *Drosophila melanogaster* Discs Large protein) and ZO-1 (zonula occludens 1 protein). http://upload.wikimedia.org/wikipedia/commons/9/96/1eul-membrane.gif

6.3 Ca^{2+} and Other Intracellular Signals

Intracellular Ca^{2+} interacts with the other intracellular signals in eukaryotic cells – IP_3 and other inositol phosphates, cyclic nucleotides (cyclic AMP and cyclic GMP), cyclic ADP ribose, NAADP, as well as NO, CO_2, H_2S, and ethylene in plants. There are three types of interactions between intracellular Ca^{2+} and other intracellular messengers:

1. Effects of the other messenger on intracellular Ca^{2+} concentrations, through effects on Ca^{2+} release into the cytosol, or removal mechanisms.
2. Effects of intracellular Ca^{2+} on the concentration of the other messenger, through effects on the enzymes responsible for their synthesis and degradation.
3. Co-regulation by Ca^{2+} and the other messenger of target molecules, such as kinases, phosphatases, channels and pumps.

6.3.1   Cyclic Nucleotides and Ca^{2+}

Interaction between cyclic AMP, cyclic GMP, cyclic ADP ribose (cADPr) and intracellular Ca^{2+} in eukaryotic cells occurs in six ways:

1. Cyclic nucleotides can allow Ca^{2+} into the cell, by opening cyclic nucleotide-gated ion channels in the plasma membrane.
2. Cyclic nucleotides can activate proteins kinases, which phosphorylate proteins that regulate intracellular Ca^{2+}. This includes plasma membrane Ca^{2+} channels, SERCA, IP_3 and ryanodine receptors, e.g. SERCA in the heart by phospholamban increases the SR Ca^{2+} store, and PKA phosphorylates the Ca^{2+} channel, increasing heart rate.

3. Cyclic AMP and cADPR increase Ca^{2+} release from the SR/ER by increasing Ca^{2+} levels in the ER/SR, and/or by regulating the ryanodine and IP_3 receptors.
4. Ca^{2+} can reduce cytosolic cyclic nucleotide concentrations via activation of phosphodiesterase by calmodulin.
5. Ca^{2+} can either activate or inhibit the cyclase responsible for the synthesis of the cyclic nucleotide, depending on the isoform.
6. Cyclic nucleotides and Ca^{2+} can work in harmony or in competition, altering the activity of the target, and changing the threshold for the cell event, or its level, e.g. phosphorylase by Ca^{2+} and PKA.

Like cytosolic free Ca^{2+}, the concentration of free cyclic AMP in the resting cell is around 0.1 μM and rises to 1–10 μM after stimulation of adenylate cyclase, via an external agonist binding on a G-protein-coupled receptor, though it can be higher than this in microdomains, such as close to cyclic nucleotide-gated ion channels in the plasma membrane. Cyclic AMP is typically an analogue signal, in contrast to intracellular Ca^{2+}, which is digital, operating a cellular switch. Cyclic AMP activates PKA, and Epac1 and 2, the latter having Ca^{2+} and phosphorylation sites, working as guanine nucleotide-exchange factor (GEF) activators on selective small GTP-binding proteins.

6.3.1.1 Ca²⁺ and Adenylate Cyclase

Ca^{2+} at concentrations above 100 μM inhibits adenylate cyclase, but this is not physiological. However, Ca^{2+} either directly, or via a Ca^{2+}-binding protein, can inhibit or activate adenylate cyclases at micromolar concentrations of Ca^{2+}, depending on cell type and isoforms. The C-terminal domain contains a calmodulin-binding site in cyclases regulated by Ca^{2+}.

Although the effects of Ca^{2+} may only affect the enzyme activity by two- or three-fold, this will still have a major effect on cyclic AMP levels in the intact cell. Ca^{2+} stimulates AC1, 3 and 8 via calmodulin, but inhibits AC5 and 6 at micromolar Ca^{2+} concentrations directly, by binding near the active centre and the Mg^{2+}-binding site. However, Ca^{2+} has no physiological effects on AC2, 4 and 7. In the hippocampus, the Ca^{2+} curve is bell-shaped, with a two- or three-fold activation from 0.1 to 1 μM, and a decrease to below basal levels from 10 to 100 μM. AC1 in the brain is high affinity and is activated by Ca^{2+}-calmodulin, with a half-maximal concentration of 0.15 μM Ca^{2+}, just above the level in a resting neurone. This is important in synaptic plasticity – the ability of a neuronal synapse to change the strength of its response to an action potential and thus the amount of transmitter released, important for memory.

Microdomains are an important feature of the interaction of Ca^{2+} with adenylate cyclase. The cyclase is often located in clusters (e.g. in or near caveolae), or as rafts in the plasma membrane. In neurones, Ca^{2+}-inhibitable adenylate cyclase can be next to ion channels opened by cyclic nucleotides, and in cells with AC1 and 8 there is a close association with SOCE. This has led to the term 'coincidence detectors'. AC9 is not a very active cyclase, but can be inhibited by Ca^{2+} through calcineurin.

6.3.1.2 Ca²⁺ and Guanylate Cyclase

Cyclic GMP plays a central role in vertebrate photoreceptors and smooth muscle, and in tissues involved in NO signalling. In vertebrate retinal rods, light triggers an increase in cyclic GMP hydrolysis, closing cGMP-gated ion channels, generating an electrical signal transmitted to the terminal. This causes a decrease in cytosolic free Ca^{2+}, which activates guanylate cyclase, stimulating the resynthesis of cGMP. Dark-adapted retinas have a very low Ca^{2+} resulting in synthesis of cyclic GMP, whereas high Ca^{2+} inhibits. Cyclic GMP works through EF-hand Ca^{2+}-binding proteins, related to calmodulin – guanylate cyclase-activating proteins (GCAP). The

GCAPs in vertebrates are GCAP1 and 2, though humans and fish have a third, GCAP3. At least five other GCAPs are found in other species. Guanylate cyclase-inhibitory proteins (GCIPs) also exist. GACPs and GCIPs have remarkable sequence similarity, each having four EF-hands, but only three are active, EF-hands 2, 3 and 4. GCAPs have a condensed structure, with the EF-hands in tandem, in contrast to the dumbbell shape of calmodulin. GCAPs are myristoylated, enabling them to insert into phospholipid bilayers. When recoverin is involved in the visual response, it leads to a Ca^{2+}/myristoyl switch. An inhibitory effect of Ca^{2+} on guanylate cyclase has been found in the protozoa *Tetrahymena* and *Paramecium*. Ca^{2+}-calmodulin inhibits. In *Paramecium* cilia both guanylate cyclase and calmodulin are located in the cilia.

6.3.2 Ca^{2+} and Protein Kinase C (PKC)

PKC is a family of serine/threonine kinases activated by DAG, the other product of PIP_2 breakdown by phospholipase C (PLC), or Ca^{2+}. There are 15 PKC enzymes in humans, divided into three subfamilies: classical or conventional, novel and atypical. The 'classical' PKC isoenzymes α, βI, βII and γ (genes *PRKCA*, *B* and *G*) need DAG, Ca^{2+} and a phospholipid, such as phosphatidyl serine, for activity. On the other hand, the 'novel' group, PKCs δ, ε, η and θ (genes *PRKCD1*, *2*, *3*, *E*, *H* and *Q*), needs DAG, but not Ca^{2+}. Both the classical and novel PKCs are activated via G-protein-coupled PLC. In contrast, the atypical PKCs Mζ, ι and λ (genes *PRKCI*, *Z*, *1*, *2* and *3*) need neither DAG nor Ca^{2+} for activity. The DAG-binding site in the C1 domain in the classical and novel PKCs is near the N-terminus. The Ca^{2+}-binding site, only found in the classical PKCs, is in the next domain, C2. Binding of DAG and Ca^{2+} causes PKC to attach to the inner surface of the plasma membrane. But, the N-terminal domain in the atypical PKCs cannot bind DAG or Ca^{2+}. PKCs often work in parallel to Ca^{2+}-activated pathways, and are an analogue component of a Ca^{2+}-activated digital cell event. PKCs play an important role in cell division and growth, cancer and inflammation. For example, the NADPH oxidase that generates superoxide in neutrophils is activated by Ca^{2+}-calmodulin and PKC, involved in the cells attaching to a surface, secreting enzymes from granules, and actin assembly/disassembly.

6.3.3   Nitric Oxide and Ca^{2+}

Nitric oxide (NO) interacts with intracellular Ca^{2+}, and at two concentration ranges of NO. Ca^{2+} can regulate the enzyme responsible for NO synthesis, or NO can activate guanylate cyclase to generate cyclic GMP, and also block oxygen uptake in mitochondria, leading to Ca^{2+} uptake. NO was discovered in living systems as an endothelial-derived relaxing factor (EDRF), that had a half-life of around 12 s and was capable of relaxing smooth muscle in the aorta. It is produced by the enzyme NO synthase (NOS; E.C.1.14.13.39), using arginine as the substrate to form citrulline and NO:

$$2L-arginine+3NADPH+2H^+ +4O_2 \rightarrow 2\,citrulline+3NADP^+ +2NO \qquad (6.22)$$

There are four isoenzymes – nNOS, eNos, iNOS and bNOS – as well as alternatively spliced forms, which are widely distributed in tissues and phylogenetically. NO is particularly important in the brain, blood vessels and the immune system. The effect of a rise in cytosolic free Ca^{2+} can be either activate or inhibit the enzyme, depending on the tissue and isoenzyme. Ca^{2+}-calmodulin can activate, whereas CaMKs can inhibit. The isoenzymes were named after the tissues where they were first found: n = neuronal, e = endothelial, i = immune and b = bacterial. All eukaryote NOS are homodimers, with two domains: an oxygenase at the N-terminus, using a haem linked via a Fe^{2+}–S bond, and a reductase at the C-terminus. Both nNOS and eNOS are activated by Ca^{2+}-calmodulin, and are virtually inactive at the nanomolar levels of cytosolic free Ca^{2+} in the resting cell. Ca^{2+}-calmodulin acts as a molecular switch, causing electron transfer from the reductase domain at its C-terminus to the haem at the N-terminus. Ca^{2+}-calmodulin binding occurs in the 'latch' domain in helices 2 and 6. CaMKI and II can have the opposite effect to Ca^{2+}-calmodulin, phosphorylating nNOS at Ser741 and Ser 847, respectively. NO plays an important role in synaptic plasticity. In contrast, in NOS2 (iNOS) calmodulin is very tightly bound, even at the low cytosolic free Ca^{2+} in the resting cell. This NOS can be activated by Ca^{2+} under certain conditions.

NO works at two levels: nano- and micro-molar. Both involve binding iron, e.g. the Fe^{2+} in haem-containing enzymes, such as guanylate cyclase or cytochrome oxidase, activating or inhibiting, respectively. Activation of guanylate cyclase causes relaxation of smooth muscle. The cyclic GMP formed activates PKG, which activates myosin light chain phosphatase by phosphorylating it, thereby dephosphorylating myosin, counteracting the action of MLCK activated via a rise in cytosolic free Ca^{2+}. This occurs at nano- to micro-molar levels of NO. But, at high micro-molar, NO binds to the Fe^{2+} in the haem of cytochrome oxidase, inhibiting it, blocking ATP synthesis and oxygen utilisation, with consequences for Ca^{2+} uptake into mitochondria. The inhibition of cytochrome oxidase by NO has an interesting role in the flash of a firefly. In invertebrates, NO also plays a role as a bacteriocidal agent. In the freshwater snail *Viviparus*, Ca^{2+} activates both the soluble and membrane-bound forms of NOS, with a half-maximum of around 90 nM Ca^{2+}. NO also plays a role in plants, with similar interaction with Ca^{2+} signalling.

6.3.4   Ca²⁺, Inositol Phosphates and Other Intracellular Signals

Intracellular Ca^{2+} has an intimate relationship with several inositol phosphates, including IP$_3$ and IP$_4$ in most cells, IP$_6$ in the nucleated erythrocytes of birds, amphibians and fish, and pyrophosphates such IP$_7$. Other interactions include those with sphingosine 1-phosphate and NAADP. The relationships between intracellular Ca^{2+} and these other signals often involve regulation of the concentration of cytosolic free Ca^{2+}, through effects on intracellular receptors that stimulate release of Ca^{2+} from the ER or other organelles.

6.4   Ca²⁺ and Monovalent Ions

There are four ways intracellular Ca^{2+} can interact with monovalent cations:

1. Opening or closing of monovalent ion channels causes a change in membrane potential that opens or closes voltage-gated Ca^{2+} channels.
2. Regulation of monovalent cation or anion channels by cytosolic free Ca^{2+}.
3. Competition of ion flux through a channel between Ca^{2+} and a monovalent ion.
4. Interaction of monovalent cations with Ca^{2+} binding sites in proteins.

Depolarising a cell through opening of Na$^+$ channels opens voltage-gated Ca^{2+} channels. Repolarisation, via inactivation, or opening of K$^+$ channels, closes these Ca^{2+} channels. In contrast, a rise in cytosolic free Ca^{2+}, close to the plasma membrane, can increase K$^+$ channel opening, hyperpolarising the membrane. Several cation channels which let Ca^{2+} into the cell, such as transient receptor potential (TRP) channels, also allow monovalent ions through, particularly Na$^+$ and Cl$^-$, the concentration of both being higher outside the cell than in the cytosol. On the other hand, cations used experimentally or clinically can affect intracellular enzymes and receptors involved in Ca^{2+} signalling. For example Li$^+$, used in the treatment of manic depression, *in vitro* inhibits the phosphoinositide cycle, thereby interfering with

Ca^{2+} signalling. Li^+ inhibits inositol monophosphate phosphatase and inositol polyphosphate 1-phosphatase, and decreases IP_3 levels in embryos.

6.4.1   Intracellular Ca^{2+} and K^+ Conductance

Repolarisation of the membranes of excitable cells, such as nerve and muscle, can occur as a result of inactivation of Na^+ channels or by opening K^+ channels. Many excitable and non-excitable cells have K^+ channels that can be opened by a rise in cytosolic free Ca^{2+}. There are six key pieces of evidence this:

1. Measurement of cytosolic free Ca^{2+}, using aequorin, arsenazo III or a fluorescent Ca^{2+} indicator, shows a correlation of Ca^{2+} concentration with the magnitude of the K^+ current and number of K^+ channels open detected using patch clamping.
2. Activation of K^+ currents by injection of Ca^{2+} into the cell.
3. Prevention of the rise in cytosolic free Ca^{2+} by microinjection of Ca^{2+} chelators, such as EGTA or BAPTA, into the cell prevents both K^+ channel opening and rapid repolarisation of the plasma membrane.
4. Prevention of the Ca^{2+} current, by removal of extracellular Ca^{2+} or Ca^{2+} channel blockers, prevents both K^+ channel opening and repolarisation of the plasma membrane.
5. Lack of effect of tetraethylammonium ions, which block voltage-gated K^+ channels.
6. Identification of the mechanism by which Ca^{2+} opens the K^+ channels, either because the K^+ channel has an intracellular domain with a recognisable Ca^{2+} or a calmodulin binding site, typically of the IQ type.

There are four classes of K^+ channels:

1. Voltage-gated: open when the membrane depolarises.
2. Inward rectifying: allowing current to pass more easily inwards than outwards.
3. Tandem pore: always or often open in the resting cell, and are the cause of the resting K^+ currents.
4. Ca^{2+}-activated: open when there is a rise in cytosolic free Ca^{2+}.

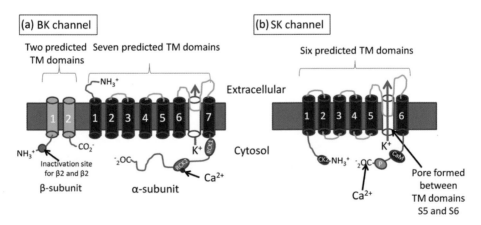

Figure 6.9 Structure of Ca^{2+}-activated K^+ channels: (a) BK channel and (b) SK channel; CK1 and 2 (= PKC site); CaM = Ca^{2+}-calmodulin binding site; P = phosphorylation site for protein kinase and protein phosphatase 2A; TM = transmembrane domain. Reproduced by permission of Welston Court Science Centre.

Activation of K⁺ efflux by intracellular Ca²⁺ was discovered in erythrocytes and molluscan neurones. The Ca²⁺-activated K⁺ channels (Figure 6.9) responsible can be distinguished in patch clamp experiments from other K⁺ channels by using specific blockers and activators. There are three types of Ca²⁺-activated K⁺ channels:

1. Big conductance (maxiK or BK, 100–300 pS), also known as slo-1 or $K_{Ca}1.1$.
2. Small conductance (SK, 2–25 pS)), also known as SK1, 2 or 3, or $K_{Ca}2.1$, $K_{Ca}2.2$ or $K_{Ca}2.3$.
3. Intermediate conductance (IK, 25–100 pS), also known as SK4 or $K_{Ca}3.1$.

The distribution of these Ca²⁺-activated K⁺ channels varies with cell type. For example, SK channels are widely expressed in neuronal tissues, where they underlie post-spike hyperpolarisations, regulate spike frequency adaptation and shape synaptic responses. These channels have six or seven transmembrane domains, and a transmembrane helix at the N-terminus. The α subunit forms homo- or hetero-trimers to form the channel. The Ca²⁺ site is a 'bowl' on the cytosolic C-terminus of the α subunit. Calmodulin can also bind some α subunits, regulating it when Ca²⁺ binds. CK2 kinase makes it more sensitive to Ca²⁺, affecting gating, as does protein phosphatase 2A. A rise in cytosolic free Ca²⁺ above 1 μM in neurones of the sea hare *Aplysia*, the snail *Helix* or cardiac Purkinje fibres increases outward K⁺ movement. In *Aplysia* such a rise in K⁺ conductance shuts off the Ca²⁺ channels as a result of membrane repolarisation. Repolarisation allows the cell to fire again by depolarisation, and Ca²⁺ channels reopen, leading to action potential oscillations.

6.4.2 Intracellular Ca²⁺ and Na⁺ Conductance

Ca²⁺ can affect Na⁺ channels in photoreceptors and insect salivary glands. Extracellular Ca²⁺, in the range 1–15 mM, affects Na⁺ channels, via an effect on surface charge, blocking the channel as the Ca²⁺ increases and affecting channel gating. However, the intracellular C-terminus of the voltage-gated $Na_v1.5$ (hH1) Na⁺ channel is regulated by intracellular Ca²⁺. There are three domains essential for this regulation:

1. The DIII–IV linker.
2. An EF-hand domain.
3. A calmodulin-binding IQ motif, which binds calmodulin in high affinity regardless of the free Ca²⁺.

Mutations in the α subunit of voltage-gated Na⁺ channels cause a range of inherited disease, several clustering around the IQ calmodulin-binding domain. In some life-threatening arrhythmias and long Q syndrome, these mutations disrupt calmodulin binding. This region exhibits five individual mutations involved in long-Q and Brugada syndromes. The Brugada mutation A1924T reduces the affinity for calmodulin. In the normal Na⁺ channel Ca²⁺-calmodulin destabilises the gate that has to shut to inactivate the channel. As a result, inactivation requires larger depolarising potentials. The Na⁺ channel is a large 220-kDa transmembrane complex with two main subunits: α and β. The α subunit forms the channel through its four main domains, and is regulated by Ca²⁺ in cardiac myocytes. Ca²⁺-calmodulin binds directly to the inactivation gate, which is a conserved 50-amino-acid region between domains III and IV. This allows access of the Ca²⁺ to the C-terminal IQ domain. In the Na⁺ channel NaChBac from the marine alphaproteobacterium HIMB114 (*Rickettsiales* sp.), millimolar extracellular Ca²⁺ blocks the channel. The outer Na⁺ selectivity filter involves Ser181 and Glu183, and carbonyls from Thr178 and Leu179, which are inside the selectivity filter where Ca²⁺ binds.

6.4.3 Ca²⁺ and Chloride Channels (CaCC)

Chloride channels are widely distributed in eukaryotes, and also occur in bacteria and Archaea. Several can be activated by a rise in cytosolic free Ca^{2+}. The Ca^{2+} binding site is typically a Ca^{2+} 'bowl' between two of the transmembrane segments. Cl^- channels are usually referred to as 'activated' or 'regulated', in contrast to 'gated' when referring to Ca^{2+} channels. In *Xenopus* oocytes, the rise in cytosolic free Ca^{2+} induced by fertilisation of a sperm opens the Cl^- channels. The outward Cl^- movement depolarises the cell. This helps to prevent fertilisation by second sperm. In vertebrate photoreceptors, Ca^{2+}-activated Cl^- channels play an important role in transmitter release. The protein involved in cystic fibrosis, CFTR, is a Cl^- channel. Another group of membrane proteins, which act as Ca^{2+}-activated Cl^- channels, are the anoctamins (ANO or TMEM16). ANO1, 2 and 6 can be activated or inhibited by Ca^{2+}-calmodulin, and have two modes of interaction, one at sub-micromolar Ca^{2+}, the other at micromolar Ca^{2+}.

Ca^{2+}-activated Cl^- channels play an important role in the secretion of electrolytes from epithelial cells, e.g. sweat. They are also involved in photoreception, sensory mechanisms, such as smell and hearing, as well as acting as secondary regulation in excitable cells, such as neurones, heart myocytes, and smooth muscle tone. Since outward Cl^- movement will tend to maintain depolarisation of the plasma membrane, this counteracts activation of K^+ channels by a rise in cytosolic free Ca^{2+}. But, if the membrane is highly depolarised, then the electrochemical potential pulls Cl^- into the cell, since the concentration of Cl^- is usually much higher extracellularly than in the cytosol. A major problem in studying Cl^- channels is the lack of pharmacological activators or blockers, either natural or synthetic.

6.5 Other Cations and Intracellular Ca²⁺

Transition metals and lanthanides, from micro- to milli-molar, interact with intracellular Ca^{2+} signalling:

1. Divalent – Co^{2+}, Mn^{2+}, Ni^{2+} Cd^{2+} – block Ca^{2+} channels.
2. Divalent – Mn^{2+}, Zn^{2+} – are enzyme activators or inhibitors.
3. Trivalent – La^{3+}, Gd^{3+} – block mechanoreceptors, Ca^{2+} channels and Ca^{2+} pumps.

Some pass through Ca^{2+} channels. Others are effective blockers of channels. This can be useful experimentally. The sensitivity to particular transition metal ions of voltage-gated Ca^{2+} channels varies. T-type LVA channels are particularly sensitive to Ni^{2+}, whereas N-type and L-type HVA Ca^{2+} channels are much less sensitive. In contrast, N-type and L-type HVA Ca^{2+} channels are sensitive to block by Cd^{2+}, whereas T-type LVA channels are resistant. In snail neurones, Ca^{2+}-activated Cl^- conductance can be prevented by Cd^{2+}, blocking the inward Ca^{2+} current.

Lanthanum (La^{3+}) has an ionic radius similar to Ca^{2+} and, as a result, inhibits Ca^{2+} signalling proteins, including Ca^{2+} channels, Ca^{2+}-MgATPases and Ca^{2+}-binding proteins. La^{3+} inhibition of Ca^{2+}-MgATPases works by La^{3+} replacing Mg^{2+} bound to ATP. $LaATP^-$ binds to the catalytic site, reducing the turnover rate and thus inhibiting Ca^{2+} transport. La^{3+} can replace Ca^{2+} as a trigger for light emission from the Ca^{2+}-activated photoprotein aequorin, but with a markedly reduced turnover rate. Another lanthanide, gadolinium (Gd^{3+}) is a particularly potent inhibitor of mechanoreceptor ion channels, which allow Ca^{2+} into the cell when the cell is touched or stretched.

The interaction between Zn^{2+} and intracellular Ca^{2+} signalling is one of particular physiological and pathological interest. Zinc is essential for normal growth and development, and several proteins have been identified in the plasma membrane, and the membranes of intracellular organelles, that are zinc transporters. Zn^{2+} binds to proteins, nucleic acids, carbohydrates and lipids. Over 2300 genes have been identified in the human genome that may bind Zn^{2+}. These include Zn^{2+} finger transcription factors, PKC and casein phosphatase II, which is inhibited by Zn^{2+}. Zn^{2+} can induce cytosolic free Ca^{2+} signals. In addition, L-type Ca^{2+} channels allow Zn^{2+} into heart cells. This induces genes driven by the metallothionein promoter, activated by Zn^{2+}-binding proteins. Zn^{2+} is stored inside the ER, and binds to many chelators used to chelate or buffer cytosolic free Ca^{2+}. It is thus important to rule out a role for Zn^{2+} when using these. Fluorescent indicators for Zn^{2+} are available, but these give punctate images, making it difficult to interpret whether the indicator is detecting the free or bound cation.

Mn^{2+} also can interact with intracellular Ca^{2+}, e.g. in the ER. Mn^{2+} is required for the maximum activity of several enzymes involved in intermediary metabolism. Experimentally, addition of extracellular Mn^{2+} can be used to test whether Ca^{2+} channels are still open after a cytosolic free Ca^{2+} signal, as Mn^{2+} quenches the fluorescence of Ca^{2+} indicators such as fura-2. The effect of other transition metals, such as Cu^{2+} and Fe^{2+}, on Ca^{2+} signalling has not been well studied.

6.6 Anions and Intracellular Ca²⁺

Several anions mimic phosphate, and are thus potent inhibitors of Ca^{2+}-MgATPases. These anions include vanadate and fluoride, in the presence of Mg^{2+}, Be^{2+} and Al^{3+}. Vanadate, at concentrations in the 10–100 nM range, was first reported to inhibit the Na^+/K^+-MgATPase in muscle. But, it was then found that vanadate, at micromolar concentrations, was also a potent inhibitor of Ca^{2+} pumps, though there is considerable variation in the concentration range necessary for inhibition. At neutral pH vanadate is mostly in the $H_2VO_4^-$ form, similar in structure and size to $H_2PO_4^-$, the main form of phosphate at pH 7, thus blocking the phosphate site on Ca^{2+}-MgATPases, inhibiting enzyme turnover, and Ca^{2+} transport. Vanadate is present as decavanadate at acid pH, which can bind both to the phosphate and ATP sites on Ca^{2+}-MgATPases.

Fluoride, in the presence of Mg^{2+}, is a potent inhibitor of SERCA Ca^{2+} pumps, via MgF_4^{2-}, which acts as a phosphate analogue similar to vanadate. Fluoride is a useful tool for uncoupling ATP hydrolysis from ion transport. As a result the energy is dissipated as heat. Aluminium and beryllium also form complexes with F^-, which inhibit Ca^{2+}-MgATPases. At pH 7 and millimolar F^-, these are BeF_3^-, AlF_3 and $AlF_3(OH)$. Acting as phosphate analogues, these bind to the Ca^{2+} free conformation of the ATPase, inhibiting it.

6.7 Summary

There are five key features that lie at the heart of how Ca^{2+} works inside cells:

1. The chemistry of Ca^{2+} is special, compared with the monovalent cations Na^+ and K^+, and the divalent cations Mg^{2+}, Mn^{2+}, Fe^{2+}, Cu^{2+} and Zn^{2+}.
2. High affinity Ca^{2+} targets inside cells trigger the cell event.
3. Sites in these targets bind Ca^{2+} in the presence of other cations, particularly Mg^{2+}.
4. Ca^{2+} binding causes the target to affect the machinery responsible for the cell event.
5. Ca^{2+} target can go wrong in a disease or pathological process.

There are seven special features about the biological chemistry of Ca^{2+}, which enabled evolution, several thousand millions of years ago, to select it as a cellular switch (Box 6.2). This special chemistry has enabled evolution to select a huge variety of proteins that bind Ca^{2+} with high affinity, in the presence of Mg^{2+}, to carry out three functions:

1. Ca^{2+} pumps and exchangers that maintain cytosolic free Ca^{2+} at sub-micromolar levels in resting cells, and return the free Ca^{2+} to sub-micromolar levels after a cell event.
2. Ca^{2+} channels that allow Ca^{2+} into the cell, or release Ca^{2+} from organelles, to raise the cytosolic free Ca^{2+} to micromolar levels.
3. High affinity Ca^{2+} binding proteins that are the intracellular Ca^{2+} target, triggering the cell event when Ca^{2+} binds, by interacting directly with the cell machinery responsible for the event.

Box 6.2 Why evolution has selected Ca^{2+} for its unique role inside cells.

1. The low free Ca^{2+} (sub-micromolar) maintained in the cytosol of all resting cells – Eukaryota, Bacteria and Archaea – resulting in a 'Ca^{2+} pressure' across the plasma membrane and across the membranes of organelles such as the ER.
2. Binding to oxygen as the main atom ligating Ca^{2+} inside cells, and not nitrogen.
3. Binding to, and coming off, proteins quickly, in milliseconds, unlike Zn^{2+}, which dissociates from ligands slowly.
4. A good charge carrier across membranes.
5. No problems with osmotic changes.
6. Only one oxidation state, unlike Fe^{2+}/Fe^{3+} and Cu^+/Cu^{2+}.
7. Precipitation is prevented, under physiological conditions, by the low free Ca^{2+} required (micromolar) to trigger the cell event, and the lack of nucleation sites for precipitates to form.

The key to how Ca^{2+} works inside cells is the target protein, whose affinity for Ca^{2+} is in the micromolar range, in the presence of millimolar Mg^{2+}. This results in a large fractional change in binding, when the cytosolic free Ca^{2+} rises to micromolar concentrations after a stimulus. The typical number of atoms in the Ca^{2+} ligand is seven or eight, Mg^{2+} preferring six. Nitrogen is a good donor for the coordination of Ca^{2+} by many synthetic compounds. But the major intracellular ligand atom for Ca^{2+} is oxygen. The oxygen comes from acidic amino acid residues, Asp and Glu, carbonyls in the peptide chain and water. There are four Ca^{2+}-binding sites in intracellular proteins (Box 6.3).

Box 6.3 Four main types of Ca^{2+} binding site in intracellular protein.

1. EF-hand, a 29-amino-acid loop, linked at each end by α-helices – high affinity.
2. C2 domain constructed between beta sheets – high affinity.
3. Clusters of multiple acidic residues, asp and glu – usually medium-low affinity.
4. Clusters of oxygens from amino acids and folding of the peptide chain – medium to low affinity.

There are three ways Ca^{2+} binding leads to a cell event:

1. Ca^{2+} binding causes the protein to bind to a target, which then triggers the event, e.g. calmodulin binding to an ion channel, the main target site for calmodulin being the amino-acid duplet IQ.
2. Ca^{2+} binding to a protein, already bound to its target, causes this to change in conformation, triggering the event, e.g. troponin C changing the troponin-tropomyosin complex which releases the inhibition of actomyosin, allowing it to contract.
3. Ca^{2+} binding activates a protein that causes covalent modification of another protein, which then changes a protein responsible for the cell event: e.g. a calmodulin kinase, that catalyses phosphorylation, or the protease calpain.

Ca^{2+} and/or Ca^{2+}-calmodulin can activate, or sometimes inhibit, a range of kinases, phosphatases and proteases. Ca^{2+} can also have effects on phospholipases and nucleases. Intracellular Ca^{2+} also interacts with other intracellular signals. These include cyclic nucleotides, NO, H$_2$S, inositides, NAADP and sphingosine 1-phosphate. The effects of these interactions are:

1. The other intracellular messenger affects cytosolic free Ca^{2+}, e.g. PKA phosphorylation of phospholamban in the ER.
2. Ca^{2+} affects the concentration of the other messenger, e.g. cyclic nucleotide phosphodiesterase activated by Ca^{2+}-calmodulin.
3. Co-activate or inhibit the same target protein, e.g. activation of phosphorylase by Ca^{2+} and cyclic AMP.

The biological chemistry of intracellular Ca^{2+} is now set to examine in detail how Ca^{2+} triggers a wide range of physiological processes in animals, protists, plants, fungi and microbes.

Recommended Reading

* A must read

Books

Campbell, A.K. (1983) Intracellular Calcium: Its universal role as regulator (monographs in molecular biophysics and biochemistry). Chichester: John Wiley & Sons Ltd. Comprehensive reviews of how Ca^{2+} works in many cell types. Fully referenced up to 1982.
*Campbell, A.K. (2015) Intracellular Calcium. Chichester: John Wiley & Sons Ltd. Chapter 6 How Ca^{2+} works in cells, pp 259–311. Fully referenced chapter.

Reviews

*Carafoli, E. (2005) The symposia on calcium binding proteins and calcium function in health and disease: An historical account, and an appraisal of their role in spreading the calcium message. *Cell Calcium*, **37**, 279–281. Introduction to a special edition on how Ca^{2+} binding proteins work.
Cheng, S.H., Willmann, M.R., Chen, H.C. & Sheen, J. (2002) Calcium signaling through protein kinases: The *Arabidopsis* calcium-dependent protein kinase gene family. *Plant Physiol.*, **129**, 469–485. The role of Ca^{2+}-activated kinases in plants.
*Denton, R.M. (2009) Regulation of mitochondrial dehydrogenases by calcium ions. *Biochim. Biophys. Acta*, **1787**, 1309–1316. How Ca^{2+} in the mitochondria regulates substrate supply for ATP synthesis by a pioneer.

*Feske, S., Rao, A. & Hogan, P.G. (2007) The Ca²⁺-calcineurin NFAT pathway. Chapter 14, pp. 365–402. In Krebs, J. and Michalak, M (Eds). *Calcium: A Matter of Life and Death, New Comprehensive Biochemistry*, Volume **42**. Amsterdam: Elsevier. Useful review of a major effect of Ca²⁺ on gene expression.

Halls, M.L. & Cooper, D.M. (2011) Regulation by Ca²⁺-signaling pathways of adenylyl cyclases. *Cold Spring Harb. Perspect. Biol.*, **3**, a004143. The interaction between Ca²⁺ and cyclic AMP synthesis.

*Hartzell, C.P.I. & Arreola, J. (2005) Calcium-activated chloride channels. *Annu. Rev. Physiol.*, **67**, 719–758. How Ca²⁺ regulates chloride channels.

*Haynes, L.P., Mccue, H.V. & Burgoyne, R.D. (2012) Evolution and functional diversity of the calcium binding proteins (CABPS). *Front. Mol. Neurosci.*, **5**, 9. The evolutionary relationship between Ca²⁺ binding proteins.

*Li, G.H., Arora, P.D., Chen, Y., Mcculloch, C.A. & Liu, P. (2012) Multifunctional roles of gelsolin in health and diseases. *Med. Res. Rev.*, **32**, 999–1025. How the Ca²⁺-regulated actin interacting protein gelsolin works.

Maki, M., Maemoto, Y., Osako, Y. & Shibata, H. (2012) Evolutionary and physical linkage between calpains and penta-EF-hand Ca²⁺-binding proteins. *FEBS J.*, **279**, 1414–1421. The structure and function of the wide ranging proteases – calpains.

Marques, R., Maia, C., Vaz, C., Correia, S. & Socorro, S. (2014) The diverse roles of calcium-binding protein regucalcin in cell biology: from tissue expression and signalling to disease. *Cellular Mol. Life Sci.*, **71**, 93–111. How a non-EF hand Ca²⁺ binding protien acts as a Ca²⁺ target in cells.

Martens, S., Kozlov, M.M. & Mcmahon, H.T. (2007) How synaptotagmin promotes membrane fusion. *Science*, **316**, 1205–1208. How Ca²⁺ triggers vesicle fusion in secretory cells.

*Perochon, A., Aldon, D., Galaud, J.P. & Ranty, B. (2011) Calmodulin and calmodulin-like proteins in plant calcium signaling. *Biochimie*, **93**, 2048–2053. The role of calmodulin in plants.

*Rusnak, F. & Mertz, P. (2010) Calcineurin: Form and function. *Physiol. Rev.*, **80**, 1483–1520. How the structure of the Ca²⁺ binding protein calreticulin in the ER explains its actions.

Research Papers

Bouron, A. & Oberwinkler, J. (2014) Contribution of calcium-conducting channels to the transport of zinc ions. *Pflugers Archiv-Eur. J. Physiol.*, **466**, 381–387. How Ca²⁺ channels can allow zinc transport.

Bozym, R.A., Patel, K., White, C., Cheung, K.H., Bergelson, J.M., Morosky, S.A. & Coyne, C.B. (2011) Calcium signals and calpain-dependent necrosis are essential for release of coxsackie virus b from polarized intestinal epithelial cells. *Mol. Biol. Cell*, **22**, 3010–3021. How Ca²⁺ can be involved in viral infection.

Fajardo, M., Schleicher, M., Noegel, A., Bozzaro, S., Killinger, S., Heuner, K., Hacker, J. & Steinert, M. (2004) Calnexin, calreticulin and cytoskeleton-associated proteins modulate uptake and growth of *Legionella pneumophila* in *Dictyostelium discoideum*. *Microbiology-Sgm*, **150**, 2825–2835. How various Ca²⁺-binding proteins are involved in the physiology of a slime mould.

*Striegel, A.R., Biela, L.M., Evans, C.S., Wang, Z., Delehoy, J.B., Sutton, R.B., Chapman, E.R. & Reist, N.E. (2012) Calcium binding by synaptotagmin's C2a domain is an essential element of the electrostatic switch that triggers synchronous synaptic transmission. *J. Neurosci.*, **32**, 1253–1260. How Ca²⁺ binding to synaptotagmin activates a switch that triggers vesicle fusion at nerve terminals.

*Sun, H.Q., Yamamoto, M., Mejillano, M. & Yin, H.L. (1999) Gelsolin, a multifunctional actin regulatory protein. *J. Biol. Chem.*, **274**, 33179–33182. The role of the Ca²⁺-binding protein gelsolin, and why it has this name.

7

How Ca²⁺ Regulates Animal Cell Physiology

Learning Objectives
1. The wide range of physiological processes that are triggered by intracellular Ca^{2+}, showing how intracellular Ca^{2+} is essential for our birth and existence.
2. How Ca^{2+} triggers nerves firing, neurotransmitter secretion at nerve terminals, muscle contraction in vertebrates and invertebrates, endocrine and exocrine secretion, release of substances during inflammation or a blood clot, and the secretion of solid material such as nematocysts and coccoliths.
3. The role, and processing, of Ca^{2+} in the endocytic pathway.
4. How Ca^{2+} works in an analogue way to regulate intermediary metabolism, to increase ATP synthesis, necessary so that a cell event has the energy required.
5. How intracellular Ca^{2+} controls stages of cell division and growth, including egg fertilisation, the cell cycle, and differentiation.
6. How Ca^{2+} is essential in the immune response to an infection.
7. How Ca^{2+} works differently in the vision of vertebrates and invertebrates.
8. How all our senses of smell, taste and touch all depend on intracellular Ca^{2+}.
9. The special case of Ca^{2+} in bioluminescence.
10. How Ca^{2+} plays a role in regulating gene expression.

7.1 Principles

A rise in intracellular Ca^{2+} is the universal signal in animal cells that causes an event. It enables us to think. It allows an Olympic athlete to run 100 m in 10 s, and a pianist play a scale of C major in just a few seconds. It enables animals with over 100 legs to coordinate these so that they move coherently in the right direction. It makes bees buzz. It is essential for a barnacle or mussel on a rock exposed at low tide to remain closed for hours, until the sea comes back in again. Ca^{2+} signals allow our eyes to see, our ears to hear, our nose to smell, our tongue to taste, and our fingers to sense touch. We are conceived, born and live on waves of free Ca^{2+} in a plethora of different cells. Ca^{2+} signals tell a nerve to fire, a muscle to contract, an insulin cell to secrete, a luminous jellyfish to flash, a photoreceptor in the eye to see, a sensor cell in the nose to smell, a cell in the ear to hear, a sperm to fertilise an egg, an embryonic cell to divide and then differentiate, a cell under attack to defend itself, and an embryonic cell between your fingers to kill itself. These can be grouped into five principle categories (Box 7.1).

Fundamentals of Intracellular Calcium, First Edition. Anthony K. Campbell.
© 2018 John Wiley & Sons Ltd. Published 2018 by John Wiley & Sons Ltd.
Companion Website: http://www.wiley.com/go/campbell/calcium

> **Box 7.1 Physiological cell events triggered by intracellular Ca²⁺.**
>
> 1. Cell excitability, e.g. nerves, bioluminescence.
> 2. Cell movement, e.g. muscle contraction, chemotaxis in inflammation.
> 3. Membrane fusion, e.g. endo- and exo-crine secretion, phagocytosis, endocytosis.
> 4. Cell division and development, e.g. fertilisation, cell cycle, differentiation.
> 5. Senses, e.g. vision, taste, hearing.

There are thus five key questions that we need to answer:

1. How is the Ca^{2+} signal generated?
2. What is the intracellular Ca^{2+} target?
3. How does the target trigger the cell event?
4. How does the cell return to rest?
5. What is the role of Ca^{2+} in generating the energy required for the event via the MgATP/ MgADP + phosphate reaction?

All of these cellular events are initiated by a primary stimulus and can be modified by a secondary regulator. Several are drug targets and are involved in disease. The energy comes from the MgATP/MgADP + phosphate reaction being far from equilibrium. The rise in cytosolic free Ca^{2+} also activates intermediary metabolism, to keep the MgATP/MgADP + phosphate reaction well on the side of ATP, away from equilibrium.

7.2 Ca²⁺ and How Nerves Work

Ca^{2+} is essential for the working of all nerves. Na^+ and Ca^{2+} channels are the two cations responsible for the generation and propagation of an action potential, and a rise in cytosolic free Ca^{2+} is the trigger for secretion at the nerve terminal. Nerves (neurones) are excitable cells that transmit an electrical pulse from one end to the other, so that a transmitter is released into the synapse from the terminal at the end of the nerve. This transmitter then provokes the next cell – typically another nerve or a muscle.

Nerves are made up of four parts (Figure 7.1):

- Dendrites
- Cell body
- Axon
- Terminal.

Binding of a transmitter to its dendritic receptor opens Na^+ and/or Ca^{2+} channels, which cause small depolarisations. Small local changes in cytosolic free Ca^{2+} also occur, as well as changes in Ca^{2+} inside the endoplasmic reticulum (ER) and mitochondria. The membrane potential of a typical nerve is −90 mV, negative inside. To fire, typically the dendritic miniature potentials need to summate to reduce the overall cell membrane potential to less than −30 mV. This generates an action potential, which travels down the axon through opening and closing of Na^+ channels. When it reaches the nerve terminal, it generates a Ca^{2+} signal that provokes fusion of vesicles with the plasma membrane, resulting in release of the neurotransmitter. A rise in cytosolic free Ca^{2+} is the trigger for vesicle fusion at the nerve terminal leading to release of a transmitter, e.g. acetylcholine, glutamate, γ-aminobutyric acid (GABA),

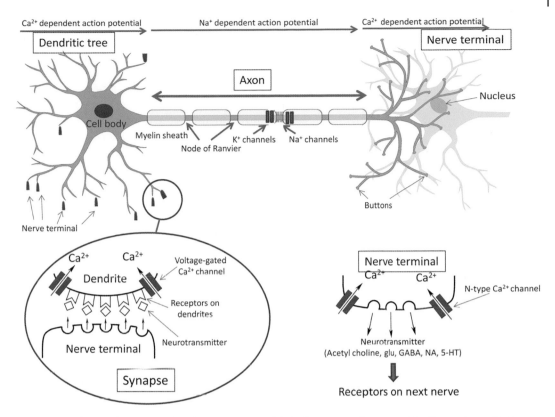

Figure 7.1 Where Ca²⁺ is involved in the action potentials in the mammalian nerve. The figure shows that receptor-operated and voltage-gated Ca²⁺ channels are opened in the dendrites. If the depolarisation of several dendrites is big enough, voltage-gated Na⁺ channels open and a Na⁺-dependent action potential flows down the axon. The action potential jumps from node to node. The Na⁺ channels are at the centre of the node, whereas the K⁺ channels are on the edge, to prevent current leakage. When the action potential reaches the nerve terminal voltage-gated Ca²⁺ channels open (e.g. N-type), Ca²⁺ moves into the cell and the rise in cytosolic free Ca²⁺ triggers neurotransmitter release. Reproduced by permission of Welston Court Science Centre.

5-hydroxytryptamine (5-HT), noradrenaline, glycine, various peptides, such as endorphins in the human brain and gut, and formylated peptides in jellyfish and other invertebrates. Secondary regulators, such as adenosine, can act at the dendrite or the nerve terminal. In some invertebrate nerves, such as those of the sea slug *Aplysia*, oscillating action potentials can occur. In this case, Ca²⁺ currents are the main cause of depolarisation to generate the action potential. The consequent rise in cytosolic free Ca²⁺ activates Ca²⁺-sensitive K⁺ channels that repolarise the cell, ready for another burst of Ca²⁺-dependent action potentials.

Nerves express a variety of Ca²⁺ channels in different parts of the cell. These can be identified by patch clamp with the use of specific channel blockers, such as nicardapine, nitredipine, ω-conotoxin and ω-agatoxin. A key Ca²⁺ channel in the nerve terminal is the N (neuronal) voltage-gated Ca²⁺ channel, which is blocked by the cone snail toxin ω-conotoxin GVIA. But there are other Ca²⁺ channels involved. Voltage-gated Ca²⁺ channels are found in the dendritic spines and nerve terminals, whereas receptor-gated Ca²⁺ channels are restricted postsynaptically to the dendritic spines. Voltage-gated Ca²⁺ channels at the presynaptic nerve terminal open rapidly, when the action potential depolarises it, allowing Ca²⁺ to enter the cell. The cytosolic free Ca²⁺ rises to several μM. This activates fusion of the vesicles with the membrane, releasing their transmitter into the synapse. The terminal then recaptures the vesicular membrane back into the cell, and pumps the Ca^{2+} out via a Ca^{2+}-MgATPase or by Na^+/Ca^{2+} exchange, ready for another action potential. If we are to escape a tiger that has entered the room, our nerves have to transmit the message to run within a few ms; for a bee to buzz, this process has to occur up to 200 times a second. The dendritic tree has hundreds of dendrites, each in a synapse with another neurone. Similarly, the terminal is a tree in synapses with hundreds of other nerves or, in the case of a motor neurone, the muscle endplate. Typical transmitters, the primary stimuli, which open Ca^{2+} channels in the dendrites are glutamate, acetylcholine and dopamine. These open either voltage-gated or ligand-gated Ca^{2+} channels, depending on the receptor type.

7.3 Muscle Contraction

There are four types of muscle contraction where a rise in cytosolic free Ca^{2+} has been fully established as the trigger for contraction, and the molecular mechanism characterised:

1. Skeletal muscle.
2. Heart muscle.
3. Smooth muscle.
4. Invertebrate muscle.

The key molecular structure causing the movement in all muscle is actomyosin (Figure 7.2). A rise in cytosolic free Ca^{2+} releases the molecular apparatus to contract, energy coming from the MgATP/MgADP + phosphate reaction, the myosin MgATPase being activated as the actin filaments slide over the myosin. The releasable Ca^{2+} store in muscle is the sarcoplasmic reticulum (SR). Up to 90% of the SR membrane protein is SERCA, which pumps Ca^{2+} into its lumen. Mammals have three SERCA genes, producing at least ten isoforms from alternative splicing, each with subtly different Ca^{2+} affinity and turnover rates. Thus fast skeletal muscle has SERCA1a, slow skeletal and heart muscle SERCA2a, and smooth muscle and non-muscle SERCA2b. SERCA2 in the heart is regulated by phospholamban, activated by protein kinase A and sarcolipin. Ca^{2+} is released by channels in the SR ryanodine receptor (RyR), RyR1 in skeletal muscle and RyR2 in the heart (Figure 7.3).

There are four common features in muscle contraction:

1. A rise in cytosolic free Ca^{2+} is the trigger.
2. The contractile complex is actomyosin.
3. The Ca^{2+} target allows the myosin to move over the actin, causing contraction.
4. Hydrolysis of MgATP to MgADP provides the energy source.

There are also several differences:

1. The primary stimulus:
   a. In skeletal and heart muscle this is an action potential, provoked by acetylcholine or the pacemaker cells, respectively.
   b. In smooth and catch muscle this a neurotransmitter or hormone, e.g. ACh.
   c. In invertebrate muscle this is a transmitter such as glutamate.
2. The source of Ca^{2+} in skeletal and heart muscle is release from the SR, but the source of Ca^{2+} in smooth muscle is via store-operated Ca^{2+} entry (SOCE), though voltage-gated Ca^{2+} plasma membrane channels can also be involved, which are the main source of Ca^{2+} in invertebrate muscle.

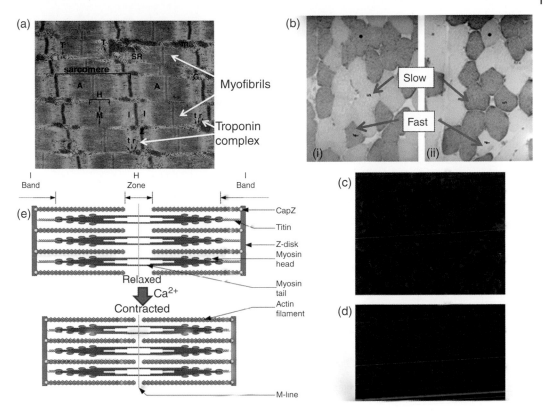

Figure 7.2 Ultrastructure of skeletal muscle. (a) Electron micrograph of skeletal muscle, showing the sarcomere. (b) Human skeletal muscle stained with antibodies, labelled with peroxidase, to fast (I) and slow (ii) myosin showing the reciprocal patchwork distribution of fast (f) and slow (s) muscle fibres. Occasionally there are hybrid fibres (●), particularly in pathology. (c) Muscle labelled with an antibody to SERCA1 showing the three fibre types: fast (type 1), slow (type 2) and intermediate (type 3). Fast fibres have more SR than slow fibres, consistent with being able to deliver Ca^{2+} faster to the troponin complex. Fibres 2A are fast glycolytic; fibres 2B are fast oxidative fibres. (d) Cardiac muscle fibres stained with antibodies to laminin β1. (e) The sarcomere with fibrils relaxed and contracted. This is a simplification of muscle structure. Titin actually stretches from the Z to the M line. Source: (a)–(d) Dubowitz & Sewry (2007). Reproduced with permission from Elsevier. (e) From http://en.wikipedia.org/wiki/Sacomere by David Richfield via Wikimedia Commons.

3. The Ca^{2+} target in skeletal, heart muscle and invertebrate muscle in barnacles is troponin C, and in smooth muscle is calmodulin activated myosin light chain kinase (MLCK).
4. The Ca^{2+} target in the catch muscle of bivalves is myosin itself.

After 80 years, our heart will have contracted and relaxed some 2500 million times. But to escape the tiger that has just entered the room, the heart beat rises immediately two to three times, as a result of noradrenaline released by nerves in the heart, and adrenaline secreted into the blood from the adrenals. When the main source is the release of Ca^{2+} from the sarcoplasmic reticulum, Ca^{2+} still has to enter from outside, if, for example, we are to keep holding our knife and fork for several minutes Otherwise Ca^{2+} would be pumped out of the cell and the muscle fibre would relax.

Mammalian skeletal muscle is triggered by acetylcholine released from nerve terminals, binding to nicotinic acetylcholine receptors. This opens their channels permeable to Na^+ and K^+, generating an action potential that moves down the muscle fibre in milliseconds through voltage

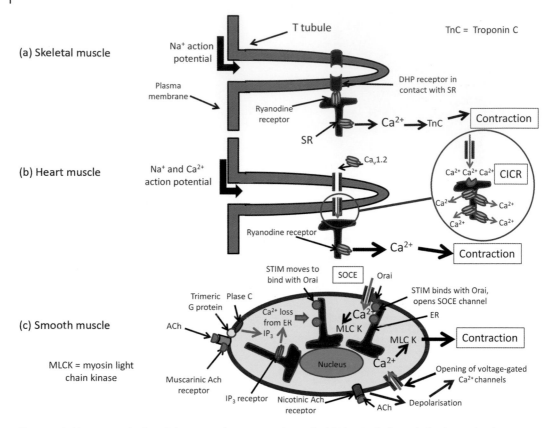

Figure 7.3 How cytosolic free Ca^{2+} rises in three types of muscle. (a) The T-tubule in skeletal muscle. The action potential generated by the acetylcholine receptor at the neuromuscular junction is dependent on the opening of voltage-gated Na^+ channels. The action potential travels down the T-tubule and affects the dihydropyridine receptor, which interacts directly with the ryanodine receptors on the SR. These generate a global Ca^{2+} signal close to the troponin C that releases the links on actomyosin to allow contraction to occur. (b) The T-tubule in heart muscle. The action potential in the myocyte, generated by the pacemaker, is dependent first on the opening of voltage-gated Na^+ channels and then voltage-gated Ca^{2+} channels which are dihydropyridine sensitive. The action potential travels down the T-tubule, opening more voltage-gated Ca^{2+} channels. These let Ca^{2+} into the cell causing an array of Ca^{2+} sparks in the space between the plasma membrane and SR. When the free Ca^{2+} is high enough, a Rubicon is crossed by the ryanodine receptors on the SR. These generate a global Ca^{2+} signal via Ca^{2+}-induced Ca^{2+} release close to the troponin C which releases the links on actomyosin to allow contraction to occur. (c) Smooth muscle. The primary stimuli for smooth muscle contraction include acetylcholine (ACh) via nicotinic or muscarinic receptors and 5-HT. Nicotinic acetylcholine receptors work by initiating Na^+-dependent action potentials, opening voltage-gated Ca^{2+} channels. Muscarinic acetylcholine and 5-HT receptors bind with the α subunit of a trimeric G-protein complex, activating phospholipase C (PLC), generating IP_3 which releases Ca^{2+} from the ER and activates the store-operated Ca^{2+} entry (SOCE) mechanism, by stimulating STIM to move close to the inner surface of plasma membrane where it binds Orai and opens its Ca^{2+} channel. Reproduced by permission of Welston Court Science Centre.

gated Na^+ channels. It traverses the T-tubules, where it causes the dihydropyridine (DHP) receptor to interact directly with the RyR1 receptors. The channels in these tetramers in the sarcoplasmic reticulum open, and release some 100–200 μM Ca^{2+} net, most of which binds within milliseconds to troponin C close by. The cytosolic free Ca^{2+} therefore only rises to a 5–10 μM. Binding of Ca^{2+} to troponin C causes a shift in troponin T and I, resulting in a change in the conformation of tropomyosin, which at rest stops actomyosin contracting (Figure 7.4). The change

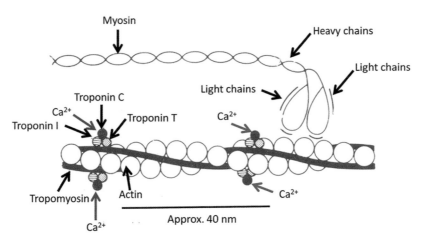

Figure 7.4 Actomyosin structure in muscle. The figure shows the close interaction between tropomyosin and the actomyosin complex. Binding of Ca^{2+} by troponin C stops the inhibitory effect of tropomyosin, allowing myosin to slide over the actin and activate its MgATPase.

in conformation of tropomyosin releases this inhibition, allowing actin to slide over the myosin, causing the muscle fibre to contract. Prolonged contraction requires continuous bursts of acetylcholine release from the nerve terminals. Continuous action potentials cause Ca^{2+} to move into the cell to prevent large loss of Ca^{2+}, which would otherwise cause the fibre to relax. Muscle tissue, such as arms and legs, is a digital system. The strength depends on the number of muscle fibres contracting, the speed of muscle movement depending on the number of slow or fast muscle fibres contracting (Figure 7.2). Once the action potentials stop, Ca^{2+} is pumped back into the SR via the SERCA1 Ca^{2+} pump, where most is bound to calsequestrin. As the cytosolic free Ca^{2+} drops, Ca^{2+} dissociates within milliseconds from the troponin C, actin slides back, tropomyosin regains its inhibitory position, and the muscle fibre relaxes. Contraction is maintained by a continuous breakdown of MgATP to MgADP and phosphate. Thus the cell has to maintain the MgATP = MgADP + $HPO_3^{2-}/H_2PO_3^-$ reaction well away from equilibrium on the side of MgATP. It does this by Ca^{2+} activating phosphorylase to stimulate glycogen breakdown, producing glucose for glycolysis, and, under aerobic conditions, activating pyruvate dehydrogenase in the mitochondria, increasing the respiratory chain and ATP synthesis there.

In heart muscle, the regular action potential generated by the pacemaker cells in mammalian heart, or nerves in neurogenic hearts, spreads across the myocytes via gap junctions, opening first voltage-gated Na^+ channels (Figure 7.5).

Once the membrane potential has dropped to less than −40 mV the voltage-gated $Ca_v1.2$ channels open, causing Ca^{2+} sparks between the inner surface of the plasma membrane and the SR. A threshold is crossed causing the RyR2 receptors in the SR to open via Ca^{2+} induced Ca^{2+} release, resulting to a rapid release of 100–200 μM Ca^{2+} net. The situation leading to contraction is now identical to skeletal muscle, though a different troponin C is used in the heart. Binding of Ca^{2+} to troponin C restricts the global cytosolic free Ca^{2+} rise to 5–10 μM, releasing the inhibition by tropomyosin of actin, and the heart myocyte contracts. Within less than a second, the voltage-gated Ca^{2+} channels close, as do the RyR2 channels in the SR. Ca^{2+} is pumped back into the SR via the SERCA2 Ca^{2+} MgATPase pump, and the precise amount of Ca^{2+} that has entered the cell is pumped back out, primarily via the $3Na^+/Ca^{2+}$ exchanger NCX. The cytosolic free Ca^{2+} drops, Ca^{2+} dissociates from troponin C, the inhibition of actin sliding over myosin recurs, and the cell relaxes. This cycle is repeated about once a second for the human heart to beat, and faster

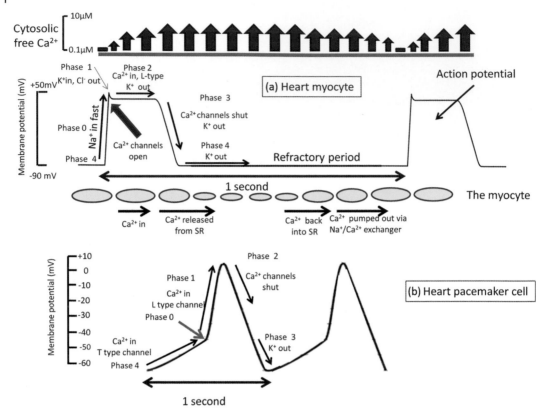

Figure 7.5 Action potentials in the heart and the heart beat. The figure shows the changes in membrane potential in two key cells in the heart: (a) the myocyte, which is responsible for the beat via muscle contraction, and (b) the pacemaker cell, which determines the periodicity of the heart beat. In mammals the pacemaker is myogenic (i.e. spontaneous), but in some animals it is neurogenic (i.e. the periodicity is controlled by nerves). A typical human heart beats once a second (e.g. when you are sitting down). The action potential lasts about 200–300 ms. Ca^{2+} enters via voltage-gated Ca^{2+} channels, causing a local rise in cytosolic free Ca^{2+}, which triggers the ryanodine receptors to release Ca^{2+} from the SR, producing the large, global rise in cytosolic free Ca^{2+} responsible for triggering contraction. The action potential is followed by a refractory period during which the Ca^{2+} is pumped back into the SR and that which entered from outside pumped out by the 3Na^{+}/Ca^{2+} exchanger (NCX). But when you run, the action potential shortens and the refractory period can almost disappear. In live cells, K^{+} channels allow K^{+} to move out of the cell, down its chemical gradient. This tends towards repolarisation of the plasma membrane. Different K^{+} channels operate at particular phases, some are designated delayed and inward rectifying. The latter means that when studied under experimental conditions the conductance inward is bigger than outward.

in small vertebrates, birds and invertebrates with hearts. When we run, or are shocked, the heart beats faster and stronger, stimulated by adrenaline released into the blood from the adrenals, and nor-adrenaline released from local nerve terminals. These bind to the β_1-adrenergic receptor, which, via a trimeric G protein, activates adenylate cyclase. Cytosolic cyclic AMP rises, activating protein kinase A that phosphorylates phospholamban on the SR, which then activates the SERCA2 pump. As a result, initially, after each beat more Ca^{2+} is pumped back into the SR, increasing the Ca^{2+} store. Then, each action potential leads to more and more Ca^{2+} being released each time, causing more troponin C to bind Ca^{2+}, and thus a stronger contraction. The increased speed of contraction is due to the effects of phosphorylation by protein kinase A on the Ca^{2+} and other ion channels in the plasma membrane, changing their gating properties.

Caffeine is an inhibitor of cyclic AMP phosphodiesterase, and activates ryanodine receptors. Acetylcholine, released by nerve terminals close to the myocytes, slows the heart down and reduces the strength of the heartbeat. The molecular mechanism of this is via effects on Ca^{2+} and K^+ channels in the plasma membrane – acetylcholine decreasing Ca^{2+} conductance and increasing K^+ conductance. The receptors are M2 muscarinic, which act via inhibitory G_i-coupled proteins, leading to inhibition of the voltage-gated Ca^{2+} channels, activation of K^+ channels and an inhibition of adenylate cyclase causing a decrease in cyclic AMP. SR Ca^{2+} release is controlled by RyR2 interacting with FKBP12.6, and triadin, junctin and calsequestrin inside the SR.

There are two Ca^{2+}-dependent mechanisms that regulate smooth muscle contraction:

- Direct effects of cytosolic Ca^{2+} via calmodulin activation of myosin light chain kinase (MLCK).
- Modulation of the sensitivity of the contractile system to Ca^{2+} by the Rho/Rho kinase signalling pathway.

Smooth muscle is provoked to contract by neurotransmitters, such as acetylcholine binding to muscarinic receptors. This activates a trimeric G protein complex. Release of the α subunit activates phospholipase C, generating IP_3 and DAG. The IP_3 binds to the ER IP_3 receptor complex, releasing Ca^{2+}. This loss of Ca^{2+} causes the protein STIM1 to aggregate and form punctae, which move towards the plasma membrane. There they bind to the protein Orai1 in the plasma membrane, which opens its Ca^{2+} channels. Ca^{2+} floods into the cell. The Ca^{2+} binds to calmodulin that activates myosin light chain kinase. This phosphorylates Ser19 on the myosin, different from skeletal or heart muscle, allowing actin to slide over the myosin, and the cell contracts. Sheets of smooth muscle cells, such as are found in arteries or the intestine, contract in synchrony and even beat, because Ca^{2+} waves can traverse between cells as a result of communication via gap junctions. Nitric oxide (NO) is an important secondary regulator, causing relaxation. Ca^{2+}-calmodulin weakens the interaction of MLCK with both actin and myosin. This enables the kinase to move along the myofibril so that it can phosphorylate a large fraction of the myosin. Telokin inhibits by competing with MLCK for the myosin.

Contraction of invertebrate adductor catch muscle closes bivalves, such as mussels, clams and oysters, when the tide goes out, or as a defence. This is provoked by a neurotransmitter or neurohormone, such as acetylcholine or glutamate, which generate a large cytosolic free Ca^{2+} signal that activates troponin C to cause the initial contraction. But these muscles have to stay contracted, sometimes for hours. After the initial contraction, cytosolic free Ca^{2+} decreases, contraction of the catch muscle being maintained by the protein twitchin, via cross-bridging of the actomyosin and very low MgATPase activity. Twitchin is activated by the phosphatase Ca^{2+}-calcineurin, which dephosphorylates it. 5-HT (serotonin) relaxes through phosphorylation of twitchin by cyclic AMP-dependent PKA. Mussels have another catch muscle, the anterior byssal retractor muscle (ABRM), a smooth muscle, responsible for keeping the bivalve attached to a rock or other substrate. This works via twitchin as well. Bivalve muscles contain another Ca^{2+} binding protein, calponin. This inhibits myosin MgATPase activity. Its phosphorylation by a protein kinase, dependent upon calcium binding to calmodulin, releases this inhibition.

An unusual contraction occurs in some ciliated protozoa that have a contractile organelle called the spasmoneme. This causes the organism to retreat when touched. Contraction is mediated by a 20-kDa EF-hand acidic Ca^{2+}-binding protein – spasmin. Thus, a rise in cytosolic free Ca^{2+} provokes contraction, the threshold being about 0.4 μM Ca^{2+}. A unique feature of this contractile system is that MgATP is not the driving force, Ca^{2+} binding is all that is required. It appears that Ca^{2+} comes from tubules, aligned along the spasmin filaments. These tubules, like the SR in muscle, can release and take up Ca^{2+} rapidly.

7.4 Chemotaxis and Ca^{2+}

Chemotaxis is the movement of cells along a chemical gradient, and occurs in animals, plants and microbes. There are two mechanisms by which cells can move: flagellate and amoeboid. Both can be regulated by a rise in cytosolic free Ca^{2+}, there being three common features:

1. An increase in cytosolic free Ca^{2+} from the SOCE mechanism.
2. Activation of a Ca^{2+} binding protein, e.g. gelsolin.
3. A change in shape via Ca^{2+} activation of the protease calpain.

   A rise in cytosolic free Ca^{2+} stimulates cytoplasmic streaming in many cells, including fertilised eggs. Chemotaxis is essential in inflammation, as macrophages, neutrophils, eosinophils and lymphocytes are attracted into a site of infection. The primary stimuli are products from the complement pathway, chemokines and substances released from lymphocytes, hormones such as insulin, and some toxins, such as mellitin in bee venom. N-formylated chemotactic peptides have been detected in patient synovial fluid, yet there is no bacterial infection, being released from mitochondrial proteins. The tripeptide f-Met-Leu-Phe (FMLP), at nanomolar, induces a rapid rise in cytosolic free Ca^{2+} in neutrophils, from submicromolar to micromolar. This stimulates amoeboid movement, but micromolar concentrations of the peptide activate release of oxygen metabolites. Activation of PLC generates IP$_3$ and DAG. IP$_3$ provokes loss of Ca^{2+} from ER activating SOCE. DAG activates protein kinase C (PKC). The neutrophil, when it touches a blood vessel, generates a local rise in cytosolic free Ca^{2+}. This provokes the cell membrane to ruffle and flatten, and then forms a spicule-type shape. The cell then sends out a pseudopod in the direction of increasing concentrations of the chemoattractant. This 'foot' pulls the cell along, until it reaches its target site. Ca^{2+} activates the gelsolin which converts the cytoskeleton from a gel into a sol, changing actin from its fibrin (F) form to the globular (G).

   Intracellular Ca^{2+} plays a key role in cell migration by chemotaxis of neurones, fibroblasts, algae and protozoa, sting and other cells in hydroids, sperm towards an egg, slime moulds attracted by cyclic AMP and nutrients, and extracellular calmodulin, the dinoflagellate *Cryptothecodinium* attracted through oxygen gradients, fibroblasts, and a variety of algae and protozoa. The rise in cytosolic free Ca^{2+} activates gelsolin, changing the actin–myosin cytoskeleton, allowing pseudopods to form and amoeboid movement to occur.

   Changes in intracellular Ca^{2+} alter the rhythm and direction of cilia and flagella, stimulated mechanically or by a chemotactic agent. Here, Ca^{2+} works through calmodulin and calaxin, inside flagella, and through tubulin with Ca^{2+} binding sites in microtubules, and contractile proteins, actin and myosin, at the base of the flagellum or cilium. The reversal of beating cilia in the protozoan *Paramecium* is induced by local rises in intracellular free Ca^{2+} as a result of Ca^{2+} channels opening in the cilia membrane, enabling it to reverse when it hits an object. Similarly, the swimming trypanosome *Crithidia oncopelti* reverses the flagellum wave at its tip when it hits an object, also enabling it to reverse. The chemoattractants progesterone in human sperm, and sperm activating and attracting factor (SAAF) in the ascidian *Ciona intestinalis*, trigger sperm to turn and swim towards the eggs. SAAF generates a Ca^{2+} transient, Ca^{2+} entering the sperm through a Ca^{2+} channel, CatSper. Ca^{2+} then regulates the wave movement of the flagella through the 'neuronal' Ca^{2+}-binding protein, EF hand calaxin, on the outer arm of the dynein motor and the microtubules in the axoneme.

   Movement of intracellular granules, phagocytosis and pinocytosis, cytoplasmic streaming, the acrosome reaction in sperm, formation of the cleavage furrow for cell division, platelets and clot retraction, and cell fusion are stimulated by a rise in cytosolic free Ca^{2+} at the right site in the cell, through kinases and Ca^{2+}-binding proteins (α-actinin, gelsolin, vincullin, villin and fragmin), forming cross-links with actin, which are broken when Ca^{2+} binds.

7.5 Intracellular Ca²⁺ and Secretion

All eukaryotic cells secrete material – solids, liquids and gases – into the surrounding milieu. Fusion of vesicles with the plasma membrane (exocytosis) releases hormones, neurotransmitters, enzymes, neurohormones in invertebrates, antibodies, cytokines and the local release of paracrines, reactive oxygen metabolites, and spermicides from certain types of fertilised eggs. Ca²⁺ is the universal trigger for all of these (Figure 7.6) binding to synaptotagmin interacting

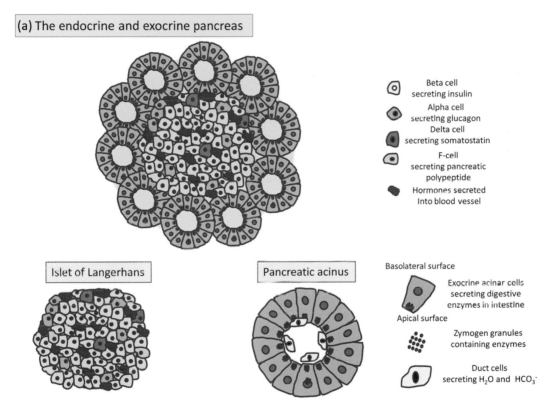

Figure 7.6 Cytosolic free Ca²⁺ changes in three types of vesicle secretion. (a) Endocrine and exocrine pancreas; (b) endocrine beta cell; and (c) pancreatic exocrine cell. In the pancreatic β-cell membrane depolarisation is provoked by a glucose-induced rise in MgATP/MgADP, which causes voltage-gated Ca²⁺ channels to open, leading to a rise in cytosolic free Ca²⁺. This causes fusion of the insulin-containing granules with the plasma membrane, releasing insulin into the blood. Around 6 mM glucose is sufficient to stimulate a human β-cell, but a mouse β-cell requires around 16 mM. Glucose is taken up into the pancreatic β-cell via the transporter GLUT1. Metabolism increases the MgATP/MgADP ratio, which blocks the K$_{ATP}$ channels in the plasma membrane. These close, resulting in membrane depolarisation, which opens T-type voltage-gated Ca²⁺ channels. This results in further depolarisation of the plasma membrane causing L-type Ca²⁺ channels and voltage-gated Na⁺ channels to open. The action potential ends with the opening of P/Q-type Ca²⁺ channels. The cell repolarises by Ca²⁺ activation of BK K⁺ channels. Restoration of the ATP/ADP ratio then allows the K$_{ATP}$ channels to reopen and the membrane becomes fully polarised again. A similar situation occurs in α-cells which release glucagon. In the exocrine pancreatic cell the global cytosolic free Ca²⁺ signal occurs at the apical surface of the cell provoked by the primary stimulus, cholecystokinin (CCK1 receptor) or acetylcholine (muscarinic M3 receptor), acting at the basal surface. The Ca²⁺ rise activates secretion at the apical surface and increases Cl⁻ conductance. Two intracellular signals are generated, IP$_3$ by acetylcholine which leads to SOCE opening, and NAADP plus cyclic ADP ribose acting on ryanodine receptors to release intracellular Ca²⁺. Reproduced by permission of Welston Court Science Centre.

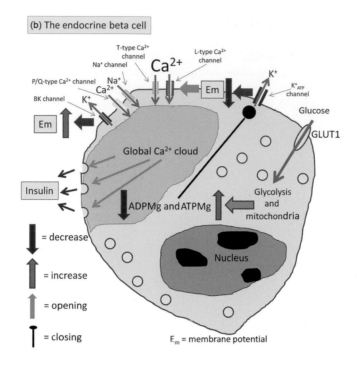

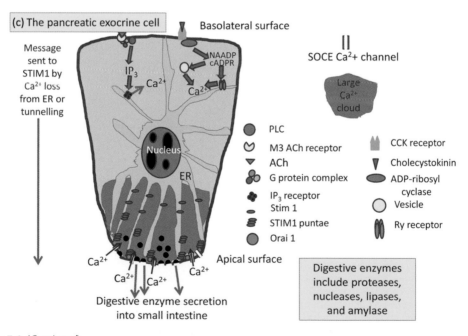

Figure 7.6 (*Continued*)

with the SNARE complex on the inner surface of the plasma membrane, allowing fusion of the vesicle membrane with the plasma membrane. In some organs the primary stimulus acts at one side of the cell, but secretion occurs at the other side; e.g., the exocrine pancreas.

The source of Ca^{2+} is a plasma membrane channel opened by:

- Depolarisation, e.g. the nerve terminal and the pancreatic beta cell.
- Store-operated Ca^{2+} entry (SOCE), e.g. the mast cell.
- A combination of receptor operated Ca^{2+} channel and SOCE, e.g. the neutrophil.

Thus the primary stimulus can be:

- An action potential, e.g. the nerve terminal.
- A hormone, neurotransmitter or immune complex, e.g. the mast cell.
- A metabolite, e.g. the pancreatic beta cell and the neutrophil.

Secondary regulators affect the level of primary stimulus required to make the cell fire, or they may regulate the amount of material secreted. The amount of insulin released by the endocrine pancreas depends on the number of β-cells that have fired. The number of vesicles containing a transmitter at a nerve terminal that fuse after an action potential can be altered by amino acids and other substances released into the synapse.

Secretion by vesicle fusion involves a sequence (Figure 7.7):

1. The vesicles are filled and matured through the ER and Golgi.
2. They then move along microtubules and microfilaments to the inner surface of the plasma membrane, to which they attach.
3. A primary stimulus triggers a rise in cytosolic free Ca^{2+}.
4. Ca^{2+} activates synaptotagmin, which interacts with the SNARE complex.

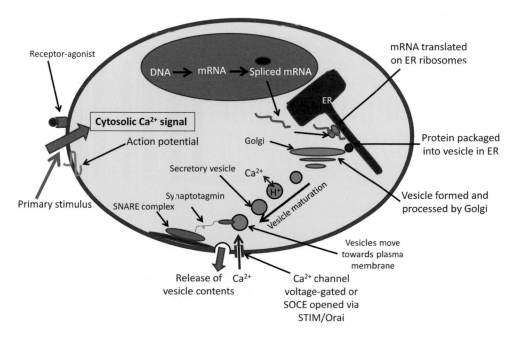

Figure 7.7 Intracellular Ca^{2+} and the sequence of secretion. Reproduced by permission of Welston Court Science Centre.

5. The vesicles 'kiss' the plasma membrane, and then can partially or fully fuse with it, releasing their contents extracellularly.
6. The empty vesicular membrane is then taken back into the cell by endocytosis and removed through the endocytic pathway.
7. The Ca^{2+} within the endocytic vesicle is pumped out.

SNARE (Figure 7.8) is an acronym from SNAp REceptor proteins, where SNAP is the soluble NSF (*N*-ethylmaleimide sensing factor) attachment protein, part of a superfamily of over 60 proteins. There are two types of SNARE: vesicle (v) and target (t). vSNAREs are involved in intracellular vesicle trafficking, whereas tSNAREs are located on target membranes. Structurally the SNARE proteins are either R or Q, depending on the role of arginine (R) or glutamine (Q) in protein binding. Key proteins in the SNARE complex of nerve terminals include syntaxin (VAMP – vesicle-associated membrane protein), SNAP-25 anchored in the plasma membrane, and synaptobrevin anchored in the vesicle membrane. NSF has MgATPase activity – important in SNARE disassembly. A key Ca^{2+} sensor is synaptotagmin with two

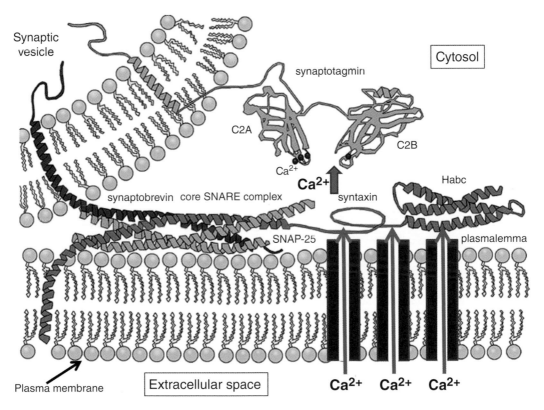

Figure 7.8 Synaptotagmin and the SNARE complex. The figure shows the SNARE complex in a nerve terminal. Ca^{2+} enters via N-type and other voltage-gated Ca^{2+} channels. SNAREs represent a large superfamily of more than 60 members in eukaryotes – another example of Darwinian molecular variation. There are two categories of SNAREs: vSNAREs in the vesicle membrane and tSNARES on the plasma membrane targeted by the vesicle. In the nerve terminal, a local rise in cytosolic free Ca^{2+} binds to synaptotagmin via C2 Ca^{2+} sites. This allows the secretory vesicle to interact with the SNARE complex via its R or Q sites and then induce fusion or 'kiss and touch' in order to release the transmitter into the synapse, so that it can trigger the next nerve or a muscle. Other secretory systems are similar in how Ca^{2+} near the inner surface of the plasma membrane causes fusion of the secretory vesicle with the plasma membrane. Source: Danko Dimchev Georgiev, M.D. [GFDL (www.gnu.org/copyleft/fdl.html) or CC-BY-SA-3.0 (http://creativecommons.org/licenses/by-sa/3.0/)], via Wikimedia Commons.

C2 Ca^{2+}-binding domains formed as loops from a β-sandwich structure – C2A binds three Ca^{2+} and C2B two Ca^{2+}. The formation of a trans-SNARE complex, with three highly helical proteins lying along each other, is essential for fusion to take place (Figure 7.8). This mechanism occurs in all endocrine and exocrine cells triggered by a rise in cytosolic free Ca^{2+} and includes:

- Nerve terminals, stimulated by an action potential.
- Salivary gland, stimulated by acetylcholine.
- Mast cells, stimulated by IgE-antigen.
- Pancreatic enzymes into the intestine, stimulated by acetylcholine or cholecystokinin.
- Platelet secretion of ADP as part of a blood clot, provoked by thrombin.
- Enzymes and reactive oxygen metabolites from phagocytes.
- Hypothalamus, e.g. dopamine, vasopressin.
- Hormones from pancreatic islet cells (alpha – glucagon, beta – insulin), the pituitary (growth hormone, TSH), the parathyroid gland (PTH), adrenal medulla (adrenaline).

In nerve terminals the synaptotagmin-SNARE mechanism is activated by Ca^{2+} entry through voltage-gated Ca^{2+} channels. On the other hand, insulin secretion is provoked by glucose, which increases the ATP/ADP ratio. This closes the K$^+$ channel (K$_{ATP}$) in the plasma membrane, causing membrane depolarisation, resulting in the opening of L-type voltage-sensitive Ca^{2+} channels. The cytosolic free Ca^{2+} rises to several micromolar (Figure 7.6a), and may oscillate, provoking fusion of the insulin-containing granules with the plasma membrane. Lowering plasma glucose reduces the intracellular ATP/ADP ratio, allowing the K$_{ATP}$ channels to reopen. The cell repolarises via delayed-rectifying and Ca^{2+}-activated K$^+$ channels, shutting the voltage-gated Ca^{2+} channels. Glucose stimulates two phases of insulin secretion: a fast transient phase, with a peak and lasting just a few minutes, followed by a slow phase. These are explained through the fusion of different types of vesicle. SOCE allows the cell to top up the ER store, and for the Ca^{2+} needed inside the secretory vesicles. Intracellular cyclic AMP amplifies insulin secretion, via phosphorylation of the voltage-gated Ca^{2+} channels and ER IP$_3$ receptors. Pancreatic islets also release glucagon from α-cells – a vesicular secretion process also provoked by a rise in cytosolic free Ca^{2+}, triggered by depolarisation and adrenaline, through a SOCE mechanism.

Like the salivary gland, the exocrine pancreas secretes both fluid and enzymes. Intracellular Ca^{2+} is the trigger for the pancreatic acini of the exocrine pancreas (Figure 7.6c) to secrete degradative enzymes – proteases, amylase, nucleases and lipases. Secretin is the major primary stimulus for fluid secretion, whereas the main primary stimuli for enzyme secretion are acetylcholine, via muscarinic receptors, from the vagal nerve, and cholecystokinin. Both induce cytosolic free Ca^{2+} transients. The primary neurotransmitter and hormonal stimuli act on the basolateral surface, and provoke secretion at the other, apical surface, into the duct that leads to the duodenum. SOCE provides the mechanism for this. There are two phases of enzyme release from the exocrine pancreas. The first phase occurs over minutes, whereas the second phase can continue for hours, necessary to digest fully a large meal. Exocrine pancreatic acinar cells have two major internal Ca^{2+} stores: the ER and the acidic secretory granules. Both may release Ca^{2+} via IP$_3$, NAADP or cyclic ADP ribose. The ER store accumulates Ca^{2+} via the SERCA pump, whereas Ca^{2+} accumulates in the granules via Ca^{2+}/H$^+$ exchange across the vesicle membrane.

Mast cells are responsible for the hypersensitivity reaction and the allergic response. A rise in cytosolic free Ca^{2+} provoked by an antigen–IgE complex, the natural primary stimulus, is the trigger for exocytosis – degranulation of mast cells with release of histamine, heparin, and the production of eicosanoids and cytokines. The synthesis of eicosanoids and platelet-activating factor (PAF), released by a non-vesicular mechanism, is activated by a rise in cytosolic Ca^{2+}, stimulating phospholipase D (PLD), producing more substrate. A rise in intracellular free Ca^{2+} also activates the transcription factors NF-κB and NFAT, leading to increased synthesis of cytokines such as interleukin (IL)-6 and -13, and tumour necrosis factor. The Ca^{2+} signal is generated via SOCE. Agents that increase cyclic AMP, such as the phosphodiesterase inhibitor theophylline and sodium cromoglycate, inhibit the rise in cytosolic free Ca^{2+} and the number of cells that secrete. Mitochondria restrict the Ca^{2+} cloud from the ER, and the Ca^{2+} entering through the SOCE channels, to a microdomain. A rise in Ca^{2+} inside the mitochondria activates ATP synthesis through substrate supply to the Krebs cycle.

Neutrophils are the first cell at the site of an infection, and provide one of the major mechanisms for killing invading microorganisms. Neutrophils are also abundant at the site of inflammatory diseases such as rheumatoid arthritis and in the lung in asthma, where they play a major role in causing pain and tissue damage. A rise in cytosolic free Ca^{2+} is required to trigger several processes in neutrophils and other phagocytes – cell spreading, chemotaxis, superoxide production, and exocytosis. They kill the invader by releasing toxic oxygen species and degradative enzymes, and by phagocytosis. A rise in cytosolic free Ca^{2+} is provoked by the primary stimuli responsible for each of these processes (Figure 7.9).

The rapid rise in cytosolic free Ca^{2+} occurs through SOCE, causing the filamentous (F) actin to break up into monomeric, globular (G) actin, via gelsolin. Ca^{2+} signals also activate the enzymatic machinery that produces reactive oxygen species, via a special oxidase using NADPH to reduce oxygen via a cytochrome *b*. Ca^{2+}-calmodulin activates this complex. There are two distinct pathways for activating this oxidase, one dependent on a rise in cytosolic free Ca^{2+}, the other not. Neutrophils isolated from the joints of patients with rheumatoid arthritis had abnormally large cytosolic free Ca^{2+} signals in the absence of extracellular Ca^{2+} (Figure 7.9b), indicating a high level of Ca^{2+} stored within the ER. Ca^{2+} signals can be localised to microdomains, close to the plasma membrane or as clouds near the ER. O_2^- can be released, or be generated inside phagosomes, where it is converted into hypochlorite (OCl^-) by the enzyme myeloperoxidase. A similar Ca^{2+}-activated mechanism generates hypobromite (OBr^-) in eosinophils, which kill eukaryotic parasites, whereas hypoiodite (OI^-), released from some fertilised eggs, prevents fertilisation by a second sperm.

Platelets vesicular secretions by thrombin, ADP and 5-HT are also all triggered by a rise in cytosolic free Ca^{2+}, via SOCE, which also activates prostaglandin and thromboxane synthesis. Platelets are activated to stick together by thrombin, produced from prothrombin as a result of tissue damage, such as a cut, by contact with collagen and vWF, and then by ADP released from activated platelets. Platelet activation leads to activation of a scramblase, shifting the negatively charged phospholipids phosphatidyl serine and phosphatidyl ethanolamine to the cell surface. This aids adhesion and catalysis of the tenase and prothrombinase complexes. There is also a receptor-operated Ca^{2+} channel, PLX1, on the platelet cell surface and the TRP channel TRPC, which can also let Ca^{2+} into the cell.

Several cell types secrete solid substances, or actual structures. In birds, calcite crystals are secreted in the oviduct in order to form the eggshell. Sting cells of jellyfish, anemones and corals have a microsyringe, contained within the nematocyst, which fires, and injects a toxin into its prey or attacker. Coccolithophores, single-celled algae found throughout the world in the sea, secrete plates, coccoliths, made of calcium carbonate, contributing half the calcite production in the sea. Intracellular Ca^{2+} plays a major role in this process. Ca^{2+} has several roles in these processes. Ca^{2+} transients are required for nematocyst discharge. Cnidocytes also have voltage-gated Ca^{2+} channels in the plasma membrane leading to Ca^{2+} entry. The capsule containing the nematocyst has a very large Ca^{2+} store. Release of this Ca^{2+} into the cytosol when the cnidocil is touched causes a very rapid change in osmotic pressure inside the nematocyst capsule, causing the nematocyst to uncoil and eject. On the other hand, the coccoliths of coccolithophores take up Ca^{2+} via a Ca^{2+}/H^+ exchanger CAX and HCO_3^-. The vesicle contains a glutamate, proline and alanine-rich (GPA) Ca^{2+}-binding protein, required for coccolith formation. The vesicle is transported to the plasma membrane, where fusion allows the coccolith to attach to the outside of the cell. This then joins up with other plates to form the complete extracellular plate structure.

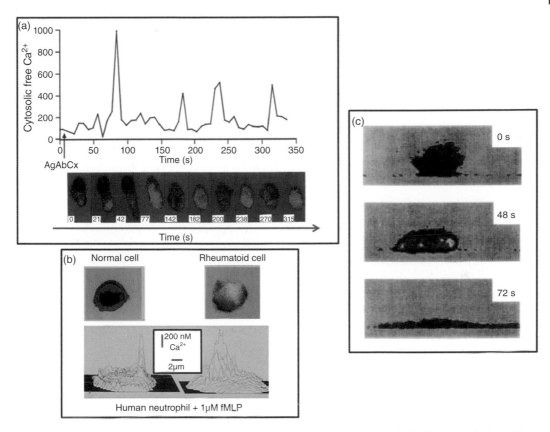

Figure 7.9 Some cytosolic free Ca^{2+} changes in neutrophils. This figure illustrates a few of the cytosolic free Ca^{2+} signals observed in neutrophils. Four types of Ca^{2+} signal induced by the peptide f-Met-Leu-Phe (FMLP), which comes naturally from the N-terminus of bacterial proteins and proteins synthesised by the mitochondrion itself, were originally seen in human neutrophils: type 1, a transient rise in Ca^{2+} occurring within 6 s; type 2, an oscillating cytosolic free Ca^{2+}; type 3, a latent Ca^{2+} transient significantly delayed (21–56 s); and type 4, no significant Ca^{2+} rise. The figure shows cytosolic free Ca^{2+} signals in neutrophils imaged using fura-2 after simulation, see Davies *et al.* (1994) *FEBS Lett.*, **291**, 135–138. (a) Fc receptors binding an immune complex. Cytosolic free Ca^{2+} oscillations are induced in a human neutrophil by soluble antibody–albumin complexes (AgAbCx) binding the Fc receptor. The mechanism involves SOCE. No cytosolic free Ca^{2+} signal was observed in the presence of EGTA. Binding of opsonised particles to C3b receptors also induces a cytosolic free Ca^{2+} signal, which activates calpain, causing freeing of β_2-integrin which enhances phagocytosis. Source: Davies *et al* (1994) *Immunology*, **82**, 57–62. Reproduced with permission from Oxford University Press. (b) FMLP in the absence of extracellular Ca^{2+}, showing a much larger cytosolic Ca^{2+} cloud in cells from a rheumatoid joint. Source: Davies *et al.* (1994) *Ann. Rheum. Dis.*, **53**, 446–449. Reproduced with permission from Wiley. (c) A cell when settling on to a surface. Source: Hallett, M.B. & Dewitt, S. (2007) *Trends Cell Biol.*, **17**, 209–214. Reproduced with permission from Wiley.

7.6 Ca²⁺ and Endocytosis

Endocytosis is the uptake of fluid, soluble substances and particles by vesiculation into cells. Invaginated vesicles move through an endocytic pathway, from plasma membrane to a lysosome, inside which proteins are degraded. Ca^{2+} plays a role in parts of this pathway. Endocytosis via clathrin-coated pits, caveolae or micropinocytosis do not typically require a rise in cytosolic free Ca^{2+}; though, in nerve terminals, there is association between the Ca^{2+}-dependent phosphatase calcineurin and dynamin 1, which is required for clathrin-mediated endocytosis, and is provoked by the rise in the cytosolic free Ca^{2+} as a result of the action potential. Ca^{2+} taken in the initial endosome is pumped into the cytosol. Ca^{2+} is required inside the lysosome, from which it can be released by NAADP, or two pore channels through phosphatidylinositol 3,5-bisphosphate (PtdIns(3,5)P2).

7.7 Ca^{2+} and Phagocytosis

Phagocytosis is the uptake of particles into cells. It occurs in vertebrate macrophages, neutrophils, eosinophils and platelets, and in invertebrate digestion. Phagocytosis involves:

1. Binding of the particle to the receptor cell surface.
2. A rise in cytosolic free Ca^{2+}.
3. Invagination of the particle into a vesicle forming a phagosome.
4. Fusion of the phagosome with a lysosome.
5. Digestion of the particle.
6. Excretion of material in the phagolysome by exocytosis.

Secondary regulators via cyclic AMP can inhibit this process, interacting with the Ca^{2+} signalling mechanism. Neutrophils, macrophages and eosinophils kill invading microbes taking up opsonised particles via Fcγ or CR receptors. Binding generates a cytosolic free Ca^{2+} signal close to the membrane forming the phagosome, which can be followed by a global Ca^{2+} signal, oscillatory, or close to the phagosome, as it fuses with the lysosome. Fc receptors lead to activation of PLCγ, activating the SOCE mechanism. In contrast, CR receptor activation activates PLD and the production of sphingosine 1-phosphate, but not IP_3. Sphingosine 1-phosphate also releases Ca^{2+} from the ER and may activate SOCE. The rise in cytosolic free Ca^{2+} activates the Ca^{2+}-binding proteins gelsolin, converting fibrous (F) actin into globular (G) actin, solubilising it close to the inner surface of the plasma membrane, enabling the phagosome to form. Ca^{2+} also activates calmodulin, PKC and annexins. These allow fusion of primary and secondary granules with the phagosome. Ca^{2+}-calmodulin activates calmodulin kinase II (CaMKII). Ca^{2+} also allows assembly and activation of the NADPH/cytochrome b complex, which generates O_2^-. Release of the negatively charged O_2^- causes the membrane to depolarise. Since Ca^{2+} entry is dependent on a negative membrane potential, concomitant with the efflux of O_2^-, Ca^{2+} activates voltage-gated proton channels, (voltage-sensing domain only protein) VSOP/H_v1 H^+ channels. This allows H^+ also to move out of the cell, compensating for the negative charge on the superoxide anion and repolarising the membrane. The rise in cytosolic free Ca^{2+} also activates K^+ channels in the plasma membrane, which hyperpolarise the cell, increasing the electrochemical force on Ca^{2+} influx. The rise in cytosolic free Ca^{2+} is essential for parts of the phagocytic pathway and eventual killing of the invading microbe.

7.8 Intracellular Ca^{2+} and Intermediary Metabolism

All processes activated by a rise in cytosolic free Ca^{2+} require MgATP, the potential energy coming from the MgATP = MgADP + phosphate reaction. Thus, in all cases, the rise in Ca^{2+} activates pathways that generate more NADH for oxidative phosphorylation in the mitochondria, and also more ATP via glycolysis, pushing the reaction well on the side of MgATP. Ca^{2+} achieves this by activating enzymes that hydrolyse the breakdown of glycogen and triglyceride, and pyruvate oxidation. Four major pathways are activated or inhibited by Ca^{2+}, depending on cell type:

1. Glycogen breakdown is activated and synthesis inhibited, generating more glucose as substrate.
2. Glycolysis is activated and gluconeogenesis inhibited, generating more ATP, and pyruvate for mitochondrial ATP synthesis.

3. Pyruvate oxidation in the mitochondria is activated, generating more NADH for ATP synthesis.
4. Fat breakdown and fatty acid oxidation are activated, generating more AcCoA in the mitochondria for the citric acid cycle.

The breakdown of glycogen to glucose 1-phosphate is catalysed by phosphorylase that is activated by AMP and covalent modification via protein kinase A or Ca^{2+}-calmodulin, the tightly bound γ subunit. This converts phosphorylase b into phosphorylase a. Ca^{2+}-calmodulin also inhibits glycogen synthesis via glycogen synthase kinase and phosphorylase kinase.

When Ca^{2+} goes up in the mammalian cell, there is an increase in oxygen uptake, because of an increase in the mitochondrial respiratory chain, leading to an increase in ATP synthesis. The rise in cytosolic free Ca^{2+} causes an increase in Ca^{2+} uptake by mitochondria, through the Ca^{2+} uniporter, leading to 5–10 µM free Ca^{2+} in the inner matrix. This activates several enzymes, resulting in an increase in NADH, and thus ATP synthesis. Pyruvate generated by glycolysis is taken up into the inner matrix of the mitochondria, where it is oxidised to generate NADH. Three matrix enzymes for this are activated by micromolar Ca^{2+}:

1. Pyruvate dehydrogenase phosphatase.
2. NAD-isocitrate dehydrogenase (IDH3).
3. α-oxoglutarate dehydrogenase.

Pyruvate dehydrogenase phosphatase dephosphorylates a subunit of the pyruvate dehydrogenase complex, pyruvate dehydrogenase being inhibited by phosphorylation. Pyruvate dehydrogenase phosphatase, is stimulated by Ca^{2+}, insulin, PEP and AMP, but competitively inhibited by ATP, NADH and acetyl-CoA. There are two isoforms of pyruvate dehydrogenase phosphatase in mammalian mitochondria: PDP1 and 2, with 55 kDa catalytic subunits designated PDP1c and 2c. Only PDP1c is activated by Ca^{2+}. Ca^{2+} also causes a large increase in affinity of isocitrate dehydrogenase 3 for the substrate *threo*-D_S-isocitrate, by decreasing the K_m. This requires ATP or ADP, the ratio influencing sensitivity to Ca^{2+}. The lower the ATP/ADP, the more sensitive the enzyme is to Ca^{2+}, fitting the need to convert more ADP into ATP. Ca^{2+} binding requires isocitrate, adenine nucleotide and Mg^{2+}. The complex can bind two Ca^{2+}, but there are no identifiable EF-hands or other obvious Ca^{2+} sites in the enzyme. Only the vertebrate enzyme is regulated by Ca^{2+}. Enzymes isolated from yeast, blowfly and locus flight muscle, potato, and the spadix of arum lily are not affected by Ca^{2+}. The three inner mitochondrial enzymes vary in their sensitivity to Ca^{2+}.

Other mitochondrial enzymes associated with ATP synthesis that are regulated by Ca^{2+} include activation of the F_o/F_1-MgATPase via a small inhibitory protein, inhibition of pyrophosphatase, an effect related to mitochondrial volume, NO synthase nNOS and, at higher Ca^{2+} concentrations, pore formation. These latter two may be particularly important in cell injury during hypoxia and changes in oxygen supply, with the generation of reactive oxygen species.

There is no calmodulin in the inner matrix of mitochondria, as it has no targeting sequence to take it in. However, calmodulin can interact with proteins in the outer mitochondrial space or membrane, and can bind to mitochondria, for example in neurones through Ca^{2+} dependent binding to hyaladherin RHAMM, the receptor for hyaluronan-mediated motility (CD168). FAD-glycerol phosphate dehydrogenase is also activated by micromolar Ca^{2+}. This enzyme works with cytosolic NAD-glycerol phosphate dehydrogenase to make the glycerol phosphate shuttle, enabling NADH generated in the cytosol to be transferred into the inner mitochondrial matrix. It is bound to the inner membrane of the mitochondrion, with its glycerol phosphate and Ca^{2+}-binding sites facing outwards. The Ca^{2+} site has a $K_{0.5}$ of 0.1 µM, ideal for responding to cytosolic changes in free Ca^{2+} in the 0.1–1 µM range. Ca^{2+} increases substantially the affinity of the enzyme for glycerol phosphate. It has two predicted EF-hand Ca^{2+} sites facing towards the cytosol – DEDEKGFITIVD and DLNKNGQVELHE – residues 1, 3, 5, 9, 12 and the carbonyl in the peptide bond of 8 providing oxygens for Ca^{2+} binding. The enzyme in yeast and plants is smaller and lacks the two predicted EF-hands, and is thus unlikely to bind Ca^{2+} physiologically.

The major stores of triglyceride in the animal body are white and brown adipose tissue. Lipolysis in brown adipose tissue is a source of body heat, while white adipose lipolysis provides fatty acid as an energy source for other tissues. Primary stimuli for lipolysis include adrenaline, via both α and β receptors, each working via trimeric G-proteins. α-Adrenergic receptors work via a rise in cytosolic free Ca^{2+}, whereas β-adrenergic receptors activate adenylate cyclase, producing cyclic AMP, which activates PKA.

7.9 Intracellular Ca^{2+} and Cell Growth

7.9.1 Principles

The development and the survival of all organisms depend on the ability of cells to double in size, and then divide into two. Intracellular Ca^{2+} plays a key role in egg fertilisation, cell growth and tissue development in all animals and plants, as well as in several bacteria. Key evidence for this has been based on experiments in model systems, and in particular on measurements and imaging of cytosolic free Ca^{2+}, first using aequorin and then fluorescent dyes. A wide variety of Ca^{2+} signals have been seen in dividing cells, including puffs, sparks and spirals, transients, oscillations, waves, and tides. Because all eukaryotic cells contain a substantial amount of Ca^{2+} in the ER, and in some cells in other organelles such as secretory vesicles, the cell has to double this Ca^{2+} before it can divide in two.

7.9.2 Cell Cycle and Ca^{2+}

Ca^{2+} transients play a key role in different parts of the cell cycle in many invertebrate and vertebrate cells (Figure 7.10), determining whether cell division occurs. Ca^{2+} is also important in taking quiescent cells (G_0) back into the active cell cycle, for example lymphocytes in antibody production. Hormones, growth factors and cytokines regulate cell proliferation in animals and plants by inducing cells from a quiescent G_0 phase into G_1, or by speeding up one or more of the phases in the ER store in the two new cells. Depletion of Ca^{2+} stores, using the ER Ca^{2+}-MgATPase inhibitor thapsigargin, stops cell division. Ca^{2+} can also be required for structural changes in chromatin. Ca^{2+} can play a role in any one of the five phases of mitosis – prophase, prometaphase, metaphase, anaphase and telophase. Each is determined by checkpoints, regulated by cell cycle-dependent kinases and cyclins. Ca^{2+} and the rise in pH in fertilised oocytes cause an increase in cyclin synthesis. Injection of the Ca^{2+} chelator BAPTA blocks cyclin phosphorylation, and arrests mitosis at entry. After mitosis, a cleavage furrow forms, and the cell divides into two – cytokinesis. Cells can leave the cell cycle from G_1 into G_0, or from G_2 into R_2. Cytosolic free Ca^{2+} signals, initiated by hormones or growth factors, trigger the cell to re-enter the cell cycle. There are three common points in yeast, mammalian and invertebrate cells where Ca^{2+} signals play a key role:

1. *Start:* the decision the cell makes between mitosis, meiosis, arrest in G_0, or conjugation.
2. *Entry*: entry into mitosis from G_2, after DNA doubling is complete.
3. *Exit*: exit from mitosis once the chromosomes have separated, leading to complete cell division.

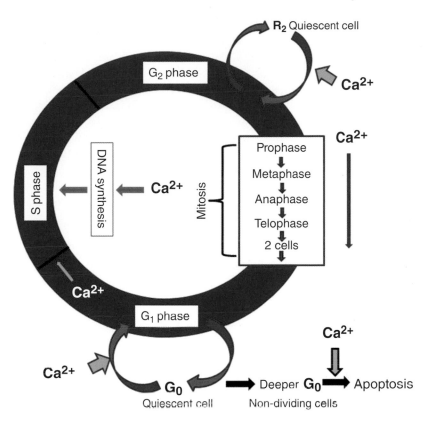

Figure 7.10 Ca²⁺ and the cell cycle. The figure shows the four main phases in the eukaryotic cell cycle: G_1, S, G_2 and M. A rise in cytosolic free Ca²⁺ is important often In G_1 and free Ca²⁺ in the nuclear region during M, after the nuclear membrane has disappeared. But cytosolic and nuclear free Ca²⁺ signals can occur in other parts of the cell cycle depending on species and cell type. Reproduced by permission of Welston Court Science Centre.

Ca²⁺ transients have been detected in different parts of the cell cycle in many invertebrate and mammalian cells. Ca²⁺ transients, which correlate with cell cycle events, include pronuclear migration, nuclear envelope breakdown, the metaphase to anaphase transition in mitosis, and the final cell division step itself. These events are prevented by injecting a Ca²⁺ chelator that stops the Ca²⁺ signal. Mitotic and other cell cycle events, including gene activation, can also be mimicked by injecting Ca²⁺ into eggs or by using Ca²⁺ ionophores. The rise in cytosolic free Ca²⁺ can be the signal telling the cell to go forward, or Ca²⁺ is simply necessary for the cell to continue in the cell cycle. Many Ca²⁺ signals during the cell cycle can be generated in the absence of extracellular Ca²⁺, the main Ca²⁺ signal being generated by release from the ER through IP_3. In sea urchin eggs, for example, IP_3 oscillations during the first two cell cycles cause oscillations in cytosolic free Ca²⁺, which lead to mitosis.

Cell cycle Ca²⁺ transients are initiated by internal programming, whereas cell proliferation can also be triggered through intracellular Ca²⁺ signalling pathways initiated by hormones, paracrines and growth factors binding to cell surface receptors, usually coupled to trimeric G-proteins. Measurements of nuclear Ca²⁺, using either targeted aequorin or fluorescent dyes such as fura-2 or fluo-3, have led to a controversy as to whether the nuclear membrane can act as a barrier to Ca²⁺. The permeability of the nuclear pore can be regulated by signalling processes inside the ER, changing during the cell cycle. Ca^{2+}-calmodulin on the mitotic spindle promotes microtubule disassembly and inhibits reassembly. Ca^{2+} signals are needed for activation of gene expression through calmodulin, activating kinases and phosphatases, and Ca^{2+} activated calcineurin phosphatase (CALNA/CALNB). Calmodulin moves into the nucleus following cell activation, and its expression is regulated during the cell cycle, doubling at the G_1/S boundary, and can be essential for G_2/M transition. Calcineurin is crucial for the activation of the T-lymphocytes provoked by a Ca^{2+} transient, activating the NFATc, by dephosphorylation in the cytosol. Activated NFATc is then translocated into the nucleus, where it up-regulates the expression of IL-2. IL-2 then stimulates the cell division and differentiation of the T cell response. CaMKI is required for regulating G_1, whereas CaMKII is essential for regulating the G_2/M transition and during mitosis. Phosphorylation of proteins catalysed by Ca^{2+}-activated CaMKII is a key part of chromosome duplication and movement to form two separate sets.

7.9.3 Fertilisation and Intracellular Ca^{2+}

Intracellular Ca^{2+} plays a key role at six steps in the fertilisation process, the fusion of a male and female gametes, and in several of the cell divisions that follow, as the fertilised egg develops into a cluster of cells:

1. Maturation of the egg, activated by hormones such as progesterone, ready to fuse with a sperm.
2. Maturation of sperm after ejaculation, by capacitation and the acrosome reaction at the sperm head, ready to fuse with an egg.
3. Chemotaxis of animal sperm, and pollen tube formation in plants.
4. Changes in cell surface proteins and the production of oxygen metabolites, both of which prevent fusion of the egg with a second sperm.
5. Activation of the pathway for cell division.
6. Activation of a differentiation pathway, as the single fertilised egg forms a cluster of cells.

Released mammalian eggs are still oocytes, arrested at metaphase II of meiosis. In contrast, in marine invertebrates and insects, the oocyte arrests with nuclei intact at the metaphase of meiosis I. Thus, after the sperm has fused and injected its DNA into the egg, the fertilised egg completes meiosis, but now has too much DNA. It gets rid of this through polar bodies, which are rejected via a sort of uneven cytokinesis, one in animals and two in invertebrates. The polar bodies then degrade over a day or so. After successful fertilisation, the egg enters the cell cycle proper, which is initiated by a large cytosolic free Ca^{2+} signal. This leads to cell division (cytokinesis), and then the formation a multicellular blastula. Once the sperm fuses with the egg, its DNA is injected into the egg, as well as a special phospholipase that generates a Ca^{2+} signal. Thus, rapidly after fusion of the sperm with the egg, there is a large rise in cytosolic free Ca^{2+}. This takes the form of a Ca^{2+} wave or tide (Figure 7.11), which moves through the egg. This Ca^{2+} rise initiates two key processes, depending on species:

1. In fish and invertebrate eggs, the Ca^{2+} signal stimulates the production of oxygen metabolites, O_2^- and HOI^-, which alters the egg surface, and inactivate other sperm, both preventing any more sperm fusing.
2. The Ca^{2+} wave or tide initiates reactions, leading to formation of the cleavage furrow, where the diploid fertilised egg divides into two.

The cytosolic Ca^{2+} signal occurs in the absence of extracellular Ca^{2+}, and thus comes from release from an internal store stimulated by IP_3. A search of the mouse genome revealed an unusual phospholipase (PLC), PLCζ, discovered by Karl Swann and Tony Lai. Injection of the

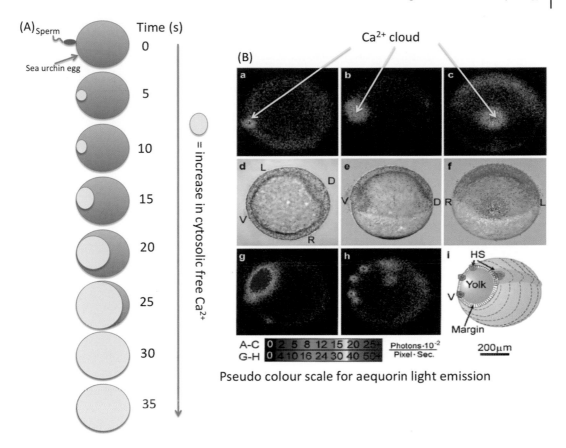

Figure 7.11 Ca^{2+} spikes, waves and tides imaged in eggs and embryos. (a) Diagrammatic image of the rise of cytosolic free Ca^{2+}, using the Ca^{2+} indicator of calcium green dextran injected into the egg, as a tide which takes about 20 s to fill a sea urchin egg (*Lytechinus pictus*) fertilised by a sperm. Data from Whitaker (2006) *Physiol. Rev.*, **86**, 25–88. (b) Localised Ca^{2+} signals in a zebrafish embryo imaging f-aequorin. Reproduced with permission from Elsevier. Three types of Ca^{2+} transient were seen: one between the epibody and ventral marginal signal (a–f), another as a yolk flash (g), and a third as marginal hot spots (h). (d, e, f) Aequorin signal superimposed on the bright field; (a, d) embryo viewed from the vegetal pole; (b, c) from the left side of the embryo; (e, f) from the ventral side of the embryo. Source: Gilland *et al.* (1999) *Proc. Natl Acad. Sci. USA*, **96**, 157–161. Reproduced with permission from PNAS. See also Webb, S.E. & Miller, A.L. (2013) *Cell Calcium*, **53**, 24–31.

cRNA coding for this initiated Ca^{2+} oscillations in mouse eggs (Figure 7.12), and Ca^{2+} transients in invertebrate eggs. PLCζ is smaller than the other PLCs β, γ, δ and ε, being around 70 kDa. Like all PLCs, PLCζ has the X and Y catalytic domains, a C2 Ca^{2+}-binding domain, and four EF Ca^{2+}-binding sites. But PLCζ is active at low Ca^{2+} concentrations (i.e. resting Ca^{2+} of around 100 nM), whereas its plasma membrane counterpart requires micromolar levels of Ca^{2+} to activate it. The target for PLCζ is PIP$_2$ on vesicles in the egg, rather than PIP$_2$ in the plasma membrane. Fertilised eggs from mice in which PLCζ has been knocked-out do not generate Ca^{2+} signals when they fuse with sperm. Hydrolysis of PIP$_2$ generates IP$_3$ and DAG. IP$_3$ releases Ca^{2+} from the ER, and DAG activates a plasma membrane Na$^+$/H$^+$ exchanger, raising the cytosolic pH from about 6.7 to 7.3. This cytoplasmic alkalinisation is necessary for stimulation of protein synthesis, but is not sufficient for cell cycle progression. Ca^{2+} is essential for this. Preventing the Ca^{2+} signal by injection of Ca^{2+} chelators stops cell cycle progression, nuclear envelope breakdown, and mitosis.

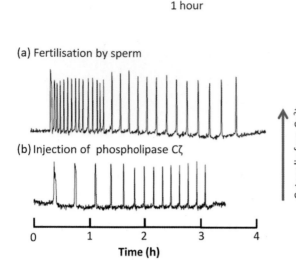

1 hour

(a) Fertilisation by sperm

(b) Injection of phospholipase Cζ

Cytosolic free Ca^{2+}

0   1   2   3   4

Time (h)

Figure 7.12 Ca^{2+} oscillations in fertilised mouse eggs. The figure shows regular spikes in cytosolic free Ca^{2+} in mouse eggs. (a) Fertilised by a sperm. (b) Induced by injection of cRNA coding for PLCζ, which uses PIP$_2$ in intracellular vesicles to generate IP$_3$ and DAG, rather than the usual plasma membrane source used by G-protein activation after binding of a primary stimulus to a receptor on the cell surface. Cytosolic free Ca^{2+} was measured using the fluorescence of the Ca^{2+} indicator Oregon Green BAPTA (OGPD)–dextran injected into the eggs. Data from Campbell and Swann (2006).

The type and timing of the Ca^{2+} signal in fertilised eggs varies between different species. In mammals the sperm induces a prolonged phase of Ca^{2+} oscillations, which last several hours (Figure 7.12), initially at intervals of about 10 min. This leads to the second meiotic division and second polar body formation. The interval between the Ca^{2+} spikes is now longer, around 20–30 min. These Ca^{2+} spikes last 3–5 h in total, and stop when the cell embryo enters interphase and renews the cell cycle, leading to formation of the male and female pronuclei. The Ca^{2+} oscillations in mammalian eggs are essential for activation of CaMKIIγ, which induces degradation of cyclin B by the proteosome, and a decrease in maturation promoting factor (MPF) and MAPK, which trigger the resumption of the cell cycle. The Ca^{2+} oscillations coincide with hyperpolarisation of the membrane, via Ca^{2+} activation of K$^+$ channels. The Ca^{2+} wave is propagated via Ca^{2+}-induced Ca^{2+} release from the ER. But, in mammalian eggs, release of Ca^{2+} from the ER activates SOCE, which is required, to maintain Ca^{2+} oscillations, by replenishing the ER Ca^{2+} store between oscillations. Ca^{2+} entry via SOCE is also necessary to activate signalling pathways upstream from CaMKIIγ, for emission of the second polar body. But Ca^{2+} oscillations maintained by SOCE are not required for the cell to resume meiosis II.

7.9.4 Differentiation and Intracellular Ca^{2+}

As the embryo develops through two, four, eight and more cells, the cells differentiate. In most animals a blastula forms with three cell layers – epiderm, mesoderm and ectoderm, though coelenterates have a mesoglea between the ectodermal and epidermal layers. As the organism develops, gene expression becomes more tightly controlled. Ca^{2+} signals play an important role in the differentiation of cells in the early embryo, evidence for this coming from two model systems: the fruit fly *Drosophila* and zebrafish.

7.10 Intracellular Ca^{2+} and the Immune Response

A rise in cytosolic free Ca^{2+} is responsible for activating cells in the immune system, enabling the animal to kill the invading microbes, signalling:

1. Transformation of T-lymphocytes to helper cells, which then activate B-lymphocytes.
2. Activation of B-lymphocytes to divide and produce antibodies.

3. Activation of cell spreading in phagocytes, so that they can pass through blood vessels.
4. Chemotaxis of cells towards the site of infection.
5. Activation of the oxidase that produces superoxide in phagocytes.
6. Secretion of digestive enzymes by phagocytes.

Antibodies (IgM, IgG, IgA, IgD and IgE) bind to an invading bacterium, protozoan or virus, which then bind to phagocytes, which remove and kill them. In order to produce lots of high-affinity antibody, both T- and B-lymphocytes have to be stimulated to divide. Ca^{2+} signals switch on the genes necessary for moving out of the dormant G_0 back into G_1 of the cell cycle. The natural mitogenic stimulus can be mimicked by the plant lectin phytohaemoglutinin (PHA), concanavalin A, pokeweed mitogen, and the Ca^{2+} ionophore A23187. These provoke a Ca^{2+} signal in lymphocytes within seconds, associated within minutes with an increase in glucose uptake, an increase in intracellular cyclic nucleotides, and an increase in phosphatidyl inositol turnover. This is followed within 15–30 min by an increase in nucleic acid synthesis, acetylation of arginine-rich histones, phosphorylation of several proteins and an increase in Na^+-dependent amino acid uptake. By 2–4 h there is an increase in protein synthesis and glycolysis, and a redistribution of lysosomal hydrolases. Activation of NFATc transcription factor by the Ca^{2+}-activated phosphatase calcineurin activates gene expression, leading to a doubling of DNA in S phase, and eventually mitosis at 48–72 h.

7.11   Intracellular Ca^{2+} and Vision

Ca^{2+} plays a key role in all eyes and complex photoreceptors (Figure 7.13). The signalling system in the eye tells the brain that photons have been received by photoreceptors in the retina. But reception of photons also alters the sensitivity to light, via light/dark adaptation in the photoreceptors, and by altering the diameter of the pupil. The brain also controls the lens so that the image on the retina is in focus. However, there are major differences in the molecular mechanisms between vertebrates and invertebrates, and even within different parts of the cell that initially absorbs a photon – the photoreceptor. Light induces a decrease in cytosolic free Ca^{2+} in vertebrate photoreceptors, whereas in invertebrate photoreceptors light induces an increase.

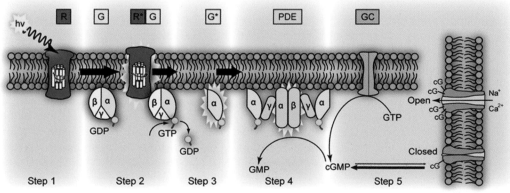

Figure 7.13 Ca^{2+} and vertebrate photoreceptors. *Source*: From http://upload.wikimedia.org/wikipedia/commons/d/de/Phototransduction.png by Jason J. Corneveaux via Wikimedia Commons.

7.11.1   Ca²⁺ and Vertebrate Vision

Vertebrate photoreceptors are modified neurones, but do not generate action potentials. The 'resting' cell is permanently activated and thus secreting continuously. The primary stimulus, light, acts to inhibit cell activity that carries on in the dark. Vertebrate photoreceptors consist of three sections: the outer and inner segments, and the synaptic terminal that secretes glutamate to excite the next neuron. Ca^{2+} plays a role in all three. Ca^{2+} as usual is the activator of exocytosis and the release of glutamate by granule fusion, using synaptotagmin and the SNARE complex. Vertebrate photoreceptors have a high cytosolic free Ca^{2+} of around 300 nM, and are thus switched on in the dark, continuously releasing glutamate to stimulate or inhibit the bipolar cells via metabotropic or ionotropic receptors, respectively, which transmit to the brain. These receptors are an analogue system. The cytosolic free Ca^{2+} decreases with increasing light intensity, down to around 50 nM, and thus the amount of glutamate released decreases in parallel with the cytosolic free Ca^{2+}. Ca^{2+} enters the synaptic terminal via L-type voltage-gated Ca^{2+} channels, which open at the depolarised voltage of –40 mV, and close when light hyperpolarises the cell. Ca^{2+} is pumped out via a plasma membrane Ca^{2+}-MgATPase type 1. In the outer segment, the primary intracellular second messenger is cyclic GMP, produced by a special retina-specific guanylate cyclase – rod outer segment guanylate cyclase (ROC-GC). Ca^{2+} enters via cyclic nucleotide-gated (CNG) channels, and exits via Na^+/Ca^{2+}-K^+ (NCKX), the transporter that exchanges four Na^+ for one Ca^{2+} plus 1 K^+. This is therefore sensitive to membrane potential, whereas plasma membrane Ca^{2+}-MgATPases (PCMCa1–4) are not. Thus, in the dark, the photoreceptor secretes glutamate as a result of a high cytosolic free Ca^{2+}. Light causes a rise in cyclic GMP, which closes cyclic GMP-gated cation channels in the outer segment membrane.

In the dark, in the retinal rods, cation channels in the plasma membrane are open as a result of binding cyclic GMP, allowing Na^+ into the cell, depolarising the membrane to –35 to –45 mV. As the photoreceptor cell is small, this depolarisation signal is transmitted electrotonically to the synapse, where voltage-gated channels open, leading to a rise in cytosolic free Ca^{2+}. This Ca^{2+} provokes the fusion of glutamate-containing granules. The depolarisation signal is graded, depending on light intensity, and is not an action potential. The Ca^{2+} channels in the presynaptic terminal of the photoreceptor form the channel with αF and αD subunits, triggering exocytosis and glutamate release. Exocytosis needs synaptotagmin, syntaxin, synaptophysin, syntaxin and SNAP-23/25, and involves a docking ribbon of vesicles. Secondary regulation involves GABA, NO, glutamate itself and dopamine, regulating the amount of glutamate released.

Light hyperpolarises the cell, producing a membrane potential of around –70 mV, and inhibiting glutamate release. High cytosolic free Ca^{2+} decreases cyclic GMP levels by activating phosphodiesterase and inhibiting guanylate cyclase. Light lowers the cytosolic free Ca^{2+}, thereby increasing cyclic GMP levels and shutting the cyclic CNG cation channels. Closing of CNG channels stops Ca^{2+} entry, but Ca^{2+} efflux continues. Bright light can reduce the cytosolic free Ca^{2+} to around 10 nM in rods and 20–50 nM in cones. Ca^{2+} effects guanylate cyclase via the EF-hand Ca^{2+}-binding proteins GCAP1 (guanylate cyclase-activating protein), and recoverin. Unlike most activating Ca^{2+}-binding proteins, GCAP is inhibited by binding Ca^{2+}. When the cytosolic free Ca^{2+} decreases from 450 to 40 nM, GACP1 activity increases and the cyclase increases. In order to recover in the dark, cyclic GMP is resynthesised by guanylate cyclase, activated by a large drop in cytosolic free Ca^{2+}.

The response of a vertebrate photoreceptor to light is an amplification cascade. Light is first absorbed by the visual pigment rhodopsin or iodopsin, a 11-*cis*-retinal made from vitamin A and bound to its protein opsin. This causes it to change to all-*trans*-retinal. The protein opsin is bound to a trimeric G-protein, transducin. This traverses the segment membrane, activating cyclic GMP phosphodiesterases on the cytosolic side. Cyclic GMP therefore drops off the plasma membrane cationic ion channel, leading to closure of the Na^+ channels and hyperpolarisation by K^+ efflux through K^+ channels. The hyperpolarisation closes the voltage-gated Ca^{2+} channels in the terminal, so the cytosolic free Ca^{2+} drops to below 100 nM, shutting off secretion at the terminal. These channels are sensitive to L-type Ca^{2+} channel blockers, such as Cd^{2+}, Co^{2+}, nifedipine and nimodipine, but are distinct from L-type channels in other excitable cells, being blocked by ω-conotoxin, which typically blocks N-type Ca^{2+} channels. The Ca^{2+} channel itself, as usual, is formed by the α_1 subunit, the gating being modulated by the β and γ/δ subunits. Dopamine, somatostatin, adenosine, glutamate via metabotropic receptors, insulin, GABA, retinoids, cannabinoids, and ions such as H^+ and Cl^- modulate inner segment Ca^{2+} channels, acting as secondary regulators of glutamate release.

7.11.2 Ca²⁺ and Invertebrate Vision

In contrast to vertebrates, invertebrate photoreceptors are switched on by light, which raises the cytosolic free Ca^{2+} to around 500 nM. The retina contains Ca^{2+}-binding proteins that take part in different aspects of the visual process. The rise in cytosolic free Ca^{2+} triggers secretion of neurotransmitters to excite the next neurone. Light results in the activation of phospholipase C (PLC), with the consequent production of IP_3 and DAG. The IP_3 binds to its receptors on the ER, provoking Ca^{2+} release into the cytosol. In *Limulus* and *Apis*, this is the main source of Ca^{2+} for the cytosolic rise, light-activated cation channels in the plasma membrane causing Na^+ influx with very little Ca^{2+} influx. However, in *Balanus* and *Drosophila*, the major source of Ca^{2+} for the cytosolic rise is via Ca^{2+} channels in the plasma membrane, opened as a result of loss of Ca^{2+} from the ER (SOCE). In the flame scallop or rough fileclam, *Lima scabra*, there is a slow inward current, similar, but not identical, to SOCE in other cells. But, the ciliary photoreceptors in scallops do not use IP_3-mediated Ca^{2+} release from the ER, there being a Ca^{2+}-independent activation of the cells by light involving cyclic GMP.

On the other hand, the fruit fly *Drosophila* photoreceptor exhibits a transient receptor potential (TRP and TRPL), caused by light-gated cation channels, the influx of Ca^{2+} and Na^+ causing a depolarisation. TRPL causes a transient depolarisation, and TRP a sustained one. Ca^{2+} is extruded from the cell in dark by Na^+/Ca^{2+} exchange. Asp621 in TRP is the Ca^{2+} sensor, and both TRP and TRPL contain calmodulin-binding sites. The cytosolic Ca^{2+} rise caused by bright light also activates CaMKII, which phosphorylates arrestin 2 (Arr2). This regulates G-protein signalling, such as receptor traffic, switches and receptor desensitisation, making the photoreceptors less sensitive to light. Ca^{2+}-calmodulin plays a negative role in light adaptation, the duration of the Ca^{2+} transients being shorter when the eye is light-adapted. Thus, as well as being a direct mechanism for responding to light, Ca^{2+} rises play a role in the adaptation of invertebrate photoreceptors to light.

7.12 Intracellular Ca²⁺ and Other Senses

All our senses of smell, hearing, taste, hot and cold, and touch also involve a rise in cytosolic free Ca^{2+} induced via TRP channels. The olfactory responsive CNG channel is highly sensitive to changes in cytosolic free Ca^{2+}. Touch and heat sensors also involve Ca^{2+} signals in the neurones which communicate to the brain.

7.13   Ca²⁺ and Bioluminescence

Bioluminescence is the emission of visible light by living organisms, occurring in 18 phyla (Figure 7.14). Some flash, some glow. In some, the light emission is intracellular, whereas in others it is secreted. There is indirect evidence that intracellular Ca^{2+} plays a key role in initiating a flash in many bioluminescent organisms. But only for a few is there direct evidence that a rise in cytosolic free Ca^{2+} is trigger for light emission. Removal of extracellular Ca^{2+} reduces light emission in several species, such as coelenterates, brittle stars, and polynoid, scale worms, particularly when stimulated by addition of KCl. KCl induces action potentials in neurones and in the photocytes themselves. Luminous hydrozoans, hydroids and jellyfish, and the ctenophore sea combs and gooseberries have Ca^{2+}-activated photoproteins with EF-hand Ca^{2+}-binding sites. Bioluminescent anthozoans, such as the sea pen *Renilla*, have Ca^{2+}-binding proteins that release the key component required for the bioluminescent reaction when the protein binds Ca^{2+}.

All bioluminescence is chemiluminescence – the emission of light from a chemical reaction. Three components are essential: luciferin, luciferase and oxygen. Bioluminescence is cold light – 'burning without fire'. The energy for light emission comes from oxidation of the luciferin by oxygen or one of its metabolites. ATP hydrolysis is not the energy source. It does not have an energy-rich bond. Five distinct families of luciferin have been found: aldehydes, flavins, imidazopyrazines, benzothiazole, and linear tetrapyrolle. But there are several others, as yet undiscovered. Primary stimuli provoking bioluminescence in various species include touch, adrenaline, melanocyte-stimulating hormone (MSH), prolactin and NO. Secondary regulators include GABA and NO. Many bioluminescent organisms exhibit a circadian rhythm, there being little point in producing light during the day.

In order to produce light, all the components of the bioluminescent reaction must mix together. In ostracods, copepods, decapods, some squids and the piddock *Pholas*, the bioluminescence is secreted into the surrounding sea water. This exocytosis of the luciferin and luciferase, usually from separate cells, will be triggered by a rise in cytosolic free Ca^{2+}. But this has yet to be demonstrated directly. A voltage-gated Ca^{2+} current triggers light emission in polynoid scale worms. Many bioluminescent organisms can be induced to flash by adding KCl, which excites their nerve net, and opens voltage-gated Ca^{2+} channels in the luminous cells (photocytes).

The hydroid *Obelia geniculata* (Figure 7.15) grows profusely throughout the world, on brown algae, particularly *Laminaria*. It produces small jellyfish. The hydroids and medusa are stimulated to flash naturally by touch, consistent with it being a mechanism to scare predators. Neurones in the epithelium generate an action potential, which is transmitted to the endodermal photocytes. This requires Ca^{2+} and gap junctions, the nerve net of coelenterates being directly electrical coupling between cells. The Ca^{2+} for triggering light emission from the photoprotein in the photocyte comes through gap junctions. The problem, however, is that the kinetics of the live photocyte emission is much faster than the isolated photoprotein. *Obelia* photocytes emit flashes with a time constant that can be as fast as 300 s^{-1}. In contrast, the maximum exponential decay of obelin provoked in the test-tube by millimolar Ca^{2+} is only 4 s^{-1}, much too slow to explain alone the fast flash *in situ*. Similar discrepancies are found in other bioluminescent species. To produce a fast flash from a Ca^{2+}-activated photoprotein, Ca^{2+} must initially saturate the Ca^{2+}-binding sites. Then, millisecond removal of the free Ca^{2+} causes Ca^{2+} to dissociate from the photoprotein, leaving the chemical intermediate left on the protein to flash rapidly.

Sea pansies and sea pens, and some scyphozoans, do not use photoproteins, but rather have a conventional luciferin–luciferase reaction. But Ca^{2+} is still involved. The luciferin, coelenterazine, is bound to an EF-hand Ca^{2+}-binding protein. When the cytosolic free Ca^{2+} rises, Ca^{2+} binds to this protein, releasing the coelenterazine so that it can bind to the luciferase, which then catalyses its oxidation by oxygen to form coelenteramide in an excited state.

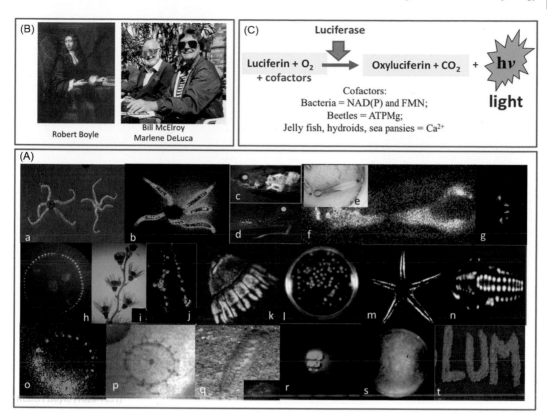

Figure 7.14 Some bioluminescent organisms. (A) A variety of bioluminescent organisms, many of which are triggered by a rise in cytosolic free Ca²⁺. (a) The brittle star *Amphipholis squamata*; the under (ventral) surface is where the photocytes are. (b) Brittle star bioluminescence. Reproduced by permission of Welston Court Science Centre. (c) A deep-sea searsiid fish secreting bioluminescent cells into the water. *Source*: Courtesy of P. Herring. (d) A deep-sea myctophid showing the fluorescence of the ventral photocytes under UV light. *Source*: Courtesy of P. Herring. (e) The piddock *Pholas dactylus* out of its burrow. Reproduced by permission of Welston Court Science Centre. (f) *Pholas* glowing. Reproduced by permission of Welston Court Science Centre. (g) The internal light organs of *Pholas* glowing. Reproduced by permission of Welston Court Science Centre. (h) GFP around the base of the umbrella in an *Obelia lucifera* jellyfish. Reproduced by permission of Welston Court Science Centre. (i) The hydroid *Obelia geniculata* showing the GFP- and obelin-containing cells. Reproduced by permission of Welston Court Science Centre. (j) The bioluminescence of the hydroid *Obelia geniculata* superimposed on the bright field showing that the photocytes correspond exactly with the GFP-containing cells. Reproduced by permission of Welston Court Science Centre. (k) The pelagic jellyfish *Periphylla*. *Source*: Courtesy of P. Herring. (l) The planktonic bioluminescent radiolarian *Thalassicolla*. Reproduced by permission of Welston Court Science Centre. (m) Bioluminescent deep-sea starfish. *Source*: Courtesy of P. Herring. (n) Bioluminescent photophores on the ventral surface of a deep-sea hatchet fish. *Source*: Courtesy of P. Herring. (o) The bioluminescence of the jellyfish *Clytia hemispherica*. Reproduced by permission of Welston Court Science Centre. (p) The jellyfish *Clytia hemispherica*. Reproduced by permission of Welston Court Science Centre. (q) A luminous scale worm. Source: Courtesy of Jean Marie Bassot. (r) The light organs glowing of the glow-worm *Lampyris noctiluca*. Reproduced by permission of Welston Court Science Centre. (s) The bioluminescence of dinoflagellates. (a, e, k, l, p, q) Bright field; (c, d, m, n, s) bioluminescence in real colour; (b, f, g, j, o, r) bioluminescence in pseudo-colour. (t) *Source*: Images courtesy of P. Herring, Welston Court Science Centre and Jean Marie Bassot. Reproduced by permission of Welston Court Science Centre. (B) Pioneers in the chemistry and biochemistry of bioluminescence. Robert Boyle (1627–1691) showed air was required for glowing wood (a fungus) and shining flesh (bacteria); that is, all bioluminescence requires oxygen – it is 'burning without fire'. Portrait of Robert Boyle © The Royal Society. William (Mac/Bill) D. McElroy (1917–1999), who discovered that ATP is required for firefly bioluminescence, with his wife Marlene DeLuca (source unknown). (C) The generic reaction responsible for all bioluminescence.

(a) *Obelia geniculata*

(b) *Obelia longissima*

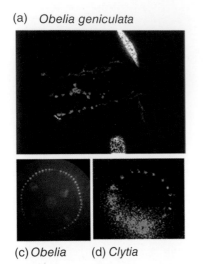

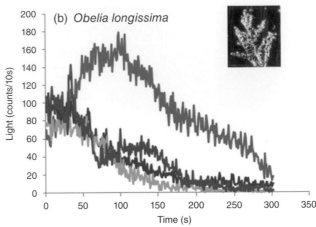

(c) *Obelia*   (d) *Clytia*

Figure 7.15 Bioluminescence of *Obelia* triggered by a rise in cytosolic Ca²⁺. The figures show the Ca²⁺-activated photoprotein obelin triggered in the hydroids *Obelia geniculata* and *Obelia longissima* by addition of 0.5 M KCl. This excites the nerve net and also the photocytes themselves. (a) Pseudo-colour image *Obelia geniculata* bioluminescence superimposed on the bright field. Hydroid 1.5 cm long. (b) Light emission from four individual photocytes in *Obelia longissima*, which locate at the base of the hydranths. The data show the variation in light signal in terms of timing and oscillations. (c) GFP fluorescence at the base of the tentacles and around the umbrella of the jellyfish *Obelia lucifera*, which was released from the hydroid *Obelia geniculata*. The GFP-containing cells are the ones which also contain obelin. Jellyfish around 1 mm in diameter. (d) Pseudo-colour image of the bioluminescence of the jellyfish *Clytia hemisphericum*, showing the photocytes at the base of the tentacle. Jellyfish around 1 cm in diameter. Jellyfish supplied by Evelyn Houliston, Villefranche, France. Experiment carried out by Campbell and Swann in Cardiff. The jellyfish were cultured in the laboratory and were thus not bioluminescent, until coelenterazine was added to activate the apoclytin. Reproduced by permission of Welston Court Science Centre.

Bioluminescence of luminous bacteria such as *Vibrio* and *Photobacterium* is controlled by an autoinducer, a homoserine lactone, produced by the bacteria themselves. This switches on the operon that produces the proteins necessary for light emission. Removal of extracellular Ca²⁺ reduces light emission (Figure 8.7a, b), which cannot be explained by an effect on cell growth.

No mammals, amphibians or higher plants are bioluminescent. But phagocytes producing reactive oxygen species chemiluminesce. This is induced by a rise in cytosolic free Ca²⁺. Fertilised eggs from invertebrates and fish also chemiluminesce, due to release of ovoperoxidase and reactive oxygen species to prevent fertilisation from a second sperm. This exocytosis is triggered by a rise in cytosolic free Ca²⁺ induced by the sperm.

7.14   Intracellular Ca²⁺ and Gene Expression

Many of the acute physiological processes in animals involve gene expression. Changes in cytosolic and organelle free Ca²⁺ are activate or down-regulate many genes. In addition, Ca²⁺-activated proteins such as calmodulin and transcription factors are translocated to the nucleus as a result of a rise in cytosolic free Ca²⁺. A good example of this is the NFATc pathway activated by the phosphatase calcineurin. Calcineurin and lysosomal Ca²⁺ are also important in sending to the nucleus the transcription factor TFEB (transcription factor EB), the master gene for lysosomal biogenesis that coordinates expression of lysosomal hydrolases, membrane proteins, and genes involved in autophagy. In addition, the ER stress pathway can be activated by depletion of the ER Ca^{2+} store, as well as malfolding of proteins within the ER. Repetitive Ca^{2+} transients cause tissue adaptations. For example, patients who stay in bed for a long while exhibit muscle weakness when they eventually try to stand up. A feedback mechanism in the muscle fibres monitors how often it gets a cytosolic Ca^{2+} rise as a result of the acetylcholine-induced action potential. There is much to be learnt about such mechanisms.

7.15   Summary

A rise in cytosolic free Ca^{2+} is a universal switch in virtually all types of animal cells, provoking nerves to fire an action potential, a muscle to contract, allowing an animal to move, or its heart to beat, exo- and endo-crine cells to secrete, a cell divide, a sperm to reach an egg and fertilise it successfully, a cell to mature, an animal to see, smell, taste, feel, and hear, and a luminous species to flash. The key evidence for this comes from:

1. Direct demonstration of intracellular Ca^{2+} signals, measured and imaged using a Ca^{2+}-activated photoprotein, such as aequorin, or a fluorescent dye such as fura-2 or fluo-3/4.
2. Identification of the mechanism causing the rise in cytosolic free Ca^{2+} using patch clamp, and pharmacological agents.
3. Identification of the Ca^{2+} target, by protein purification, protein mapping, antibody location, inhibitors of Ca^{2+} binding proteins.
4. Effect of knock-outs, using transgenic mice and siRNA in tissue culture.
5. Genome sequences and microarray, and the identification of protein variants from alternative splicing.

The reason Ca^{2+} is able to trigger such a diverse set of phenomena is the universality of the Ca^{2+} pressure in all eukaryotic cells across the plasma membrane, and between the ER and the cytosol. A relatively small absolute release of Ca^{2+} from an internal store, or across the plasma membrane, will cause a large fractional rise in cytosolic free Ca^{2+}, typically in the range $1–10\,\mu M$ (total Ca^{2+} release *ca* 100 micromolar). This is sufficient to cause a significant change in Ca^{2+} binding to target proteins, which then activate the cell event. There are eight key intracellular targets for cytosolic free Ca^{2+} (Box 7.2).

Box 7.2   Key intracellular Ca²⁺ targets.

- Troponin C in skeletal and heart muscle.
- Myosin light chain kinase in smooth muscle, via calmodulin.
- Myosin itself in many invertebrate muscles.
- Synaptotagmin in exocytotic secretion, in nerve terminals, endocrine and exocrine cells.
- Calmodulin in a wide range of metabolic processes.
- Covalent protein modification by Ca^{2+} activated kinases, phosphatases, and proteases.
- Ca^{2+} binding proteins that interact with actin and the cytoskeleton.
- NFATc and other transcription factors in regulating gene expression.

The mechanism of release of Ca^{2+} into the cytosol can involve:

- Opening of voltage-gated or receptor operated Ca^{2+} channels in the plasma membrane.
- Release from the SR/ER, via ryanodine receptors directly or Ca^{2+} induced Ca^{2+} release, or IP_3 generated as a result of activation of phospholipase C.

- Opening of store-operated Ca^{2+} channels in the plasma membrane as a result of loss of Ca^{2+} from the ER (SOCE).
- Release from organelles, such as lysosomes or intracellular vesicles.
- Movement of signals or action potentials through cell layers via gap junctions.

The Ca^{2+} signal can be regulated and restricted to a micro-domain by mitochondria. Depending on cell type, the Ca^{2+} is removed from the cytosol by:

- Pumping back into the SR/ER via a SERCA MgATPase.
- Pumping out of the cell by the plasma membrane MgCaATPase (PMCA).
- Pumping out of the cell by the $3Na^+/Ca^{2+}$ exchanger NCX, or $4Na^+/Ca^{2+}+K^+$ NCKX in eyes.

Intracellular Ca^{2+} also interacts with other intracellular signals such as cyclic nucleotides, NO and H_2S.

Box 7.3 Darwinian molecular variations in Ca^{2+} signalling.

1. The type of Ca^{2+} signal (e.g. transient, plateau, oscillation).
2. The location of the Ca^{2+} signal (e.g. microdomain, wave, tide).
3. How the Ca^{2+} signal is generated (e.g. ER Ca^{2+} release via IP_3 or ryanodine receptors, voltage-gated Ca^{2+} channels, SOCE).
4. How Ca^{2+} is removed after the cell event (e.g. plasma membrane Ca^{2+} pump, Na^+/Ca^{2+} exchange, SERCA).
5. The key Ca^{2+} target protein (e.g. troponin C, calmodulin, synaptotagmin, gelsolin).
6. The type of Ca^{2+}-binding site (e.g. EF-hand, C2, acidic clusters, three-dimensional oxygens).
7. The target protein for the Ca^{2+}-binding protein (e.g. kinase, phosphatase, ion channel, SNARE complex, myosin, actin).

Darwinian variations in Ca^{2+} signalling components between different cell types (Box 7.3) enable Ca^{2+} to have a specific action of events in different cell types. There is also molecular variation in the level of expression of proteins in Ca^{2+} signalling pathways, and subtle differences in their Ca^{2+} affinities and kinetic properties from slightly different protein sequences generated by different genes, alternative splicing or covalent modification. This results in the individual organism having a selective advantage in a particular situation, to escape a tiger, to digest food faster, to think quicker, and so on. Darwinian–Wallace Natural Selection depends on this molecular biodiversity within the population of a species.

Recommended Reading

*A must.

Books

*Campbell, A.K. (2015) Intracellular Calcium. Chichester: John Wiley & Sons Ltd. Chapter 7. How Ca^{2+} regulates animal cell physiology. Fully referenced chapter.

Reviews

*Capiod, T. (2011) Cell proliferation, calcium influx and calcium channels. *Biochimie*, **93**, 2075–2079. Review of how changes in intracellular Ca²⁺ can regulate cell division.

*Halestrap, A.P. (2009) Mitochondrial calcium in health and disease. *Biochim. Biophys. Acta*, **1787**, 1289–1290. How Ca²⁺ regulates key enzymes in the mitochondria.

Hernandez-Ochoa, E.O., Pratt, S.J.P., Lovering, R.M. & Schneider, M.F. (2016) Critical role of intracellular RYR1 calcium release channels in skeletal muscle function and disease. *Frontiers Physiol.*, 6.420. How the ryanodine receptor releases Ca²⁺ in skeletal muscle.

Hong, F., Haldeman, B.D., Jackson, D., Carter, M., Baker, J.E. & Cremo, C.R. (2011) Biochemistry of smooth muscle myosin light chain kinase. *Arch. Biochem. Biophys.*, **510**, 135–146. How Ca²⁺ activates the protein that triggers smooth muscle contraction.

Kahl, C.R. & Means, A.R. (2003) Regulation of cell cycle progression by calcium/calmodulin-dependent pathways. *Endocr. Rev.*, **24**, 719–736. How Ca²⁺/calmodulin can regulate parts of the cell cycle.

Lederer, W.J., Bers, D.M. & Eisner, D.A. (2013) Calcium signaling in heart: Multiscale, diverse, rapid, local, and remarkable. *J. Mol. Cellular Cardiol.*, **58**, 3–4. How various aspects of Ca²⁺ are crucial to the heart.

Luzio, J.P., Bright, N.A. & Pryor, P.R. (2007) The role of calcium and other ions in sorting and delivery in the late endocytic pathway. *Biochem. Soc. Trans.*, **35**, 1088–1091. Evidence of a role for Ca²⁺ in endocytosis.

Nunes, P. & Demaurex, N. (2010) The role of calcium signaling in phagocytosis. *J. Leukoc. Biol.*, **88**, 57–68. The evidence for the role of Ca²⁺ in phagocytosis.

Pan, Z., Brotto, M. & Ma, J. (2014) Store-operated Ca²⁺ entry in muscle physiology and diseases. *BMB Rep.*, **47**, 69–79. The role of SOCE in muscle.

Ramakrishnan, N.A., Drescher, M.J. & Drescher, D.G. (2012) The SNARE complex in neuronal and sensory cells. *Mol. Cell. Neurosci.*, **50**, 58–69. How the SNARE complex works through the Ca²⁺ signal to trigger secretion.

*Swann, K. & Lai, F.A. (2016) Egg activation at fertilization by a soluble sperm protein. *Physiol. Rev.*, **96**, 127–149. How the Ca²⁺ signal in a fertilised egg is generated by injection of a special phospholipase C by the discoverers.

Tu, M.K., Levin, J.B., Hamilton, A.M. & Borodinsky, L.N. (2016) Calcium signaling in skeletal muscle development, maintenance and regeneration. *Cell Calcium*, **59**, 91–97. How Ca²⁺ signals can control muscle tissue.

Webb, S.E. & Miller, A.L. (2013) Ca2(+) signaling during activation and fertilization in the eggs of teleost fish. *Cell Calcium*, **53**, 24–31. The role of intracellular Ca²⁺ in development using a model fish system.

Whitaker, M. (2008) Calcium signalling in early embryos. *Philos. Trans. R. Soc. Lond. B Biol. Sci.*, **363**, 1401–1418. Ca²⁺ signals are involved in early embryos.

Research Papers

Bonfanti, D.H., Alcazar, L.P., Arakaki, P.A., Martins, L.T., Agustini, B.C., De Moraes Rego, F.G. & Frigeri, H.R. (2015) ATP-dependent potassium channels and type 2 diabetes mellitus. *Clin. Biochem.*, **48**, 476–482. The role of ATP K⁺ channels in insulin secretion to open voltage-gated Ca²⁺ channels.

Campbell, A.K. & Hallett, M.B. (1983) Measurement of intracellular calcium-ions and oxygen radicals in polymorphonuclear leukocyte–erythrocyte 'ghost' hybrids. *J. Physiol.*, **338**, 537–550. First evidence that a rise in cytosolic free Ca²⁺ is required to generate oxygen metabolites by neutrophils.

Del Pilar Gomez, M. & Nasi, E. (2009) Prolonged calcium influx after termination of light-induced calcium release in invertebrate photoreceptors. *J. Gen. Physiol.*, **134**, 177–189. Evidence for the role of Ca^{2+} in invertebrate eyes.

Demaurex, N. & Nunes, P. (2016) The role of STIM and Orai proteins in phagocytic immune cells. *Am. J. Physiol.-Cell Physiol.*, **310**, C496–C508. Evidence for the mechanism generating Ca^{2+}signals in phagocytes.

Liu, M., Vhen, T.Y., Ahamed, B., Li, J. & Yau, K.W. (1994) Calcium-calmodulin modulation of the olfactory cyclic nucleotide-gated cation channel. *Science*, **266**, 1348–1354. Evidence for a role for cytosolic Ca^{2+} in smell.

Montell, C. (2005) Trp channels in *Drosophila* photoreceptor cells. *J. Physiol.*, **567**, 45–51. The role of transient receptor channels in fruit fly eyes.

Saunders, C.M., Larman, M.G., Parrington, J., Cox, L.J., Royse, J., Blayney, L.M., Swann, K. & Lai, F.A. (2002) PLC zeta: A sperm-specific trigger of Ca^{2+} oscillations in eggs and embryo development. *Development*, **129**, 3533–3544. Evidence for a new phospholipase C in generating the crucial Ca^{2+} signals in eggs after sperm fertilisation.

*Shi, L., Shen, Q.T., Kiel, A., Wang, J., Wang, H.W., Melia, T.J., Rothman, J.E. & Pincet, F. (2012) SNARE proteins: One to fuse and three to keep the nascent fusion pore open. *Science*, **335**, 1355–1359. How the SNARE complex works with Ca^{2+} to trigger secretion.

*Skelding, K.A., Rostas, J.A. & Verrills, N.M. (2011) Controlling the cell cycle: The role of calcium/calmodulin-stimulated protein I and II. *Cell Cycle*, **10**, 631–639. How Ca^{2+}-calmodulin can control parts of the cell cycle.

Webb, S.E. & Miller, A.L. (2013) Ca^{2+} signaling during activation and fertilization in the eggs of teleost fish. *Cell Calcium*, **53**, 24–31. Evidence for the role of Ca^{2+} in fish egg activation.

Wissenbach, U., Philipp, S.E., Gross, S.A., Cavalie, A. & Flockerzi, V. (2007) Primary structure, chromosomal localization and expression in immune cells of the murine Orai and Stim genes. *Cell Calcium*, **42**, 439–446. Key properties of the two proteins that lock together to generate store-operated Ca^{2+} entry.

8

Intracellular Ca²⁺, Microbes and Viruses

Learning Objectives
1. What Eukaryote, Bacterial and Archaean microbes are.
2. What these microbes do in response to primary stimuli or stress.
3. Indirect evidence for a role of Ca²⁺.
4. Direct evidence in bacteria with the vital measurement of cytosolic free Ca²⁺.
5. Identification of bacterial proteins with the potential for regulating cytosolic free Ca²⁺, and the problem of mistakes using sequence similarities and experiments with vesicles.
6. Identification of a non-proteinaceous Ca²⁺ channel in bacteria.
7. The small amount of evidence of a role for intracellular Ca²⁺ in archaens.
8. The role of Ca²⁺ in virus infection and growth in eukaryotic and bacterial cells.
9. Direct evidence in eukaryote microbes, with the vital measurement of cytosolic free Ca²⁺, together with proteins.

8.1   The Puzzle

Microbes are single-celled micro-organisms that are so tiny, thousands, or even millions, can fit on the top of a pin. They are the dominant organisms on our planet. All contain calcium, which plays a role in regulating the behaviour of many species. They occur in all three domains of life – Bacteria, Archaea and Eukaryota. They are found in the sea, fresh water, soil, clouds, hot springs, high salt, acid or alkaline lagoons, and in the bodies of organisms, including our own. Some events in microbes are regulated by changes in cytosolic free Ca²⁺ (Figure 8.1). In our body we have some 10^{13}–10^{14} cells, but we have over ten times as many bacterial cells, representing several hundred species, in our gut alone compared with the eukaryotic cells in the rest of our body. Microbes are usually single celled, but many can associate together to form multicellular structures. Viruses need a live cell to reproduce, and infect all these three cell types. Viruses include bacteriophages that infect bacteria, and those that cause animal and plant diseases.

   Just like multicellular animals and plants, microbes maintain a cytosolic free Ca²⁺ in the micromolar to sub-micromolar range, even in the presence of 1–10 mM external Ca²⁺. So the Ca²⁺ pressure is there. It was once thought that Ca²⁺ was simply toxic to bacteria, and the only interest the cell had was to get rid of it. A key discovery was that several physical, chemical and biological agents induce Ca²⁺ transients inside many microbes. Unfortunately, the literature contains many artefactual conclusions about Ca²⁺ in microbes based on indirect evidence. For

Fundamentals of Intracellular Calcium, First Edition. Anthony K. Campbell.
© 2018 John Wiley & Sons Ltd. Published 2018 by John Wiley & Sons Ltd.
Companion Website: http://www.wiley.com/go/campbell/calcium

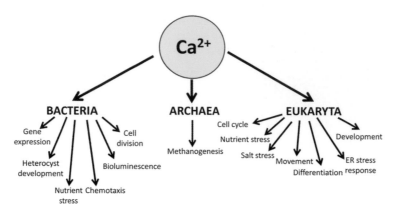

Figure 8.1 Possible roles for intracellular Ca^{2+} in microbes. Reproduced by permission of Welston Court Science Centre.

example, three proteins, ChaA, YrbG and PitB, have been reported to be major transporters of Ca^{2+} in *Escherichia coli* and other bacteria. Yet direct measurement of cytosolic free Ca^{2+} showed that none of these proteins was responsible for maintaining the Ca^{2+} gradient across the plasma membrane of normal, live cells.

There are four questions that must be answered if a role of intracellular Ca^{2+} in a microbe is to be established:

1. What is the molecular mechanism responsible for maintaining the very low cytosolic free Ca^{2+}?
2. Are there primary stimuli that cause a rise in cytosolic free Ca^{2+}?
3. What ion channels are there which can be opened by these stimuli to cause a rise in intracellular Ca^{2+}?
4. Are there Ca^{2+} targets, high-affinity Ca^{2+}-binding proteins, inside microbes, and what do they do?

8.2 What Microbes Do

In order to survive and evolve successfully, microbes do a lot or things (Table 8.1), where intracellular Ca^{2+} can play a key role. Microbes also defend themselves against a physical, chemical and biological attack, and adapt to a wide range of environments, such as different osmolarities, temperature, pH, nutrient supply, and the availability of water.

8.3 Indirect Evidence of a Role for Intracellular Ca^{2+} in Bacteria

Bacterial growth occurs in three phases: dormant, exponential and stationary, though some cells are still dividing slowly in the stationary phase. The role of Ca^{2+} may be different in these phases. Indirect evidence shows that intracellular Ca^{2+} may regulate bacterial behaviour:

- Removal of external Ca^{2+} reduces or blocks the bacterial process.
- Blockers of Ca^{2+} channels or pumps in eukaryotes (e.g. local anaesthetics, conotoxin, amiloride) reduce or block the bacterial process.

Table 8.1 Examples of what microbes do.

Activity	Examples	Bacteria	Archaea	Protozoa	Algae
Growth	Cell size	All	All	All	All
Cell division	Generation time	All	All	All	All
Movement	Chemotaxis, phototaxis, thermotaxis, gliding, swarming	E. coli, myxobacteria, *Bacillus* and many others		*Paramecium*	
Infection	Virulence; competence (ability to take up DNA)	*Yersinia*, *Streptomyces*	*Methanobrevibacter*	*Trypanosomes*	
Structural formation	Heterocyst swarming	*Anabena*, *Bacillus*		Yeasts	
Defence	Spore formation, spore germination, antibiotics, prevention of killing by phagocytes	*Bacilli*, Enterobacteriaceae, *Salmonella*, *Yersinia*		*Trypanosomes*	
Adaptation	Thermal vents, high salt	Thermophiles	*Haloferax*	Yeasts	
Symbiosis	Bacteria + fungus, bacteria or alga, intestine of mammals	*Bacterioides*	*Methanobrevibacter*	*Candida*	Lichens, *Hydra*
Evolution	Mutation	All	All	All	All

- Manipulation of intracellular Ca^{2+}, using ionophores or caged Ca^{2+} compounds has predicted effects on the bacterial process.
- Putative Ca^{2+} transporters, with sequence similarities to eukaryotic Ca^{2+} pumps or exchangers, identified from searches of bacterial genomes.
- Ca^{2+}-binding proteins identified from genome searches, using sequence similarities with eukaryotic proteins and Ca^{2+}-binding motifs, such as the EF-hand.

But none of these provides definitive proof that intracellular Ca^{2+} is the signal for a particular process or phenomenon. This requires direct correlation of changes in intracellular free Ca^{2+} with the cellular event. The literature on the role of Ca^{2+} in microbes is thus full of misleading speculation and artefact, leading to flawed hypotheses and conclusions. Five components are necessary in bacteria if intracellular Ca^{2+} is to play a role in cell signalling:

1. Low cytosolic free Ca^{2+}, sub-µM to µM in the presence of mM extracellular Ca^{2+}.
2. Ca^{2+} influx, activated by primary stimuli, leading to a rise in cytosolic free Ca^{2+}.
3. Ca^{2+} efflux mechanisms.
4. Ca^{2+}-binding proteins, with µM affinity as Ca^{2+} targets or sinks inside the cell.
5. Targets for the Ca^{2+}-binding protein mediating the cell event when there is a rise in cytosolic free Ca^{2+}.

As with mitochondria, in bacteria only a few hundred Ca^{2+} ions have to move to increase the intracellular Ca^{2+} by tens of µM, as opposed to the millions of Ca^{2+} in a eukaryotic cell.

8.4 Direct Evidence for a Role of Intracellular Ca^{2+} in Bacteria

The key discovery is that the bacterial membrane, like that of eukaryotes, always maintains a very low cytosolic free Ca^{2+}, even in the presence of mM Ca^{2+} extracellularly. The only way to measure free Ca^{2+} inside live bacteria at present is to use a Ca^{2+}-activated photoprotein, such as aequorin, whose gene can be engineered into a plasmid. The use of fluorescent dyes or proteins produces artefacts. Free Ca^{2+} has been measured in *E. coli*, *Streptococcus pneumonia*, the cyanobacterium *Anabaena*, and *Bacillus subtilis*. In all cases the resting cytosolic free Ca^{2+} was in the sub-µM to µM range, in the presence of mM Ca^{2+} outside the cells. So there has to be a Ca^{2+} efflux mechanism. But clear evidence that changes in free Ca^{2+} trigger events in bacteria is weak. Free Ca^{2+} transients, measured using recombinant aequorin, are induced in *E. coli* by metabolic toxins such as methylglyoxal, chemorepellents, monosaccharides, such as glucose and fructose, and agents that attack the plasma membrane, such as complement. Thus, there must be a mechanism for a regulated Ca^{2+} influx, analogous to receptor-mediated Ca^{2+} transients in eukaryotes. But bacterial membranes have little or no cholesterol or phosphoinositides. Thus, bacteria cannot have an inositol trisphosphate (IP_3)/diacylglycerol (DAG) Ca^{2+} signalling system, nor a store-operated Ca^{2+} entry (SOCE) mechanism. Furthermore, there is no evidence for action potentials in bacteria that might open voltage-gated Ca^{2+} channels, and there is no vesicular SNARE excretion system, analogous to eukaryotic cells. The electrochemical potential that drives Ca^{2+} into the cell has two components: the concentration gradient of Ca^{2+}, and the membrane potential across the plasma membrane, which, under aerobic conditions can be some −200 mV, negative inside, maintained by a Mitchell mechanism through the respiratory chain. However, definitive evidence for high affinity intracellular Ca^{2+} targets is weak.

Bacteria exist in a wide range of ionic environments. *E. coli*, for example, can survive in fresh water, where the Ca^{2+} may be as low 1 µM, in the gut where the extracellular free

Ca²⁺ is around mM, and in sewage effluents in sea water where the extracellular Ca²⁺ is as high as 10 mM. Ponds in the Antarctic have been found where bacteria can live in 50 mM Ca²⁺ or higher. Mechanosensitive channels play an important role in protecting bacteria from osmotic damage. Gram-negative bacteria, such as *E. coli*, have an inner membrane, and an outer membrane, with a periplasmic space between, where the free Ca²⁺ may be higher than that in the surrounding fluid because of a Donnan potential of 10–20 mV, caused by large anions, such as membrane-derived oligosaccharides (MDOs) facing into the periplasm.

Marine blue-green algae (cyanobacteria), and a few non-marine bacteria, require Ca²⁺ for growth. Removal of extracellular Ca²⁺ increases the generation of *E. coli* by about 10%. Such small changes in generation time for bacteria can be highly significant. Spore formation in Gram-positive *Bacillus* requires the accumulation of large amounts of Ca²⁺ (Figure 8.8 below). This Ca²⁺ has to be pumped out if the spore is to germinate. However, demonstrations of effects of removal of external Ca²⁺ on other bacterial processes are scant.

8.5   How Much Ca²⁺ is There in Bacteria?

Since the number of free Ca²⁺ ions inside a single bacterium is very small, the molecular mechanisms for moving Ca²⁺ in and out of bacterial cells are likely to be quite different from those in the plasma membrane of eukaryotes. Element analysis has shown that bacteria contain significant total amounts of Ca²⁺. In spores, large amounts of Ca²⁺ are bound to dipicolinic acid. In bacterial cells, Ca²⁺ binds to DNA, the complex poly-(R)-3-hydroxybutyrate (PHB)–polyphosphate (PP), and anionic groups on proteins and carbohydrates facing outside the cell. Acidocalcisomes and PHB-containing vesicles contain Ca²⁺ in bacteria.

A typical *E. coli* cell is a cylinder about 1 µm long and 0.5 µm wide, having a total volume ($\pi r^2 h$) of about 0.2 fl (2×10^{-16} l). This compares with a hepatocyte in the liver, 20 µm in diameter having a volume ($4/3\pi r^3$) of about 4 pl (4×10^{-12} l). The periplasm of *E. coli* accounts for about 20% of its total volume. Thus, the cytoplasm has an estimated volume of about 0.16 fl. Assuming most of this is water, then, at a cytosolic free Ca²⁺ of 0.1 µM, there are less than 10 Ca²⁺ ions free in the cell cytosol. When a Ca²⁺ transient occurs, raising the cytosolic free Ca²⁺ to say 10 µM, then only 1000 Ca²⁺ ions have to move into the cell. This compares with a hepatocyte, which would have 250 000 Ca²⁺ ions in the cytosol when the free Ca²⁺ is 0.1 µM and 25 million when the cytosolic free Ca²⁺ is 10 µM. These numbers are a major problem, since an ion channel with a conductance of just 10 pS would allow several million ions to move into a cell within less than 1 s. Yet Ca²⁺ transients in bacteria take several minutes to rise from the resting level of 0.1 to 10 µM, measured as a mean of several million cells (Figure 8.2). Furthermore, a typical plasma membrane Ca²⁺-MgATPase, responsible for pumping Ca²⁺ out of a eukaryotic cell, has a turnover number of about 10 s⁻¹, with a K_m for Ca²⁺ of 1 µM. Thus, one single eukaryotic pump molecule could remove 10 µM Ca²⁺ from inside a bacterium within less than 1 s. Yet the Ca²⁺ transients in bacteria takes several minutes to return back to resting levels (Figure 8.2). At pH 7, there would be only eight free H⁺ ions in the cytosol of a single *E. coli* cell, but 150 000 K⁺ ions and 10 000 Na⁺ ions, at cytosolic concentrations of 150 mM K⁺ and 10 mM Na⁺, respectively. The number of ATP molecules at a concentration of 3 mM in *E. coli* would be 29 000. Yet the number of adenosines in the chromosome of 3×10^6 base pairs is about 130 000. When *E. coli* divides it has to produce at least four times its initial adenine content in just 20–30 min.

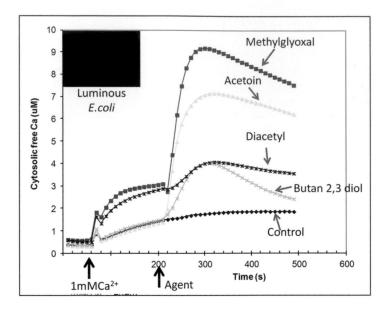

Figure 8.2 Examples of cytosolic free Ca²⁺ transients in *Escherichia coli*. The *E. coli* (JM109) were transformed with a plasmid containing the DNA coding for the Ca²⁺-activated photoprotein aequorin. The figure shows the effect of adding 1 mM CaCl₂ on the cytosolic free Ca²⁺, which rises from submicromolar to about 3 µM and plateaus at this value. Addition of methylglyoxal, acetoin, butane-2,3-diol or diacetyl all induced transient rises in cytosolic free Ca²⁺. Methylglyoxal produced the biggest rise, the peak in cytosolic free Ca²⁺ being about 9 µM. Source: Campbell *et al.* (2007). Reproduced with permission from Elsevier.

8.6   How Bacteria Regulate Their Intracellular Ca²

Since all bacteria maintain a cytosolic free Ca²⁺ in the µM to sub-µM range, and several agents induce changes in intracellular Ca²⁺, there must be mechanisms for regulating Ca²⁺ efflux and influx. There have been many claims for putative Ca²⁺ channels, pumps and exchangers in bacteria, based on genome searches, or Ca²⁺ movement in vesicles isolated from broken cells. However, measurement of free Ca²⁺ in live cells, or patch clamping of spheroplasts, where the outer membrane of Gram-negatives is removed, have not confirmed these hypotheses. The only clear evidence is that there are Ca²⁺ pumps dependent on ATP in both Gram-positive and -negative bacteria. The best evidence for a Ca²⁺ channel in bacteria is not a protein, but rather a complex of PHB and PP.

Several conditions cause transient rises in cytosolic free Ca²⁺ in *E. coli* and other bacteria (Figure 8.2). These are dependent on external pH and monovalent cation concentration (Figure 8.3). The effects of pH and monovalent cations are explained by inhibition of Ca²⁺ influx by H⁺, Na⁺ and K⁺. Some sugars, chemorepellents, methylglyoxal, fermentation products such as butane-2,3-diol, antibiotics and complement cause a rise in cytosolic free Ca²⁺, up to tens of µM. In contrast, chemoattractants caused a small decrease in cytosolic free Ca²⁺. The potency of the bacterial fermentation products was methylglyoxal > acetylmethylcarbinol > diacetyl > butane-2,3-diol > propane-1,2-diol. These substances are produced by gut bacteria under anaerobic conditions and can be produced by eukaryotic cells. The potency of sugars in cytosolic free Ca²⁺ transients was glucose = fructose > galactose > lactose > ribose.

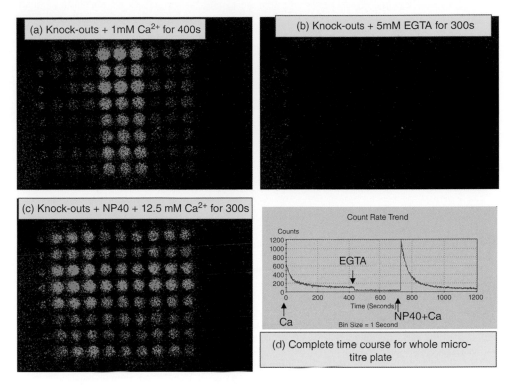

Figure 8.3 Screening of *E. coli* mutants using the Ca²⁺-activated photoprotein aequorin. Individual mutants of *E. coli* from the Keio knock-out collection, National BioResource Project (NIG, Japan): *E. coli* were transformed with a plasmid containing the DNA coding for the Ca²⁺-activated photoprotein aequorin and the full photoprotein formed by addition of coelenterazine. Each mutant was placed in a microtitre well and 1 mM extracellular Ca²⁺ added. The rise in cytosolic free Ca²⁺ was imaged. (a) Ca²⁺ at 1 mM for the first 400 s; (b) EGTA (5 mM final) was then added to remove the extracellular Ca²⁺ in order to image Ca²⁺ efflux; (c) 1% NP40 + 12.5 mM Ca²⁺ were then added to expose the remaining active aequorin in the cells to high Ca²⁺ in order to convert the light emission into absolute cytosolic free Ca²⁺ (see Chapters 4 and 5); (d) complete time course showing light emission measured from the entire microtitre plate. Mutants defective in Ca²⁺ influx or efflux were then examined in detail. Reproduced by permission of Welston Court Science Centre.

8.6.1 Ca²⁺ Influx into Bacteria

There are several potential mechanisms for Ca²⁺ influx in bacteria:

1. Ca²⁺ exchanger, analogous to the Ca²⁺/Na⁺ or Ca²⁺/H⁺ exchangers in eukaryotes, e.g. ChaA and YrbG.
2. Ca²⁺-phosphate symport, e.g. PitB.
3. Ca²⁺ channel, analogous to Ca²⁺ channels in eukaryotes.
4. Ca²⁺ importer, analogous to mitochondria.
5. Ca²⁺ ionophore, analogous to ionomycin.
6. Multiple leaks through proteins in the plasma membrane.
7. Non-proteinaceous Ca²⁺ channel.

The most convincing evidence is for a non-proteinaceous Ca²⁺ channel causing Ca²⁺ influx in bacteria. ChaA, YrbG and PitB are not significant transporters of Ca²⁺ in intact cells. Knock-out mutants showed no difference in the rise and fall of cytosolic free Ca²⁺ when

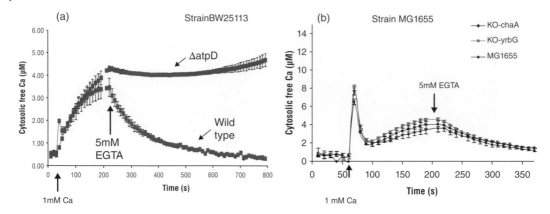

Figure 8.4 *Escherichia coli* mutant defective in Ca²⁺ efflux. (a) Individual mutants of *E. coli* from the Keio knock-out collection, National BioResource Project (NIG, Japan): *E. coli* containing the DNA coding for the Ca²⁺-activated photoprotein aequorin and the full photoprotein formed by addition of coelenterazine. After addition of CaCl₂ (final concentration 1 mM Ca²⁺) the cytosolic free Ca²⁺ rose to just over 4 μM and then plateaued. EGTA (final concentration 5 mM) was then added to observe Ca²⁺ efflux in the absence of influx, since there was now no extracellular Ca²⁺. Ca²⁺ efflux in the wild-type strain (BW25113) was rapid, the cytosolic free Ca²⁺ decreasing to less than 0.5 μM within 5 min. In contrast, in the Δ*atpD* mutant, where the gene coding for the Fₒ/F₁-MgATPase in the plasma membrane had been knocked-out, there was no detectable Ca²⁺ efflux. This defect in efflux was explained by a drop in intracellular ATP necessary for Ca²⁺ efflux. Source: Naseem *et al.* (2009). Reproduced with permission from Elsevier. (b) Knock-outs of the putative Ca²⁺ transporter genes *chaA* and *yrbG* in strain MG1655 were tested for Ca²⁺ influx and efflux. No significant differences were observed between the wild-type and the knock-outs, showing that neither ChaA nor Yrbg, alone at least, were responsible for Ca²⁺ influx or efflux under these conditions. A similar negative result was seen when using a knock-out of *pitB*. Source: Naseem *et. al.* (2008) Cytosolic free Ca²⁺ regulates protein expression in *E. coli* through release from inclusion bodies. *Biochim. Phys. Acta*, **1778**, 1415–1422. Reproduced with permission from Elsevier.

extracellular Ca²⁺ was increased or removed, compared with wild-type, since there were no changes in cytosolic free Ca²⁺ in cells in which the genes encoding these proteins had been knocked-out (Figures 8.3 and 8.4). Similarly, the multidrug transporter, LmrP, in *Lactococcus lactis* has been identified as a potential Ca²⁺ transporter. This has a predicted high-affinity EF-hand Ca²⁺-binding motif, $K_d^{Ca} = 7.2$ μM in isolation, with two critical acidic residues involved in Ca²⁺-binding, and was capable of transporting ⁴⁵Ca into membrane vesicles, measured using entrapped fura-2. But, once again, there is no evidence that this caused Ca²⁺ transport in live cells!

Cation and anion channels have been found in bacteria. These include K⁺, Na⁺ and Cl⁻, the K⁺ channel potentially being gated by Ca²⁺. But, no convincing homologues of eukaryotic Ca²⁺ channels have yet been found in bacteria. Many bacteria have mechanosensitive ion channels that open at different pressures, and are relatively non-selective for both cations and anions. These have large conductances, in the nano-siemens range, and would therefore be expected to allow Ca²⁺ into cells down its electrochemical gradient. They exhibit subconductance states, which can be explained by portals opening and closing under appropriate conditions. However, knock-outs of the major mechanosensitive channels in *E. coli*, MscL (L = large conductance) or MscS (S = small conductance), as with knock-outs of ChaA, YrbG and PitB, showed no apparent defect in either Ca²⁺ influx or efflux. Even a triple knock-out of three mechanosensitive ion channels in *E. coli* – MscL, MscS and MscK – still showed a large Ca²⁺ influx and efflux.

The best candidate for a Ca^{2+} *influx* channel is in fact not a protein, but is a helical complex of polyhydroxybutyrate (PHB) and polyphosphate (PP) (Figure 8.5) – a hypothesis pioneered by Rosetta Reusch in the United States. In *E. coli*, high levels of PHB accumulate in the stationary phase, and correlate with a large rise in cytosolic free Ca^{2+}, and vice versa. PHB acts as an energy storage molecule, and has been found in human tissues, in plasma and atherosclerotic plaques, in rat mitochondria, and associated with the plasma membrane Ca^{2+} pump, increasing in several tissues in diabetes induced by streptozotocin. This suggests PHB may be involved in Ca^{2+} transport in eukaryotes and pathogenesis. Four pieces of evidence support PHB–PP as a Ca^{2+} channel in bacteria:

1. The complex binds Ca^{2+}.
2. The complex occurs in the plasma membrane of Gram-positive and Gram-negative bacteria.
3. Insertion of purified PHB–PP complexes into artificial bilayers produces ion channels that conduct Ca^{2+}, with a conductance for Ca^{2+} 18 times that of Na^+. Like eukaryotic Ca^{2+} channels, PHB-PP also conducts Sr^{2+} and Ba^{2+}, and is blocked by La^{3+}, Co^{2+} and Cd^{2+}.
4. Membrane PHB–PP levels correlate with the size of cytosolic free Ca^{2+} transients.

Interestingly, the rise in cytosolic free Ca^{2+} in *E. coli* is higher at alkaline pH, consistent with the pH sensitivity of PHB–PP.

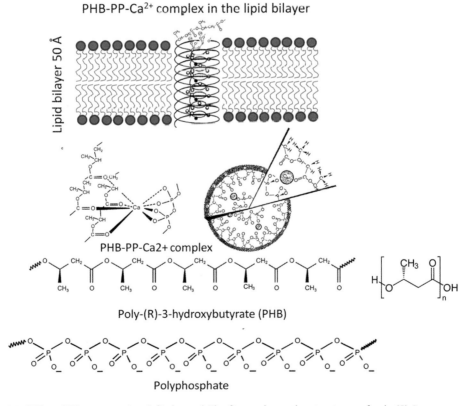

Poly-(R)-3-hydroxybutyrate (PHB)

Polyphosphate

Figure 8.5 PHB and PP as a putative Ca^{2+} channel. The figure shows the structures of poly-(*R*)-3-hydroxybutyrate (PHB), polyphosphate (PP) and the PHB–PP–Ca^{2+} complex in the lipid bilayer as a putative Ca^{2+} channel in bacterial membranes. Reusch (2000). *Source*: Courtesy of R.N. Reusch.

8.6.2 Ca²⁺ Efflux out of Bacteria

The best candidate for Ca^{2+} efflux, in Gram-negative and Gram-positive bacteria, is a MgATPase. Several P-type MgATPases have been found. These include the *yloB* gene in *Bacillus subtilis*, coding for a Ca^{2+}-MgATPase with sequence similar to those in the ER and Golgi of eukaryotes, the *pacL* gene coding for a Ca^{2+}-MgATPase found in *Synechococcus*, and putative Ca^{2+}-MgATPases in *Streptococcus faecalis, Flavobacterium odoratum*, and the cynobacterium *Anabaena variabilis*. A BLAST analysis of *mgtA*, which appears to code for a P-type MgATPase in *E. coli*, shows a sequence homology with the human pMR1 Ca^{2+}-MgATPase. However, a knock-out of *mgtA* in *E. coli* showed no defect in Ca^{2+} efflux in live cells. In *E. coli* a dramatic reduction in Ca^{2+} efflux was found in a knock-out of the F_0/F_1-MgATPase subunit AtpD (Figure 8.4), due to a low ATP level in the *atpD* knock-out. Restoration of ATP to that of the wild-type, by adding glucose, restored Ca^{2+} efflux. In contrast, lowering ATP in the wild-type, using the 2,4-dinitrophenol (DNP), reduced Ca^{2+} efflux to a level similar to that in the *atpD* knock-out, supporting the hypothesis that ATP is required for Ca^{2+} efflux.

8.7 Regulation of Bacterial Events by Intracellular Ca²⁺

8.7.1 Ca²⁺ and Growth of Bacteria

Several marine bacteria, blue-green algae (cyanobacteria), *Azotobacter*, and a few non-marine bacteria such as *Streptococcus pneumonia*, require Ca^{2+} for optimal growth. What is not clear is whether this requirement is extra- or intra-cellular, nor whether changes in intracellular Ca^{2+} can signal effects on growth, and thus affect generation time. In contrast, many bacteria grow quite happily in EGTA (Figures 8.6 and 8.7). But, *E. coli* grown in 5 mM EGTA have a generation time in exponential phase some 5 min longer than cells grown in 1–10 mM Ca^{2+}. In addition, cells in EGTA go into the stationary phase earlier, the growth curve flattening off sooner (Figure 8.6a and b). A 10% reduced generation time in a mixed culture with another population, dividing 10% slower, would mean that after 24 h there will be some 20 000 times more of the faster growing cells than their competitors (Figure 8.6d). The cytosolic free Ca^{2+} measured using recombinant aequorin was consistent with an important role for Ca^{2+} in the growth cycle. Furthermore, in the presence of Ca^{2+} in the growth medium, 41 genes were up-regulated and 69 genes down-regulated, compared with cells grown in the absence of Ca^{2+}, detected by microarray.

8.7.2 Ca²⁺ and Bacterial Movement

There are four ways bacteria move:

1. Brownian motion.
2. Pushing aside by newly dividing cells.
3. Flagellar dependent movement – swimming in suspension.
4. Translocation over surfaces; of particular interest are 'swarming' and two forms of gliding.

The most convincing evidence that intracellular Ca^{2+} plays a role in bacterial movement is in the flagella-based chemotaxis of *E. coli*, phototaxis in *Halobacterium*, the gliding of cyanobacteria, and the 'swimming' motion of *Synechococcus*.

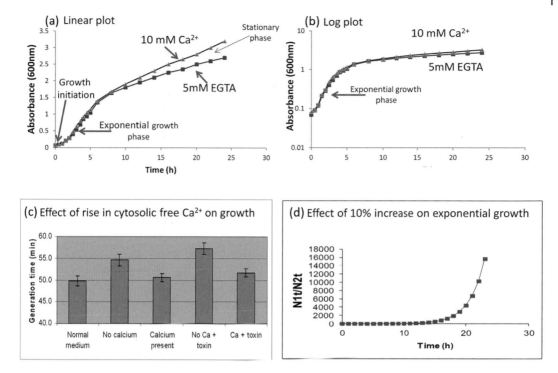

Figure 8.6 Effect of Ca²⁺ in cell growth of *E. coli*. (a and b) The effect of 10 mM extracellular Ca²⁺ versus 5 mM EGTA on cell growth measured by light scattering at 600 nm (A_{600}). The exponential and so-called stationary phases are indicated by the arrows. The putative Ca²⁺ channel PHB–PP is highest in the stationary phase. Small effects of Ca²⁺ are more easily seen when plotting linear growth curves (a), rather than with a conventional microbiologist's log plot (b). A rise in cytosolic free Ca²⁺ causes just a 10% reduction in generation time (c). (d) The consequence of two cell populations competing for the same nutrients and one population (N1) is dividing just 10% faster than the other (N2). Within 24 h there will be nearly 20 000 times more N1 than N2. A small change by small change Darwinian effect. Reproduced by permission of Welston Court Science Centre.

8.7.2.1 Chemotaxis and phototaxis

Chemotaxis is the movement of cells along a chemical gradient. It plays a crucial role in the gut, for organisms such as *Helicobacter pylori*, the cause of stomach ulcers; *Campylobacter jejuni*, a major cause of food poisoning; *Vibrio cholerae*, the cause of cholera; *Vibrio anguillarum*, a fish pathogen; *Salmonella*, the most common cause of food poisoning; *Pseudomonas aeruginosa*, a cause of chronic lung infections; *Brachyspira hyodysenteriae*, a gut pathogen; *Borrelia burg-dorferi*, the cause of Lyme disease; *Leptospira interrogans*, attracted by blood through haemo-globin; and *Treponema denticola*, as oral spirochaete attracted by serum and some sugars and amino acids. In plants, chemotaxis is important to the pathogens *Agrobacterium tumefaciens*, which are attracted by phenols released by wounds in the plant, as well as being involved in the symbiosis of *Rhizobium*, which is attracted by substances released by nodules in plant roots. While *Azospirillium* and *Halobacterium* are attracted by light at wavelengths required for photosynthesis.

Bacterial swimming is driven by flagella with motors, whose energy comes from the pro-ton motive force. Rotation of the flagellum causes the bacterial cell to move in one direction. Nutrients, such as amino acids, attract some bacteria, while other chemicals, which are poten-tially toxic, repel. On the other hand, phototaxis occurs when light of particular wavelengths attracts an organism, such as the cyanobacterium *Synechocystis*. It is easy to calculate that it would be impossible for any receptors on the bacterial cell surface to sense the difference in concentration or light intensity between one end of the bacterium and the other. A bacterium senses gradients by tumbling. In *E. coli* serine increases the time between tumbles, and thus attracts bacteria towards the source. Toxic agents, such as Ni^{2+}, decrease the time between tumbles, and thus repel the bacteria. Removal of external Ca^{2+} reduces chemotaxis in *E. coli*, and blocks chemotaxis in *Bacillus subtilis*. Furthermore, the ionophore A21387 can increase chemotaxis – an effect that depends on the presence of external Ca^{2+}. Various Ca^{2+} antagonists, such as conotoxin and cations, inhibit chemotaxis in *E. coli*. While the ionophore A23187, or release of caged Ca^{2+} from nitr-5, increases the rate of tumbling in *E. coli*. But raising intracellular Ca^{2+} using A23187 in *Bacillus subtilis* causes the cells to tumble continuously, thereby inhibiting chemoattraction. Nitrendipine, verapamil, La^{3+} and conotoxin inhibit chemorepulsion in *E. coli*, conotoxin blocking Ca^{2+} entry and a rise in free Ca^{2+}, thereby causing tumbling to stop. Chemorepellents cause an increase in cytosolic free Ca^{2+} and chemoattractants a small decrease, measured using recombinant aequorin. The veracity of measurements using fura-2 in bacteria is in doubt. But there does appear to be a Ca^{2+} sensor inside chemotactic bacteria, which affects the direction of flagellar rotation, and the time between tumbles. In order to undergo chemotaxis three components are required:

1. A motor driven by an energy source in the cell that can be regulated by signals inside the cell.
2. A receptor facing out of the cell that can communicate to the inside.
3. A signal transduction pathway that can regulate the motor.

Flagella rotation is either clockwise or anti-clockwise. The normal rotation causes the bacterium to move in one direction. When the rotation is reversed, briefly, the bacterium 'tumbles', a sort of spin around, and then it moves off in another direction. Ca^{2+} in *E. coli* reduces the time between tumbles, and thus mediates the action of a chemorepellent. *E. coli* can sense a variety of amino acids, sugars and dipeptides, as well as pH, temperature, or oxygen. The chemoattractant or -repellent binds to a receptor, which is a methyl-accepting protein dimer or multimer facing into the periplasm. The receptor transmits a signal to the domain facing into the cytoplasm, which signals to a histidine kinase, CheA, through an adaptor protein, CheW. This results in autophosphorylation of CheA, followed by transfer of the phosphate to CheY, the fourth member of the signalling pathway. Phospho-CheY then interacts with the flagella motor, where it induces an immediate change in rotation of the flagella. This triggers the bacterium to tumble, so that it moves off in another direction. Conversely non-phosphorylated CheY allows the flagella motor to continue to rotate in one direction only. The result is less frequent 'tumbling', allowing the bacteria to be attracted to one direction. A rise in cytosolic free Ca^{2+} somehow reduces the time it takes for the flagellum to change its direction of rotation, and thus the bacterium tumbles more often. CheA, CheW and CheY are not required for the induction of tumbling when cytosolic free Ca^{2+} is increased, e.g. by photorelease from a caged Ca^{2+} compound. Chemorepellents, such as Ni^{2+}, cause a rise in cytosolic free Ca^{2+} in *E. coli*, whereas chemoattractants such as serine causes a small decrease.

8.7.2.2 Gliding
Bacterial gliding occurs in cyanobacteria, and in the soil Gram-negative myxobacterium *Myxococcus xanthus*, which has no flagella. The gliding machinery in myxobacteria requires extracellular Ca^{2+} (0.1–0.3 mM), and can be induced by 1 mM Ca^{2+}. But there is no convincing evidence that changes in intracellular Ca^{2+} play a role in this movement. Cyanobacteria

photosynthesis accounts for some 50% of carbon fixation in the sea, and marine mats of cyanobacteria contain microprecipitates of $CaCO_3$ with Mg^{2+}. The cyanobacterium *Synechocystis* has evolved a highly efficient CO_2 concentrating machinery. A key component of this is the ABC importer CmpABCD. This concentrates CO_2 1000-fold. CmpABCD consists of four proteins:

1. CmpA, a periplasmic high-affinity solute-binding lipoprotein binding bicarbonate, and apparently Ca^{2+} cooperatively.
2. CmpB, a membrane permease.
3. CmpD, a cytoplasmic MgATPase.
4. CmpC, a second ABC MgATPase/solute-binding fusion protein, with an additional domain highly homologous to CmpA that regulates transport.

Since CmpA binds Ca^{2+} cooperatively, it is an unproven candidate for a Ca^{2+} importer. Evidence indicating the presence of Ca^{2+} channels in cyanobacteria has been obtained using the fluorescence of fluorescein-labelled dihydropyridine, a Ca^{2+} channel binder in eukaryotes. The rod-shaped cyanobacterium *Synechococcus* can move rapidly at up to 25 μm s⁻¹ without a change in shape. Extracellular Ca^{2+} is essential for this. The Ca^{2+} target is a secreted 130-kDa protein, with one EF-hand Ca^{2+}-binding site, one haemolysin-type Ca^{2+}-binding site, and 12 GGXGXD Ca^{2+}-binding motifs. Speed increases as the extracellular Ca^{2+} concentration increases over the range 0.9–2 mM. Movement is blocked by the voltage-sensitive Ca^{2+} channel blockers verapamil and nitrendipine, but not by the Ca^{2+} channel blockers conotoxin and diltazem. Nor does the P-type Ca^{2+}-MgATPase blocker orthovanadate inhibit movement. Furthermore, 4-bromo-A23187 at 10 μM does not inhibit motility, but at 100 μM causes cells to clump. Interestingly, photosynthesis is not required directly for motility, thus the effect of Ca^{2+} cannot be on photosystem II, the cells continuing to move in cyanide – a potent blocker of cytochrome oxidase. Motility is inhibited by a low concentration of terbium – a lanthanide and a known competitive inhibitor of Ca^{2+}. The gliding machinery in myxobacteria also requires extracellular Ca^{2+} (0.1–0.3 mM), 1 mM Ca^{2+} inducing construction of the gliding apparatus. Similarly, movement of the cyanobacterium *Synechocystis* attracted by light is blocked by EGTA, La^{3+} and the calmodulin inhibitor trifluoperazine, and partially inhibited by agents that inhibit Ca^{2+} signalling in eukaryotes including pimozide (a voltage-gated L-type Ca^{2+} channel inhibitor), orthovanadate (a blocker of Ca^{2+} efflux via the Ca^{2+}-MgATPase) and A23187.

A wave of depolarisation is induced during gliding of the filamentous cyanobacterium *Phormidium uncinatum* in response to a light/dark transition, a photophobic response, or by chemorepellents. The light 'receptor' occurs in the first third of the filamentous trichome structure. The polarisation wave propagates along the entire length of the filament. Extracellular Ca^{2+} is required for both the photophobic response, and production of the action potential. This is also associated with the uptake of ⁴⁵Ca, implicating a Ca^{2+} channel in the depolarisation. The extracellular Ca^{2+}-binding protein oscillin is the Ca^{2+} target. Ruthenium Red, an inhibitor of electrogenic Ca^{2+} transport, and the Ca^{2+} channel blocker La^{3+}, inhibit movement reversal, but do not affect the speed of movement.

Thus, there are many examples where extracellular Ca^{2+} is required for gliding, the process being inhibited by agents that block Ca^{2+} channels and signalling events in eukaryotes. The evidence supports a role for Ca^{2+} extracellularly, but there is little evidence to support a role for changes in cytosolic free Ca^{2+} being a signal for gliding. What is needed is measurement of cytosolic free Ca^{2+} in live bacteria, correlating this with movement, and the use of knock-outs to identify how intracellular Ca^{2+} is regulated, and how it works.

8.7.2.3 Swarming

Swarming of bacteria depends on flagella, and the presence of a surfactant. The bacteria migrate cooperatively on mass over, for example, an agar surface. Major changes in gene expression occur. Swarming has been studied best in the genera *Aeromonas, Bacillus, Escherichia, Proteus, Pseudomonas, Salmonella, Serratia, Vibrio* and *Yersinia*. Different genes are expressed in different parts of the swarm, dependent on nutrients and medium viscosity. But, as with gliding, although removal of extracellular Ca^{2+} may disturb bacterial swarming in some species, there is as yet little direct evidence to support a role for intracellular Ca^{2+} as a signal for this process.

8.7.3   Quorum Sensing and Gene Expression

Walk along the beach at night, preferably when there is no moon, and you may be lucky enough to see a piece of rotting fish or crab glowing in the dark. This is caused by luminous bacteria, typically *Vibrio harveyi* or *Photobacterium phosphoreum*. Yet individual bacteria free floating in the sea are not luminous. In the 1960s, this puzzle led to one of the most important discoveries in microbiology – quorum sensing. These particular bacteria, as they grow, release a homoserine lactone that induces the bacterial *lux* operon encoding the proteins necessary for light emission. Each bacterium contributes to the whole, leading to a build-up of the quorum-sensing compound in the surrounding fluid. When the concentration reaches a critical level, the operon is switched on. The bacteria therefore require a 'quorum' before they can produce visible light.

Removal of extracellular Ca^{2+} markedly inhibits the development of bioluminescence in the marine *Vibrio harveyi* and the soil *Photorhabdus luminescens* (Figure 8.7), which cannot be explained by an effect of Ca^{2+} on cell growth. *Photorhabdus luminescens* (*Xenorhabdus luminescens*) is a soil bacterium, which infects dead flesh, and is famous for glowing wounds and bodies on battlefields, having been recorded in Chinese literature over 2000 years ago, in the American Civil war. Even Florence Nightingale is reputed to have seen luminous wounds in the Crimean war, which did not become gangrenous because *Photorhabdus* produces a potent antibiotic. It is not yet clear whether the effect of Ca^{2+} is intra- or extra-cellular, or whether Ca^{2+} acts to regulate the production or the activity of the autoinducer, or on gene expression. In *Serratia liquefaciens*, quorum sensing compounds induce changes in intracellular Ca^{2+}, concomitant with an increase in protein kinase. However, there is little evidence that quorum sensing generally is induced by changes in cytosolic Ca^{2+}. On the other hand, a class of anti-quorum metabolites has been identified. These are carbohydrate metabolites – methylglyoxal, butane-2,3-diol, propane-1,3-diol, acetoin and dimethylglyoxal. These induce Ca^{2+} transients in *E. coli* (Figure 8.2), and can inhibit growth, although the Ca^{2+} signal does not appear to play a major role in growth inhibition. Interestingly, quorum-sensing compounds can affect host cells through intracellular Ca^{2+}, when an animal or plant is infected by bacteria.

8.7.4 Ca^{2+} and Bacterial Metabolism

Cyanobacteria are photosynthetic bacteria found in great abundance in the sea, and also occur in terrestrial ecosystems. Cyanobacteria produce oxygen, but many also fix nitrogen, reducing it to ammonia. The latter process requires the enzyme nitrogenase. Removal of extracellular Ca^{2+} inhibits several metabolic processes in the non-heterocyst cyanobacterium *Gloeopcapsa*, such as acetylene reduction caused by nitrogenase. This enzyme is highly sensitive to oxygen, and is inactivated by it. Superoxide scavengers and Ca^{2+} prevent this. These cyanobacteria can fix nitrogen in the absence of oxygen, but Ca^{2+} only protects nitrogen fixation in oxygen.

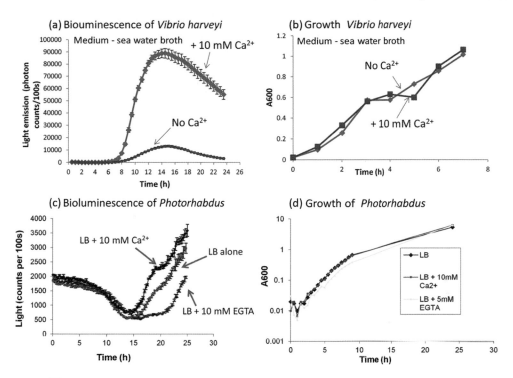

Figure 8.7 Effect of Ca²⁺ on the bioluminescence and growth of *Vibrio harveyi* and *Photorhabdus luminescens*. (a and b) *Vibrio harveyi*, bioluminescence being measured from cells in microtitre plates imaged using an intensified CCD (ICCD) camera (Photek): (a) bioluminescence; (b) growth of cells. Campbell and Naseem (unpublished). (c and d) In the soil bioluminescent bacterium *Photorhabdus luminescens*. Kind gift from Dr Clark, University of Cork, Republic of Ireland. Temperature = 20 °C. Bioluminescence measured from cells in microtitre plates imaged using an ICCD camera (Photek). (c) Bioluminescence and (d) growth of cells. Noman and Campbell (unpublished). LB = Luria–Bertani medium. The results show for both species of bacteria a large increase in bioluminescence when cells were grown in medium containing Ca²⁺, but no detectable effect of Ca²⁺ on growth.

8.7.5 Bacterial Defence – Dormancy, Spore Formation and Germination

Bacteria have to defend themselves against a variety of environmental hazards. These include changes in osmolarity, temperature, pH, lack of nutrients, lack of water, products of competing bacteria, such as antibiotics and methylglyoxal, and eukaryotic attack by components of the immune system. *Bacilli*, *Clostridia* and *Sporosarcina*, when short of water or key nutrients, defend themselves by forming spores – a process that requires a large influx of Ca²⁺ and its binding to dipicolinic acid. These genera contain several important human and animal pathogens, including anthrax (*Bacillus anthracis*), and bacteria that cause severe diarrhoea, such as *Clostridium difficile*. The spores contain a large amount of Ca²⁺-dipicolinate (Figure 8.8), essentially in crystalline form. This Ca²⁺ has to be released if the spore is to germinate. These spores are very stable, and can remain dormant for decades in the soil.

Spore formation starts with an asymmetric cell division, signalling nucleoids with no nuclear membrane, the two nuclear bodies producing the mother cell and the prespore. Progressively, a membrane, and a thick external coat, the exosporium, develop around the spore nucleoid. The nascent spore then accumulates large amounts of dipicolinic acid, and the cell takes up large amounts of Ca²⁺, which binds to the dipicolinate anion. The developing spore becomes

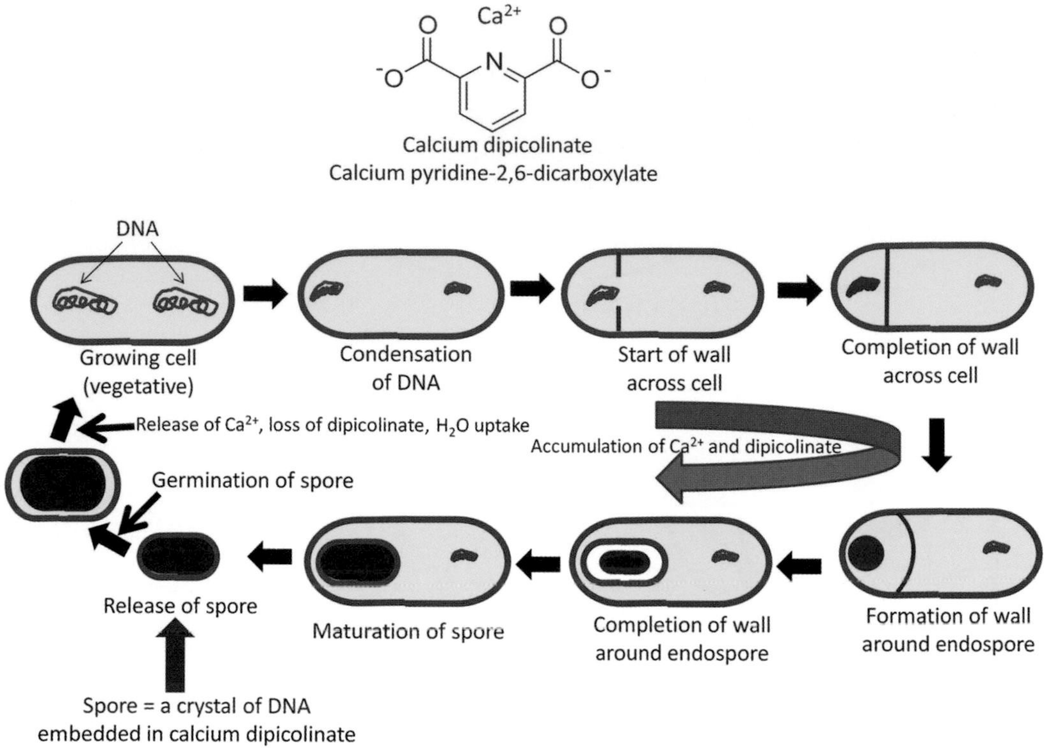

Figure 8.8 Ca^{2+} and bacterial spore formation. The figure shows the formation of bacterial spores after nutrient deprivation or drought, involving the accumulation of calcium dipicolinate. Water or nutrients activate the pumping out of Ca^{2+} and loss of dipicolinate, so that the cell can return to a vegetative stage. Reproduced by permission of Welston Court Science Centre.

dehydrated, forming a semi-crystalline state. The mature spore is released by lysis of the bacterial cell. Spores can form in low Ca^{2+} media, but these spores have a low dipicolinic acid, and thus low Ca^{2+} content, making them unstable. Dipicolinic acid, an aromatic dicarboxylic acid (Figure 8.8) poorly soluble in water, binds Ca^{2+} using oxygen as the ligand atom. However, none of this interesting chemistry explains why evolution has selected this particular molecule for Ca^{2+} binding. Germination of a spore follows the opposite course, and can start within seconds of the stimulus, the DNA becoming active again. The stimuli for spore germination include water, and nutrients such as serine. Rapid rehydration is followed by loss of dipicolinic acid and extrusion of Ca^{2+}.

8.7.6   Bacterial Infection – Virulence, Competence and Defence

In order to infect a host successfully, bacteria have to prevent being killed by the host defence. *Yersinia* and *Salmonella* switch off the ability of phagocytes to ingest and kill them. In contrast, *Streptococcus pneumoniae* has to exchange DNA between cells in order to become virulent. Several *Yersinia* are highly pathogenic to man and animals, *Yersinia pestis*, causing bubonic plague, and exhibit a 'low calcium response'. Virulent *Yersinia* contain a special 70-kb plasmid, pCD1 in *Y. pestis*, that codes for three anti-host proteins, *Yersinia* outer proteins (Yops). When the *Yersinia* cell binds to a phagocyte, which aims to kill it, a unique trans-envelope structure is assembled – the type III secretion system or the 'injectisome'. This injects three anti-host proteins into the phagocyte cytosol. These Yops act in different ways to prevent the bacterium being engulfed by phagocytosis, and to block the release of cytokines that would normally enhance the inflammatory response. Secretion of the *Yersinia* proteins is blocked in the presence of mM external Ca²⁺. At 37 °C, protein secretion of Yops by the type III system is activated in the absence of Ca²⁺, while growth ceases – growth restriction. When taken from 26 to 37 °C virulent strains of *Yersinia*, within two generations, stop growing in Ca²⁺ free media, the cells switching more quickly into the stationary phase. Thus, *in vitro*, the presence of 2.5 mM Ca²⁺ and 1.5 mM Mg²⁺ allows vegetative growth, but represses the synthesis of the virulence proteins encoded by the Lcr (Low calcium response) plasmid. A similar story occurs in *Pasteurella*, a rabbit pathogen. But, as yet, there is no convincing evidence that the virulence or growth of either genus involves Ca²⁺ signals within the bacteria themselves. A βγ-crystallin-type Ca²⁺-binding protein has been isolated from *Yersinia pestis* that could be involved. This has two Ca²⁺-binding domains of the Greek key motif type. Binding of Ca²⁺ induces structural changes. Direct measurements of cytosolic free Ca²⁺ must be performed in *Yersinia* during its interaction with the host phagocyte, and mutagenesis studies carried out, to confirm the importance of Ca²⁺-binding to putative targets. *Salmonella* defends itself against phagocyte attack by inhibiting the generation of toxic oxygen species, a process requiring external Ca²⁺. But, once again, there is no direct evidence this involves Ca²⁺ signals within the bacterial cell.

The Gram-positive *Streptococcus pneumoniae* causes pneumonia, intracellular Ca²⁺ playing a role in growth and competence, requiring at least 0.15 mM Ca²⁺ in the external medium. Higher mM Ca²⁺ induces a stress response, which is regulated by a small protein released into the external medium. As bacterial growth reaches the stationary phase, cells lyse, releasing fragments of DNA. This DNA can be taken up by viable cells, but only if they are 'competent' to do so. Ca²⁺ is required to make the viable cells competent. Ca²⁺ uptake occurs, and the process can be inhibited by an amiloride derivative that inhibits 3Na⁺/Ca²⁺ exchange in eukaryotes. But, the genome of *Streptococcus pneumoniae* has not revealed any obvious candidates homologous to such eukaryotic proteins. Cytosolic free Ca²⁺ has been measured using recombinant aequorin, but changes in intracellular free Ca²⁺ have yet to be correlated with competence or growth.

8.7.7 Development of Bacterial Structures

Bacteria grow well in free suspension, or in colonies on surfaces, largely as individual cells. But some form extraordinary structures. These include:

1. Filaments, with differentiating heterocysts.
2. Fruiting bodies.
3. Clusters.
4. Swarms and biofilms.
5. Mats.

Marine mats composed of cyanobacteria contain intracellular micro-precipitates of CaCO₃ with magnesium. These must therefore accumulate intracellular Ca²⁺. But the best evidence for a role of intracellular Ca²⁺ in bacterial structures is in heterocysts of cyanobacteria.

8.7.7.1 Heterocysts
Cyanobacteria are photosynthetic and contain fluorescent proteins, phycobiliproteins, which are dichroic: blue in daylight, hence cyano-, and red fluorescent under UV light. They represent one of the few instances where free Ca²⁺ measurement has been correlated with a process in live bacteria. Many cyanobacteria, such as *Anabaena*, grow as long filaments or chains of cells attached to each other. Every tenth cell or so differentiates into a heterocyst (Figure 8.9).

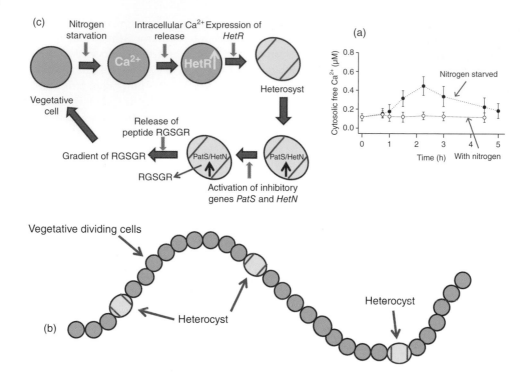

Figure 8.9 The role of Ca^{2+} in heterocyst formation provoked by nitrogen starvation. Typically cyanobacteria in the laboratory are supplied with nitrogen, 10 mM $NaNO_3$ and 5 mM NH_4Cl in the culture medium. (a) Removal of this nitrogen source causes a rise in cytosolic free Ca^{2+}, monitored using cells transformed with the Ca^{2+}-activated photoprotein aequorin. This forces the cells to fix atmospheric N_2. Data from Torrecilla *et al.* (2004) *Microbiology*, **150**, 3731–3739. (b) The pathway involves activation of the HetR gene, and then the inhibitory genes PatS and HetN, which generate the peptide RGSGR as a gradient, so that the heterocysts occur about every ten cells in the growing filament. (c) The cycle from vegetative cell through heterocyst and back to vegetative cell.

These develop to fix nitrogen. Intracellular Ca^{2+} is involved in this differentiation process, to protect its nitrogenase from oxygen. Heterocysts form in a semi-regular pattern in response to lack of nitrogen. Calmodulin inhibitors, and manipulation of extra- and intra-cellular Ca^{2+}, suggest that Ca^{2+} plays a key role. But the crucial data were obtained when cytosolic free Ca^{2+} was measured using aequorin, allowing changes in cytosolic free Ca^{2+} to be correlated with the occurrence and timing of the differentiation process. Furthermore, evidence for Ca^{2+} efflux mechanisms by a Ca^{2+}/H^+ exchanger or Ca^{2+}-MgATPase was also obtained. A potential Ca^{2+}-binding protein target for intracellular Ca^{2+} was also identified.

The differentiation process involves specific gene expression changes, which require Ca^{2+}. A key gene is *het*R, which codes for an unusual cytosolic serine protease, activated by Ca^{2+}. Nitrogen deprivation, induced by removal of nitrate or ammonium from the growth medium, induces a slow cytosolic free Ca^{2+} transient, rising from a resting level of 0.1–0.2 μM after 45 min, and reaching a peak of 0.4 μM within 2–3 h. The Ca^{2+} transient differs markedly from transients initiated in *Anabaena* by heat and cold shock, acid shock, salt stress or light/dark transitions. For example, cold or hyperosmotic shock provoke a peak in the cytosolic free Ca^{2+} of 2–3 μM, some ten-fold more than that induced by nitrogen deprivation. The Ca^{2+} transients caused by other stimuli also last for a much shorter time, typically 3–4 min or at most 1 h. Furthermore, in the nitrogen-deprived cells, the source of the Ca^{2+} rise is intracellular, since it still occurs in EGTA, but is lowered when the cells are loaded with the Ca^{2+} chelator BAPTA.

This also inhibits heterocyst differentiation. The calmodulin inhibitor trifluoperazine inhibits differentiation. Furthermore, suppression, amplification, or inappropriate regulation of the Ca^{2+} signal inhibits heterocyst differentiation. Two important intracellular proteases have been identified in heterocysts. One is Ca^{2+}-activated. The other is specific for phycobiliproteins. Proteolysis supplies amino acids for the synthesis of new proteins required in the differentiation process. *het*R activation starts about 2 h after nitrogen deprivation, consistent with Ca^{2+} being the signal. Typically, *Anabaena* filaments only have 10–20% of the cells as heterocysts. Thus, the exact cells in which the Ca^{2+} signals are generated is uncertain, and the Ca^{2+} level in each cell may even be much higher than first thought, if it is only produced in cells designated to be differentiating heterocysts.

8.7.7.2 Inclusion Bodies

Bacteria store a range of substances in inclusion bodies, a few nanometres in diameter. These are made up of proteins, small organic molecules, polymers such as PHB, or inorganic ions such as PP. These are dynamic structures, substances being released from, and taken up by, them. In *E. coli*, raising the cytosolic free Ca^{2+} to high micromolar for 1–2 h leads to substantial intracellular release of protein from the inclusion bodies, paralleled by a decrease in generation time.

8.7.8   Ca²⁺ and Gene Expression in Bacteria

Increases in cytosolic free Ca^{2+} have been correlated with changes in gene expression in several bacteria – *E. coli*, *Bacillus subtilis*, *Anabaena* and *Actinobacillus pleuropneumoniae*. Interestingly, removal of Ca^{2+} from the growth medium of the luminous bacteria *Vibrio harveyi* and *Photorhabdus luminescens* inhibited the development of the bioluminescence, but there was no detectable effect of Ca^{2+} on growth (Figure 8.7). In *Vibrio harveyi*, the increase in bioluminescence in the presence of Ca^{2+} was associated with increases in the level of several proteins, suggesting that Ca^{2+} is a stress regulator in bacteria. In the case of *E. coli*, transcriptome analysis using microarray technology showed that over 100 genes responded to an increase in cytosolic free Ca^{2+} induced by adding extracellular Ca^{2+}, expression of 41 being elevated and 69 depressed. Three of these were transport proteins, and four others were membrane proteins, being potential candidates for Ca^{2+} transport. The expression of a further 943 genes was changed.

8.7.9   Uptake of Nucleic Acid by Bacteria

There are three ways by which bacteria can take up DNA or RNA:

1. Conjugation.
2. Transduction.
3. Transformation.

In conjugation, bacterial cells fuse together and DNA is transferred directly from one cell to the other. Transduction is performed through bacterial viruses – bacteriophages. Transformation, on the other hand, involves the uptake of DNA from the external medium, through the wall and membrane of the cell into to cytoplasm. Ca^{2+} appears to play a role in all three mechanisms, but there is as yet no direct evidence that changes in intracellular Ca^{2+} are essential. In order to be transformed, a bacterium must be 'competent'. This can be induced artificially by cold/heat shock in the presence of a high concentration of $CaCl_2$, a standard technique in genetic engineering.

8.7.10 Bacterial Metabolic Toxin Hypothesis

Bacteria release gases, ions, metabolites, proteins and nucleic acids, and a range of peptide toxins, several of which are highly pathogenic. But, there is another type of toxin produced by bacteria, which interacts with Ca^{2+} signalling. Gas, and metabolic toxins-aldehydes, ketones and acids-are produced by bacteria in the large intestine when sugars are not absorbed in the small intestine. These play a key role in producing the systemic symptoms of lactose and food intolerance – severe headache, muscle and joint pain, heart palpitations, increased micturition, various allergies including eczema, sinusitis, and infertility. The metabolic toxins regulate the balance of microflora in the large intestine through Ca^{2+} signals and cell growth. Several, including methylglyoxal, diacetyl and butane-2,3-diol, provoke Ca^{2+} transients (Figure 8.2). The metabolic toxins also affect eukaryotic cells, through effects on ion channels and Ca^{2+} signalling, including reducing the heart rate and causing arrhythmias. A further consequence of bacterial metabolic toxins is that they can covalently modify neurotransmitters and hormones, likely to play a major role in cancer, type 2 diabetes, Alzheimer's and Parkinson's diseases.

8.7.11 Intracellular Ca^{2+} in Bacteria – Conclusions

Bacteria, like all eukaryotic cells, maintain a very low cytosolic free Ca^{2+}, sub-micromolar to low micromolar, in the presence of mM Ca^{2+} outside the cell. Cytosolic free Ca^{2+} rises transiently when the external Ca^{2+} is increased, or when the cells are exposed to a stimulus. Thus, bacteria have both Ca^{2+} influx and efflux mechanisms, capable of rapidly regulating the level of intracellular Ca^{2+}. Ca^{2+}-MgATPases, CaCA, Ca^{2+}/H^+ exchangers and the non-proteinaceous PHB–PP complex have been identified in several bacterial species. But definitive identification of the mechanisms responsible for Ca^{2+} influx and efflux remains elusive, as potential candidates have not been confirmed to regulate cytosolic free Ca^{2+} in live cells. The number of Ca^{2+} ions that have to move into the cell to cause significant changes in intracellular Ca^{2+} is in the tens to hundreds. Mechanosensitive ion channels exist in bacteria with a conductance measured by patch clamp in the nano-siemens range. A single channel would therefore be expected to allow tens of millions of Ca^{2+} ions to enter in 1 s, enough to raise the intracellular Ca^{2+} by many micromolar. Yet, in live cells, the changes in cytosolic free Ca^{2+} occur over minutes. Ca^{2+} plays a role in several bacterial processes, but whether Ca^{2+} acts as an intracellular signal, or whether it plays a passive role inside or outside the cell is not yet clear. Inhibitory effects of removal of extracellular Ca^{2+} can be explained by Ca^{2+}-binding proteins, with Ca^{2+}-binding sites facing the external medium. Definitive identification of Ca^{2+}-binding proteins in bacteria, analogous to those in eukaryotes, also remains elusive, as does the mechanism by which Ca^{2+} regulates gene expression. Calmodulin is not found in bacteria. Genomic searches, using Ca^{2+}-binding motifs, have identified candidates in several bacteria, but not convincingly in *E. coli*. Ca^{2+}-binding proteins, such as crystallin, with EF-hand or Greek key-type Ca^{2+}-binding domains have been found. But most are extracellular, or require high micromolar or even millimolar Ca^{2+} to bind. The fact that the cytosolic free Ca^{2+} can rise to tens of micromolar in *E. coli*, and the cells remain viable, suggests that the intracellular Ca^{2+} targets in bacteria may have a lower affinity for Ca^{2+} than the Ca^{2+}-binding proteins in eukaryotes. The evidence that Ca^{2+} plays a role in some bacterial phenomena can be summarised as follows:

1. Removing external Ca^{2+}, followed by a Ca^{2+} dose response curve, shows a requirement for Ca^{2+} in the process.
2. Manipulation of intracellular Ca^{2+}, using ionophores or caged Ca^{2+} compounds, is consistent with changes in intracellular Ca^{2+} regulating the process.

3. Blockers of eukaryotic Ca^{2+} channels or transporters have the predicted effect in bacteria.
4. Some cytosolic Ca^{2+}-binding proteins have been identified.
5. ATP is required for Ca^{2+} efflux.
6. In a few cases, increases in cytosolic free Ca^{2+} have been induced by agents provoking the cellular process, which correlate with the physiological process.
7. A non-proteinaceous putative Ca^{2+} channel has been identified, PHB–PP.

Proteins initially identified as Ca^{2+} influx and efflux mechanisms have now been discounted, in *E. coli* at least. Furthermore, it has been a huge mistake not to examine the effect of knockout and other mutants of putative Ca^{2+} transporters and Ca^{2+} target proteins on Ca^{2+} influx and efflux in live cells, and cell physiology and growth. High ångström resolution crystal structures have revealed the potential molecular basis of Ca^{2+}/H^+ exchange of Yfke in the Gram-positive bacterium *Bacillus subtilis*. Two of the transmembrane helices in putative antiporters form a hydrophilic cavity, providing the pathway for cation exchange. Ca^{2+} binding is typically via glutamate residues. However, what is still required is a correlation of Ca^{2+} exchange with cytosolic free Ca^{2+} measured in the live cell and its loss in knock-out mutants. Cation exchange in vesicles is encouraging, but not sufficient proof that it occurs in the live cell. The use of blockers of eukaryotic Ca^{2+} channels is also flawed, given the fact that no homologues of eukaryotic Ca^{2+} channels have been convincingly found in bacteria. The most convincing cases for a role of intracellular Ca^{2+} as a signal in bacteria are:

1. Heterocyst formation in the cyanobacterium *Anabaena*.
2. Chemotaxis of *E. coli* and *Bacillus subtilis*.
3. Regulation of growth and gene expression in *E. coli*.
4. Spore formation in *Bacillus* and related species.

For the remainder of bacterial processes, including growth and movement, evidence is more convincing that Ca^{2+} has an extracellular role, or is required in the periplasm. A further possibility is that Ca^{2+} is required to maintain structures within the cytoplasm, such as the chromosome. There is still much to be learned about the role of Ca^{2+} in bacteria, requiring a new experimental and conceptual approach. What is needed is a method for monitoring free Ca^{2+} and ATP in individual bacteria.

8.8 Role of Intracellular Ca²⁺ in Archaea

Archaea, formally known as archaebacteria, were discovered by Carl Woese and George Fox in 1977, based on ribosomal RNA sequences. They are the third domain of life, the other two being Bacteria and Eukaryota. Microfossil Archaea have been found in rocks some 3500 million years old, or in extremes of acid and alkaline pH. Archaea have been found in the human body – gut, mouth, vagina and teeth. Evidence of a signalling role for intracellular Ca^{2+} in Archaea is weak. But, in one species, there is evidence internal Ca^{2+} can be regulated, and that cytosolic free Ca^{2+} is in the sub-micromolar to micromolar range (Figure 8.10). Other evidence that Ca^{2+} has a role in Archaea is based on ^{45}Ca fluxes, effects of manipulation of external and internal Ca^{2+}, and the effects of Ca^{2+} on isolated proteins. A major problem is the lack of evidence for a physiological or pathological phenomena activated by primary stimuli, that would require an intracellular signalling pathway based on Ca^{2+}. Searches of Archaea genomes for sequence similarities with eukaryotic Ca^{2+} signalling proteins have not been very successful.

Archaea are the only cells that produce methane. Measurement of breath methane in patients with irritable bowel syndrome (IBS) or inflammatory bowel disease (IBD) provided the first

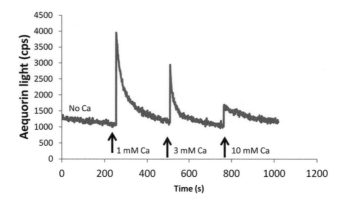

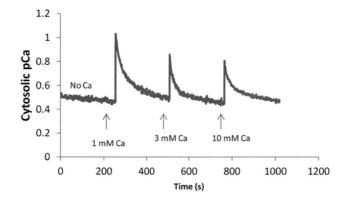

Figure 8.10 Cytosolic free Ca^{2+} in *Haloferax volcanii*. The halophile *Haloferax volcanii* was transformed with an archaean plasmid containing the gene for the Ca^{2+}-activated photoprotein aequorin, with the codon frequency for this organism. Light emission from the cells was then converted into a rate constant and this converted into cytosolic free Ca^{2+} using a standard curve of aequorin. However, these estimations were much lower than the real free Ca^{2+} inside the cells, as the cytosolic concentration of potassium in *Haloferax* is some 4 M. This will reduce the apparent affinity of aequorin for Ca^{2+}. Nevertheless, the results still show that this archaean can maintain a very low cytosolic free Ca^{2+} in the presence of millimolar extracellular Ca^{2+}. Campbell, Lestini and Myllikallio (unpublished).

evidence of a role for Archaea in human disease. Methanogenesis requires external Ca^{2+}, and Ca^{2+} transport has been observed in vesicles made from Archaea. Na^{+}, K^{+}, choline and respiratory chain uncouplers affect ^{45}Ca influx and efflux, evidence that the Archaean *Methanobacterium thermoautotrophicum* has H^{+}/Ca^{2+} and Na^{+}/Ca^{2+} exchangers (e.g. MaX1). Experimental conditions expected to lower or raise intracellular Ca^{2+} affected cell growth. Furthermore, methane production is dependent on the external Ca^{2+} concentration (25–125 μM), and is increased substantially by loading the cells with Ca^{2+}. Methane production is inhibited by Co^{2+} and Ni^{2+}, which block Ca^{2+} channels in eukaryotes. The possibility that a key enzyme in methanogenesis, methenyltetrahydromethanopterin cyclohydrolase, might be regulated by Ca^{2+} is supported by X-ray fluorescence showing that this enzyme in *Methanopyrus kandleri* has 0.5 mol Ca^{2+} bound per subunit. *Methanobacterium thermoautotrophicum* also contains a protein, MTH 1880, having a highly acidic domain with aspartates and glutamates, a Ca^{2+}-binding motif. External Ca^{2+} also induces aggregation of Archaean bipolar tetraether liposomes. Mechanosensitive channels are found in Archaea, with the potential to act as Ca^{2+} channels. Interestingly, one of the few natural uses of Ba^{2+} is found in the methanol dehydrogenase of an archaean.

Several extracellular and intracellular proteins isolated from Archaea bind, and are activated by, Ca^{2+}. These include the inducible alkaline phosphatase in the halophile *Haloarcula marisma*, the recombinase Rad51 from *Methanococcus voltae*, similar to the Ca^{2+}-stimulated DNA polymerases in humans, yeast and *E. coli*, and analogues of intracellular human lens βγ-crystallin in *Methanosarcina acetivorans*. Crystallins have two β-sheet-type Ca^{2+}-binding sites: one with moderate affinity, the other with low affinity. But, these proteins require millimolar Ca^{2+} concentrations for full Ca^{2+} binding, far higher than would be expected for cytosolic free Ca^{2+}. A 24-kDa calmodulin-like protein has been reported in *Halobacterium salinarium*. This activates cyclic nucleotide phosphodiesterase, and is inhibited by calmodulin inhibitors, Ca^{2+} binding being confirmed by ^{45}Ca. No genuine EF-hand proteins were detected in the first Archaeal genome sequences available. A Na^+/Ca^{2+} exchanger similar to NCX in eukaryotic cells has been found in the archaean *Methanococcus jannaschii*, designated as NCX-Mj. This has ten transmembrane helical domains, with four cation-binding sites at its centre, one for Ca^{2+}, and three for Na^+, fitting the three Na^+ for one Ca^{2+} exchange necessary for NCX. The 3D structure also has channels for ion movement. NCX-Mj is a member of the super CaCA family, exhibiting sequence similarities between eukaryotes, bacteria and archaeans.

The red archaean *Haloferax volcanii* (originally *Halobacterium volcanii*) was discovered by the French microbiologist Benjamin Volcani (1915–1999) and is a halophile, living in high salt, such as in the Dead Sea and the Camargue. *Haloferax volcanii* is the cause of the pink colour in flamingos. In the laboratory it is grown in a medium containing 3M NaCl, and astonishingly has an estimated intracellular K^+ of some 3–4 M. By expressing aequorin, it has been possible for the first time to measure free Ca^{2+} inside a live archaean (Figure 8.10). As with all cells, *Haloferax volcanii* maintains a cytosolic free Ca^{2+} in the micromolar to submicromolar range. *Haloferax* produces PHB, which now needs investigating as a potential Ca^{2+} channel, and can also produce methylglyoxal. If a role for intracellular Ca^{2+} can be established in the methanogens, then this has major therapeutic potential as a drug target. Nothing is known about the pharmacology of Archaea. They are not affected by antibiotics.

8.9 Intracellular Ca²⁺ and Viruses

8.9.1 Eukaryotic Viruses

Intracellular Ca^{2+} plays a role in the infection, replication, assembly and disassembly of several animal and plant viruses. Good evidence has been found in measles, HIV, Herpes, enteroviruses, coxsackievirus B, Epstein–Barr virus (EBV), hepatitis B, rotavirus, Dengue virus, papillomavirus (HPV), cytomegalovirus, pseudorabies virus and several invertebrate viruses. Viruses that infect Archaea, and bacteriophages that infect bacteria, can be affected by Ca^{2+}. Eukaryotic viruses affect cell metabolism, and can exploit, hijack, or disrupt the Ca^{2+} signalling system in order to take over the biochemistry of a cell, and replicate. Many viruses interfere with normal Ca^{2+} signalling, involving effects on Ca^{2+} channels in the plasma membrane, and Ca^{2+} movement in the ER and mitochondria. Viruses can induce cytosolic free Ca^{2+} signals through SOCE, voltage-gated Ca^{2+} channels, and metabotropic Ca^{2+} channels. They can also form ion channels and pores from viroporin proteins. Viruses can also affect gene expression and replication through activation of Ca^{2+}-calmodulin kinases (CaMKs) and calreticulin. The alteration in Ca^{2+} signalling either enhances viral replication, or prevents the cell going into apoptosis, which would otherwise leave the virus stranded in mid-stream. Intracellular Ca^{2+} also plays a crucial role in the immune system, in a viral infection.

Patch clamping of cells infected with measles virus showed reduced voltage-gated Ca^{2+} channel activity and increased metabotropic glutamate currents. These effects on Ca^{2+} signalling are important in encephalitis, and other perturbations of the central nervous system, a major concern in measles virus infection. Intracellular Ca^{2+} also plays an important role in HIV infection (AIDS). HIV alters voltage-dependent Ca^{2+} channels, glutamate receptor channels and membrane transporters, disrupting Ca^{2+} signalling in the ER and plasma membrane. Two HIV-1 proteins are responsible for this, gp120, a viral coat protein, and Tat, a transcription regulator. Both are toxic to neurones. Ca^{2+} overload induced by HIV leads to disruption of mitochondria and oxygen metabolite production. These play a key role in dementia and encephalitis induced by HIV infection. Cdk9 T-loop phosphorylation regulates Ca^{2+}-calmodulin-dependent protein kinase ID (CaMKID), and is involved in the activation of Tat. Extracellular Ca^{2+} can also interact with proteins such as neuraminidase on the surface of viruses.

Herpes simplex activates a rapid cytosolic Ca^{2+} transient through the production of IP_3. This is necessary for the activation of focal adhesion kinase, which occurs within 5 min of infection. Similarly, coxsackievirus B activates phospholipase C (PLC) in endothelial cells, IP_3 releasing Ca^{2+} from the ER, thereby activating SOCE. The rise in cytosolic free Ca^{2+} activates the protease calpain 2, which is required for correct trafficking of the virus in the cell. Epstein–Barr virus (EBV) immortalises lymphocytes, altering Ca^{2+} signalling by increasing ER Ca^{2+}, affecting the resting cytosolic free Ca^{2+} and SOCE. LMP-1 is a key EBV protein in this process, increasing SOCE through expression of STIM and Orai. EBV also reduces release of cytochrome c from mitochondria, by interfering with the voltage-dependent anion protein 1. This reduces apoptosis, important in cancer. Enteroviruses also have an anti-apoptotic mechanism involving the protein enterovirus 2B, which can form pores in the ER and Golgi membranes, causing Ca^{2+} to leak out and disturbing the exchange of Ca^{2+} between the ER and mitochondria.

Hepatitis B increases cytosolic free Ca^{2+} via SOCE, through the protein HBVx, which increases expression of ER calreticulin. This increases virus replication by inhibiting the pathway induced by interferon (IFN)-α, through a reduction of the translocation of its IFN-regulatory factors, using the JAK–STAT signalling pathway. Calreticulin also reduces the antiviral effect of IFN by inhibiting, and reducing expression of, protein kinase R, 2′,5′-oligoadenylate synthetase and STAT1 phosphorylation.

Viruses can also initiate autophagy via Ca^{2+} signals. For example, rotavirus-induced cytosolic Ca^{2+} signals result in the hijacking of membranes involved in autophagy. This enables the viral proteins to be moved to the site of virus replication. The modification of Ca^{2+} signalling by rotavirus involves the protein NSP4 (non-structural protein 4), which activates the SOCE pathway, and the Na^+/Ca^{2+} exchanger NCX in the reverse of its usual action, i.e. it lets Ca^{2+} in. NSP4 is a glycoprotein that traverses the ER membrane and increases cytosolic free Ca^{2+} via a mechanism independent of PLC activation. The rise in cytosolic free Ca^{2+} induced by rotavirus is essential for effective viral replication, assembly and release. Similarly, prevention of apoptosis favours the replication of Dengue virus. This virus has a capsid protein (DENV C) that binds the protein cyclophilin-binding ligand. This alters Ca^{2+} signalling in the host cell. It is this binding that prevents apoptosis, through effects on mitochondrial membrane potential and prevention of caspase-3 activation.

Activation of calmodulin by Ca^{2+} plays a role in the replication of several viruses. For example, human papillomavirus (HPV) produces a protein HPV-E2 that binds nuclear receptor interaction protein (NRIP). NRIP has an IQ domain that binds Ca^{2+}-calmodulin, the NRIP–Ca^{2+}-calmodulin complex activating the phosphatase calcineurin. This stabilises protein E2 by dephosphorylating it, reducing ubiquitination. Furthermore, infection of cells by human cytomegalovirus causes a rise in cytosolic free Ca^{2+}, having major effects on intermediary metabolism, including glycolysis, the citric acid cycle, fatty acid and nucleotide synthesis.

A key target is CaMK kinase 1 (CaMKK1), which is important for viral replication and activation of glycolysis, a process not found with many other viruses. CaMKII contributes to pelvic pain in neurogenic cystitis induced by the Bartha's strain of pseudorabies virus.

Ca^{2+} also plays a role in infection by the invertebrate viruses, such as pathogenic viruses in mosquito, the occluded viruses baculoviruses and cypoviruses, and the non-occluded viruses densoviruses and the iridoviruses. Transmission orally of these viruses is inhibited by Ca^{2+}, but enhanced by Mg^{2+}. In plants, viruses induce cytosolic free Ca^{2+} signals as an important part of the viral response. Virus-infected plants regulate cytosolic free Ca^{2+} via Ca^{2+}-MgATPases in a way that prevents Ca^{2+} overload and damage to chloroplasts, thereby reducing cell damage induced by oxygen metabolites.

Intracellular Ca^{2+} is also involved in cell fusion provoked by certain viruses. Thus, many viruses induce Ca^{2+} signals or disturb Ca^{2+} signalling – crucial to viral replication and the prevention of cell death by apoptosis of infected cells. Ca^{2+} signalling is a good target for drug discovery in the treatment of viral infections.

8.9.2 Bacterial Viruses – Bacteriophages

Ca^{2+} is involved in the infection of bacteria by several bacteriophages, through binding to extracellular sites, and not through intracellular Ca^{2+} signals. It has been known since the mid-1920s that Ca^{2+} in the medium can influence the ability of a phage to infect and replicate inside the bacterium, and to kill it. Our gut bacteria are full of them, as are the intestines of many animals; 70% of marine bacteria are infected by phage. Infection of a bacterial host involves attachment, injection of the DNA or RNA into the cell, expression of the bacteriophage genes, then lysis to release new phage, and consequent death of the bacterium. Plaque formation, an indicator of bacterial lysis, caused by phage φμ4, requires external Ca^{2+} in the medium, and is inhibited by citrate, a Ca^{2+} chelator. Similarly, divalent cations, particularly Ca^{2+} at around 5–10 mM, are required for streptococcal phage survival in milk. T5 replication in *E. coli* requires at least 0.1 mM Ca^{2+} in the medium. Ca^{2+} and ATP have been shown to bind to the mycobacteriophage, each phage binding some 118–148 ATP molecules and about 3000 Ca^{2+} ions. When Ca^{2+} is required by a phage for attachment, removal of Ca^{2+} can irreversibly inactivate the phage. Thus, requirement for Ca^{2+} in phage replication can be for:

1. Attachment of the phage to the cell.
2. Phage penetration.
3. Synthesis, once the nucleic acid is inside the bacterium.

Only in the latter could there be a role for intracellular Ca^{2+}. For example, formation of Ph1 phage in *Lactobacillus casei* requires Ca^{2+}, yet the adsorption phase does not. In this case, Ca^{2+} is required for penetration of the phage into the bacterium. Host cells with adsorbed phage in the absence of Ca^{2+} show no penetration and are not killed. Once again, the key experiment of measuring the effect of a bacteriophage on cytosolic free Ca^{2+} in a bacterial host has not been carried out.

8.10 Intracellular Ca²⁺ and Eukaryotic Microbes

Changes in intracellular Ca^{2+} play a key role in the life of many eukaryotic microbes. A rise in cytosolic free Ca^{2+} can cause:

- A change of movement, such as the internal signal for chemotaxis.
- Changes in circadian rhythm.
- Effects on cell division.

- Effects on reproduction.
- The formation and germination of spores.
- Control of genes that enable a cell to adapt to changes in environmental conditions, such as nutrient supply and starvation, salt, pH, temperature, drought and light.

Intracellular Ca^{2+} has been particularly well studied in the ciliates *Paramecium* and *Tetrahymena*, the yeasts *Saccharomyces* and *Candida*, the slime moulds *Physarum* and *Dictyostelium*, the pathogen *Trypanosoma*, as well as the unicellular green alga *Chlamydomonas*, the red bread mould *Neurospora*, the unicellular protozoan/alga *Euglena*, the giant unicellular alga *Acetabularia* and the amoeboflagellate *Naegleria*. Trypanosomes played an important role in the discovery of the mitochondrial Ca^{2+} uniporter (MCU). All eukaryotic microbes contain the machinery for Ca^{2+}-calmodulin- and calcineurin-activated pathways, as well as Ca^{2+} importers and Ca^{2+} efflux mechanisms. But many eukaryotic microbes also contain special organelles associated with the regulation of intracellular Ca^{2+}, such as the vacuole and calcisome (acidocalcisome).

8.10.1   Yeast

Yeasts are unicellular organisms that are widespread in nature, and play a major role in our lives. As result of protein and DNA sequencing, yeasts are now regarded as fungi. They are crucial in the production of bread and alcoholic drinks. Intracellular Ca^{2+} plays an important role in many aspects of yeast physiology. Measurement of cytosolic free Ca^{2+} in yeast, using recombinant aequorin, and fluors such as fura-2 and fluo-3, has established that yeasts:

1. Maintain a resting Ca^{2+} concentration around 50–200 nM.
2. Generate cytosolic free Ca^{2+} transients when stimulated by nutrients, such as glucose and galactose.
3. Adapt to changing environmental conditions, including temperature, nutrients, salts, lack of oxygen and drought, through changes in cytosolic free Ca^{2+}.
4. Regulate their cell cycle through changes in cytosolic free Ca^{2+}.
5. Exhibit a stress response when Ca^{2+} is dumped out of the ER.

Yeasts have played an important role in unravelling parts of the Ca^{2+} signalling system:

- The Golgi MgATPase (SPCA), which pumps Ca^{2+} and Mn^{2+} into the Golgi, was discovered in yeast. Ca^{2+} and Mn^{2+} are necessary for maximum activity of several Golgi enzymes, and Ca^{2+} is useful for secretory granule formation.
- The ER stress mechanism was discovered in yeast, involving the sensor Ire1p and the unfolded protein response, activated by prolonged loss of Ca^{2+} from the ER.
- Exploitation of yeast genetics by producing mutants, knock-outs and the expression of Ca^{2+} signalling components from other organisms has provided important mechanistic information about Ca^{2+} signalling.

Yeasts divide in one of two ways: budding or fission. Their DNA is 50% similar to our own. Thus, yeasts have long been used as model systems to work out conserved molecular mechanisms and biochemical pathways. The most popular yeasts for experimental study have been *Saccharomyces cerevisiae* and *Schizosaccharomyces pombe*, and *Candida albicans*, the cause of thrush, the Ca^{2+}-activated phosphatase calcineurin being an important intracellular Ca^{2+} target.

Yeasts exhibit several similarities in Ca^{2+} signalling components compared with multicellular eukaryotes. But there are also important differences. For example, in ER Ca^{2+} uptake and release, and the SPCA Golgi Ca^{2+}/Mn^{2+} pump. As in animal cells, IP_3 plays a role in generating cytosolic Ca^{2+} signals. But IP_3 generated by activation of PLC, rather than working via the ER IP_3 receptor, is rapidly converted into IP_4 and IP_5 by a dual kinase, and then into IP_6, which activates ER receptors to release Ca^{2+} into the cytosol. This mechanism plays a role in growth of yeast on glucose. The cytosolic free Ca^{2+} transient induced by glucose comes mainly from Ca^{2+} influx, and requires the proteins Plc1, the Gpr1/Gpa2 G-protein complex and hexokinase. Yeast mitochondria contain small Ca^{2+}-binding mitochondrial carriers (SCaMCs), such as the protein SAL1 as a Ca^{2+}-dependent carrier of MgATP, MgADP and phosphate across the mitochondrial membrane. A neat feature of the evidence for this has been the use of firefly luciferase targeted to the mitochondria, enabling intra-mitochondrial ATP to be monitored in intact cells. Interestingly, mitochondrial membranes contain the $PHB–PP–Ca^{2+}$, a putative Ca^{2+} channel in bacteria, and, in human cell lines, engineered reduction of mitochondrial PPi affects mitochondrial bioenergetics and transport.

Unlike animal cells, yeasts have a large intracellular vacuole, which has a Ca^{2+}/H^+ exchanger, Vcx1p/Ha1p and Pmc1p. This stores a large amount of Ca^{2+}, in addition to that in the ER. These two stores make the total cell Ca^{2+} around 2 mmol kg^{-1} cell water, 90% of which can be in the vacuole, where it complexes with PP_i. The Ca^{2+} pump Vcx1p has a high Ca^{2+}capacity, but low affinity, causing the vacuole to take up Ca^{2+} rapidly. This attenuates Ca^{2+} signals, when there is a sudden rise in cytosolic free Ca^{2+}. Pmc1p is less efficient. A P5A-type MgATPase, Cta4, plays a role in Ca^{2+} uptake into the ER of fission yeast, whereas Pmr1 is a Ca^{2+}-MgATPase in the Golgi. Null mutants showed only a 50% loss of Ca^{2+} uptake into isolated Golgi, indicating there is another Golgi Ca^{2+} pump. A third Ca^{2+}-MgATPase exists in the ER. Disruption of Pmr1 has major effects on the activity of Golgi enzymes, such as NADPH cytochrome *c* reductase, dolichyl phosphate mannose synthase, a membrane Ca^{2+}-MgATPase, and on proteins in the secretory pathway. Ca^{2+} in the Golgi is important for correct protein sorting and secretion.

Glucose produces a rise in yeast cytosolic free Ca^{2+} through glucose metabolism:

1. Glucose enters the cell via the plasma membrane glucose transporter.
2. Glucose is then phosphorylated to glucose 6-phosphate by hexokinase, and converted into glucose 1-phosphate by phosphoglucomutase.
3. Glucose 1-phosphate activates PLC, producing IP_3 and DAG from membrane phosphatidyl inositol 4,5-bisphosphate (PIP_2).
4. IP_3 open a plasma membrane Ca^{2+} channel, which lets Ca^{2+} into the cell.
5. IP_3 is then phosphorylated to IP_6, which releases Ca^{2+} from the ER.

The Ca^{2+} channel is sensitive to nifedipine and gadolinium, and may be a SOCE-type mechanism. Sal1p is a target for the glucose-induced Ca^{2+} signal, and is essential in cells where transport of ATP in mitochondria is compromised. The cytosolic free Ca^{2+} transient generated by glucose activates the H^+-MgATPase in the vacuole membrane, causing extrusion of protons.

The Ca^{2+}-calmodulin-activated protein phosphatase calcineurin plays an important role in Ca^{2+} signalling in yeast. This activates trans factors Crz1 and 2, which cause Ca^{2+} to be pumped into the vacuole, and activates the NFATc transcription pathway. Cell division requires two established signalling pathways: TORC1 and PKA, both of which are activated by intracellular signals. Rises in cytosolic free Ca^{2+} are involved in both. TORC1 is activated by amino acids and PKA by glucose, which generates cytosolic free Ca^{2+} transients. If little glucose is available, then the expression of many genes is changed, via transcription factor activation. For example, the high-affinity glucose transporter in yeast, HXT2, is only expressed when glucose is limiting. Induction of HXT2 occurs rapidly on alkinisation of the medium, which requires both the pathways activated by Snf1 and calcineurin. Calcineurin activates transcription factor Crz1, which moves to the nucleus when the cell is stressed by a change in pH. Calcineurin activation of transcription regulates carbohydrate gene expression. Calcineurin also plays a key role in adaption to environmental stresses, such as high salt and high temperature. Activated calcineurin dephosphorylates the zinc finger transcription factor Crz1p/Tcn1p, causing it to concentrate in the nucleus. On the other hand, PKA stimulates phosphorylation of Crz1p, cyclic AMP acting in competition with Ca^{2+}. However, reports of Ca^{2+} affecting cytosolic or nuclear yeast proteins at mM concentrations must be ignored. Yeasts grow better at acid pH, than neutral or alkaline pH. An alkaline medium activates a stress response through a rise in cytosolic free Ca^{2+} and activation of the calcineurin pathway.

Homologues to animal transient receptor potential (TRP) channels have been identified in yeast, and may play a role in increasing cytosolic free Ca^{2+}, for example, the vacuolar membrane protein homologue Yvc1p. Interestingly, intracellular Ca^{2+} has played a role in understanding an important problem in industry, because of the need to be able to freeze *Saccharomyces cerevisiae* in dough, and then reactivate it. Damage during freezing is caused by osmotic stress. Overexpression of the *CRZ1/TCN1* gene, whose product is the target for calcineurin, improved tolerance to both salt and freezing.

8.10.2 *Paramecium* and Related Ciliates

Paramecium is a unicellular ciliated eukaryotic microbe whose cilia are regulated by changes in cytosolic free Ca^{2+}. *Paramecium* is common in many lakes, ponds and other types of fresh water, but species are also found in brackish and marine environments. They feed on microbes such as bacteria, algae and yeasts, which they engulf through a mouth-like structure leading to an intracellular vacuole, which gets rid of water and unwanted ions, including Ca^{2+}. They buzz around with the aid of their cilia, which can number as many as 15 000. Two processes are triggered by a rise in cytosolic free Ca^{2+}:

- Reversal of ciliate movement.
- Trichocyst secretion.

Intracellular Ca^{2+} may also regulate:

- Phagocytosis for digestion.
- Osmoregulation.
- Gene expression.
- The cell cycle.

A key primary stimulus is touch. But also changes in osmolarity and light regulate the behaviour of cells. *Paramecium* contains all the necessary molecular machinery for Ca^{2+} signalling:

- Internal Ca^{2+} stores.
- Ca^{2+} release channels.
- SOCE and voltage-gated Ca^{2+} channels in the plasma membrane.
- The SNARE fusion complex, required for the docking and fusion of the vacuole with the plasma membrane.
- Ca^{2+} target proteins, such as calmodulin, Ca^{2+}-activated kinases, and the phosphatase calcineurin.

The *Paramecium tetraurelia* genome contains 34 distinct Ca^{2+} channels, divided into six subfamilies, based on domain structure, pore type, selectivity filter (e.g. amino acid sequences GVGD and GIGD), and how they are activated and opened. There are IP_3 receptors, and some receptors related to ryanodine receptors. In addition, the SOCE mechanism causes a large rise in cytosolic free Ca^{2+}. All these Ca^{2+} signalling components play a role in the reversal of cilia and exocytosis.

When a *Paramecium* hits an object, such as a stone, the cilia reverse the direction of their beat, so that the cell can reverse and swim away. Touch provokes an action potential in the cilia membrane, due to voltage-gated Ca^{2+} channels, each having a conductance of about 0.1 pS. Mutants with no voltage-gated Ca^{2+}-dependent depolarisation are unable to reverse. The inward current in normal cilia is not blocked by the Na^+ channel blocker tetrodotoxin. But interestingly it is not easy to block it with Mn^{2+}, Co^{2+}, La^{3+} or the verapamil derivative D-600, which block many other types of voltage-gated Ca^{2+} channel. Ca^{2+} enters the cilia, resulting in a local rise in free Ca^{2+}, which causes Ca^{2+} channel inactivation. The Ca^{2+} current is sufficient to produce a rise in free Ca^{2+} within each cilium of about 10 µM. This rise was confirmed using aequorin injected into a single cell, and is required for ciliary reversal to take place. At a cytosolic free Ca^{2+} of 10 nM to 1 µM the cell moved forwards, but when the cytosolic free Ca^{2+} was raised to between 1 and 100 µM the *Paramecium* moved backwards. But the molecular mechanism causing this reversal is not clear. The volume of a cilium means that each only has a handful of individual Ca^{2+} ions. Interestingly, changes in free Ca^{2+} in animal cilia have not been detected.

A global rise in cytosolic free Ca^{2+} is the signal for cortical morphogenesis and exocytosis, the primary stimulus producing a wave of free Ca^{2+} in the cytosol. *Paramecium* contains dense secretory vesicles called trichocysts. These are also found in dinoflagellates. The trichocyst contains a hair-like structure that is secreted, and plays a role in defence and obtaining food. *Paramecium* contains several thousand trichocysts, which secrete their long thin threads simultaneously, triggered by a rise in cytosolic free Ca^{2+}. Exocytosis in other protozoa is also triggered by a rise in cytosolic free Ca^{2+}. For example, in the pathogen *Toxoplasma gondii*, exocytosis of the special organelle, the microneme, involves activation of a Ca^{2+}-dependent kinase, TgCDPK1. This is not found in mammals and thus is a good target for drug discovery. The release of the microneme is required for motility, invasion of the host and egress, which are lost when TgCDPK1 is suppressed. Another related ciliate to *Paramecium* is the kidney-shaped *Colpoda*, often found in hay infusion. It divides by forming cysts, which release between two and eight individual cells. The signal for encystment is a rise in cytosolic free Ca^{2+}, which activates appropriate kinases. Some eukaryotic microbes have flagella, which are larger than cilia, and are controlled by the protein dynein. The beat of *Chlamydomonas* flagella is severely reduced by removal of extracellular Ca^{2+}, and by Ca^{2+} channel- and mechanosensitive channel-blockers. Mechanical stimulation of the cell increases Ca^{2+} influx and beat frequency, Ca^{2+} activating the dynein arm of the flagellum.

8.10.3 Slime Moulds

Slime mould is a broad term for organisms that appear as a gelatinous slime, and produce spores to reproduce. They are distinct from fungi, and occur in a variety of colours. Two slime moulds have been extensively studied, where intracellular Ca^{2+} plays a key role in their physiology: *Physarum* and *Dictyostelium*. When there is plenty of food, slime moulds exist as single cells. But lack of nutrients causes them to aggregate and form fruiting bodies, which release a large number of spores. The single cells are amoeboid and haploid. They feed on phagocytosed bacteria and other waste material. Mating with another cell enables them to grow into plasmodia. Four physiological processes involve intracellular Ca^{2+}:

- Chemotaxis.
- Differentiation.
- Cytoplasmic streaming.
- Phagocytosis.

Cytosolic free Ca^{2+} transients are required for chemotaxis, and differentiation involves Ca^{2+} oscillations. These are regulated by two internal stores:

- The IP_3-sensitive ER.
- Acidocalcisomes.

Ca^{2+} stores, Ca^{2+} channels, IP_3 receptors, Ca^{2+} pumps, calmodulin, calcineurin, penta EF-hand Ca^{2+}-binding proteins, and all other components of the Ca^{2+} signalling machinery are found in slime moulds. Novel EF-hand Ca^{2+}-binding proteins (CBP5–9) are encoded in the *Dictyostelium* genome. But slime moulds contain a special, vesicular Ca^{2+} store: the acidosome or acidocalcisome. This contains an H^+/Ca^{2+} antiporter that concentrates Ca^{2+} in the acidosome. Cytosolic free Ca^{2+} increases markedly at the aggregation stage, dropping at the migration stage, but rising again at the culmination stage. At all stages, except the vegetative stage, the chemotactic stimulus cyclic AMP (50 nM) provokes transients in cytosolic free Ca^{2+}.

Physarum and *Dictyostelium* also have a vegetative structure called a plasmodium, enveloped by an outer membrane, and containing hundreds of nuclei. The fan-shaped plasmodium flows out in an amoeboid fashion, engulfing microbes and decaying plant material. This occurs spontaneously, but can also be activated by chemotactic stimuli. Cyclical changes in cytosolic free Ca^{2+}, measured with injected aequorin, correlate with the polarity and activation of cytoplasmic streaming. Injection of EGTA inhibits this, the active range of cytosolic free Ca^{2+} being 0.5–10 μM. Amoeboid movement occurs through the production of pseudopods from the plasma membrane. Ca^{2+} activates gelsolin and calmodulin, causing a gel–sol conversion in the actomyosin microfilaments through the transition of helical, filamentous F-actin to globular G-actin. This gel–sol conversion may also aid movement and fusion of intracellular vesicles with the plasma membrane. Myosin and tropomyosin bind to F-actin, but not G-actin, the myosin MgATPase being activated when it binds. The activation of cytoplasmic movement by a rise in cytosolic free Ca^{2+} occurs in amoebae, slime moulds, foraminifera, mammalian macrophages and neutrophils, mammalian and invertebrate nerve, fertilised eggs, and the giant plant cell *Cara*.

The cell differentiation factor (DIF) for stalk formation in slime moulds works by raising cytosolic free Ca^{2+}. Germination of spores from slime moulds requires autoactivation, swelling of the spore, followed by emergence of the amoeboid form. These require Ca^{2+}-calmodulin to activate Ca^{2+} efflux, analogous to the germination of bacterial spores. This uptake reverses in the late event of amoeboid emergence, resulting in Ca^{2+} uptake by the cell. Novel EF-hand Ca^{2+}-binding proteins, such as CAF-1, have been shown to be involved in the differentiation process, the gene being activated by starvation. Cytosolic free Ca^{2+} has been measured in the slime moulds *Dictyostelium* and *Physarum*, using aequorin, fura-2 and indo-1, showing that the cytosolic free Ca^{2+} is different in prestalk and prespore cells. Manipulation of cytosolic free Ca^{2+}, using EGTA or ionophore A23187, has predictable effects on morphology, differentiation and shape.

Extracellular cyclic AMP provokes cytosolic free Ca^{2+} transients in *Dictyostelium discoideum*, regulating gene expression, through Ca^{2+}-calmodulin. The calmodulin-binding protein nucleomorphin, NumA1, is located in intranuclear patches, next to the nuclear envelope and in the nucleolus. Ca^{2+} influx provoked by cyclic AMP is mediated by non-esterified fatty acids generated as the result of activation of phospholipase A2, addition of arachidonic or linoleic acid provoking a cytosolic free Ca^{2+} transient. The *number* of cells which produce a Ca^{2+} transient rise as the fatty acid concentration increases, the level of the Ca^{2+} rise also being dependent on stimulus concentration. The Ca^{2+} rise is via a SOCE mechanism, requiring initial release of Ca^{2+} from internal stores, both ER and acidosomes. Gradients of extracellular cyclic AMP or Ca^{2+} act as chemoattractants, the two being stronger together than on their own. There are also free Ca^{2+} gradients and gradients of vesicular store Ca^{2+} along the differentiating slime mould. Cyclic GMP inside the cell may also interact with the Ca^{2+} signalling machinery. Sporulation is triggered by Ca^{2+} and malate.

8.10.4 Luminous Radiolarians

Radiolarians are protozoans, which can be large, and even multinucleate. Several are biolumi-nescent, as the result of Ca^{2+}-activated photoproteins similar to those in luminous hydroids and jellyfish, for example *Thalassicolla* in the surface sea water off the west coast of Ireland. *Thalassicolla* cells are several millimetres across, and look like frog spawn (Figure 7.14A, part l). Shaking the cells produces blue flashes of light, as a result of mechanosensitive cation channels in the plasma membrane that let Ca^{2+} into the cell from the surrounding sea water. However, internal Ca^{2+} store release cannot be ruled out. Interestingly, there can be some four times as much apophotoprotein as active photoprotein, suggesting that coelenterazine luciferin is limiting.

8.11 Summary

All microbes – Bacteria, Archaea or Eukaryota – maintain a free Ca^{2+} inside in the sub-micro-molar to micromolar range, even when the extracellular Ca^{2+} is 1–10 mM. Thus, there have to be mechanisms for Ca^{2+} influx and efflux. Many experiments implicate Ca^{2+} in the multiplica-tion or activity of bacterial and archaeal microbes, and viruses. Only in some is the evidence convincing that changes in cytosolic free Ca^{2+} act as the signal (Box 8.1).

Box 8.1 Where the best evidence is for a role of intracellular Ca²⁺ in regulating microbes.

- Bacteria – chemotaxis, gene expression, growth, spore formation.
- Yeast – growth.
- Moulds – growth and differentiation.
- Protozoa – ciliate movement.
- Archaea – methanogenesis.
- Viruses – replication and defence.

Investigation of the role of intracellular Ca^{2+} in bacteria was hampered for many years by the assumption Ca^{2+} was simply toxic in the cell. So all bacteria had to do was to get rid of it. Iden-tification of Ca^{2+} influx triggered by external stimuli, the effect of this on gene expression and growth, and the need for ATP in Ca^{2+} efflux, argue for an active role for Ca^{2+} in bacteria. But, the key experiment of correlating measurement of cytosolic free Ca^{2+} with changes in a physi-ological process in a bacteria and archaeans has been carried out in just a few cases. Cytosolic free Ca^{2+} is best measured using a Ca^{2+}-activated photoprotein, such as aequorin or obelin, so that changes in free Ca^{2+} inside the microbe can be correlated with gene expression, cell growth and replication. The small molecular fluors, which have been such a brilliant success in eukary-otes, are not suitable for measuring free Ca^{2+} inside bacteria. A major step forward would be a method for measuring and imaging free Ca^{2+} and ATP in individual microbes.

Bacteria and Archaea are very small. So to cause a large change in cytosolic free Ca^{2+} only a small number of Ca^{2+} ions, in the tens or hundreds, have to move across the outer membrane.

This is much smaller than the tens of thousands necessary in eukaryotes, although organelles such as mitochondria are also similarly small. Thus, the Ca^{2+} influx and efflux mechanisms in microbes are likely to be different from those in eukaryotes. If a role for intracellular Ca^{2+} can be firmly established in a bacterial or archaeal species, then this would be a novel target for drug discovery, independent of antibiotics.

Sequence similarities have been fraught with problems, although Ca^{2+} exchangers in the CaCA family may be important. High ångström resolution crystal structures have revealed the potential molecular basis of Ca^{2+}/Na^+ and Ca^{2+}/H^+ exchange of NCX-Mj in the archaean *Methanococcus jannaschii*, CAX-Af in the archaean *Archaeoglobus fulgidus*, and Yfke in the Gram-positive bacterium *Bacillus subtilis*. Two of the transmembrane helices in these putative antiporters form a hydrophilic cavity, providing the pathway for exchange. Ca^{2+} binding is typically via glutamate residues. These three-dimensional structures shed light on the evolution of the CaCA family of Ca^{2+} transporters and exchangers. However, what is still required is a correlation of Ca^{2+} exchange with cytosolic free Ca^{2+} measured in the live cell, and its loss in knock-out mutants. Cation exchange in vesicles is encouraging, but not sufficient proof that it occurs in the live cell. Bacterial proteins, 30–50% similar to Ca^{2+} signalling proteins in eukaryotes, have turned out to be red herrings. The failure to follow up potentially important proteins that could regulate Ca^{2+}-dependent processes, with measurements of free Ca^{2+} on live cells, together with molecular genetic and mutational studies, to establish their precise functional role has seriously held back the field. However, there is substantial evidence that a rise in cytosolic free Ca^{2+}, particularly in microdomains, regulates eukaryotic microbes, such as protozoa, yeast, and slime moulds, and viral infection.

A key issue is the electrochemical potential that will drive Ca^{2+} into a microbial cell. This has two components:

- The concentration gradient of Ca^{2+}.
- The potential across the plasma membrane.

In eukaryotes, the membrane potential is typically −20 to −90 mV, negative inside, and is established by the K^+ gradient, maintained through the sodium pump. However, in bacteria and archaeans, under aerobic conditions, the membrane potential is set by a Mitchell-type mechanism through the respiratory chain, and is typically around −200 mV, negative inside. Under anaerobic conditions little is known about the membrane potential in bacteria and archaeans. Functional Ca^{2+} channels have been identified in the plasma membrane of many eukaryotic microbes, but none have been convincingly found in bacteria or archaeans, in spite of the fact that K^+, Na^+ and Cl^- channels have been found in certain species.

There is still much to learn about Ca^{2+} in microbes of all types, and its significance in their evolution. The key Darwin–Wallace question for Bacteria and Archaea is: What is the *selective advantage* of intracellular Ca^{2+} regulation, rather than what is its *function*?

Recommended Reading

*A must.

Books

*Campbell, A.K. (2015) Intracellular Calcium, Chichester: John Wiley & Sons Ltd.. Chapter 8, pp 389–442. Intracellular Ca^{2+} and microorganisms. Fully referenced article with the evidence for a role of intracellular Ca^{2+} in bacteria, archaens, viruses and eukaryotic microbes.

Reviews

*Campbell, A.K., Matthews, S., Vassel, N., Cox, C., Naseem, R., Chaichi, J., Holland, I.B., Green, J. & Wann, K. (2010) Bacterial metabolic 'toxins': A new mechanism for lactose and food intolerance, and irritable bowel syndrome. *Toxicology*, **278**, 268–276. How metabolites produced by gut bacteria interact with host cell Ca²⁺signalling.

Dominguez, D.C. (2004) Calcium signalling in bacteria. *Mol. Microbiol.*, **54**, 291–297. General review of how Ca²⁺ is regulated and acts inside bacteria.

*Higgins, D. & Dworkin, J. (2012) Recent progress in *Bacillus subtilis* sporulation. *FEMS Microbiol. Rev.*, **36**, 131–148. How intracellular Ca²⁺ is crucial to spore formation and fertilisation.

Hyser, J.M. & Estes, M.K. (2015) Pathophysiological consequences of calcium-conducting viroporins. *Annu. Rev. Virol.*, **2**, 473–496. How intracellular Ca²⁺ is involved in viral infection.

*Reusch, R.N. (2012) Physiological importance of poly-(R)-3-hydroxybutyrates. *Chem. Biodivers.*, **9**, 2343–2366. Evidence for the role of a non-proteinaceous channel as a Ca²⁺ in bacteria.

Shemarova, I.V. & Nesterov, V.P. (2014) Ca²⁺ signaling in prokaryotes. *Microbiology*, **83**, 431–437. Review of the evidence for intracellular Ca²⁺ acting as signal in bacteria.

*Zhou, Y., Frey, T.K. & Yang, J.J. (2009) Viral calciomics: Interplays between Ca²⁺ and virus. *Cell Calcium*, **46**, 1–17. How intracellular Ca²⁺ interacts with viral infection in cells.

Research Papers

*Campbell, A.K., Naseem, R., Holland, I.B., Matthews, S.B. & Wann, K.T. (2007) Methylglyoxal and other carbohydrate metabolites induce lanthanum-sensitive Ca²⁺ transients and inhibit growth in E. coli. *Arch. Biochem. Biophys.*, **468**, 107–113. First evidence for Ca²⁺ transients in bacteria caused by various bacterial metabolites.

Cheshenko, N., Del Rosario, B., Woda, C., Marcellino, D., Satlin, L. M. & Herold, B.C. (2003) Herpes simplex virus triggers activation of calcium-signaling pathways. *J. Cell Biol.*, **163**, 283–293. Ca²⁺ signalling pathways activated by a human virus.

Cubitt, A.B., Firtel, R.A., Fischer, G., Jaffe, L.F. & Miller, A.L. (1995) Patterns of free calcium in multicellular stages of *Dictyostelium* expressing jellyfish apoaequorin. *Development*, **121**, 2291–2301. Free Ca²⁺ changes in a slime mould.

Jones, H.E., Holland, I.B. & Campbell, A.K. (2002) Direct measurement of free Ca²⁺ shows different regulation of Ca²⁺ between the periplasm and the cytosol of *Escherichia coli*. *Cell Calcium*, **32**, 183–192. First measurement of free Ca²⁺ in the periplasm of a Gram negative bacterium.

*Knight, M.R., Campbell, A.K., Smith, S.M. & Trewavas, A.J. (1991) Recombinant aequorin as a probe for cytosolic free Ca²⁺ in *Escherichia coli*. *FEBS Lett.*, **282**, 405–408. First genuine measurement of free Ca²⁺ in live bacteria.

Meng, Y.H., Dong, G.R., Zhang, C., Ren, Y.Y., Qu, Y.L. & Chen, W.F. (2016) Calcium regulates glutamate dehydrogenase and poly-gamma-glutamic acid synthesis in *Bacillus natto*. *Biotechnol. Lett.*, **38**, 673–679. Evidence of a role for intracellular Ca²⁺ in a bacterium.

*Naseem, R., Wann, K.T., Holland, I.B. & Campbell, A.K. (2009) ATP regulates calcium efflux and growth in E. coli. *J. Mol. Biol.*, **391**, 42–56. Evidence for a Ca²⁺ pump in E. coli, and clear evidence that three previously proposed Ca²⁺ regulators, ChaA, YrbG and PitB, are not involved, using Keio knock-outs.

*Torrecilla, I., Leganes, F., Bonilla, I. & Fernandez-Pinas, F. (2004) A calcium signal is involved in heterocyst differentiation in the cyanobacterium *Anabaena* sp PCC7120. *Microbiology*, **150**, 3731–3739. Evidence that Ca²⁺ signals trigger differentiation in a cyanobacterium.

Vancek, M., Vidova, M., Majernik, A.I. & Smigan, P. (2006) Methanogenesis is Ca^{2+} dependent in *Methanothermobacter thermautotrophicus* strain deltah. *FEMS Microbiol. Lett.*, **258**, 269–273. Evidence that intracellular Ca^{2+} is involved in methane production by archaeans.

Watkins, N.J., Knight, M.R., Trewavas, A.J. & Campbell, A.K. (1995) Free calcium transients in chemotactic and non-chemotactic strains of *Escherichia coli* determined by using recombinant aequorin. *Biochem. J.*, **306**, 865–869. Evidence of a role for Ca^{2+} in affecting tumbling in a chemotaxic bacterium.

9

Role of Intracellular Ca²⁺ in Plants and Fungi

> **Learning Objectives**
> 1. What plants do, and how this can be triggered by primary stimuli.
> 2. The intracellular Ca²⁺ stores in plants, with two key differences from animals cells – the vacuole and the chloroplast.
> 3. Identification of the components of Ca²⁺ signalling in plants – Ca²⁺ channels, Ca²⁺ stores, release mechanisms and intracellular target proteins.
> 4. Regulation of intracellular Ca²⁺ by the vacuole tonoplast and chloroplast.
> 5. Evidence of a role for intracellular Ca²⁺ by direct measurement of the effects of light, gas exchange, gravity, fertilisation and metabolism.
> 6. The role of intracellular Ca²⁺ in the response of plants to stress and cell death.
> 7. Evidence for a role for intracellular Ca²⁺ in mosses, liverworts and ferns.
> 8. Evidence for a role for intracellular Ca²⁺ in fungi, yeast and lichens.

9.1 Role of Ca²⁺ in Plants

It has been known since the early years of the twentieth century that calcium is essential for the healthy growth of plants, and that plants store large amounts of calcium in the form of calcium oxalate. Every gardener knows that calcium is needed for healthy plant growth, and every vigneron knows that calcareous soils can have a beneficial effect on the growth of vines. Ca²⁺ is absorbed through the roots, typically at the apical tip region, and plays a structural role in the middle lamella of plant cell walls. Two polymers of polygalacturonic acid hold Ca²⁺ in an 'egg box' structure as a major component of pectin in the cell wall. Every cook who makes jellies, jams and chutneys will recognise the effect of Ca²⁺ on pectin, necessary for a firm gel. This can be solubilised from the cell wall of fruit by hot water and/or a Ca²⁺ chelator. But, just as in animals, intracellular Ca²⁺ has a major role to play in the biology of all higher plants. The evidence for this has been based on:

1. Effects of manipulating extracellular or intracellular Ca²⁺ on plant physiology.
2. Measurement of cytosolic free Ca²⁺ in live plants cells, and intact plants, correlating this with the cell or plant event.
3. Identification of the components of the Ca²⁺ signalling system in plants, including Ca²⁺ pumps and exchangers, Ca²⁺ channels, internal Ca²⁺ stores and Ca²⁺-binding proteins.
4. Experiments showing how changes in intracellular Ca²⁺ occur and how they act to cause a physiological event.

Fundamentals of Intracellular Calcium, First Edition. Anthony K. Campbell.
© 2018 John Wiley & Sons Ltd. Published 2018 by John Wiley & Sons Ltd.
Companion Website: http://www.wiley.com/go/campbell/calcium

9.2   What Stimulates Plants?

9.2.1   Plant Cell Stimuli

Plants have many cellular events that are similar to those in animal cells, and which can be triggered by a range of physical, chemical or biological stimuli (Table 9.1). Charles Darwin (1809–1882), carrying out pioneering experiments with his son Francis, set the scene, showing that plants grow towards light (phototropism), particularly blue light. They showed that it was the tip of the coleoptile that had the light sensor, which transmits a message to a site lower down in the plant, where the bending takes place. Darwin wrote several books on plants, including orchids, climbing plants and insectivorous plants. Plants also played an important part in the life of Charles' grandfather, Dr Erasmus Darwin (1731–1802), one of the great geniuses of the eighteenth century, a polymath. In *Phytologia*, Erasmus writes an amazing account of what we now call photosynthesis. The current chapter highlights again how the key to Charles' BIG idea – small variations within and between species – has played a crucial role in the evolution of the Ca^{2+} signalling system in plants, as we have already seen in animals.

Plants grow towards light because cells on the darker side elongate more than those exposed directly to the light. Physical stimuli include light, dark, temperature, gravity and wind, and, like animals, plants have hormones. These all can generate cytosolic free Ca^{2+} signals in particular cell types. And, like animals, intracellular Ca^{2+} plays a key role in fertilisation – seed germination. Plants also have defence mechanisms against stress agents. These utilise intracellular Ca^{2+} signals. Stresses include wind, cold shock, heat, drought, salinity, fungal elicitors and oxidative

Table 9.1 Cellular events in plants signalled by intracellular Ca^{2+}.

Type of event	Example
Movement	Growth affected by gravity
	Flowers closing at night
	Growth affected by wind
	Insectivorous plants
Transpiration/respiration	Stomata and guard cells opening and closing
Secretion	Scent
	Nectar
	Hormones
Metabolism	Photosynthesis
Circadian rhythm	Light/dark cycle
Fertilisation	Seed germination by pollen
	Formation of the pollen tube
Defence	Wind
	Cold shock
	Drought
	High salt
Death	Loss of leaves in autumn
	Development

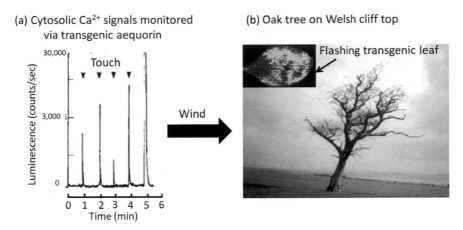

(a) Cytosolic Ca²⁺ signals monitored via transgenic aequorin

(b) Oak tree on Welsh cliff top

Flashing transgenic leaf

Touch

Wind

Figure 9.1 The effect of touch on plant cytosolic free Ca²⁺ and growth. (a) Cytosolic free Ca²⁺ spikes induced by wind in a tobacco seedling made transgenic with the Ca²⁺-activated photoprotein aequorin. Source: Knight *et al.* (1991). Reproduced with permission from Nature. (b) The predicted effect of recurrent cytosolic free Ca²⁺ spikes induced by wind, causing the cells on one side of the oak tree to grow faster than the other, so the tree is 'bent' by the wind.

stress. Furthermore, cell death by apoptosis plays an important role in plants. For example, it is the mechanism responsible for leaves dropping in autumn. Changes in intracellular Ca²⁺ play a role in all of these cell events. A good example is wind. This produces dramatic effects on growth, development, and distribution. Wind can reduce plant size and leaf area, reducing markedly crop yields. However, the shortest plants often have the strongest stems, trees have stronger trunks. The stems, petioles, cuticles and cell walls of plants are thickened by wind. Wind also affects the distribution of cell types, in structures such as stomata, sclerenchyma and root hairs. In Wales, we are familiar with trees and bushes on the cliff tops having been bent by the wind (Figure 9.1). This is a not a mechanical blowing, but rather a signalling mechanism, whereby the cells on one side of the tree or bush grow at a slightly different rate from the other, so that the plant is protected against destruction by the prevailing wind. Wind generates regular intracellular Ca²⁺ signals that switch on the calmodulin gene and defence genes. Intracellular Ca²⁺ also plays a role in many pathological events in plants. However, plants have several features that are not found in animal cells. Plant cells have a wall, and they have two organelles not found in animal cells:

- Chloroplasts, responsible for trapping sunlight.
- The vacuole.

These play an important role in Ca²⁺ signalling.

9.2.2 Plant Hormones

Like animals, plants produce substances released by one tissue that signal reactions in another – hormones. The first plant hormone to be discovered, between 1913 and 1926, was an auxin based on indole acetic acid (IAA). Other auxins were discovered, but IAA is the most abundant. It promotes cell elongation in the coleoptile and stem, increases cell division in callus tissue in the presence of cytokinins, and regulates various growth and development processes, such as roots from detached leaves and stems.

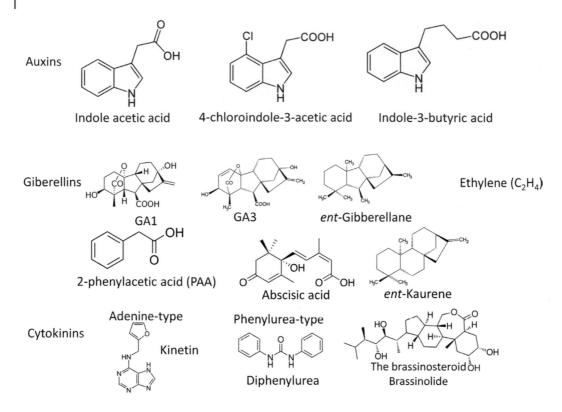

Figure 9.2 Some major plant hormones. Many of these can induce cytosolic free Ca^{2+} signals under particular circumstances.

Six classes of plant hormone are now known (Figure 9.2): auxins, gibberellins, cytokinins, the fruit-ripening hormone ethylene, brassinosteroids and abscisic acid. Several other substances also have hormone-like properties, and include jasmonic acid, salicylic acid, strigolactone, some flavonoids and some peptides. Hormones regulate the physiological processes in plants, and several are involved in the response to stress or in pathological processes. As in animals, extracellular ATP can regulate plant cells. Many processes involve intracellular Ca^{2+}. Such physiological processes include:

1. Cell growth and size, such as stems, leaves and roots.
2. Cell development and differentiation, such as xylem, phloem, guard cells and root hairs.
3. Nodule formation.
4. Seed and pollen formation.
5. Seed germination and pollen tube formation.
6. Flowers growing, and opening and closing.
7. The light/dark cycle, including photosynthesis, metabolism and flower opening and closing.
8. Effects of gravity.
9. Defence against stress.

Ca^{2+} plays a vital role in tip growth and branching of hyphae, dimorphism, the cell cycle, and defence against stress. Cytosolic Ca^{2+} signals can be provoked by auxins, gibberellins and abscisic acid. Other agents that induce Ca^{2+} signals include glutathione and oxygen metabolites.

Model systems have been important in providing the evidence for a role for intracellular Ca^{2+} in plant physiology and pathology. These include:

1. The tobacco plant *Nicotiana plumbaginifolia*.
2. The common weed *Arabidopsis thaliana*.
3. The pea *Pisum*.
4. The unicellular large alga *Chara*.
5. Rice *Oryzias*.
6. Quince for saline-resistant species.
7. The seaweed *Fucus*.
8. The small brown alga *Pelvetia*.

9.2.3 Intracellular Signals in Plants

As with animal cells, plants use several different intracellular messengers to trigger and regulate cellular processes; Ca^{2+}, cyclic AMP, inositol trisphosphate (IP_3) and reactive oxygen species. Plant cells maintain a resting membrane potential higher than animal cells, around 100–200 mV, negative inside. They have Ca^{2+} channels in their plasma membrane that can be voltage-gated, and are regulated by external stimuli and intracellular molecules. Plant cells have IP_3 and ryanodine receptors on the endoplasmic reticulum (ER), which cause Ca^{2+} release into the cytosol, as in animal cells. Plant mitochondria also regulate cytosolic free Ca^{2+} signals. But unique to plants is Ca^{2+} regulation in and by chloroplasts and the vacuole, with its tonoplast membrane. The vacuole can accumulate large amounts of calcium oxalate, and can be a major source for cytosolic free Ca^{2+} signals. On the other hand, movement of Ca^{2+} into chloroplasts can regulate photosynthesis and metabolic processes within this organelle. Ca^{2+}-binding proteins have been found in plants. The ubiquitous calmodulin regulates several processes in plants, including gene expression, while a calcineurin-like phosphatase plays an important role in guard cells. The study of intracellular Ca^{2+} in plants was opened up in the late 1980s and early 1990s by the ability to measure, and image, free Ca^{2+} in the cytosol of live plant cells, intracellular organelles, and even intact plants, using fluorescent dyes and recombinant aequorin. The cytosolic free Ca^{2+} in resting plant cells is submicromolar, as in all animal cells, and cytosolic free Ca^{2+} signals play a key role in triggering responses of plants to external and internal stimuli, or stress.

9.3 Requirement of Plants for Ca²⁺

It has long been known that Ca^{2+} is required for the normal growth and survival of all plants, monocotyledons usually requiring less than dicotyledons. Removal or manipulation of external Ca^{2+} affects many processes in plants (Table 9.2). Ca^{2+} is required for flowering, and for the formation of nodules in leguminous plants when infected by rhizobia. The bacteria in the nodules fix nitrogen, and are thus important in crop rotation. Removal of external Ca^{2+} also affects organ and cell development, the formation of the pollen tube, guard cells and the size of the pore in stomata, abscission, and motility of unicellular algae. Lack of Ca^{2+} also damages cell wall formation, particularly when stimulated by plant hormones such as auxins. This can affect cell elongation, organelle structure, and membrane permeability. Ca^{2+} may also act as an antagonist to toxic metals, such as Al^{3+}, Ni^{2+}, Zn^{2+} and Mn^{2+}, found in the soil, and high Ca^{2+} can counteract the toxic effects of high salinity. The latter is particularly important for plants that grow near the sea and in salt marshes. The key to establishing that Ca^{2+} has a role as an intracellular signal in plant

Table 9.2 Evidence of a role for Ca^{2+} in plants based on the effects of manipulating extracellular Ca^{2+}.

Effect of Ca^{2+}
Growth
Root tip growth and development
Growth and development of *Fucus*
Organelles
Accumulation in isolated chloroplasts
Ca^{2+} accumulation in pollen tips
Phenomena
Induces contraction in green algae
Differentiation of *Nitella*
Auxin-stimulated cell wall deposition
Ca^{2+} transport in aquatic plants
Serine transport in tobacco
Modulation in leguminous plants
Electrical activity in *Nitella*
Inhibits abscission
Inhibits cytoplasmic motility in *Chara*
Enzymes
Ca^{2+}-calmodulin activates NAD kinase in the pea *Pisum*

cells was the measurement of cytosolic free Ca^{2+} in live plant cells, as well as the identification of the proteins responsible for regulating intracellular Ca^{2+} and its intracellular targets.

Ca^{2+} outside plant cells is typically in the range 1–10 mM, whereas cytosolic free Ca^{2+} is in the nano- to micro-molar range. In nature, calcium deficiency is not common. But when it does occur, it leads to stunted growth and structural changes in the leaves. Ca^{2+} is also required for proper production of fruit, such as apples. Severe calcium deficiency can lead to bitter-pit in apples, blossom end rot, and tip burn. Plants require Ca^{2+} during the synthesis of new cell walls, particularly the middle lamellae. Ca^{2+} also plays a role in the mitotic spindle, and is required extracellularly for the normal functioning of the plasma membrane. Ca^{2+} deficiency leads to necrosis in various parts of the plant, particularly young meristems, root tips, and young leaves where cell division is active. The meristem consists of undifferentiated plant cells where growth takes place, and can give rise to various organs, such as roots, leaves and flowers. Ca^{2+} deficiency can also lead to yellowing of leaves, as with nitrogen deficiency. Ca^{2+} exchanges other cations, and is the counterion for several organic and inorganic anions across the plasma membrane or intracellular organelles, such as the vacuole. Intracellular Ca^{2+} can induce movement of chloroplasts.

9.4   Where Ca^{2+} is Stored in Plants

The Ca^{2+} content of plants varies between 0.1 and over 5% of the dry weight. Ca^{2+} is found extra-cellularly in the aplastic space, and bound to pectin in the cell wall, and intracellularly in several organelles. As in animal cells, Ca^{2+} is found in the ER, and can be taken up by mitochondria and chloroplasts. But a major difference from animal cells is the large vacuole that is found in all plant cells. This is typically 30% of the cell volume, but can be as much as 80%. The inside is acid, and contains a large amount of Ca^{2+}, either free or in the form of calcium oxalate. The vacuole is surrounded by a membrane called the tonoplast. This contains Ca^{2+}/H^+ exchangers, and other proteins associated with Ca^{2+} signalling. Estimates for the total Ca^{2+} concentration in the vacuole range from 0.1 to 10 mM. This is maintained by the Ca^{2+}/H^+ exchanger in the tonoplast, which utilises the low pH inside the vacuole. This low pH is established by a V-type H^+ pump and a pyrophosphatase, which also pumps H^+ across the tonoplast membrane. Stimulation of plant cells by physiological stimuli or stress can cause release of Ca^{2+} from the vacuole, leading to a rise in cytosolic free Ca^{2+}. But Ca^{2+} signals can also occur from other plant organelles, such as mitochondria, chloroplasts, the ER and nucleus, as well as from the opening of Ca^{2+} channels in the plasma membrane.

9.5   Measurement of Cytosolic Free Ca²⁺ in Plants

The detection and imaging of cytosolic free Ca^{2+} signals in plants has established beyond doubt that intracellular Ca^{2+} does play a key role as a signal of many cellular processes in plants (Figure 9.3). Plant cells, like animal cells, can exhibit rapid cytosolic Ca^{2+} transients lasting just a few seconds, or produce several spikes, repetitive transients, or oscillations. Some Ca^{2+} signals last minutes or even hours. For example, cytosolic free Ca^{2+} oscillates in guard cells over 30–60 min, and imaging has identified Ca^{2+} tides and waves in plant cells. However, the array of puffs,

Figure 9.3 Examples of free Ca^{2+} signals in live plant cells. (a) Cytosolic free Ca^{2+} spikes in tobacco seedlings (*Nicotiana*) induced by wind, cold shock or a fungal elicitor, measured using transgenic aequorin. Source: Knight *et al.* (1991). Reproduced with permission from Nature. (b) Increase in cytosolic free Ca^{2+} in plant leaf guard cells induced by 1 μM abscisic acid. The plant leaf was from the broad bean *Vicia faba*, the cytosolic free Ca^{2+} being measured using the fura-2, the membrane potential being held at –40 mV and the intracellular solution including 10 mM potassium glutamate. Source: Schroeder & Hagiwara (1990). Reproduced with permission from PNAS and the author, Prof. J. Schroeder.

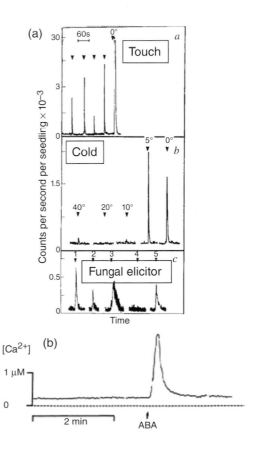

sparks and other types of Ca^{2+} signals seen in animal cells have not yet been well reported in plant cells. Cytosolic free Ca^{2+} signals have been measured and imaged following stimulation of plant cells, or intact plants, by cold, wind, light, cell division and differentiation, fertilisation, oxidative stress, and cell death. Red light induces micromolar changes in cytosolic free Ca^{2+} in plants. The free Ca^{2+} indicators used have been aequorin, small organic fluors, and energy transfer indications such as the cameleons.

The first measurement of cytosolic free Ca^{2+} in a plant cell used aequorin injected into the large unicellular alga *Chara*, a rise in cytosolic free Ca^{2+} affecting cytoplasmic streaming. Fura-2, transgenic aequorin and cameleons allowed changes in cytosolic free Ca^{2+} to be imaged in different parts of intact plants, such as the roots, cotyledons, leaves and pollen tubes, as well as individual cells, such as root hairs and guard cells. Measurement, and imaging, of cytosolic free Ca^{2+} in intact plants, using transgenic tobacco seedlings containing aequorin, showed that touch, mimicking the natural stimulus of wind, cold shock at around 5 °C, and fungal elicitors from yeast all provoked cytosolic free Ca^{2+} transients (Figure 9.3). The fastest were induced by touch, lasting just a few seconds, whereas those induced by the fungal elicitor lasted 1 min or more. Imaging of these Ca^{2+} signals showed that in seedlings the roots were more sensitive to cold than the cotyledons (Figure 9.4), and that there was considerable heterogeneity in the timing and duration of the Ca^{2+} signals between individual seedlings. Furthermore, Ca^{2+} signals

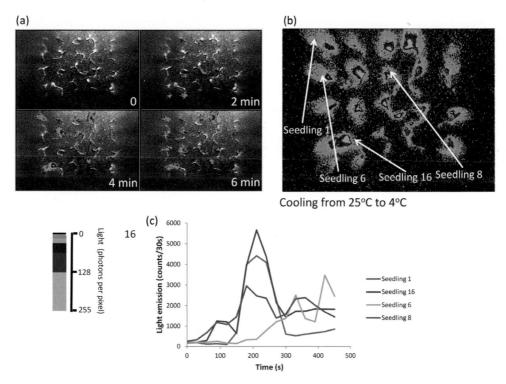

Figure 9.4 Individuality of cytosolic free Ca^{2+} signals in tobacco seedlings. Seedlings were transgenic with the Ca^{2+}-activated photoprotein aequorin, placed in front of an intensified CCD (ICCD) photon counting imaging camera (Photek) and the seedling cooled from 25 to 4 °C. (a) Ca^{2+} signals detected at 2, 4 and 6 min, the Ca^{2+} signals appearing in the roots at a higher temperature than the cotyledons. (b) The total light emitted from each of the 20 seedlings integrated over 10 min as the seedlings were cooled. (c) The time-course of light emission from four selected seedlings, showing major differences in the timing of the Ca^{2+} signals between individual seedlings, even though these were isogenic. Furthermore, some seedlings exhibited Ca^{2+} signals which oscillated, others did not. Reproduced by permission of Welston Court Science Centre.

in whole leaves start at the tips, and move into the centre. The Ca^{2+} channel blocker La^{3+} only inhibits the Ca^{2+} signals induced by cold shock, and not by touch or wind, suggesting that the main source of the Ca^{2+} induced by wind was intracellular release, and that by cold was through Ca^{2+} channels in the plasma membrane. However, mechanical stimulus of *Arabidopsis* roots induced cytosolic free Ca^{2+} signals within 1–18 s, detected and imaged by the fluorescence resonance energy transfer (FRET) yellow cameleon Ca^{2+} indicator, which were blocked by La^{3+}, suggesting these were through channels in the plasma membrane. Acidification of the cytosol, and alkalinisation of the external apoplastic space occurs through the Ca^{2+}/H^+ exchanger. The Ca^{2+} signals activate the NADPH oxidase responsible for generating O_2^- and other reactive oxygen species. But conversely H_2O_2 induces Ca^{2+} transients in tobacco detected by transgenic aequorin. Although gravity did not induce Ca^{2+} signals in *Arabidopsis* roots, changes in cytosolic free Ca^{2+} have been observed in seedlings induced by gravity. Thus there is variation in how intracellular Ca^{2+} is regulated between different plant species.

The trapping of photons by chlorophyll in the chloroplasts, converting this into ATP synthesis, is not the only response of plants to light. Phytochromes are proteins found in many leaves and flowers that enable them to trap sunlight at wavelengths not suitable for chlorophyll. However, phytochromes are signalling molecules, sensitive to far-red light around 650–750 nm, which flowering plants use to set their circadian rhythms, and regulate their time of flowering. Phytochromes are also used to regulate seed germination (photoblasty), seedling elongation, the size and shape of leaves, chlorophyll synthesis, and the straightening of seedling stems. Absorption of red light by plant phytochromes induces cytosolic free Ca^{2+} signals, which activate Ca^{2+}-calmodulin signalling.

Cold shock induces cytosolic free Ca^{2+} signals that switch on defence genes. But there is another type of response to cold – vernalisation, which also causes cytosolic free Ca^{2+} signals. The term vernalisation is derived from the Greek meaning 'spring', and describes the process where cold temperatures are required for some plants to flower. This is distinct from the defence system that enables plants to defend themselves against severe drops in temperature, such as frost. But both processes have been shown to induce cytosolic free Ca^{2+} signals. Cold acclimatisation causes Ca^{2+} influx into the cytosol, and is mediated by a cold sensor. Ca^{2+} comes from the apoplast – the free diffusional space outside the cell – that includes the cell wall. Ca^{2+} is also released into the cytosol from the ER and the vacuole. Imaging of intact mature leaves and seedlings (Figure 9.4 and 9.5) showed that the roots were more sensitive to cold, producing a Ca^{2+} signal at a lower temperature than the cotyledons. Furthermore, cooling the roots sends a signal to the cotyledons or leaves, which then generate their own cytosolic free Ca^{2+} signal (Figure 9.6). The Ca^{2+} transients generated by low temperature induce genes coding for proteins that protect the plant against freezing. Seeds and fungal spores can survive incredibly low temperatures, even down to liquid nitrogen. In autumn, cold acclimatisation causes water to be withdrawn from the xylem, stopping, for example, woody stems splitting when they freeze. There are also cryoprotective proteins. Low temperatures also increase membrane rigidity. Genes are also induced by water shortage and changes in salinity.

Plant responses to red light through phytochrome, gravity, touch, cold shock, fungal elicitors, drought and salinity are all accompanied by transient elevations of cytosolic free Ca^{2+}. The concept of a Ca^{2+} memory was first put forward by Knight and coworkers, based on the observation that repetitive stimulation (e.g. by mechanical stimulus) led to decreasing Ca^{2+} spikes. After a few minutes the plant recovered, and large Ca^{2+} spikes could be generated again. In the case of H_2O_2, it can take more than 1 h for the plant to recover, and again generate a cytosolic free Ca^{2+} signal following further addition of H_2O_2. Down-regulation in the Ca^{2+} signals is likely to be caused by loss of Ca^{2+} from a releasable Ca^{2+} store, which has to be replenished during a resting phase.

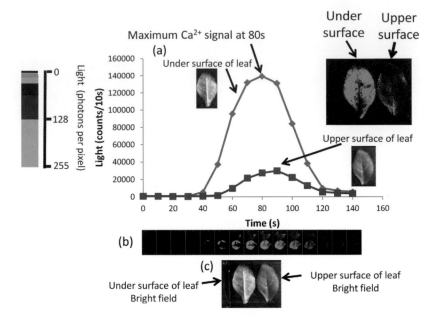

Figure 9.5 Imaging of cytosolic free Ca²⁺ in leaves of a tobacco plant. Tobacco plants were transgenic with the Ca²⁺-activated photoprotein aequorin. Two leaves, showing the under and upper surfaces, respectively, were placed in front of a ICCD photon counting imaging camera (Photek) and the leaves cooled from 25 to 4 °C. (a) Light emission plotted for each leaf. (b) The time-course of the imaged leaves. (c) Bright field picture of the two leaves. Campbell, Trewavas and Knight (unpublished experiment). Reproduced by permission of Welston Court Science Centre.

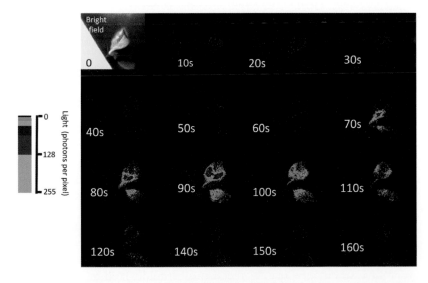

Figure 9.6 Effect of cooling plant roots on cytosolic free Ca²⁺ in the leaves. Tobacco plants were transgenic with the Ca²⁺-activated photoprotein aequorin. Two leaves, showing the under and upper surfaces, respectively, were placed in front of a ICCD photon counting imaging camera (Photek) and the roots cooled from 25 to 4 °C. The figure shows cytosolic free Ca²⁺ generated in the leaves as a result of cooling the root. Campbell, Trewavas and Knight (unpublished); see Campbell *et al.* (1996) *Cell Calcium*, **19**, 211–218 for other data. Reproduced by permission of Welston Court Science Centre.

Several chemical stimuli have been shown to induce cytosolic free Ca^{2+} signals in plants. These include plant hormones and glutathione. For example, auxin can induce Ca^{2+} signals in protoplasts isolated from wheat (*Triticum aestivum*) seedlings. Changes in cytosolic pH and oxygen metabolites often accompany changes in cytosolic free Ca^{2+}. Thus an important question is: What is cause or consequence?

9.6   Components of Ca^{2+} Signalling in Plants

The complete toolbox of Ca^{2+} signalling proteins has been identified in plants by purification and genome analysis:

- Ca^{2+} channels in the plasma membrane.
- Ca^{2+} storage organelles – ER, chloroplast, vacuole.
- Organelles that regulate cytosolic free Ca^{2+} – mitochondria, chloroplasts, ER, vacuole.
- High affinity Ca^{2+} binding proteins.
- Ca^{2+} sensitive genes.

9.6.1 Ca^{2+} Pumps and Exchangers

Plants have two types of protein that take Ca^{2+} across the plasma membrane, and those of intracellular organelles:

- Ca^{2+}-MgATPases.
- Ca^{2+}/H^+ exchangers.

Ca^{2+}-MgATPases are type II MgATPase, and usually have a high affinity for Ca^{2+} with K_d^{Ca} in the range 0.5–15 μM, but with low capacity. In contrast, the Ca^{2+}/H^+ exchangers tend to be of lower affinity, with K_d^{Ca} in the range 10–20 μM, but with high capacity. In keeping with Darwinian variation, all plants express several different genes coding for a Ca^{2+} pump or exchanger with subtly different properties. There are two types of Ca^{2+}-MgATPases: IIA and B. These can be regulated by changes in cytosolic free Ca^{2+}, either through Ca^{2+}-calmodulin or phosphorylation. This is because they have an autoregulatory site at the N-terminus, which has a Ca^{2+}-calmodulin-binding site, and a serine that can be phosphorylated by Ca^{2+}-dependent protein kinase. These MgATPases can be located in different parts of the cell. For example, AtACA1 is found in the plasma membrane, whereas AtACA4 is found in the tonoplast and inner plastid membrane. Ca^{2+}-MgATPase accounts for 0.1% of the protein in the plasma membrane, and is some 30–100 times less abundant than H^+-MgATPases.

Arabidopsis also has 11 genes coding for Ca^{2+}/H^+ exchangers (AtCAX1–11). Different isoforms are located in different membranes, and changes in cytosolic free Ca^{2+} can regulate their gene expression.

9.6.2 Ca^{2+} Channels in the Plasma Membrane

Two types of voltage-sensitive Ca^{2+} channels have been identified in plants:

- Depolarisation-activated (DACC).
- Hyperpolarisation-activated (HACC).

There are also voltage-independent Ca^{2+} channels (VICC). Cyclic nucleotide-gated channels also exist. Calmodulin down regulates channel activity. The *Arabidopsis* genome identifies many potential channels for monovalent and divalent cations. At least six families of cation channel have been identified. Three are K⁺ channels (shaker). The others are cyclic nucleotide-gated, two-pore domain TPK/VK and a K_{ir}-like protein. Twenty proteins are poorly ion selective, such as cyclic nucleotide-gated channels with putative Ca^{2+}-binding domains. Some 20 proteins are Ca^{2+} permeable (e.g. glutamate receptors), and at least one, TPC1, has a two-pore cation-selective channel.

9.6.3 Plant Organelles and Ca^{2+}

As with animal cells, plant ER and mitochondria play an important role in regulating cytosolic free Ca^{2+}. But cytosolic free Ca^{2+} is also regulated by chloroplasts and the vacuole. A typical plant cell has 10–100 chloroplasts per cell, 2–10 μm in diameter and 1–2 μm thick. Chloroplasts make plants green, trapping light, converting the energy into ATP, and NADP into NADPH for carbohydrate synthesis. Like mitochondria, chloroplasts contain their own circular DNA, coding for some 60–100 genes, often with more than one DNA copy per chloroplast. Also like mitochondria, chloroplasts have an outer and an inner membrane, with a space between. Inside the inner membrane is an aqueous fluid, the soma, within which is the thylakoid membrane. This is a complex internal structure that houses the photosynthetic system, and itself has a luminal space within it. Embedded within the thylakoid membrane, with photosystems I and II to make NADPH and ATP via the proton motive force, are the pigments that trap the light – chlorophyll, carotenoids, and phycobiliproteins in algae utilising energy transfer to extend the wavelengths of light trapped by the chloroplast. Changes in free Ca^{2+} have been detected in chloroplasts using targeted aequorin.

The vacuole is a membrane-bound organelle found in all plants and fungi, as well as some protists, animal cells and bacteria. The tonoplast is the membrane that surrounds the vacuole. The vacuole has two main functions:

- To get rid of harmful substances.
- To provide turgor pressure to maintain the strength of the plant.

The tonoplast transports H⁺ into the vacuole, making it acid, and can also remove toxic heavy metals. It contains aquaporins to regulate water movement in and out of the vacuole. Importantly, the tonoplast membrane contains a chloride channel with a conductance of some 50 pS, which can be activated by Ca^{2+}-calmodulin, acting on the cytosolic side.

Ca^{2+} is regulated in, and by, both chloroplasts and the vacuole, the latter via the tonoplast. Once again it has been measurement of free Ca^{2+} inside living plant cells and in intact plants that has been crucial in providing the evidence for this. Both organelles can contain large amounts of Ca^{2+}. Chloroplast total Ca^{2+} can be as high as 4–23 mM. Similarly, the Ca^{2+} content in tonoplasts is high, stored as calcium oxalate. Light can induce Ca^{2+} uptake in isolated chloroplasts from spinach at much lower free Ca^{2+} concentrations. The inhibitor DCMU, a specific and sensitive blocker of the photosynthetic electron transport chain and thus ATP synthesis from light, inhibits this Ca^{2+} uptake, consistent with the driving force being the membrane potential, negative inside. Modification of the membrane potential, using lipophilic cations, ionophores and K⁺, disrupts the Ca^{2+} influx.

Using aequorin, targeted either to the cytosol or chloroplasts, in the weed *Arabidopsis* and the tobacco plant *Nicotiana plumbaginifolia*, oscillations in cytosolic free Ca^{2+} have been observed during the normal 24 h light/dark cycle, mimicked by changes in free Ca^{2+} inside the chloroplast soma. The peak in cytosolic free Ca^{2+} was 0.5–0.7 μM, with a trough of 0.1–0.15 μM. Like mitochondria, chloroplasts accumulate a higher free Ca^{2+} than the cytosol, rising from 0.15 μM to a peak of 5–10 μM. Switching off the light causes a large Ca^{2+} influx into the stroma of the chloroplasts, with a lag of about 5 min, reaching a peak by 20–30 min. The magnitude of the Ca^{2+} influx is proportional to the duration of exposure to the light prior to the dark. In the normal 24 h light/dark cycle a burst of stromal free Ca^{2+} has been observed in the chloroplasts every time the light was switched off. Microelectrodes have also shown that chloroplasts take up Ca^{2+} on illumination, giving a Ca^{2+} store potentially dischargeable in the dark.

Ca^{2+} regulates several processes inside chloroplasts. Ca^{2+} plays a vital role in the thylakoid membrane, which houses a Ca^{2+}/H^+ exchanger. It also regulates several chloroplast enzymes. But chloroplasts also modify cytosolic free Ca^{2+} signals in plants. Movement of Ca^{2+} into chloroplasts increases ATP synthesis and affects photosynthesis. Two Ca^{2+}-binding sites have been identified in isolated chloroplasts – K_d^{Ca} of 8 and 51 µM. Binding the low-affinity Ca^{2+} site correlates with a change in chlorophyll *a* fluorescence, being involved in inhibiting the spill-over of energy from photosystem II to I during photosynthesis. Ca^{2+} is also regulated in and by the tonoplast.

Like animal cells, Ca^{2+} can be released from the ER into the cytosol via IP_3. However, the bulk of total Ca^{2+} inside plant cells is stored in the central vacuole, into which it is pumped via a Ca^{2+}-MgATPase and a Ca^{2+}/H^+ antiporter, of which there are two genes, *CAX1* and *CAX2*. The driving force for uptake is the electrochemical potential across the tonoplast membrane created by the acid pH within the vacuole. The tonoplast Ca^{2+}/H^+ exchanger in higher plants is voltage sensitive. Specialised plant cells can accumulate large amounts of calcium oxalate, thought to be a defence against herbivores. Although the vacuole can receive and mediate Ca^{2+} signals, it is not clear what role calcium oxalate stores play in acute Ca^{2+} signalling.

9.6.4   Ca²⁺-Binding Proteins in Plants

An array of Ca^{2+}-binding proteins as targets for rises in cytosolic free Ca^{2+} have been discovered in plants. Some are EF-hand Ca^{2+}-binding proteins, which include calmodulin, calmodulin-like proteins, Ca^{2+}-activated protein kinases dependent or independent of calmodulin, calcineurin B like proteins, and the protein SOS3/CBL. Non-EF-hand Ca^{2+}-binding proteins in plants include calreticulin, phospholipase D (PLD), annexins and a Ca^{2+}-binding protein discovered in pistils, PCB.

The ubiquitous four EF-hand Ca^{2+}-binding binding protein calmodulin has been implicated in the response of plants to light, gravity, mechanical stress, osmotic stress, phytohormones, heat shock and several pathogens. A response element to calmodulin has also been identified that controls Ca^{2+}-sensitive genes. Calmodulin-like proteins differ from calmodulin, that has 148 amino acids, in that they are usually longer, have sequence similarities below 75%, and have between one and six EF Ca^{2+}-binding sites. They have been implicated in development, and responses to environmental and pathological stress.

Some 1085 genes coding for kinases have been identified in the *Arabidopsis* genome; 34 have been identified as Ca^{2+}-dependent protein kinases (CDPKs), independent of calmodulin. Plant calmodulin-dependent kinases are highly expressed in rapidly growing cells, such as roots and flowers. Ca^{2+}-calmodulin regulates their kinase activity, which can also autophosphorylate, activating the kinase further. There are also kinases (CaMKs) that need to bind both Ca^{2+}-calmodulin and Ca^{2+}, at separate sites, to be fully activated.

Kinases regulated by Ca^{2+} independently of calmodulin (CDPKs) are also widely distributed, being found in plants, protozoa and algae. The N-terminal protein kinase is joined to a C-terminus, and has an autoregulatory domain and a calmodulin-like domain, with usually four EF-hand Ca^{2+}-binding sites, which is the Ca^{2+} sensor, activating the N-terminal serine/threonine protein kinase when Ca^{2+} is bound. The N-terminus is highly variable, giving Darwinian variation for its selective advantage in different cell types and plant processes. These processes are diverse, and include metabolic effects on starch and protein accumulation (e.g. in immature rice seed), phytohormone signalling and gene expression regulated by light, gravitropism, thigmotropism, the cell cycle, and development of tissues such as roots and the pollen tube (e.g. in *Petunia*), the ability to tolerate cold, salt and drought (e.g. in tobacco), and responses involving abscisic acid, nodule number in leguminous plants and defence against pathogens. CDPKs are not integral membrane proteins but 24 of the 34 in the *Arabidopsis* genome do appear to have the potential for binding to membranes, since they have a N-myristylation motif. This may explain why they can be found in multiple sites in cells, including the cytosol, plasma membrane, nucleus, ER, peroxisome, outer membrane of mitochondria and intracellular oil droplets. An important Ca^{2+} target in animal cells is the phosphatase calcineurin. Calcineurin itself appears not to occur in plants. However, plants do contain calcineurin-like proteins (CBLs) that may play an important role in stress responses to cold and salt.

There are significant Ca^{2+}-binding proteins in plants whose Ca^{2+}-binding sites are not of the EF-hand type. These include phospholipase D (PLD), which is regulated through a C2 domain involving a Ca^{2+}-phospholipid site. This phospholipase plays a role in responses to ethylene, abscisic acid, α-amylase secretion, stomatal closure, leaf senescence, and stress responses to drought and pathogens. Other modulators of PLD include substrates and products of phospholipase C (PLC), which generates IP_3 for release of Ca^{2+} from the ER. Other non-EF-hand Ca^{2+}-binding proteins include the annexins, with four to eight repeats of some 70 amino acids, some with MgATPase or peroxidise activity, and which may be involved in secretion. Calreticulin is the major Ca^{2+}-binding protein in the ER and also is a stress protein. A novel Ca^{2+}-binding protein, PIP, has been found in the pistil and anthers of some plants, and thus may play a role in pollen–pistil interactions and pollen development.

9.6.5 Ca^{2+}-Sensitive Genes in Plants

An increase in cytosolic free Ca^{2+} induces changes in gene expression in many plants. In *Arabidopsis*, the transcriptome has revealed some 230 Ca^{2+}-sensitive genes, 162 up-regulated and 68 down-regulated, when there is an increase in cytosolic free Ca^{2+}. These include saline stress genes, and ion channels. The mitogen-activated protein kinase (MAPK) pathway and genes coding for enzymes that generate reactive oxygen species also interact with Ca^{2+} signalling. External peptide signals (AtPeps) in *Arabidopsis* have a plasma membrane receptor (AtPepR1) that activates an intracellular signalling pathway involving Ca^{2+}. This leads to expression of defence genes in *Arabidopsis*, such as PDF1, MPK3 and WRKY33. AtPeps provokes an inward Ca^{2+} current in mesophyll cells, causing a rise in cytosolic free Ca^{2+}. The receptor, AtPepR1, has a guanylate cyclase activity and a cyclic nucleotide-gated ion channel (CNGC2). This pathway is important in the defence against pathogens. A key target for Ca^{2+} spikes is CDPK1. This regulates the transcription factor 'repression of shoot growth' (RSG), through a kinase activated by Ca^{2+}, phosphorylating S114 in RSG. This causes it to bind the cytosolic signalling protein 14-3-3, thereby preventing it from entering the nucleus. RSG regulates the expression of genes involved in gibberellin synthesis.

9.7   How Intracellular Ca^{2+} Provokes Cellular Events in Plants

The cellular events that can be regulated by changes in intracellular Ca^{2+} include:

- Opening and closing of stomata.
- Cell cycle and cell division.

- Pollen tube formation.
- Seed pollination.
- Seed germination.
- Response to light, gravity and touch.
- Legume formation.
- Intermediary metabolism.
- Phloem transport.
- Defence to the stresses of wind, cold and high salt.

9.7.1 Light and Intracellular Ca²⁺ in Plants

Light does much more to plants than provide the energy for ATP and carbohydrate synthesis, via photosynthesis. Plants grow towards light, seen in forests and etiolated seedlings. Flowers open in the light, and may close in the dark. Plants contain light-sensitive proteins, such as phytochromes, sensitive to red light, and phototropin, sensitive to blue light. These light sensors generate cytosolic free Ca^{2+} signals. Thus, morphogenetic effects of growth and development provoked by red and blue light are mediated by Ca^2, via Ca^{2+}-sensitive genes.

9.7.2 Control of Opening and Closing of Stoma/Stomata

All leaves contain tiny pores that regulate their water (transpiration), and gas content, particularly O_2 and CO_2. These are called stoma (plural = stomata) (Figure 9.7). These pores are in the epidermis of the leaf, and are surrounded by 'guard cells' and subsidiary cells, which are regulated by changes in cytosolic free Ca^{2+}. The walls of guard cells are not symmetrical. As a result, changes in size of the guard cell affect the size of the pore. Movement of cations and anions across the guard cell membrane causes water to move osmotically, in or out of the cell. Reduction in guard cell size, as a result of water loss, makes the pore smaller, and may even close it, and vice versa. Uptake of water makes the guard cell bigger, and because of the asymmetry, the pore opens and gets bigger. Stoma open and close as part of the daily light/dark cycle. They thereby regulate gas exchange, and the water content of the leaf. Guard cells are sensitive to light intensity and quality, temperature, leaf water status, and intracellular CO_2. But a major primary stimulus for closing stomata is abscisic acid (Figure 9.2).

Abscisic acid is a plant hormone that regulates growth and stomatal closure, particularly under environmental stress, but it also can regulate seed maturation and dormancy. Abscisic acid acts on the guard cells by increasing cytosolic free Ca^{2+} from 50 nM to some 1 µM, which precedes stomatal closure (Figure 9.7). This rise in cytosolic free Ca^{2+} is generated from influx through the plasma membrane and release from internal stores by IP_3. Abscisic acid also causes an increase in cytosolic pH, mediated at least partly by Ca^{2+}-mediated inhibition of the plasma membrane H^+-MgATPase, 1 µM Ca^{2+} producing virtually 100% inhibition. Two intracellular mechanisms can cause stomata to be opened or closed through the guard cells: Ca^{2+}-dependent and Ca^{2+}-independent. The intracellular Ca^{2+} targets are calmodulin and a phosphatase calcineurin B-like protein. By interacting with the microfilaments that are arranged radially within the guard cell, conformation changes in the Ca^{2+}-binding proteins alter the shape of the cell so that the stoma closes. The Ca^{2+} signal also opens slow (S-type) anion channels in the plasma membrane, and can activate rapid (R-type) anion channels. Abscisic acid also increases the cytosolic pH to around 7.9. This activates K^+ efflux channels, and increases the number of K^+ channels available for activation. Prolonged opening of anion channels leads to a large loss of Cl^- and malate^{2-}, which causes a large depolarisation from an initial membrane potential of around −100 to −200 mV. This also opens voltage-gated K^+ channels, leading to a large loss of K^+ from

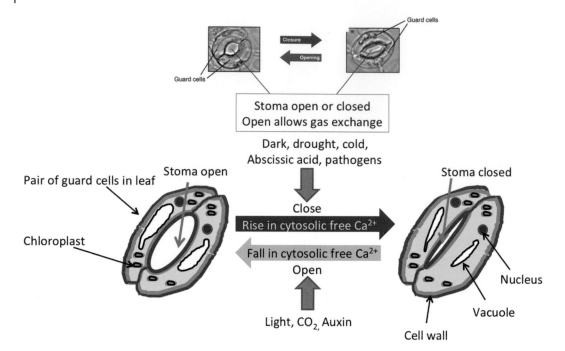

Figure 9.7 Effect of cytosolic free Ca^{2+} on guard cell closure. The figure shows the effects of hormones, drought and light on stoma opening and closing through changes in cytosolic free Ca^{2+}. *Source*: Bright field picture of guard cells from June Kwak and Pascal Mäser (2010) via Wikipedia Commons. http://upload. wikimedia.org/wikipedia/commons/thumb/2/20/Guard_cells_signals.png/800px-Guard_cells_signals.png.

the cells. The total cation plus anion loss can be as much as 0.3 M, leading to osmotic water loss from the guard cell, which can then contract by up to 50%. A rise in cytosolic free Ca^{2+} blocks K^+ inward channels, increasing the membrane depolarisation.

9.7.3 Wind and Mechanical Stimulation of Plants

It has been known for centuries that many plants are sensitive to mechanical stimuli, such as touch or wind. Darwin documents many species that respond to mechanical stimulation via sensitive stems, petioles, flower peduncles, stamens and roots. Measurement of cytosolic free Ca^{2+} using transgenic aequorin in tobacco, *Arabidopsis* and the moss *Phycomitrella patens* has established that mechanical stimulation induces rapid Ca^{2+} transients. These Ca^{2+} signals are not affected by removal of external Ca^{2+} using EGTA, nor are they inhibited by the animal cell Ca^{2+} channel blockers La^{3+} or Gd^{3+}. Therefore the source of Ca^{2+} for the cytosolic free Ca^{2+} signals is release from internal stores – the ER and/or vacuole.

9.7.4 Gravity Sensing and Growth

The stems of plants grow upwards, and the roots grow downwards, because plants have gravity sensors. Changes in intracellular Ca^{2+} play an important role in signalling the growth of plant cells in the right direction. Some plants, such as vines, runner beans and climbers, can use Ca^{2+} signals to make them grow around objects such as rocks, tree trunks and poles. A key extracellular molecule is thigmotropin. Using Ca^{2+}-sensitive dyes, changes in cytosolic free Ca^{2+} have been seen in thigmotropism. Thigmotropism (Greek *thigmo* = touch) is a movement in which an organism moves or grows in response to touch or contact stimuli. In *Arabidopsis* roots, mechanical stimulation causes a rapid increase in cytosolic free Ca^{2+}, an increase in oxygen metabolites, and a decrease in the pH gradient across the plasma membrane, through alka- linisation of the apoplastic space and acidification of the cytosol. The apoplast, or apoplastic space, is the space outside the plasma membrane in plants that aids water and solute transport across plant tissues.

On the other hand, gravity affects gravity sensors in the roots and shoots of plants, detecting if they are out of alignment with the growth axis. Within a few minutes of changing the position of a plant relative to gravity, there is an increase in cytosolic pH from around 7.2 to 7.6, with a concomitant decrease in apoplastic pH from 5.5 to 4.5, and an increase in cytosolic free Ca^{2+}. The changes in pH are mediated via activation of the plasma membrane H^+-MgATPase. Grav- ity can induce the production of the plant hormone ethylene, important in tomato ripening, resulting in a change in net Ca^{2+} distribution in the roots. Ca^{2+} controls growth by secreting cell wall structure.

9.7.5 Fertilisation, Germination and Differentiation

A rise in cytosolic Ca^{2+} is the trigger for several events involved in plant germination:

- Polarisation of fertilised eggs.
- Formation of the pollen tube.
- Fertilisation of the ovum by pollen.
- Seed germination after fertilisation.

Fertilisation of the eggs of the seaweeds *Fucus* and *Pelvetia* by the male gamete leads to polarisation of the cell in which cytosolic free Ca^{2+} plays a crucial role. Fertilisation first stimu- lates the secretion of a rigid, but adhesive, wall. This enables the fertilised egg to stick to a substratum, such as a rock. After about 12 h, the egg 'germinates', changing from a symmetrical sphere to a pear shape, from which it grows into a seaweed. The cell's polarity is maintained by a current of around 60 pA, flowing through the egg from the growing tip to the fixed end. Thermoelectric equilibrium is maintained by a counter-current, in the opposite direction, car- ried by ions from the external medium. Ca^{2+} currents and cytosolic free Ca^{2+} play a key role in this process. There is a gradient of cytosolic free Ca^{2+} from the tip to the fixed end. This Ca^{2+} gradient is maintained by differences in the influx of Ca^{2+} between one end of the egg and the other. In *Fucus*, the cytosolic free Ca^{2+} gradient across the embryo determines the asymmetry of the first cell division, distinguishing the shoot (thallus) from the rhizoid, Ca^{2+} entering at the rhizoid pole and exiting via the thallus pole. Ca^{2+} plays a role in the polarisation of other plants, such as the giant marine plant cell *Acetabularia*.

Seed germination is the process whereby a plant emerges from its seed and begins to grow. It can be triggered by water, oxygen, a change in temperature and light. When a pollen grain hits a flower head, it sticks to the stigma, part of the pistil, the female part of a flower. But the DNA in the pollen still has to reach the ovum (egg), at the base of the stamen, if it is to successfully fertilise the egg and a new seed produced. So, the pollen stimulates production of the 'pollen tube', which grows to find its way the egg. Pollen tube growth involves slow-moving Ca^{2+} waves, regulated by IP_3. The direction of pollen tube growth is towards areas where the free Ca^{2+} is raised (Figure 9.8). Changes in cyclic AMP may also be involved. The cytosolic free Ca^{2+} signals activate calmodulin, and then gene expression, which controls the growth.

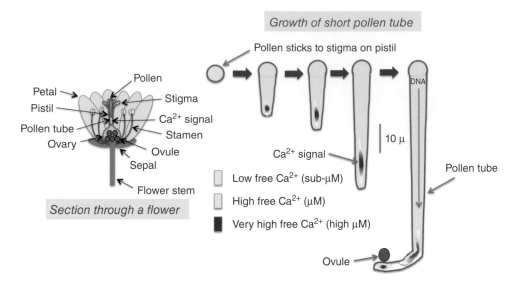

Figure 9.8 When a pollen grain arrives at a flower, via the wind or a bee, the pollen sticks to the pistil. This triggers the growth of a pollen tube – shown here. In maize this can be as long as 6 cm. Calcium signals shown here, through a fluorescent calcium indicator, affect the growth and direction of the tube. The tube carries the DNA from the pollen to the ovule, fertilising it to start the formation of a seed, which has the capacity to grow into a new plant. Note: the yellow in the pollen shown here is not a Ca^{2+} signal. Based on pioneering work of Malho and Trewavas (1996), *Plant Cell*, **8**, 1935–1949. Reproduced by permission of Welston Court Science Centre.

9.7.6 Legumes

Legumes are plants that have nodular structures in their roots that harbour symbiotic bacteria – rhizobia. They fix nitrogen, converting it into ammonium, the reason for crop rotation. Typical legumes in the family Leguminosae are the pea (*Pisum*), the broad bean (*Vicia*), the lentil (*Lens*), the soybean (*Glycine*), the kidney bean (*Phaseolus*), the peanut (*Arachis*) and the southern pea (*Vigra*). Legumes have receptors for Nod factors, such as lipochitin oligosaccharides, that are extracellular signalling molecules. The Nod factors provoke oscillations in cytosolic free Ca^{2+} in root epidermal cells. These activate calmodulin-activated kinase, leading to the CYCLOPS (Copy number alterations Yielding Cancer Liabilities Owing to Partial losS) protein, and activation of the symbiotic pathway for nodule formation. The initiation of intracellular infection of legume roots, by symbiotic rhizobia bacteria and arbuscular mycorrhiza fungi, is preceded by the induction of specific calcium signals in and around the nucleus of root epidermal cells.

9.7.7 Intermediary Metabolism

Ca^{2+} is an essential component of the sucrose-signalling pathway, leading to an increase in fructan synthesis. However, the role of changes in cytosolic free Ca^{2+} in the regulation of glucose metabolism and mitochondrial oxidation of pyruvate is not as well documented as they are in animal cells. Nevertheless, when a rise in cytosolic free Ca^{2+} activates a plant cell process, a rise in Ca^{2+} would also activate the intermediary pathways necessary for the generation of ATP needed for the cell event.

9.7.8   Transport by Phloem

Phloem consists of sieve tubes that transport metabolites, such as sugars made by photosynthesis in the leaves, to other parts of the plant, e.g. immature leaves, roots, developing flowers, fruit and seeds. Phloem also transports amino acids, other solutes, proteins, viruses and signalling molecules. Ca^{2+} can play a role in determining whether these substances are transported or not. Phloem has large pores some 1–2 μm in diameter, and is lined inside by former intracellular organelles such as the ER, mitochondria and plastids, and is surrounded by active cells. Transport is passive and non-selective, using pressure gradients to drive it. There are two types of plastid, S and P, the latter containing P-protein. Release of Ca^{2+} into the sieve elements of legumes stimulates rapid dispersal of crystalloid P-protein, which is reversible. This explains why the concentration of Ca^{2+} in the sieve tubes is greater than that in the surrounding tissue. Calmodulin, protein kinases and Ca^{2+} channels are found there.

9.7.9   Defence Against Stress

Plants are susceptible to a wide range of physical, chemical and biological stresses. These include wind, drought or water excess, high salinity and cold shock. They also have circadian rhythms that enable the plant to adapt to changes in light and temperature during the daily cycle. Cytosolic free Ca^{2+} signals are involved in helping the plant survive these stresses. Herbivores, such as caterpillars, also generate waves of cytosolic free Ca^{2+}, the leaves responding to vibrations set up by munching, releasing glutamate, opening Ca^{2+} channels that generate the Ca^{2+} waves.

9.7.9.1 Wind
Where I live, close to a cliff top looking across the Severn estuary, there is a large oak tree (Figure 9.1b). But, the trunk is not straight. Rather, it is bent away from the prevailing wind. Many bushes and trees on the cliff tops of Wales are shaped similarly. But, the bending is not induced mechanically. Rather, wind generates regular cytosolic and nuclear Ca^{2+} signals. These activate genes that cause cells on one side of the tree to grow slightly faster than the other, causing the tree to look bent by the wind. This reduces the chance of the tree or bush being blown over and killed by a strong wind. This is somewhat speculative for an oak tree. But measurements of cytosolic free Ca^{2+} in whole plants using transgenic aequorin have shown that wind does indeed induce regular cytosolic free Ca^{2+} transients. These induce genes which affect cell growth.

9.7.9.2 Water – Drought or Excess
Plants have to endure a wide range of water supply. Desert plants may have to survive weeks or even months without any rain, whereas plants that live near or in water have to cope with continual excess. Thus, plants have evolved structures to keep their water content in the right balance. Signalling mechanisms adapt plants to changes in water supply. Changes in intracellular Ca^{2+} play an important role in these signalling mechanisms. Plants have an abscisic acid response element activated by an increase in intracellular Ca^{2+}, induced by drought and cold shock. In wetland plants, an increase in cytosolic Ca^{2+} leads to apoptosis in the aerenchyma – the gas-filled space separating cells in roots.

9.7.9.3 High Salinity
High level of saline (NaCl) in the soil is a worldwide problem, causing a decrease in crop yield. Changes in intracellular Ca^{2+} play an important role in the ability of plants to survive high salinity. Some 20% of irrigated land around the world is affected by high levels of salinity. NaCl

damages plants through osmotic effects, or toxicity of either Na^+ or Cl^-, though much less is known about the role of Cl^- than Na^+ in salinity stress. There are two types of plants so far as saline is concerned: halophytes and glycophytes. Halophytes grow naturally in high salt, whereas glycophytes have a low tolerance to high levels of saline in the soil.

In order to combat the toxic effects of high salinity, the plant has a saline sensor, and a signalling pathway that leads either to removal of Na^+ from cells, or sequesters it so that it is inactive. To combat effects of increased osmolarity caused by high salinity, plants synthesise a range of organic solutes such as proline, glycine, betaine, sorbitol, mannitol, pinitol and sucrose. These raise the osmotic pressure in the cytosol, thereby combating the raised extracellular osmotic pressure.

Changes in cytosolic free Ca^{2+} have been detected in both halophytes and glycophytes. There are differences in the type and magnitude of the Ca^{2+} signal between the two types. Increases in cytosolic free Ca^{2+} may be transient, prolonged or oscillatory. In the quince (*Cydonia oblonga*), which is salt-resistant, NaCl induces Ca^{2+} oscillations, after which the cytosolic free Ca^{2+} decreases. Na^+ induces an influx of Ca^{2+} through opening of Ca^{2+} channels in the plasma membrane, with a simultaneous increase in cytosolic pH. In contrast, in rice, which is sensitive to salt, NaCl decreases cytosolic free Ca^{2+}, with a corresponding reduction in growth. In *Arabidopsis*, which is also salt-sensitive, transgenic aequorin showed an increase in cytosolic free Ca^{2+} in response to either NaCl or drought.

There are both transmembrane osmotic and salt sensors in the plasma membrane of many plant cells. These open Ca^{2+} channels, but also cause rapid increases in IP_3 and diacylglycerol, via activation of phospholipase C (PLC) induced by osmotic stress. Thus, release of Ca^{2+} from internal stores, such as the ER and vacuole, is important. In addition, regulation by chloroplasts, and independent regulation in the nucleus, may occur. An increase in cytosolic free Ca^{2+} is sensed by the protein SOS3. This binds to SOS2, so that the SOS2–SOS3 complex is a kinase, which activates SOS1 by phosphorylation. SOS1 is a Na^+/H^+ antiporter, so activation leads to efflux of Na^+ and influx of H^+. SOS1 has its C-terminus facing the cytosol, and thus may also sense Na^+ directly. The most abundant Na^+/H^+ exchanger in the vacuole membrane, the tonoplast, of *Arabidopsis* is AtNHX1. Binding of a calmodulin-like protein, AtCaM15, regulates this transporter. Ca^{2+}-dependent binding reduces Na^+/H^+ exchange. Mechanosensitive channels in plant membranes can also cause a rise in cytosolic free Ca^{2+}. Salt also induces expression of the homeobox gene ATHB7, but this appears to be independent of the rises in cytosolic free Ca^{2+} induced by salt. Thus saline stress, through Na^+ toxicity or osmotic stress, can lead to increases or decreases in cytosolic free Ca^{2+} depending on the plant and cell type.

9.7.9.4 Low Temperature

Low temperatures induce Ca^{2+} signals in plants. Different parts of the plant can have different sensitivities (Figures 9.4 and 9.5). For example, the roots of seedlings transgenic with aequorin generated a Ca^{2+} signal at 17–18 °C, whereas the leaves required a lower temperature of less than 10 °C to produce a Ca^{2+} signal. Low temperatures induce defence genes such as LT178. Modulation of cooling rates changes the Ca^{2+} signals, which correlate with the patterns of LT178 expression. Reduction of the Ca^{2+} transients also reduces LT178 gene expression. In rice, overexpression of CDPK13 and calreticulin-interacting protein (CRTinP1) showed that they, and calreticulin itself, may be involved in the cold stress response.

9.8   Fungal Elicitors

Fungal elicitors are substances produced when a fungus or yeast invades another organism, such as a plant. They induce cytosolic free Ca^{2+} signals in plants (Figure 9.3). The elicitors 'draw out' a signalling response in the host organism. These substances are produced by the plant host and also by the fungus. They include oligosaccharides, oligogalacturonides, phytoalexins and carotenoids. The complete response of the plant to fungal elicitors involves hypersensitive reactions around the local cell death induced by the fungus. This is followed by generation of reactive oxygen species and nitric oxide (NO), strengthening of the cell walls of the plant cells through deposition of callose and lignin and hydroxyproline-rich proteins, synthesis of salicylic and jasmonic acids, which induce defence genes, and induction of 'pathogenesis-related' (PR) proteins. The whole process leads to local and/or systemic resistance of the plant to the fungal infection. The selective advantage of this complex elicitor response is that it acts as a defence against the fungus, as well as being anti-fungal, restricting fungal growth. The question arises: What role do the Ca^{2+} signals play in this?

9.9 Apoptosis

Cell death by apoptosis plays an important role in the life of most plants. In autumn, and at other times, a layer of cells that connect the leaf of many plants to the stem is signalled to commit suicide. The result of this apoptosis is that the leaf falls off. As with animal cells, Ca^{2+} signals are involved in such apoptotic mechanism in plants.

9.10 Intracellular Ca²⁺ and Plant Pathology

A number of elements found in soil can be toxic to plants. These include sodium, arsenic, cadmium, copper, nickel, zinc and selenium. Plants have evolved two strategies to defend themselves against this toxicity, both potentially involving Ca^{2+}:

- Ion exclusion.
- A mechanism of tolerance.

Ca^{2+} deficiency has serious effects on plant growth and the development of fruit. And there are also a range of bacteria and viruses that attack plants. Unlike animals, plants do not have an immune system to combat infections. However, they do have signalling systems that defend plants against toxins and infectious agents. Cytosolic free Ca^{2+} plays a role in regulating genes, enabling the plant cell to combat thus pathology. Cell death by anoxia or ozone, inducing reactive oxygen species, both increase cytosolic free Ca^{2+}.

9.11 Ca²⁺ in Mosses, Liverworts and Ferns

Mosses, liverworts and ferns are small, soft plants that are typically 1–10 cm tall, though some species are much larger. Although the data on Ca^{2+} in these organisms is much less than in higher plants, there is clear evidence that cytosolic free Ca^{2+} does play a role in the biology of these as well. Many live in hostile environments, where intracellular Ca^{2+} would be expected to play a role in their survival. Mosses, liverworts and ferns can do damage to structures on which they are growing. They can extract Ca^{2+} from these. For example, the Ca^{2+} content is higher when grown on calcified substrates such as marble. Calcium requirement for ferns varies widely, and they can accumulate Ca^{2+} in their cell wall. Measurement of free Ca^{2+} has been made using transgenic aequorin in several moss species. Cold shock, mechanical perturbation and pH changes, as well as blue light, but not red light, induce calcium transients in the moss *Physcomitrella patens*. Like higher plants, CDPKs play a key role in mediating effects of cytosolic free Ca^{2+} on growth and other cellular processes.

9.12 Ca^{2+} in Fungi

9.12.1   Fungi and Intracellular Ca^{2+}

Much less is known about the role of intracellular Ca^{2+} in fungi than in plants. Transgenic aequorin, and live cell imaging of fluorescent Ca^{2+} indicators, has established that, as with all live cells, cytosolic free Ca^{2+} in fungi is maintained in resting cells at sub-micromolar levels. Some data are also available on Ca^{2+} channels and Ca^{2+}-binding proteins, but few studies have been carried out relating changes in intracellular free Ca^{2+} in fungi to physiological or pathological events.

Fungi represent a separate Kingdom from animals and plants, with an estimated number of species of 1.5 million worldwide. As a result of DNA analysis fungi now include yeasts and moulds (see Chapter 8 for Ca^{2+} in yeasts), as well as mushrooms and toadstools, and lichens, which are symbionts of a fungus with an alga or cyanobacteria. Fungi themselves are not photosynthetic, and do not contain chloroplasts, but do have vacuoles similar to plants. Fungal cells contain all the usual eukaryote intracellular organelles – nucleus, ER, Golgi, mitochondria, lysosomes and vesicles. Fungal cells have walls, made of chitin and glucans.

Fungi are of major importance economically. Yeasts are used in making bread, wine and beer. Moulds give us antibiotics, such as penicillin. And of course many fungi are delicious to eat, including the expensive truffles, though several are extremely poisonous. Several are bioluminescent. Some produce narcotics, such as LSD. Others are pathogens; *Candida*, for example, causes thrush. Fungi also can be very damaging, causing building materials to disintegrate. They can infect crops and vines, where they cause immense damage and loss of yield. They reproduce both sexually and asexually, yeasts undergoing division by fission or budding.

Extracellular primary stimuli and secondary regulators are not as well characterised in fungi as they are in animals and plants. But, several processes in fungi require intracellular signalling pathways. These include growth and differentiation at the tips of hyphae, formation of fruiting bodies and spores, and germination of the spore. In addition, degradative enzymes are secreted to digest and decompose matter that provides the nutrients for the fungus. And intermediary metabolism has to be regulated. Indirect evidence that intracellular Ca^{2+} regulates these processes in fungi includes:

- Responses to nutritional and environmental stimuli requires Ca^{2+}.
- Branching of hyphae in *Neurospora* and *Achlya* induced by a Ca^{2+} ionophore.
- Effects on the circadian rhythm of *Neurospora*.
- Inhibition of sporulation in *Penicillium* by removal of external Ca^{2+}.
- Loss of budding in a yeast mutant that has a defect in Ca^{2+} transport in the vacuole.

Key components of the Ca^{2+} signalling system have been found in fungi:

- Calmodulin.
- Ca^{2+} pumps and H^+ exchangers.
- Release of Ca^{2+} by IP_3 from vacuoles in *Neurospora*.

A search of the genome in *Neurospora* has identified proteins associated with Ca²⁺ signalling, but there are significant differences from animals and plants. Whilst fluorescent dyes have been used successfully to measure cytosolic free Ca²⁺ in some fungi, these overload into intracellular organelles, particularly intracellular vesicles, and also leak out of cells, making data difficult to interpret properly. However, high levels of expression of transgenic aequorin have been obtained in *Neurospora* and *Aspergillus* by optimising codon usage. Three external stimuli – mechanical, hypo-osmotic shock and high external Ca²⁺ – caused transient increases in cytosolic free Ca²⁺ lasting a few hundred seconds. The vacuole is a major Ca²⁺ store in fungi, using polyphosphates as the Ca²⁺ ligand. As in plants, this acts as a source for induced cytosolic free Ca²⁺ rises, but is also a way of removing cytosolic free Ca²⁺ if it stays above micromolar levels for long, protecting the fungal cell against Ca²⁺ toxicity. Ca²⁺ uptake by the vacuole is driven by an H⁺/Ca²⁺ exchanger.

As with animal and plant cells, there is an intimate interaction between intracellular Ca²⁺ and cyclic nucleotides in plants. For example, in fungi, protein kinase A (PKA) activates Ca²⁺ channels. The transition from polar to apolar growth correlates with a cytosolic free Ca²⁺ transient mediated by PKA phosphorylation. The reverse process of apolar to polar growth does not give rise to a Ca²⁺ transient. But measurement of cytosolic free Ca²⁺, using transgenic aequorin in *Aspergillus niger*, showed that activation of PKA leads to Ca²⁺ transients mediated by phosphorylation of Ca²⁺ channels in the plasma membrane. In other organisms, cytosolic Ca²⁺ can regulate the level of cyclic AMP by activation or inhibition of adenylate cyclase, or by calmodulin activation of cyclic AMP phosphodiesterase. Sucrose and light cause changes in cyclic AMP in *Aspergillus*, which mediates fungal virulence through the pathogenicity of dimorphic switching. Cyclic AMP, via PKA, also regulates intermediary metabolism. Hypo-osmosis also increases cyclic AMP.

9.12.2 Intracellular Ca²⁺ and Yeast

Yeasts are now considered as fungi, based on their reproduction and genome. A particularly important mechanism, first worked out in yeast, is the ER stress response (see Chapter 8). The production of malfolded proteins, or the prolonged release of Ca²⁺ from the ER, activates this stress response. GRP78 (BiP), a protein in the ER lumen that binds Ca²⁺, has an ATPase activity, and binds hydrophobic domains exposed in malfolded proteins. As a result BiP drops off a domain of protein IRE1P, which crosses into the nucleus. This autophosphorylates, leading to the activation of a transcription factor, which binds to the 'unfolded protein response' (UPR) element on the DNA. This induces defence genes. If these are successful, the cell survives. If not, the cell crosses the Rubicon and undergoes apoptosis. A similar mechanism is found in animal cells. The protein Cch1 restores intracellular Ca²⁺ in fungal cells during ER stress. Cch1, coupled to its subunit Mid1, forms a highly efficient Ca²⁺ channel in the plasma membrane opened by loss of Ca²⁺ from the ER, in a similar way to the store-operated Ca²⁺ channel opened through the Orai/STIM mechanism in animal cells.

Several fungi, including *Candida* and *Cryptococcus*, are pathogenic, particularly in people whose immune system is weakened. For example, *Cryptococcus neoformans* causes life-threatening meningitis in patients with a weak immune system. Mutants defective in CCH1 or MID1 are not viable in limiting extracellular Ca²⁺. The anti-fungal agent azide induces Ca²⁺ influx. These promote ER stress by blocking ergosterol synthesis. Transfection of Cch1/Mid1 into HEK293 animal cells forms a Ca²⁺ channel detected by patch clamp, producing an I_{CRAC}-like current similar to SOCE.

9.12.3 Lichens

Lichens are symbionts of a fungus, alga and/or cyanobacterium. They occur all over the world and play a major role in soil formation. They are able to grow on sites that appear quite inhospitable, such as bare soil and rocks, wood, tree bark and leaves, shells, and barnacles. As a result they can live in extreme environments, such as the polar, alpine and semi-arid desert areas, where they are often the dominant species. Little is known about the role of intracellular Ca^{2+}.

9.13 Ca^{2+} and Slime Moulds

Slime moulds get their common name as a result of their appearance as gelatinous slime. They are protists that use spores to reproduce. They were once classified as fungi, but they are now regarded as a separate group (see Chapter 8). They are found world-wide feeding on microorganisms that live on dead plant material. As a result they are commonly found on forest floors, particularly logs of deciduous trees, soil and lawns. In tropical areas they are found in fruit and inflorescence. A well-known slime mould is the yellow *Physarum polycephalum*, in which signalling mechanisms involving cyclic AMP and Ca^{2+} have been studied.

9.14 Summary

It is clear from measurements of cytosolic free Ca^{2+} in plants and fungi that these follow the universal principle of all cells:

1. They maintain a sub-micromolar free Ca^{2+} even in the presence of mM extracellular Ca^{2+}.
2. Cytosolic free Ca^{2+} rises to several micromolar when the cell is activated or stressed.

In addition to the ER and mitochondria, plants contain two organelles that can act as Ca^{2+} stores, and regulators of cytosolic free Ca^{2+}:

- Chloroplasts.
- Vacuole.

Box 9.1 Ca^{2+} signals shown to initiate a cell and whole plant event.

- Response to light, gravity and touch.
- Opening and closing of stomata via guard cells.
- Cell cycle and cell division.
- Pollen tube formation.
- Seed pollination.
- Seed germination.
- Legume formation in the roots.
- Intermediary metabolism.
- Phloem transport.
- Defence to the stresses of wind, cold, drought and high salt.
- Response to fungal elicitors
- Cell death by apoptosis, e.g. leaves falling in autumn.

Thus, many physiological stimuli and stresses induce cytosolic Ca²⁺ signals in plants, which induce the response of the plant cell (Box 9.1). These are caused, and regulated, by Ca²⁺ channels, pumps and exchangers in the plasma membrane and/or the membranes of organelles. Cytosolic free Ca²⁺ signals also occur in fungi, including yeasts, which correlate with the cell and organism event (Box 9.2).

Box 9.2 Examples of Ca²⁺ signals initiating events in fungi.

- Responses to nutritional and environmental stimuli.
- Branching of hyphae in *Neurospora* and *Achlya*.
- Circadian rhythm in *Neurospora*.
- Sporulation in *Penicillium*.
- Budding in yeast.
- ER stress and unfolded protein response in yeast.
- Intermediary metabolism.

Plants and fungi, like animal cells, contain an array of Ca²⁺-binding proteins and enzymes regulated by Ca²⁺ that covalently modify other proteins. This argues strongly for a universal role for Ca²⁺ as an intracellular regulator in plants. A frequent result of activation of Ca²⁺ signalling pathways in plants is the regulation of gene expression. But there is still a lot to learn about the sources of Ca²⁺ for internal release, and how this occurs. Changes in intracellular pH are found in developing cells, such as root tips, nodules, and in response to elicitors, hormones, cold, saline and other stresses. Thus, changes in cytosolic free Ca²⁺ are often accompanied by changes in cytosolic pH, and production of reactive oxygen species. The question therefore arises whether these are a cause or consequence of the rise in cytosolic free Ca²⁺? The central question is: How does a rise in plant cell cytosolic free Ca²⁺ trigger a cellular event?

Much less is known about the role of intracellular Ca²⁺ as a signal of cell events in fungi. Nevertheless, here too, there is evidence that extracellular agents can induce increases in cytosolic free Ca²⁺ that play a key role in activating the proteins responsible for the cell event.

Recommended Reading

*A must.

Books

*Campbell, A.K. (2015) Intracellular Calcium. Chichester: John Wiley & Sons Ltd. Chapter 9 Role of intracellular Ca²⁺ in plants and fungi. Fully referenced evidence.
*Luan, S.E. (Ed) (2011) Coding and Decoding Calcium Signals in Plants. Berlin: Springer. Useful multiauthor book on how Ca²⁺ works in plants.

Reviews

*Dodd, A.N.K.J. & Sanders, D. (2010) The language of calcium signaling. *Annu. Rev. Plant Biol.*, **61**, 593–620. How Ca²⁺ signals are interpreted in plant cells.
Hetherington, A.M. & Brownlee, C. (2004) The generation of Ca²⁺ signals in plants. *Annu. Rev. Plant Biol.*, **55**, 401–427. How intracellular Ca²⁺ are produced in plants.

McAinsh, M.R. & Pittman, J.K. (2009) Shaping the calcium signature. *New Phytol.*, **181**, 275–294. How the particular type of Ca^{2+} signal is generated.

Peiter, E. (2011) The plant vacuole: emitter and receiver of calcium signals. *Cell Calcium*, **50**, 120–128. The role of the plant vacuole in regulating cytosolic free Ca^{2+}.

Perochon, A., Aldon, D., Galaud, J.P. & Ranty, B. (2011) Calmodulin and calmodulin-like proteins in plant calcium signaling. *Biochimie*, **93**, 2048–2053. What calmodulin and its analogues do in plants.

Steinhorst, L. & Kudla, J. (2013) Calcium – a central regulator of pollen germination and tube growth. *Biochim. Biophy. Acta-Mol. Cell Res.*, **1833**, 1573–1581. How Ca^{2+} signals allow pollen DNA to reach the ovum.

*Trewavas, A.J. (2011) Plant cell calcium, past and future. In Luan, S. (Ed.) *Coding and Decoding of Calcium Signals in Plants*. Berlin: Springer, Chapter 1, pp. 1–6. Wide ranging review of the role of Ca^{2+} signalling in plant by a pioneer.

Research Papers

*Braam, J. & Davis, R.W. (1990) Rain-induced, wind-induced, and touch-induced expression of calmodulin and calmodulin-related genes in *Arabidopsis*. *Cell*, **60**, 357–364. How touch and wind regulate cell development through intracellular Ca^{2+}.

*Campbell, A.K., Trewavas, A.J. & Knight, M.R. (1996) Calcium imaging shows differential sensitivity to cooling and communication in luminous transgenic plants. *Cell Calcium*, **19**, 211–218. Pioneering imaging of Ca^{2+} signals in whole plants using recombinant aequorin.

Defalco, T.a.B.K.W. & Snedden, W.A. (2010) Breaking the code: Ca^{2+} sensors in plant signalling. *Biochem. J.*, **425**, 27–40. The Ca^{2+} targets in plant cells.

*Haley, A., Russell, A.J., Wood, N., Allan, A.C., Knight, M., Campbell, A.K. & Trewavas, A.J. (1995) Effects of mechanical signaling on plant-cell cytosolic calcium. *Proc. Natl Acad. Sci. USA*, **92**, 4124–4128. How wind and touch generate Ca^{2+} signals in plants.

*Knight, M.R., Campbell, A.K., Smith, S.M. & Trewavas, A.J. (1991) Transgenic plant aequorin reports the effects of touch and cold-shock and elicitors on cytoplasmic calcium. *Nature*, **352**, 524–526. Pioneering evidence for Ca^{2+} signals evoked by various stresses in the tobacco plant, a classic model system.

*Plieth, C. & Trewavas, A.J. (2002) Reorientation of seedlings in the earth's gravitational field induces cytosolic calcium transients. *Plant Physiol.*, **129**, 786–796. How gravity can regulate the direction of growth in plants through Ca^{2+} signals.

Sai, J.Q. & Johnson, C.H. (2002) Dark-stimulated calcium ion fluxes in the chloroplast stroma and cytosol. *Plant Cell*, **14**, 1279–1291. How Ca^{2+} moves between the cytosol and chloroplast after dark.

Watahiki, M.K., Trewavas, A.J. & Parton, R.M. (2005) Dynamic localization of calmodulin domain protein kinase (CDPK) and its relationship to calcium signalling in growing pollen tubes. *Plant Cell Physiol.*, **46**, S28. How the Ca^{2+} target locates in pollen tubes as they take their DNA towards the ovum.

Whalley, H.J. & Knight, M.R. (2013) Calcium signatures are decoded by plants to give specific gene responses. *New Phytologist*, **197**, 690–693. How particular Ca^{2+} signals cause responses in different genes.

Yano, K., Yoshida, S., Muller, J. *et al.* (2008) CYCLOPS, a mediator of symbiotic intracellular accommodation. *Proc. Natl Acad. Sci. USA*, **105**, 20540–20545. How the protein CYLOPS (Copy number alterations Yielding Cancer Liabilities Owing to Partial losS) and Ca^{2+} are involved in legumes and nodule formation.

10

Pathology of Intracellular Ca²⁺

Learning Objectives
1. The difference between physiology and pathology.
2. When intracellular Ca²⁺ is foe, rather than a friend.
3. How intracellular Ca²⁺ is involved in the four types of cell death – necrosis, apoptosis, autophagy and lysis.
4. Genetic abnormalities in Ca²⁺ signalling proteins and pathology.
5. How Ca²⁺ is involved in oxygen toxicity.
6. Inappropriate Ca²⁺ signalling.
7. How intracellular Ca²⁺ can help cells defend themselves against stress or attack.

10.1 What is Pathology?

Pathology is the science of abnormalities in any living organism, provoked by internal or external agents. These abnormalities occur because:

- The cell has been attacked by a physical, chemical or biological agent.
- The cell finds itself in a hostile environment.
- The organism has a genetic disorder.
- Cells have suffered stress induced within the organism.
- The cell has simply run out of steam.

Intracellular Ca²⁺ is involved in some way in all pathological processes (Figure 10.1), either as a cause or consequence of damage to a cell or tissue. When a cell is injured, Ca²⁺ inevitably floods into the cell down its electrochemical gradient. If the rise in cytosolic free Ca²⁺ that follows is severe enough, this damages intracellular proteins and nucleic acids, as well as organelles such as the mitochondria. This may then lead to cell death by necrosis or apoptosis. But, the rise in cytosolic free Ca²⁺ in an injured cell also activates defence mechanisms, which may save the cell from death. Furthermore, there are defence mechanisms in organelles such as the endoplasmic reticulum (ER), which also lead the cell to make a Rubicon decision between survival and death. Changes in intracellular Ca²⁺ occur in a wide range of pathological processes, and inevitably in cell death. Intracellular Ca²⁺ is also involved in the unwanted side effects of many drugs.

Fundamentals of Intracellular Calcium, First Edition. Anthony K. Campbell.
© 2018 John Wiley & Sons Ltd. Published 2018 by John Wiley & Sons Ltd.
Companion Website: http://www.wiley.com/go/campbell/calcium

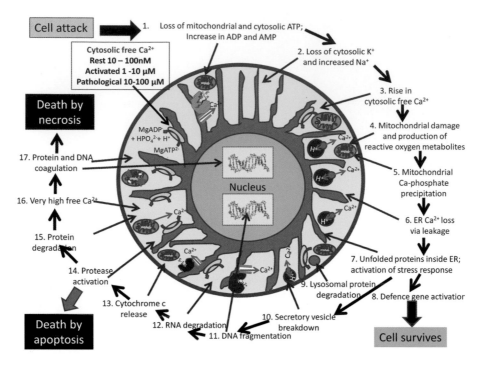

Figure 10.1 Intracellular Ca²⁺ and cell pathology. The figure shows pathways leading to cell death and where intracellular Ca²⁺ can be involved. Parts of the cell are identified which are damaged when the free Ca²⁺ in the cytosol or within organelles is too high. The numbers refer to the order in which these are damaged when a cell is injured and then follows a pathway of cell death or recovery. DNA helix. Brian0918 (2009).

10.2   Types of Pathology

Cell pathology can be classified in two ways:

- The cause of the insult.
- The type of response and mechanism that ensues.

The insult may be inherited or acquired. As with physiological primary stimuli, pathological agents provoking cell injury can be:

- Physical – mechanical damage, change in temperature (hot or cold), physical pressure, a change in osmotic pressure and electric shock.
- Chemical – toxic metals, inorganic and organic poisons, too much or too little oxygen, various oxidants, organic solvents and detergents.
- Biological – infectious eukaryotes, bacteria, archaeans and viruses, and biological toxins.

Pain is the most obvious medical problem that causes someone to go to the doctor. This involves nerves in the nociception pathway. Important voltage-gated Ca²⁺ channels in this pathway are Ca$_v$2.2 of the N and T type, which are particularly important in chronic pain, and Ca$_v$2.3 of the R type. Pain relievers block these channels. For example, Ziconotide (also known as SNX-111; Prialt), is a synthetic derivative of the cone snail toxin ω-conotoxin and blocks N-type Ca²⁺ channels in presynaptic nerve terminals, thereby preventing transmitter release. It is used to alleviate postoperative pain. Thus, Ca²⁺ can play a major role in the action of many drugs that are used to treat disease (see Chapter 11).

Pathological changes in Ca^{2+} occur in many human and animal diseases, cell injury, cell stress and cell death. Darwinian 'small change by small change' mechanisms can be very important in disease. Experimentally, we study cell physiology and pathology in the laboratory over periods of minutes, hours and days. Yet a rampant cancer can take many months to kill an unlucky patient. Just a small percentage change in the rate of cell division over the rate of cell death will lead to a very large change in total cell number within a few months. Such differences are almost impossible to detect directly in the laboratory.

There are several ways changes in intracellular Ca^{2+} can play a role in a pathological process:

1. Large accumulation of Ca^{2+} inside the cell, leading to irreversible injury and death.
2. Damage to the Ca^{2+} signalling system, with consequent disturbance of cytosolic free Ca^{2+} and intracellular Ca^{2+} signalling mechanisms, leading to cell malfunction.
3. Genetic mutations in Ca^{2+} signalling proteins, such as Ca^{2+} channelopathies, Ca^{2+} pumpopathies and malignant hyperthermia, either inherited or induced.
4. Inappropriate activation or inhibition of the Ca^{2+} signalling system in one or more cell types, such as neutrophil activation in rheumatoid arthritis.
5. Cell death or irreversible cell damage, induced by an increase in intracellular Ca^{2+}, such as apoptosis, leading to activation of, or damage to, intracellular proteins, other molecules, and damage to organelles; e.g. Ca^{2+} damage to mitochondria after a heart attack.
6. Abnormal changes in the expression of Ca^{2+} signalling genes and proteins.
7. Activation of defence mechanisms by intracellular Ca^{2+}, such as defence against unfolded proteins in the ER, or membrane pore formers such as complement.

10.3   Intracellular Ca²⁺ – Friend or Foe?

Since all living cells maintain a cytosolic free Ca^{2+} in the sub-micromolar range against a Ca^{2+} concentration outside some 1000–100 000 times this, when things go wrong there are inevitable changes in Ca^{2+} inside cells. Damage to the plasma membrane leads to a rise in intracellular Ca^{2+}, as will a decrease in ATP, caused, for example, by a loss of oxygen. There are thus four ranges for cytosolic free Ca^{2+} in cells (Box 10.1):

Box 10.1 The four ranges of cytosolic free Ca²⁺ in eukaryotic cells.

1. Resting cells: sub-micromolar, typically 10–100 nM.
2. Cells activated by a physiological stimulus: typically 1–10 μM.
3. Cells injured by a physical, chemical or biological agent: typically 10–100 μM.
4. Dead or dying cells: typically 0.1–1 mM.

These ranges are a guide, reversible or irreversible damage depending critically on how long the cytosolic free Ca^{2+} remains high. Transient changes in cytosolic free Ca^{2+} of tens of micromolar can be coped with if they only last 1 s or so, e.g. in a fast muscle twitch. But even a few μM free Ca^{2+} can be pathological if it stays at this level for minutes without compensatory mechanisms to stop mitochondrial damage, or complete loss of the intracellular Ca^{2+} stores by pumping Ca^{2+} out of the cell. Ca^{2+} oscillations are a mechanism allowing the mean cytosolic free Ca^{2+} to remain at 1–10 μM, without net Ca^{2+} being lost from the cell. But, a persistent cytosolic free Ca^{2+} in the high micromolar range will activate degradative enzymes, such as proteases and nucleases, as well as leading to Ca²⁺ overload in the mitochondria, phosphate precipitation, and irreversible damage to the oxidative ATP-generating machinery. A cytosolic free Ca²⁺ in the 0.1–1 mM range leads to protein coagulation, nucleic acid precipitation, and cell death. Interestingly, it is a persistent **low** Ca²⁺ inside the ER that is pathological, activating stress pathways. Cells have to survive a wide range of stresses. Thus, all cells have defence mechanisms, several of which can be activated by a pathological change in intracellular Ca²⁺. So, two key questions arise:

1. Is a rise in intracellular Ca²⁺ a cause or consequence of cell injury?
2. Is a pathological rise in intracellular Ca²⁺ friend or foe?

Many changes in intracellular Ca²⁺ will be a consequence of injury to particular cells. But Ca²⁺ also plays an important role in initiating damage to cells or, conversely, in defending them against attack. A key issue is whether damage to the cell is reversible, and if not whether it inevitably leads to cell death (Box 10.2).

Box 10.2 Sustained, high cytosolic free Ca²⁺ leading to cell injury – reversible or irreversible.

- Cell death by necrosis, autophagy or apoptosis – irreversible.
- High Ca²⁺ influx after ischaemia and the calcium paradox – irreversible.
- Mutations in Ca²⁺ signalling proteins that regulate cytosolic Ca²⁺ – irreversible.
- Inappropriate activation of intra- or extra-cellular toxic oxygen species – reversible.
- Bacterial metabolic toxins on gut bacteria and host cells – reversible.

10.4   Intracellular Ca²⁺ and Cell Death

Intracellular Ca²⁺ plays a key role in all forms of cell death. Cell death occurs when the biochemical and physiological functions that maintain a living entity cease. A dead cell cannot divide, or carry out its specialised functions. Ultimately a dead cell disappears, by fragmentation or digestion, either internally or externally by a phagocyte. Without a low cytosolic free Ca²⁺ the biochemical processes within cells cannot work. Ca²⁺ can initiate the pathway to cell death or, under other circumstances, intracellular Ca²⁺ can activate defence processes enabling the cell to survive attack or stress. Once the cytosolic free Ca²⁺ is over 10 μM or so, it can damage the cell irreversibly. At high micro- or milli-molar Ca²⁺:

- Calcium phosphate precipitates.
- DNA and RNA are disrupted.
- Proteases and nucleases are activated.
- Proteins are denatured.
- Enzymes, not normally regulated by physiological concentrations of cytosolic free Ca²⁺, are activated or inhibited.
- Mitochondria are damaged by Ca²⁺ overload, and Ca²⁺ precipitation.

In contrast, prolonged release of Ca²⁺ from the ER activates a stress response, whereby the cell attempts to protect itself against malfolded proteins. If this is inadequate, the cell crosses the Rubicon and kills itself.

An international commission has defined 12 types of cell death. These are distinguished by differences in structures seen microscopically, and by biochemical differences in the death

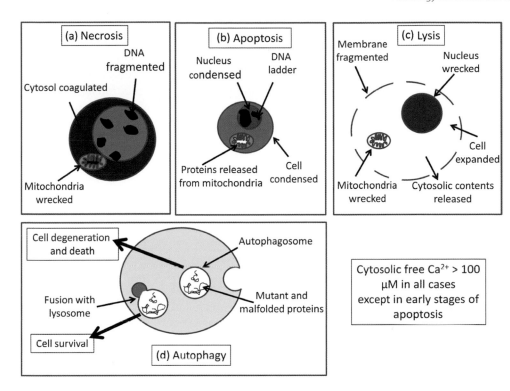

Figure 10.2 The four main types of cell death: (a) necrosis, (b) apoptosis, (c) lysis and (d) autophagy. Defence mechanisms are activated by a rise in cytosolic free Ca²⁺ that can allow the cell to survive. However, when the cytosolic free Ca²⁺ remains in the high micromolar for too long mitochondria are damaged, phosphate precipitates, and proteins and nucleic acids are degraded and coagulate, leading to cell death. Reproduced by permission of Welston Court Science Centre.

sequence. Four are designated as typical (common), with eight others being uncommon (atypical). Essentially, there are only four types of cell death (Figure 10.2):

1. Necrosis.
2. Programmed, usually called apoptosis.
3. Autophagy.
4. Lysis.

Changes in intracellular Ca²⁺ occur in all these types of cell death, and can be central to initiating irreversible damage to the cell.

10.4.1 Necrosis

Necrosis comes from the Greek *necro* = dead and was originally defined by cell structures seen under the microscope, particularly in the nucleus. It involves coagulation of intracellular proteins and degradation of nucleic acids. Necrosis is provoked by severe stress caused by:

- Lack of oxygen or key nutrients.
- Toxins.
- Physical trauma.
- Infections.

Seven morphological types of necrosis have been identified: coagulative, liquefactive, gummatous, haemorrhagic, caseous, fatty and fibrinoid. Necrosis inevitably involves a catastrophic permeabilisation of the plasma membrane, with a consequent flooding in of Ca^{2+} and flooding out of ions, such as potassium, and small organic molecules, and then macromolecules. High cytosolic free Ca^{2+} causes calcium phosphate precipitation in the mitochondria, and condensation of DNA into clusters. Protein precipitation and coagulation occur, induced by Ca^{2+}. But necrosis can also occur independent of Ca^{2+}, e.g. induced experimentally by CCl_4. Necrotic cells rarely send chemical signals to phagocytes, unlike apoptotic cells. Thus, necrotic cells can lead to unwanted dead tissue, and break up, triggering a damaging inflammation.

10.4.2 Apoptosis

Apoptosis is an internally programmed type of cell death, which prevents cell debris being strewn all over the place, stopping havoc around the dead cell. It was first observed during development, in lymphocytes after recovery from an infection such as a cold, and in tumours. Apoptosis is signalled, and involves activation of specific genes. But apoptotic cell death is not always pathological to the whole organism. It is essential in the formation of tissues in the embryo, and in development:

- Neurones that do not find partners during brain development kill themselves via apoptosis.
- The tail of a tadpole disappears to produce an adult frog.
- We have separate fingers because intervening cells are killed in the embryo.
- A leaf falls in autumn.

Apoptosis is essential in developing organs, determining shape, and getting rid of cells in the wrong place. It also gets rids of redundant cells, e.g. lymphocytes after a cold. This type of cell death is signalled from within the cell. The sequence can be initiated by an inappropriate rise in intracellular Ca^{2+}, and involves condensation of the nucleus and expression of cell surface proteins, so that the dead cell can be removed by phagocytes, without the chaotic inflammation that can follow cell lysis or necrosis. The name 'apoptosis' was given by Kerr, Wyllie and Currie, who noticed a large number of dead cells in stained cancer sections with condensed nuclei. This term was suggested by their colleague at the University of Aberdeen, Professor James Cormack of the Department of Greek. In ancient Greek, apoptosis (απoπτώσϊσ) is used to describe the dropping off of petals from leaves from a tree, though in Modern Greek it is used to describe baldness! Note that the 'p' should be silent, as in other Greek-derived words such as pterodactyl. In a tumour there is a constant battle between cell division and cell death. It has been estimated that in adults up to 70 billion cells die every day by apoptosis, and up to 30 billion in a child. A special form of programmed cell death is 'anoikis', derived from the Greek meaning 'the state of being without a home'. It is induced by cells detaching from holdfasts in the matrix that surrounds them. Anoikis is required in many organs to maintain the right cell balance, and is important in normal tissue and tumour development. Apoptosis is an attractive drug target. A difference of just 1% death versus division will cause the tumour to disappear within weeks. *In situ* there are two primary stimuli initiating apoptosis:

- Intrinsic.
- Extrinsic.

The latter includes TNF and steroids in animals, and hormones for leaf abscission in plants. Morphological changes in apoptotic cells include:

- Condensation of chromatin in the nucleus, followed by fragmentation of the DNA detected as a ladder.
- Cell shrinkage.
- Blebbing of the plasma membrane, followed by the formation of cell fragments that can be taken up more easily by phagocytes than the whole cell.

Apoptosis is a highly active process involving gene expression, and expression of phospholipids on the outside of the cell. A pathological rise in cytosolic free Ca^{2+} causes movement of phosphatidyl serine from the inner leaflet of the lipid bilayer to the outer, where it is recognised by receptors on the surface of macrophages. Ca^{2+} can be involved at several stages of apoptosis, and can be the intracellular signal initiating the apoptotic pathway. A prolonged increase in cytosolic free Ca^{2+} activates a range of intracellular enzymes, including caspases, calpain and other proteases, endonucleases, phospholipases, transglutaminases, kinases and phosphatases. An abnormal rise in cytosolic free Ca^{2+} can activate inappropriately Ca^{2+}-activated cation and anion channels, and also increase the production of reactive oxygen metabolites. Several of these may have a relatively low affinity for Ca^{2+}, and are not activated by physiological Ca^{2+} signals.

A crucial step in the initiation of the pathway to complete apoptosis is the activation of a cascade of proteases – caspases (cysteine-aspartic proteases). They are synthesised as pro-enzymes, analogous to proteases in the intestine, a peptide having to be cleaved off them if they are to be active. There are several types of caspases, with subtly different affinities and substrate specificities, and they are thought of as 'executioner' proteins. At least 12 human caspases have been identified, and are grouped as either *initiator* caspases (e.g. caspase-2, -8, 9 and -10), which start the pathway off, or *effector* caspases (e.g. caspase-3, -6 and -7), which go on to mediate the effect on other cell components. Caspases cleave many cytoskeletal, cytosolic and nuclear proteins, leading to nuclear condensation and blebbing of the plasma membrane. They can also inactivate inhibitors of apoptosis by cleavage. Two key steps in initiating the apoptotic pathway leading to caspase activation are:

1. Release of cytochrome *c* from mitochondria.
2. Changes in the distribution of phospholipids across the bilayer of the plasma membrane.

Release of cytochrome *c* from mitochondria is triggered by overload of Ca^{2+} inside the mitochondria. Cytochrome *c* is loosely bound to cardiolipin on the surface of the mitochondrial inner membrane. Its normal function is to pass electrons, through its haem moiety, from the cytochrome bc_1 complex to cytochrome oxidase in the respiratory chain. But it also very water soluble. So, once there is a large pore in the inner mitochondrial membrane, it leaks out into the cytosol. Here it binds and activates caspase-9, which then sets up a caspase cascade involving caspase-3 and -7. This, in turn, activates enzymes that degrade the components causing the morphological and biochemical cellular changes that make a cell apoptotic. The rise in cytosolic free Ca^{2+} also activates calmodulin, Ca^{2+}-dependent kinases, nitric oxide (NO) synthase, and transglutaminases. In the nucleus, high Ca^{2+} disrupts DNA by competing with Mg^{2+}, aiding nuclear change. Loss of Ca^{2+} from the ER initiates apoptosis via activation of caspase-12, but released cytochrome *c* also binds to the ER inositol trisphosphate (IP_3) receptor causing a further increase in cytosolic free Ca^{2+}. The BCL-2 family is important in apoptosis. Furthermore, the IP_3 receptor can be cleaved by calpain or caspase-3. But, the key step is a sustained, large increase in Ca^{2+} within the inner matrix of the mitochondria. This reduces ATP synthesis, and provokes the formation of large protein complex that forms a pore. This pore acts as a voltage-operated channel, allowing movement of ions and small molecules less than 1500 kDa into the cytosol. This is followed by massive protein release. Formation of this pore can also be stimulated by oxidative stress, pyridine nucleotides, alkalinisation and a reduction in the membrane potential across the inner mitochondrial membrane.

An important step at the end of the apoptotic sequence is loss of phospholipid symmetry in the plasma membrane, as a result of activation of a scramblase by the large rise in cytosolic free Ca^{2+}. Normally, all the phosphatidyl serine is found in the inner leaflet of the lipid bilayer of the plasma membrane. Activation of the scramblase causes the phosphatidyl serine to flip into the outer leaflet, providing a receptor for macrophages to engulf the apoptotic cell.

Other key proteins in apoptosis released with cytochrome *c* from permeabilised mitochondria include AIF and the BCL-2 family that control entry into the caspase pathway. AIF is normally anchored to the inner membrane of the mitochondria. In some cells, when released with cytochrome *c* into the cytosol, it moves into the nucleus, where it helps to provoke large-scale DNA fragmentation. Caspase cleaves AIF, leading to further permeabilisation of the mitochondrial membrane. Movement of AIF to the nucleus leads to the classic DNA ladder seen on gel electrophoresis of DNA isolated from apoptotic cells. The evidence for a role of AIF comes from its down-regulation using small interfering RNA (siRNA). This suppresses apoptotic stimuli such as glutamate, low oxygen or *N*-methyl-d-aspartate (NMDA)-independent neuronal cell death. In the nucleus, a rise in intranuclear Ca^{2+} activates:

- Endonucleases.
- Transcription factors leading to gene expression.
- Disorganisation and unfolding of chromosomes.

Genes include calmodulin, c-*fos*, *gad1* (growth arrest genes), c-*jun* and NF-κB, resulting in an arrest in G_2 of the cell cycle. Ca^{2+} in the nucleus is also required for accumulation of clusterin. Isolated nuclei undergo DNA fragmentation when incubated with Ca^{2+} and ATP, as occurs in apoptosis. In addition, caspase-3 cleaves several proteins associated with Ca^{2+}, including the IP_3 receptor, the plasma membrane Ca^{2+}-MgATPase and the Na^+/Ca^{2+} exchanger, as well as the sodium pump, the Na^+/K^+-MgATPase. The latter is important in apoptosis, as cell shrinkage occurs as a result of inhibition of the sodium pump.

10.4.3 Autophagy

Autophagy is the process where a cell degrades itself from inside using lysosomal enzymes. It was first described in the 1960s, and does not inevitably lead to cell death. Intracellular Ca^{2+} plays a role under particular circumstances. Autophagy involves sequestering of part of the cytosol and organelles, which are then delivered to the lysosome or intracellular vacuole for degradation and recycling of the degradation products. Autophagy can occur in any eukaryotic cell, and is regulated by kinases, phosphatases and MgGTPases. But when it gets out of hand, it leads to cell death. Autophagy interacts with the endocytic pathway, and is carefully regulated so that the cell forms a membrane within itself, degrading the contents inside the membrane. The nutrients derived from this are then used by other parts of the cell. Autophagy is involved in some developmental processes, and in diseases such as nerve loss, cancer and infection. It is particularly important when a eukaryotic cell is starved of nutrients. There are eight types. Unlike apoptosis, autophagy does not cause condensation of the nucleus. Instead, a large vacuole forms in the cytoplasm containing the intracellular component or an engulfed bacterium. Autophagy can be triggered by ER stress, and can be a survival mechanism. Two Ca^{2+}-activated enzymes are usually essential:

- Ca^{2+}-dependent activation of AMP-activated protein kinase.
- Ca^{2+}-calmodulin kinase DAPK (death-associated protein kinase).

As in necrosis, precipitates of calcium phosphate can occur. A reduction in cytosolic free Ca^{2+} stops calpain-mediated changes of the protein Atq5.

10.4.4 Lysis

Cell lysis occurs when the cell membrane bursts open and the cell releases its contents, as a result of catastrophic increases in the permeability of the plasma membrane. This can occur as a result of osmotic shock or attack on the plasma membrane, either by internal or external agents. In all cases, a rapid rise in cytosolic free Ca^{2+} occurs prior to lysis. Internal attack occurs, for example, when a virus is released. External agents include pore formers:

- The terminal membrane attack complex of complement.
- T-cell perforin.
- Bacterial toxins such as alfatoxin and streptolysin.
- Cell wall or membrane degradation, e.g. when bacteria are attacked by antibiotics or are engulfed by neutrophils, macrophages and other phagocytes.

The first thing that happens when the plasma membrane is damaged or attacked is a large rise in cytosolic free Ca^{2+}, as it pours into the cell. This can activate defence mechanisms, whereby the cell attempts to remove the potentially lethal pores. Or it can lead to irreversible damage to organelles, such as mitochondria, and activate degradative enzymes – proteases, nucleases, phosphatases and phospholipases. In many cases the cell simply explodes, causing local chaos, with cell fragments that will induce inflammation, as macrophages and other phagocytes attempt to clean up the mess.

The membrane attack complex of complement is a good model for the sequence of Rubicons that occurs when pores form in the plasma membrane of attacked cells (Figure 10.3). The membrane attack complex consists of five proteins – C5b, C6, C7 and C8 plus several C9 molecules. The C5b678 complex is bound to the plasma membrane. As C9s bind, they insert into the membrane and cause Ca^{2+} to enter the cell. The cytosolic free Ca^{2+} rises to several micromolar within less than 1 min. This Ca^{2+} rise activates processes within the cell, such as the production of superoxide from neutrophils, and also activates a protection mechanism, whereby the cell tries to protect itself by removing the potentially lethal complexes in the membrane by vesiculation, i.e. endocytosis or budding. If the cell removes the complexes successfully it recovers fully within a few minutes. If not, then the cell crosses the next Rubicon as the result of further C9s forming a larger pore. This is big enough to let ATP and other small organic molecules out of the cell. As a result, the cell ATP drops markedly. However, the cell can still survive if it can remove the complexes via vesiculation fast enough. If it does not achieve this in time, the cell lyses and releases all its contents into the surrounding fluid. When erythrocytes release haemoglobin, or other cells release lactate dehydrogenase, this reflects the number of cells irreversibly lysed, and is not a graded release of protein from all of the cells.

10.4.5 Cell Death Conclusions

Although there are common principles in each type of cell death, each cell type under different conditions of stress exhibits differences in the precise biochemical pathways and structural changes that ultimately lead to destruction and removal of the cell. This is seen in apoptosis, where there is a cytochrome *c*-dependent and -independent activation of caspases. These, in

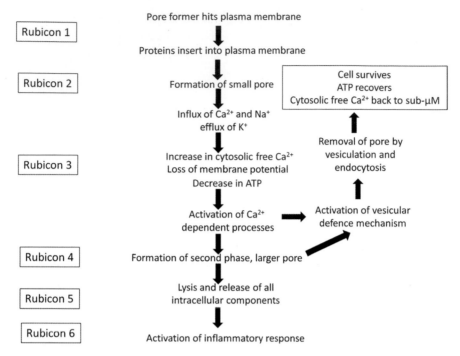

Figure 10.3 The sequence of Rubicons after attack by a pore former. Reproduced by permission of Welston Court Science Centre.

turn, exhibit subtle variations in specificity and enzymatic activity. Cell death is a classic example of the Rubicon principle. The cell is attacked, stressed or programmed, but only sets off on the pathway to cell death and destruction if the necessary threshold is crossed. Take a mammalian egg fertilised by a sperm, which then divides approximately every 24 h. The embryo starts out with a volume of just 1 nl (10^{-9} l). If it and its progenitors divide once every day, by 1 week the embryo will be the size of a pea. But within 1–2 months it will be the size of an asteroid, and in 9 months it will be the size of the Milky Way. Quite a large baby! This tells us that in life, cell death is occurring all the time. So, when considering a cancer, if the death rate is just 1% a day less than the division rate then within a few months the tumour will be huge. Apoptosis is a great drug target for treating cancer. But unless this small change by small change issue is taken into account then the approach will be flawed.

A rise in intracellular Ca^{2+} occurs prior to all forms of cell death. This rise can initiate the pathway to cell death, or under other circumstances Ca^{2+} can activate defence processes enabling the cell to survive attack or stress. Once the cytosolic free Ca^{2+} rises to more than 10 µM, the cell can be damaged irreversibly. At high micro- or milli-molar Ca^{2+}, calcium phosphate precipitates, Ca^{2+} disrupts DNA and RNA, Ca^{2+} activates proteases and nucleases, and causes proteins to denature and precipitate, as well as activating or inhibiting many enzymes not normally regulated by physiological concentrations of cytosolic free Ca^{2+}. In the heart, following myocardial infarction, Ca^{2+} uptake into mitochondria precipitates Ca^{2+} phosphate, irreversibly damaging the organelle, leading to cell death. In contrast, prolonged release of Ca^{2+} from the ER activates a stress response, whereby the cell attempts to protect itself against malfolded proteins. If this is inadequate the cell crosses the Rubicon and kills itself.

> **Box 10.3 Mutations in Ca²⁺ signalling genes causing inherited conditions.**
>
> - Ca²⁺ channels: voltage-gated or store-operated; e.g., night blindness.
> - Ca²⁺ pumps: plasma membrane, ER and Golgi; e.g., Darier's, Hailey–Hailey.
> - ER Ca²⁺ release proteins: ryanodine and IP₃ receptors – e.g. malignant hyperthermia.
> - Ca²⁺ targets: calmodulin, troponin C, phosphatases, kinases and proteases; e.g. CPVT.

10.5   Genetic Abnormalities in Ca²⁺ Signalling Proteins

Mutations in genes coding for Ca^{2+} signalling proteins are responsible for a number of inherited conditions (Box 10.3). Mutations in exons may result is a truncated protein by introducing a stop codon, or they may change the protein sequence through missense or nonsense mutations. In some cases, mutations in introns can change the protein sequence by changing splicing. Interestingly, most are not immediately lethal, but only cause major problems when the person carrying the mutation is stressed. An example of this is malignant hyperthermia, caused by a mutation in the ryanodine receptor in the sarcoplasmic reticulum (SR) of skeletal muscle, or in some cases in the voltage-gated Ca^{2+} channel $Ca_v1.1$. This condition is important clinically, since it is provoked by certain anaesthetics, such as halothane, and muscle relaxants, such as succinyl choline. If not treated quickly with the ryanodine receptor blocker dantrolene, the high temperature that ensues can be lethal.

Although many mutations causing inherited conditions involving Ca^{2+} signalling proteins have been identified, the precise way in which these mutations cause the symptoms of the disease have rarely been established. The use of knock-out mice has provide some information about how mutations can be linked to symptoms, but compensating mechanisms different from those in the human disease can obscure the precise mechanisms. There have been some surprises, for example Darier's disease. As well as direct effects on Ca^{2+} signalling through mutations in Ca^{2+}-signalling genes, there are conditions involving mutations in genes coding for proteins that interact with Ca^{2+} signalling, and thus have indirect effects on cell regulation through intracellular Ca^{2+}. An example of an inherited pathological condition involving Ca^{2+} signalling without mutations in Ca^{2+} signalling genes is one where antibodies to receptors or Ca^{2+} channels are produced. Thus, there are four main groups of mutations in Ca^{2+} signalling genes that cause, or are associated with, inherited conditions:

1. Ca^{2+} channels in the plasma membrane: voltage-gated or store-operated.
2. Ca^{2+} pumps: plasma membrane, ER and Golgi.
3. ER Ca^{2+} release proteins: ryanodine and IP₃ receptors.
4. Ca^{2+} targets: calmodulin, troponin C, phosphatases, kinases and proteases.

Most mutations in Ca^{2+} signalling genes are autosomal recessive, one or two are X-linked, and a few, such as Darier's disease, are autosomal dominant. Although the prevalence of the diseases associated with mutations in specific Ca^{2+} signalling genes is relatively rare, Darwin teaches us that they must be common enough in the population to have a selective advantage, particularly in heterozygotes. We can only guess at what that might be. Interestingly, one inherited disease involving immune deficiency has had a major impact on Ca^{2+} signalling since it led to the discover of Orai1 – the plasma membrane protein which forms the store-operated Ca^{2+} entry (SOCE) channel after interacting with the ER protein STIM.

10.5.1   Ca²⁺ Channelopathies

Several calcium channelopathies have been identified (Table 10.1). The conditions are associated with tissues that have high expression of the Ca^{2+} channel concerned. Thus, $Ca_v1.1$ is the key Ca^{2+} channel in skeletal muscle, which fits its tendency for malignant hyperthermia, though the more common cause of this condition involves mutations in the ryanodine receptor. $Ca_v1.4$ is found particularly in rods in the eye, and thus mutations cause night blindness. In contrast, mutations in key neuronal Ca^{2+} channels, P/Q and T, cause epilepsy and other problems in the brain. An important mutation in a non-voltage-gated channel, found in most non-excitable cells, causes immunodeficiency. This led to the discovery of Orai1 – the plasma membrane protein that binds to STIM1 in the ER, thereby opening SOCE channels in the plasma membrane.

Another important channelopathy is caused by mutations in the gene coding for the lysosomal Ca^{2+} channel mucolipin 1 (TRPML1 = MCOLN1). This produces a nasty lysosomal storage disease – mucolipidosis type IV.

10.5.2   Ca²⁺ Pumpopathies

Several conditions are associated with mutations in plasma membrane Ca^{2+} pumps (PMCA), or Ca^{2+} pumps in the ER/SR and Golgi (Table 10.2).

Table 10.1 Ca^{2+} channelopathies due to mutations in voltage-gated Ca^{2+} channels.

Ca²⁺ channel type	Ca_v	α₁ Subunit type	Inherited disease
L	1.1	S	Hypokalaemic periodic paralysis, malignant hyperthermia
L	1.2	C	Timothy syndrome, severe arrhythmia syndrome
L	1.3	D	Not known
L	1.4	F	Incomplete X-linked congenital stationary night blindness, retinal disorder with autism
P/Q	2.1	A	Type 2 episodic ataxia, type 6 spinocerebellar ataxia, type 1 familial hemiplegic migraine, epilepsy
N	2.2	B	Not known
R	2.3	E	Not known
T	3.1	G	Not known
T	3.2	H	Epilepsy – childhood absence and idiopathic generalised, autism spectrum disorder
T	3.3	I	?

Table 10.2 Ca^{2+} pumpopathies.

Ca²⁺ pump	Location	Condition
PMCA	Plasma membrane	Deafness
SERCA 1	SR	Brody myopathy
SERCA2	ER	Darier's disease
SPCA	Golgi	Hailey–Hailey disease

Knock-outs of PMCAs have produced some interesting, if surprising results. Knocking-out or mutating a major cell protein is not lethal, and often only leads to relatively minor impairments. For example, the knock-out of PMCA2 causes deafness in mice and balance problems. Whereas mice with knock-outs of PMCA4 appear normal, though their sperm are abnormal, leading to male sterility. This is because some 90% of all PMCA in sperm is type 4. Pathological conditions where there may be defects in Ca^{2+} pumping include Alzheimer's disease. Surprisingly, knocking-out the SERCA2 pump only causes a moderate dysfunction of the heart, despite a large decrease in the SR Ca^{2+} store. The myocytes compensate by increasing Ca^{2+} influx through L-type Ca^{2+} channels in the plasma membrane and the Na^+/Ca^{2+} exchanger, and by increasing the response of the myofibril system to Ca^2. But the ATP2A2 gene coding for SERCA2 is mutated via missense and other mutations in Darier's disease. Yet the only pathological effects are skin lesions and in some cases psychological problems. More than 150 mutations have been found in the ATP2A gene of patients with Darier's disease. On the other hand, in heart failure small ubiquitin-related modifier (SUMO)-ylation of SERCA2A at Lys480 and Lys580 may be important, suggesting a role for degradation of a Ca^{2+} signalling protein.

10.5.3 Mutations in ER Ca²⁺ Release Proteins

Four conditions are associated with mutations in proteins that are responsible for releasing Ca^{2+} from the SR/ER into the cytosol:

- Huntington's disease – mutations in the IP_3 receptor.
- Malignant hyperthermia – mutations in the RyR1 receptor or $Ca_v 1.1$.
- Central core disease – mutations in the RyR1 receptor.
- Cardiac arrhythmia and catecholaminergic polymorphic ventricular tachycardia (CPVT) – mutations in RyR2 receptor.

Mutations in the IP_3 receptor have been found in Huntington's disease. On the other hand, mutations in the ryanodine RyR1 receptor produce malignant hyperthermia or central core disease. In contrast, mutations in RyR2 produce cardiac arrhythmias, which can lead to sudden death. Malignant hyperthermia (hyperpyrexia) is an autosomal dominant condition that shows no obvious abnormalities until the person is exposed to large amounts of caffeine or a volatile anaesthetic, such as halothane, isoflorone or succinylcholine. In people with malignant hyperthermia, these drugs induce a massive and uncontrollable increase in oxidative metabolism in skeletal muscle, overwhelming the ability of the body to regulate its O_2/CO_2 balance and body temperature. Malignant hyperthermia is related to central core disease, both phenotypically and genetically. More than 80 missense mutations have been detected in this condition and which explain over 50% of cases. Each mutation increases the sensitivity of the ryanodine receptor to caffeine and volatile anaesthetics. Administration of the anaesthetic causes a rapid opening of the mutant ryanodine receptor, leading to a large release of Ca^{2+} from the SR, a subsequent muscle contraction, and a rise in body temperature due to the massive turnover of the SERCA MgATPase. Treatment with dantrolene, to inhibit the ryanodine receptors, can cause a successful decrease in morbidity and mortality.

On the other hand, central core disease, usually an autosomal dominant inherited condition, but can be recessive, is caused by another mutation in RyR1. This mutation results in myopathy early in life, with consequent hypotonia, proximal muscle weakness and delays in the motor sensory system. Intracellularly, it is characterised by regions where there are no mitochondria, known as 'cores'. Similar cores are present in minicore disease – an inherited condition seen at birth, leading to hypotonia, distal joint laxity, and later scoliosis and respiratory insufficiency. In contrast, mutations in Ry2 are the cause of catecholamine polymorphic arrhythmogenic syndrome, leading to tachycardia (catecholaminergic polymorphic ventricular tachycardia (CPVT)), which can cause sudden death. More than 20 mutations in the RyR2 gene have been detected, occurring in 1 : 10 000 europeans. The mutations make the heart ryanodine receptor sensitive to phosphorylation by protein kinase A, leading to increased Ca^{2+} release from the SR, and thus an increase in cytosolic free Ca^{2+}. Some people with CPVT have mutations in calsequestrin 2, leading to loss on ER Ca^{2+}, and disruption of its interaction with the luminal side of the ryanodine receptor. Mutations in calmodulin and triadin can also cause CVPT.

10.5.4 Mutations in Ca^{2+} Target Proteins

Mutations in Ca^{2+} target proteins associated with disease include calmodulin, troponin C, calcineurin and calpain. Mutations in calmodulin can be lethal, causing sudden cardiac arrest in the young individual. Heterozygous missense mutations in calmodulin (CALM1) have been found to cause severe cardiac arrhythmia such as the rare inherited disorder, CPVT, which can cause sudden cardiac arrest during exercise or sudden emotion, and syncope, resulting in death in young individuals. Similarly, mutations in cardiac troponin C cause heart problems, in particular hypertrophic cardiomyopathy or dilated cardiomyopathy. Mutations in the phosphatase calcineurin may be associated with Alzheimer's and Huntingdon's diseases. Mutations have also been found in the protease calpain, and are involved in the pathogenesis of the autosomal recessive disorder limb-girdle muscular dystrophy type 2A (LGMD2A), which shows selective atrophy of muscle in the proximal limbs.

10.5.5 Proteins Associated with Ca^{2+} Signalling

Abnormalities and changes in expression of several proteins closely associated with Ca^{2+} signalling have been found. An important example is annexinopathy, which plays an important role in leukaemia. Disruption of Ca^{2+} signalling can also occur in lipid storage diseases such as Gaucher's, due to mutations in glucocerebrosidase, Sandhoff's disease, due to mutations in hexosaminidase, and Niemann–Pick disease type C, which can lead to neuronal dysfunction. In Sandhoff's disease, Ca^{2+} may decrease via the SERCA pump, whereas in Niemann–Pick disease type C there is an impairment of Ca^{2+}-dependent fusion of endosomes with lysosomes, leading to accumulation of cholesterol in the cell.

10.6 Oxygen and Cell Pathology

Most eukaryotic cells require oxygen for ATP synthesis in the mitochondria, though there are cells, such as erythrocytes and the kidney medulla, which rely on substrate-level phosphorylation. Lack of oxygen – hypoxia – leads to a large drop in MgATP, reducing the activity of Ca^{2+} pumps, causing Ca^{2+} to leak into the cell and out of the ER. A sustained rise in cytosolic free Ca^{2+} causes overload in the mitochondria. If the oxygen supply is not restored quickly, then the cell will die by necrosis. Typical conditions include stroke or myocardial infarction. But, too much oxygen too quickly is toxic, because it causes production of oxygen metabolites. This is the 'oxygen paradox'. A further problem is the so-called 'calcium paradox'.

10.6.1 Ca^{2+} Paradox

The puzzle arose in the 1960s. When organs, such as the heart, were perfused with a Ca^{2+}-free solution, followed by a solution containing Ca^{2+}, irreversible damage occurred. This was because removal of extracellular Ca^{2+} made the cell membrane permeable to Ca^{2+}. So when external Ca²⁺ was restored it flooded into the cell, causing irreversible damage to the mito-chondria. This was then called the 'calcium paradox'. It was discovered by a medical student called Ariaen Zimmerman in perfused rat heart, but was then found to occur in many tissues, including the kidney. In a student experiment Zimmerman asked his friend, Willem Hulsmann, a physician and biochemist, if he could replace glucose in a perfused heart preparation with acetoacetate. To their surprise it was fatal. The heart stopped beating. Thinking this might have something to do with citrate and the Krebs cycle, they took Ca²⁺ out of the perfusion medium. This was also fatal. After a few strong contractions, the heart stopped beating and died. The calcium paradox had been discovered! Just 2 min without extracellular Ca²⁺ was enough to cause massive release of intracellular enzymes and the necrosis. It was called the calcium para-dox because of apparent similarities with a previous idea called the potassium paradox. The importance of Ca²⁺ in the cell injury process was supported by protection against cell injury using Ca²⁺ channel blockers such as verapamil. Is the calcium paradox simply an experimental artefact? Tissues *in situ* never see blood containing no Ca²⁺! But a key question was whether it had any relevance to the storage of human hearts for transplantation, or when isolated during a heart operation, or indeed when natural insults hit the heart, such as ischaemia or infarction.

When the blood supply to a tissue is restricted the tissue becomes ischaemic. This stress, involving lack of oxygen and nutrients, may not be immediately fatal. But it will be if the blood flow stops completely as a result an obstruction caused by a blood clot (embolus or thrombus) or a plaque. This is an infarction. This is what happens in a heart attack or stroke, and will lead to cell death by necrosis if the blood supply is not resumed within a minute or so. The calcium paradox is important because it highlighted the role of massive Ca²⁺ influx in cell injury and death, the mechanisms of injury inside the cell being relevant to reperfusion injury in the presence of Ca²⁺ and oxidative damage. It also had a major influence on the develop-ment of solutions used for the heart during operations. Removal of extracellular Ca²⁺ causes morphological and functional disruption. Cells begin to separate at their gap junctions and intercalated discs separate, normally held together by cadherin complexes, cadherin being a Ca²⁺-dependent adhesion protein. In addition, some glycocalyx is lost. There is also a large increase in Ca²⁺ permeability of the plasma membrane. When extracellular Ca²⁺ is restored, after a few contractions, ATP synthesis is severely impaired, and Ca²⁺-phosphate precipitates in the mitochondria. The result is loss of coupling between adjacent cells, and irreversible cell injury, all initiated by catastrophic influx of Ca²⁺.

There are some similarities between the calcium paradox mechanism and the 'oxygen para-dox'. But, there are also some important differences. First, loss of oxygen, followed by reoxy-genation of tissues such as the heart, leads to a rapid influx of Na⁺, then Ca²⁺, the Na⁺ influx not appearing to be so significant in the Ca²⁺ paradox. A rise in intracellular Na⁺ will lead to an increase in intracellular Ca²⁺ via the Na⁺/Ca²⁺ exchanger. Secondly, injury following loss of oxygen can be reversed, but not so with the calcium paradox. Thirdly, anoxia-reoxygenation occurs naturally in heart attacks and strokes, and in operations when organs are isolated. Com-plete loss of extracellular Ca²⁺ does not occur naturally. However, there is a cardioplegic solu-tion known as Bretschneider's (histidine-tryptophan-ketoglutarate (HTK)) that was developed during the 1960s, which is used in central Europe in heart transplants and operations. This medium nominally contains no Ca²⁺. However, there is a trace of Ca²⁺, and the solution is used cold at 4–8 °C. Cardioplegia is the deliberate, but temporary, stopping of cardiac activity (e.g. in heart surgery and transplantation). It is therefore essential to have a medium that keeps the heart alive, allowing the heart to function normally once the blood supply is restored. Most countries, however, use cardioplegic solutions containing Ca²⁺. The calcium paradox is thus a laboratory artefact, and does not occur under natural circumstances. Nevertheless, it has provoked research that provided insights into when intracellular Ca²⁺ can be a foe rather than a friend.

Intracellular Ca^{2+} plays an important role in cell injury after a stroke, caused by a blood clot in a cerebral blood vessel, and is a leading cause of death. Ca^{2+} overload is a major factor in cell death, entering neurones via the NMDA glutamate receptor and voltage-gated Ca^{2+} channels. Ca^{2+} is also released from the ER via ryanodine receptors prior to cell death. The accumulation of unfolded proteins activates the ER stress response, and presenilin 1 may allow Ca^{2+} to leak out of the ER. Transient anoxia also activates the NF-κB transcription factor leading to increased expression of the Na^+/Ca^{2+} exchanger in the plasma membrane, increasing Ca^{2+} into the cell.

10.6.2 Oxidative Damage and Intracellular Ca^{2+}

Reduction of O_2 to H_2O requires four electrons:

$$O_2 + 1e \rightarrow O_2^- + 1e + 2H^+ \rightarrow H_2O_2 + 1e + H^+ \rightarrow H_2O + OH^· + 1e + H^+ \rightarrow H_2O \qquad (10.1)$$

Oxygen metabolites damage proteins, lipids and nucleic acids. When they attack membranes, these can become permeable to Ca^{2+} causing Ca^{2+} to leak into the cytosol. Reduction in oxygen supply followed by excess oxygen, especially in the mitochondria, causes the production of oxygen metabolites – superoxide (O_2^-) and hydrogen peroxide (H_2O_2) – from enzymes such as cytochrome oxidase. Another damaging oxygen metabolite is singlet oxygen ($^1O_2$). There are also oxidases that produce oxygen metabolites normally. Furthermore, production of oxygen metabolites is a key part of the mechanism by which phagocytes kill invading microbes. A rise in cytosolic free Ca^{2+} in neutrophils, eosinophils and macrophages activates an NADPH oxidase via calmodulin, producing superoxide. In the phagosome, this dismutates to hydrogen peroxide, which then reacts with Cl^- in neutrophils and Br^- in eosinophils to produce OCl^- and OBr^-, respectively, catalysed by a peroxidase. These really do kill all known germs! In contrast, some fertilised fish and invertebrate eggs secrete an ovoperoxidase that forms OI^- as a spermicide, to prevent a second sperm fusing with the egg. Secretion is triggered by a wave of cytosolic free Ca^{2+}. Attack of the plasma membrane of microbes or sperm by hypohalides leads to a large flow of Ca^{2+} into the cell. The short half-life of several oxygen metabolites means that they only diffuse a short distance from their site of formation. Oxidative cell damage and disruption of intracellular Ca^{2+} arises from excess oxygen supply following damage to blood vessels or the blood–brain barrier, or changes in oxygen scavengers. This may play a role in the way amyloid-β peptide plaques damage neurones in Alzheimer's disease. Experimental addition of enzymes that generate superoxide, or addition of H_2O_2, to cells causes an influx of Ca^{2+} into the cell, which damage intracellular proteins, mitochondria, the ER and nucleic acids, and can affect gene expression. But the concentration of oxidant required is often very high, well above physiological. Furthermore, it is not clear whether the superoxide or H_2O_2 is acting on the outer surface of the attacked cell, or whether they penetrate into the cell.

10.7 Inappropriate Ca^{2+} Signalling

Inappropriate effects on Ca^{2+} signalling occurs when the Ca^{2+} signalling system is activated under pathological circumstances. A good example is in hyperventilation. This causes a reduction in free Ca^{2+} in the blood because of an increase in plasma pH. If free Ca^{2+} drops below about 0.9 mM, nerves fire spontaneously, leading to unwanted muscle contractions. A similar unwanted activation of the Ca^{2+} signalling system in skeletal muscle occurs in cramp. Such inappropriate effects on Ca^{2+} signalling play a role in several diseases.

10.7.1   Cramp

Most people will have experienced pain in the legs or arms as a result of unwanted muscle contraction. Such cramp in skeletal muscle occurs as a result of fatigue, or lack of salts such as Na^+, K^+ and Mg^{2+}, for example from intensive sweating, seen in athletes. This can occur at night in bed, and can be very painful. Cramp can also occur in smooth muscle during menstruation or gastroenteritis. In skeletal muscle cramp, if the SR Ca^{2+} pump is not working properly, because of osmotic disturbance, the cytosolic free Ca^{2+} remains high, so Ca^{2+} remains bound to troponin. Failure of ATP to drop off myosin, can also prevent muscle relaxing.

10.7.2   Immune System and Other Organs in Disease

In immune-based and inflammatory diseases, defence cells are inappropriately activated by a rise in cytosolic free Ca^{2+}. For example, neutrophils are activated in the rheumatoid joint to release oxygen metabolites (Figure 10.4), myeloperoxidase and proteases, which are the main cause of damage and pain in the synovial joint. Similarly, in inflammatory bowel disease (IBD), phagocytes are attracted into the gut, where they are inappropriately activated, causing similar damage. In both cases, phagocytes are activated by complement fragments and Fc receptors,

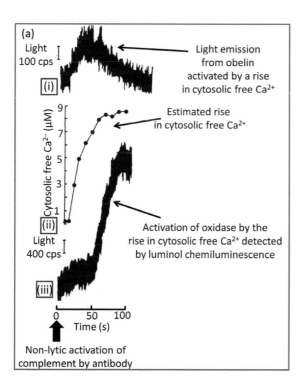
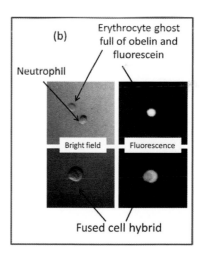

Figure 10.4 Activation of oxygen metabolite production by the non-lytic action of complement. The figure shows the rapid increase in cytosolic free Ca^{2+} in neutrophils activated by the non-lytic action of complement. (a) (i) Light emission from the Ca^{2+}-activated photoprotein obelin. (ii) Cytosolic free Ca^{2+} estimated from (i). (iii) Light emission from luminol, detecting the production of oxygen metabolites by the neutrophils activated by the rise in cytosolic free Ca^{2+}. Cytosolic free Ca^{2+} was measured using the Ca^{2+}-activated photoprotein obelin, entrapped within sealed erythrocyte ghosts and then fused with a live neutrophil using Sendai virus (b). From Hallet and Campbell (1982) Measurement of changes in cytoplasmic free Ca^{2+} in fused cell hybrids. *Nature*, **295**, 155–158. Reproduced with permission from Nature.

and mitochondrial-derived chemotactic formylated peptides, via a rise in cytosolic free Ca^{2+}. In contrast, in eczema and asthma, it is mast cells that are inappropriately activated by IgE receptors, causing a rise in cytosolic free Ca^{2+} that triggers the release of histamine. In the lung, this causes contraction of smooth muscle through an inappropriate rise in cytosolic free Ca^{2+} in these cells. Inappropriate rises in cytosolic free Ca^{2+} activate lymphocytes, phagocytes, mast cells and other cells of the immune system in neurogenerative disease, and any condition where the immune system is activated.

An inappropriate rise in cytosolic free Ca^{2+} also may provoke demyelination from oligodendrocytes in the brain in multiple sclerosis and in neurones in Alzheimer's disease. Experimentally, the build-up of extracellular amyloid-β peptide in neurones causes the resting free Ca^{2+} to rise to some 300 nM. When this is prolonged, it leads to irreversible cell damage and death by apoptosis or necrosis. Other neurodegenerative diseases where Ca^{2+} signalling is impaired, and has a role in the disease mechanism, include peripheral neuropathies, amyotrophic lateral sclerosis, and Parkinson's and Alzheimer's diseases.

Parkinson's disease is characterised by degeneration, and loss of, monoaminergic neurones in the brain stem and basal ganglion, with consequential large loss of dopaminergic neurones crucial for motor control. A key question is whether mitochondrial stress induced by Ca^{2+} causes the dopaminergic cells to die.

There are two types of Alzheimer's disease. The common, sporadic, late-onset type is found in people older than 65 years. Less common is the familial or early-onset type involving mutations in presenilin 1 or 2. Alzheimer's is characterised by plaques of amyloid-β peptide, neurofibrillary tangles and hyperphosphorylation of the τ protein, which lead to cell death. Patients suffer long-term emotional stress, and progressive and irreversible memory loss. Disruption of Ca^{2+} signalling is associated with early onset Alzheimer's, through presenilin and Ca^{2+} leaking from the ER. Reactive oxygen species may alter Ca^{2+} in the ER. Amyloid-β peptide disrupts ER Ca^{2+} and decreases RyR3 expression.

Disruption of Ca^{2+} signalling also occurs in Gaucher's, Sandhoff's, and Niemann–Pick diseases. In Sandhoff's disease Ca^{2+} may decrease via the SERCA pump, and in Niemann–Pick disease type C there is an impairment of Ca^{2+}-dependent fusion of endosomes with lysosomes, leading to accumulation of cholesterol in the cell. The Ca^{2+} signalling system is therefore a good target for drug discovery in all conditions where intracellular Ca^{2+} is activated inappropriately.

10.7.3 Bacterial Metabolic Toxins

Intracellular Ca^{2+} plays a vital role in the production and action of metabolites produced by microbes in the gut (Figure 10.5). These metabolites play a key role in causing the symptoms of lactose and food intolerance, irritable bowel syndrome (IBS), and inflammatory bowel disease (see Chapter 8, Section 8.7.10). They may also play a key role in diseases around the body, including type 2 diabetes, Parkinson's and Alzheimer's diseases. They do this by interacting with Ca^{2+} signalling in host and bacterial cells, and through the covalent modification of hormones and neurotransmitters. Metabolic toxins are produced by bacteria digesting sugars and other foods not absorbed in the small intestine, under the anaerobic conditions of the large intestine. Metabolites include alcohols, diols, aldehydes, ketones and acids (Figure 10.5). Several generate Ca^{2+} signals in bacteria, affecting gene expression and cell growth. A crucial metabolite is methylglyoxal, which has effects on both bacteria and eukaryotic cells. Methylglyoxal generates Ca^{2+} signals in bacteria, stimulates contraction of smooth muscle in isolated ileum, causes a negative inotropic effect on the heart, inhibits K^+ channels and can provoke apoptosis in tissue culture cells. These effects are mediated by a rise in intracellular Ca^{2+}.

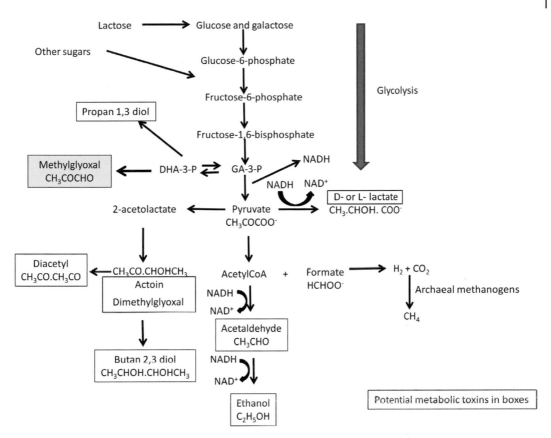

Figure 10.5 Bacterial metabolic toxins. The figure shows the production of metabolic toxins from sugars such as lactose, which, in the absence of oxygen, is needed to oxidise NADH to NAD⁺ to allow glycolysis to generate ATP. This leads to the generation of metabolic toxins which are alcohols, aldehydes and ketones. These can be oxidised further to short-chain acids. GA-3-P = glyceraldehyde-3-phosphate; DHA-3-P = dihydroxyacetone-3-phosphate. The metabolic toxins are highlighted in boxes; a particularly important one is methylglyoxal. All are capable of causing a rise in cytosolic free Ca^{2+} in both bacteria and eukaryotic cells. Reproduced by permission of Welston Court Science Centre.

The importance of this bacterial metabolic toxin hypothesis is seen when examining the symptoms exhibited by patients with IBS, as well as lactose and food intolerance. These patients exhibit gut pain and distension as a result of gas produced by gut microbes. But the metabolic toxins affect cells around the body via Ca^{2+} signalling. And they can activate diarrhoea in some patients, while others suffer from constipation. The main systemic symptoms around the body are headache, loss of concentration and fuzzy thinking, muscle and joint pain, heart palpitations, allergies, increased maturation, and decreased fertility. These can be explained by the metabolic toxins acting directly through intracellular Ca^{2+} in nerves, smooth and skeletal muscle, the heart, cells of the immune system such as neutrophils and mast cells, and other cells around the body. Methylglyoxal can also covalently modify amine transmitters and hormones such as 5-hydroxytryptamine (5-HT), dopamine, adrenaline and noradrenaline (Figure 10.6). This inactivates them, and also potentially generates inhibitors. Methylglyoxal can also inactivate protein hormones such as insulin, relevant to the pathogenesis of diabetes. These covalent modifications prevent normal Ca^{2+} signalling being activated.

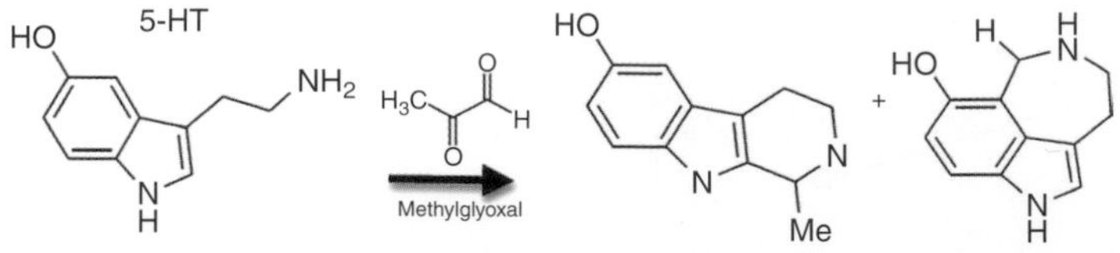

Figure 10.6 Reaction of amine transmitters and hormones with methylglyoxal. The Pictet–Spengler reaction, which inactivates the transmitter 5-HT after reaction with methylglyoxal. The reaction also occurs with other amine transmitters, including dopamine, adrenaline and noradrenaline. This reaction has the potential to explain how the gut effects nerves in different parts of the body, including the intestine, brain and periphery, and also explain how the gut is involved in Parkinson's and Alzheimer's diseases. Reproduced by permission of Welston Court Science Centre.

10.8   ER Stress Response

Release of Ca^{2+} from the endoplasmic reticulum (ER) activates a stress response via gene expression in the nucleus. The nucleus is usually regarded ultimately as the central controller of all the activities in the cell. But, the ER plays an equally important role in determining whether a cell fires, divides, defends itself against stress, or dies by apoptosis. Ca^{2+} inside the ER plays a key role in this decision-making. The ER is a three-dimensional spider's web wrapped around the nucleus and extending to the plasma membrane. It is also a dynamic structure, constantly on the move, and generating small vesicles that transport proteins to the Golgi. The ER has four main functions in eukaryotic cells:

1. Regulation of intracellular Ca^{2+}, being the major releasable Ca^{2+} store in cells.
2. Maturation of proteins, involving folding and processing, ready for secretion or trafficking to the plasma membrane, as they move towards the Golgi.
3. Lipid synthesis.
4. Response to stress.

The ER is able to communicate with the cytosol through release of Ca^{2+}, and proteins that cross the ER membrane and interact with proteins facing or in the cytosol. One of these proteins, protein kinase R-like ER kinase (PERK), interacts with caspases, and determines whether the cell crosses the Rubicon towards apoptosis. The ER also communicates with the mitochondria through Ca^{2+} release and the protein BCL-2. There is also a communication pathway between the ER and nucleus. It is this that controls the ER stress response. Changes in Ca^{2+} within the ER and cytosol play a central role in the communication of the ER to the rest of the cell. Crucial to Ca^{2+} signalling is the movement of the ER transmembrane protein STIM towards the internal junction with the plasma membrane, where it interacts with Orai, opening its Ca^{2+} channel so that Ca^{2+} flows into the cell, causing a large rise in free Ca^{2+} in the cytosol. This is SOCE.

The ER stress response occurs when proteins do not fold correctly inside the ER, or when there is a sustained, large loss of Ca^{2+} from the ER. The lumen of the ER is essentially an oxidising environment enabling S–S bonds to form between cysteines in proteins, unlike the cytosol, which has a reducing environment. The ER has an array of chaperones, foldases and carbohydrate-processing enzymes as part of its protein-processing factory. Several are Ca^{2+}-binding proteins including GRP94 and GRP78 (BiP) (GRP = glucose response protein), calnexin, calsequestrin in muscle, and calreticulin. GRP78 and GRP94 are chaperones and heat shock proteins. BiP (GRP78) is a major Ca^{2+}-binding protein with ATPase activity in the lumen or the ER, using ATP hydrolysis to aid protein folding. GRP94 has 15 Ca^{2+} sites. GRP78 binds nascent proteins and hands them on to GRP94. Calreticulin and calsequestrin are Ca^{2+}-binding lectin proteins. Calreticulin is a negative regulator of SOCE. The unfolded protein response is triggered in one of four ways:

- Glucose/energy deprivation.
- Disturbance of ER Ca^{2+}.
- Oxidative stress.
- Ischaemia.

These activate BiP, and stimulate degradation of the misfolded proteins via the proteasome. Crucially, BiP communicates to the nucleus, to activate the unfolded protein response element and NF-κB. Failure to fold proteins correctly inside the ER causes induction of chaperone genes in the nucleus: immunoglobulin heavy chain binding protein, BiP (GRP78) and GRP94. This 'unfolded protein response' is a major stress pathway in all animal, fungal and plant cells. Failure to fold proteins correctly inside the ER occurs naturally when too much protein is expressed, or when cells are attacked by particular viruses. It can be induced experimentally by tunicamycin. Dumping Ca^{2+} out of the ER also induces this stress pathway. The experimental way of doing this is to inhibit the ER SERCA pump, using thapsigargin or cyclopiazonic acid. The core of the ER stress response is a triad of stress-sensing proteins:

- Inositol-requiring enzyme 1 (IRE1).
- PERK.
- Activating transcription factor 6 (ATF6).

All three regulate portions of the transcriptional unfolded protein response element. PERK also attenuates protein synthesis during ER stress, and IRE1 interacts directly with the c-Jun N terminal kinase stress kinase pathway. BiP (GRP78) binds to the luminal N-terminal domains of two key transmembrane proteins in the ER – Ire in yeast and IRE1α in mammals, and PERK. BiP (GRP78) also activates the ER attached protein ATF6. PERK crosses the membrane between the ER lumen and the cytosol. On the other hand, Ire1 crosses the membrane between the ER lumen and the nucleus. Exposure of hydrophobic domains of proteins being processed in the ER causes activation of Ire1 and PERK. When BiP (GRP78) loses its Ca^{2+}, normally bound when the ER free Ca^{2+} is hundreds of micromolar, or when the hydrophobic domain in BiP (GRP78) binds to unfolded protein, BiP (GRP78) drops off the Ire site facing the lumen of the ER. Ire, having lost its BiP (GRP78) inside the ER, dimerises, activating its kinase domain, resulting in autophosphorylation. This converts it into an endonuclease, which, in turn, in yeast, causes alternative splicing of the mRNA from the gene *HAC1*. In mammalian cells, IRE activates the transcription factor X-box binding protein-1 (XBP1). The transcription factor binds to the unfolded protein response element, inducing genes, whose products protect the cell against the initial stress. At the same time, activation of the kinase PERK facing the cytosol results in phosphorylation, and thus inactivation, of the translational initiation factor eIF-2α. This helps the cell cope with stress, because it reduces the amount of general protein synthesis. Another protein changed by ER stress is ATF6, which initially is anchored on the ER, but which moves to the Golgi under these stress conditions. There, ATF6 is cleaved to produce a soluble transcription factor that moves to the nucleus. Thus, there are three phases in the unfolded protein response in the ER:

1. Signalling from the ER to the nucleus occurs via IRE1 and ATF6, and the cytosol via PERK. IRE induces chaperone synthesis, which attempts to remedy the unfolded or malfolded proteins.
2. PERK attempts to reduce global protein synthesis, and to increase the degradation of unfolded and malfolded proteins in the ER.

3. Activation of transcription factors induces expression of defence genes, whose proteins help to get rid of 'bad' proteins, which are moved out of the ER and degraded by the proteasome. If this does not happen adequately the cell takes a decision to die by apoptosis.

Thus, if the stress response is unsuccessful, an alarm signal is generated that signals the mitochondria to cross the Rubicon and induce apoptosis via release of cytochrome *c* and activation of the caspase cascade. There is also a cytochrome *c*-independent route to apoptosis. Overexpression of BiP (GRP78) alleviates the ER stress response.

The response of ER to stress is important in several pathological situations. Ca^{2+} in the ER plays a key role in this. These include obesity, proteinuric kidney disease, muscle disease, pancreatic β-cells and diabetes, retinal degeneration and death, as a result of mutation that causes misfolding of rhodopsin, which accumulates in the ER inducing the stress response, amyotrophic lateral sclerosis, ischaemia, liver disease, neurodegenerative diseases, including Alzheimer's and Parkinson's diseases, ageing, tumour progression, and several immune- and inflammatory-based disease. ER stress in enlarged fat tissue induces inflammation, modifies adipokine secretion, and saturated fat induces ER stress in muscle and in pancreatic β-cells. ER stress impairs insulin secretion, leading to apoptosis. ER stress is involved also in some genetic disorders, such as Darier's disease.

A further mechanism for activating the ER stress response is through oxidative damage. Addition of reactive oxygen species to cells in tissue culture generates slow rises in cytosolic free Ca^{2+}, often taking over 1 h. The source of the Ca^{2+} is increased permeability of the plasma membrane. However, release from the ER is also a source, through inhibition of the SERCA pump. This can activate the ER stress response, which may also be activated as a result of oxidation of ER proteins, or by damage to ER chaperones. Ca^{2+} release from the ER increases the level of the stress response induced by reactive oxygen species.

Mitochondria are also involved in the ER stress response. When the cytosolic free Ca^{2+} rises, mitochondria take up Ca^{2+} in preference to ATP synthesis. This Ca^{2+} uptake depolarises the inner mitochondrial membrane, and activates metabolic enzymes that produce NADH for the respiratory chain, such as pyruvate dehydrogenase. Overloading of Ca^{2+} in mitochondria occurs when there is a prolonged, pathological rise in cytosolic free Ca^{2+} and in oxidative stress. The model for this is glutamate excitotoxicity in neurones, which activates NO synthase. A combination of Ca^{2+} and NO causes a collapse to the mitochondrial membrane potential, leading to cell death.

ER stress plays a role in the response of plants to stress. Thus, like yeast and animal cells, plants have a similar mechanism for defending their cells against stresses, such as pathogens and heat shock through signals in the ER.

10.9   Summary

Pathology is the science of abnormalities in any living organism provoked by internal or external agents:

- Physical, chemical or biological agent.
- Hostile environments.
- Genetic disorder.
- Stress.
- Ageing.

Intracellular Ca^{2+} is involved in some way in virtually all these pathological processes. Because of the calcium pressure across the plasma and ER membranes, intracellular Ca^{2+} changes when a cell is injured. There are four ranges for cytosolic free Ca^{2+} in cells (see Box 10.1). Intracellular Ca^{2+} is involved in:

- The four types of cell death – necrosis, apoptosis, autophagy and lysis.
- Genetic abnormalities in Ca^{2+} signalling proteins.
- Oxygen toxicity.
- How bacterial metabolic toxins affect bacteria and host eukaryotic cells.
- Inappropriate Ca^{2+} signalling.

A key issue is whether damage to the cell is reversible, and if not whether it inevitably leads to cell death. Pathological changes in intracellular Ca^{2+} can be a friend, activating defence mechanisms:

- Removal of potentially lethal pores in the plasma membrane.
- The ER stress response.

Recommended Reading

*A must read.

Books

*Campbell, A.K. (2015) Intracellular Calcium. Chichester: John Wiley & Sons Ltd. Chapter 9, Pathology of intracellular Ca^{2+}, pp 473–498. Fully referenced.
*Krebs, J. & Michalak, M.E. (Eds) (2007) Calcium: A Matter of Life and Death. Amsterdam: Elsevier. Contains several articles describing genetic abnormalities in Ca^{2+} signalling proteins.

Reviews

*Berridge, M.J. (2014) Calcium regulation of neural rhythms, memory and Alzheimer's disease. *J. Physiol.-London*, **592**, 281–293. Potential role of intracellular Ca^{2+} in brain injury.
*Campbell, A.K. & Luzio, J.P. (1981) Intracellular free calcium as a pathogen in cell damage initiated by the immune system. *Experientia*, **37**, 1110–1112. Novel hypothesis about the role on intracelllar Ca^{2+} in protecting cells against immune attack.
*Campbell, A.K., Matthews, S.B., Vassell, N., Cox, C., Naseem, R., Chaichi, M.J., Holland, I.B. and Wann, K.T. (2010). Bacterial metabolic 'toxins'. A new mechanism for lactose and food intolerance, and irritable bowel syndrome. *Toxicology*, **278**, 268–276. Key hypothesis explaining the symptoms of food intolerance and IBS.
Deftereos, S., Papoutsidakis, N., Giannopoulos, G., Angelidis, C., Raisakis, K., Bouras, G., Davlouros, P., Panagopoulou, V., Goudevenos, J., Cleman, M.W. & Lekakis, J. (2016) Calcium ions in inherited cardiomyopathies. *Med. Chem.*, **12**, 139–150. Role of Ca^{2+} in inherited heart disease.
Ivanova, H., Vervliet, T., Missiaen, L., Parys, J.B., De Smedt, H. & Bultynck, G. (2014) Inositol 1,4,5-trisphosphate receptor-isoform diversity in cell death and survival. *Biochim. Biophys. Acta-Mol. Cell Res.*, **1843**, 2164–2183. Role of IP_3 receptors in pathology.

Duchen, M.R. (2000) Mitochondria and calcium: from cell signalling to cell death. *J. Physiol.*, **529**, 57–68. How Ca²⁺ in mitochondria can lead to cell death.

Feske, S. (2007) Calcium signalling in lymphocyte activation and disease. *Nature Rev. Immunol.*, 7, 690–702. The role of intracellular Ca²⁺ in the immune response, and the discovery of Orai, a key protein in store-operated Ca²⁺ entry.

Hovnanian, A. (2007) SERCA pumps and human diseases. *Subcell. Biochem.*, **45**, 337–363. Role of SERCA pumps in disease.

MacLennan, D.H. (2000) Ca²⁺ signalling and muscle disease. *Eur. J. Biochem.*, **267**, 5291–5297. How Ca²⁺ is involved in muscle disease by the discoverer of calsequestrin and calreticulin.

Magenta, A., Dellambra, E., Ciarapica, R. & Capogrossi, M.C. (2016) Oxidative stress, microRNAs and cytosolic calcium homeostasis. *Cell Calcium*, **60**, 207–217. Interaction between intracellular Ca²⁺ and oxygen metabolite injury.

Piper, H.M. (2000) The calcium paradox revisited: an artefact of great heuristic value. *Cardiovasc. Res.*, **45**, 123–127. A useful analysis of what has been learnt from an artifical overloading of heart cells by Ca²⁺.

Medina, D.L., Di Paola, S., Peluso, I., Armani, A., De Stefani, D., Venditti, R., Montefusco, S., Scotto-Rosato, A., Prezioso, C., Forrester, A., Settembre, C., Wang, W. Y., Gao, Q., Xu, H.X., Sandri, M., Rizzuto, R., De Matteis, M.A. & Ballabio, A. (2015) Lysosomal calcium signalling regulates autophagy through calcineurin and TFEB. *Nat. Cell Biol.*, **17**, 288–299.

Pla, A.F., Kondratska, K. & Prevarskaya, N. (2016) STIM and ORAI proteins: crucial roles in hallmarks of cancer. *Am. J. Physiol.-Cell Physiol.*, **310**, C509–C519. How Ca²⁺ can be involved in cancer.

Striessnig, J., Bolz, H.J. & Koschak, A. (2010) Channelopathies in Caᵥ1.1, Caᵥ1.3, and Caᵥ1.4 voltage-gated L-type Ca²⁺ channels. *Pflugers Arch. Eur. J. Physiol.*, **460**, 361–374. A review of genetic mutations in three Ca²⁺ channels.

Tempel, B.L. & Shilling, D.J. (2007) The plasma membrane calcium ATPase and disease. *Subcell. Biochem.*, **45**, 365–383. How the pump responsible for maintaining the sub-micromolar free Ca²⁺ in all cells is involved in disease.

*Zhivotovsky, B. & Orrenius, S. (2011) Calcium and cell death mechanisms: a perspective from the cell death community. *Cell Calcium*, **50**, 211–221. A review of the role of Ca²⁺ in cell death.

Research Papers

Crotti, L., Johnson, C.N., Graf, E., *et al.* (2013) Calmodulin mutations associated with recurrent cardiac arrest in infants. *Circulation*, **127**, 1009. How mutations in then ubiquitous Ca²⁺ binding protein may explain death in babies and small children.

Davies, E.V., Campbell, A.K., Williams, B.D. & Hallett, M.B. (1991) Single cell imaging reveals abnormal intracellular calcium signals within rheumatoid synovial neutrophils. *Br. J. Rheumatol.*, **30**, 443–448. How cells from rheumatoid patients have a larger releasable ER store than healthy people.

Hu, Z.L., Bonifas, J.M., Beech, J., Bench, G., Shigihara, T., Ogawa, H., Ikeda, S., Mauro, T. & Epstein, E.H. (2000) Mutations in ATP2C1, encoding a calcium pump, cause Hailey–Hailey disease. *Nat. Genet.*, **24**, 61–65. Identification of the mutations in the SERCA2 Ca²⁺ pump responsible for a genetic disease.

Lahat, H., Pras, E., Olender, T., Avidan, N., Ben-Asher, E., Man, O., Levy-Nissenbaum, E., Khoury, A., Lorber, A., Goldman, B., Lancet, D. & Eldar, M. (2001) A missense mutation in a highly conserved region of Casq2 is associated with autosomal recessive catecholamine-induced polymorphic ventricular tachycardia in Bedouin families from Israel. *Am. J. Human Genetics*,

69, 1378–1384. A mutation that messes up a major Ca^{2+} binding protein in the SR may be the cause of very high heart beat induced by exercise, which can be lethal.

Li, J., Huang, R., Liao, W., Chen, Z. & Zhang, S. (2012) Dengue virus utilizes calcium modulating cyclophilin-binding ligand to subvert apoptosis. *Biochem. Biophys. Res. Commun.*, **418**, 622–627. How an RNA virus that causes dengue fever uses Ca^{2+} to avoid the cell it is growing commiting suicide.

Pinton, P., Giorgi, C., Siviero, R., Zecchini, E. & Rizzuto, R. (2008) Calcium and apoptosis: ER-mitochondria Ca^{2+} transfer in the control of apoptosis. *Oncogene*, **27**, 6407–6418. The role of movement of Ca^{2+} from the ER to mitochondria in programmed cell death.

*Sakuntabhai, A., Ruiz-Perez, V., Carter, S., *et al.* (1999) Mutations in ATP2A2, encoding a Ca^{2+} pump, cause Darier disease. *Nat. Genet.*, **21**, 271–277. Evidence that mutations in the SERCA2 Ca^{2+} pump are the cause of an inherited disease that only shows as skin lesions and some psychiatric problems.

11

Pharmacology of Intracellular Ca²⁺

Learning Objectives
1. Pharmacological targets for intracellular Ca²⁺ – direct and indirect.
2. How clinical and/or experimental drugs affect tissue function through intracellular Ca²⁺.
3. How pharmaceuticals are used to investigate Ca²⁺ signalling.
4. How natural toxins and poisons affect Ca²⁺ signalling.
5. How drugs affect the Ca²⁺ receptor.
6. The agents that affect Ca²⁺ in bacteria.
7. How antibodies interact with intracellular Ca²⁺.
8. How other ions interact intracellular Ca²⁺.

11.1 Background

Many pharmacological agents affect the Ca²⁺ signalling system directly or indirectly, activating or inhibiting a component of the Ca²⁺ toolkit. There are three main classes:

1. Those used clinically to treat or prevent pain, heart arrhythmias, angina, heart failure, hypertension, diabetes and mental problems (Table 11.1).
2. Those used experimentally as activators or inhibitors of cellular mechanisms, in order to identify specific receptors, pumps, organelles or intracellular proteins involved in regulating cytosolic free Ca²⁺, generating a Ca²⁺ signal, or its target in the cell response.
3. Hazardous substances, including natural toxins and man-made poisons.

Many substances have their origin in natural compounds. We now have a wide range of substances that affect Ca²⁺ channels, receptors that provoke or regulate intracellular Ca²⁺ signals, Ca²⁺ pumps, transporters and Ca²⁺-binding proteins (Figure 11.1). These have played a major role in unravelling the Ca²⁺ signalling system.

The original compounds were small organic molecules. However, several peptide toxins have been found in animals. Some have been mimicked by the pharmaceutical industry. For example, ziconotide (SNX-111; Prialt) is based on the peptide ω-conotoxin from the cone snail *Conus magnus*, and is used to treat severe and chronic pain. It acts on N-type voltage-sensitive Ca²⁺ channels. Extracts from plants, fungi, animals and microbes have been used for thousands of years to treat human ailments, as well as being used to stimulate or

Fundamentals of Intracellular Calcium, First Edition. Anthony K. Campbell.
© 2018 John Wiley & Sons Ltd. Published 2018 by John Wiley & Sons Ltd.
Companion Website: http://www.wiley.com/go/campbell/calcium

Table 11.1 Examples of pharmacological substances used clinically that interact directly with intracellular Ca^{2+} or Ca^{2+} movement.

Class of compound	Type	Specific example	How it alters Ca^{2+}
Anaesthetic	General	Urethane	Nerve terminal Ca^{2+} block?
	Local	Lignocaine	Ca^{2+} channels
Pain reliever	Opiate	Morphine	Nerve excitability
	Anti-migraine	5-Hydroxytryptamine (5-HT) blocker	
Ca^{2+} antagonist	Ca^{2+} channel blocker	Amlodipine	Reduces Ca^{2+} entry
	Phenothiazine	Chlorpromazine	Calmodulin inhibitor
Anti-psychotic		Trifluoperazine	Calmodulin inhibitor
Cardiovascular	Cardiac glycoside	Digoxin (digitalin)	Blocks Na^+ pump followed by Na^+/Ca^{2+} exchange
Anti-allergy	Anti-histamine	Disodium chromoglycate	Block Ca^{2+} entry into mast cells
Stimulant	Methylxanthine	Caffeine, LSD, amphetamine	Activates ryanodine receptors in the SR
Muscle relaxant		Succinylcholine	Prevents action potential rise in Ca^{2+}

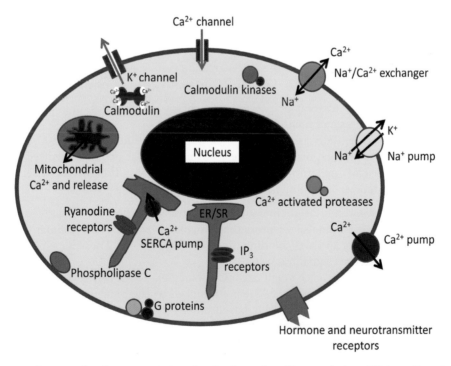

Figure 11.1 Ca^{2+} targets for pharmaceuticals and toxins. Reproduced by permission of Welston Court Science Centre.

repress mental activity. Many interact with the Ca^{2+} signalling system. One of the first such drugs used clinically was digitalis (Figure 11.2), extracted from foxgloves by Erasmus Darwin (1731–1802), the grandfather of Charles Darwin, and William Withering (1741–1799) (Figure 11.2), who had learnt of the use of foxglove extracts to treat dropsy, now known as congestive heart failure, from an old woman who was a herbalist in the English county of Shropshire. Darwin gave his results to his eldest son, the first Charles, a brilliant medical student at Edinburgh, who used this data for a thesis, which gained him the prestigious Aesculian Prize. Tragically, soon after, Charles cut himself while dissecting the body of a former boy patient. He got blood poisoning and died. Erasmus was distraught, but went on to publish the foxglove data in a journal produced by the Royal College of Physicians, with the title 'An account of the successful use of foxglove in some dropsies and in pulmonary consumption'. Withering went on to publish his own data in his book, *An Account of the Foxglove and Some of its Medical Uses* (Figure 11.2). But the 'plagiarism' by Darwin led to an estrangement between these two brilliant men. Digitalis works by inhibiting the sodium pump, the plasma membrane Na^+/K^+-MgATPase, an increase in cytosolic Na^+ leading to Ca^{2+} entry via the $3Na^+/Ca^{2+}$ exchanger NCX.

The crucial problem about pharmacological compounds is how specific they are for the target they are aimed at.

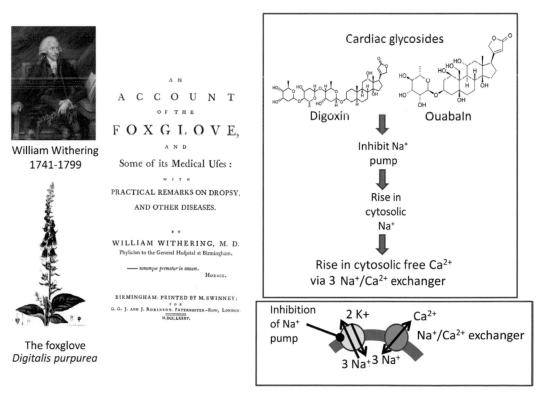

Figure 11.2 The foxglove and cardiac glycosides, with its discoverer and mechanism. Picture of a foxglove. Withering (1778). Picture of William Withering drawn and engraved by W. Bond after a painting by Carl Frederik von Breda (1759–1818). Portrait of William Withering © The Royal Society. The mechanism shows how inhibiting the Na^+ pump causes a rise in cytosolic free Ca^{2+} via the $3Na^+/Ca^{2+}$ exchanger (NCX).

11.2   Pharmacological Targets for Intracellular Ca²⁺

Agents that work directly on Ca²⁺ signalling affect a particular component of the Ca²⁺ signalling system, whereas agents acting indirectly affect a mechanism that interacts with the Ca²⁺ signalling system (Table 11.1). The nomenclature is a little confusing, since calcium antagonist or calcium blocker are used by different people to describe the same compound. Antagonist implies an analogue, graded inhibition, whereas blocker implies a digital mechanism. Each voltage-gated Ca²⁺ channel is digital – it is either open or closed. So the channel is either blocked or opened by a drug – digital. But, the drug acts to change the **number** of Ca²⁺ channels open and closed at any one time. This can alter the cell in an analogue manner, affecting the strength and/or timing of the cell event. Alternatively, the action of the agent can have a digital effect on a cell, causing it to stop firing at all or to fire continuously. But the effect of a pharmacological agent on an enzyme will be graded – analogue, as the concentration of the agent is changed. There are at least nine main targets, in the plasma membrane and intracellularly, for substances that interfere directly with intracellular Ca²⁺ (Box 11.1).

Box 11.1   Protein targets for pharmaceuticals that affect Ca²⁺ signalling.

1. Ca²⁺ channels in the plasma membrane.
2. Ca⁷⁺ channels, and their receptors, in intracellular organelles.
3. Ca²⁺ pumps and exchangers in the plasma membrane.
4. Ca²⁺ pumps and exchangers in intracellular organelles.
5. Cell surface receptors that lead to changes in intracellular free Ca²⁺.
6. Enzymes that lead to changes in intracellular Ca²⁺.
7. Proteins and enzymes that interact with Ca²⁺ signalling.
8. Ca²⁺-binding proteins.
9. Genes that code for, or control, proteins involved in Ca²⁺ signalling.

Pharmacological agents that have direct effects on Ca²⁺ signalling include Ca²⁺ channel blockers such as dihydropyridines and conotoxins, and trifluoperazine as an inhibitor of calmodulin. Agents that act indirectly on Ca²⁺ signalling include Na⁺ or K⁺ channel blockers. These affect membrane potential, and as a consequence voltage-gated Ca²⁺ channels. Agents that act indirectly on Ca²⁺ signalling also include those that activate or inhibit enzymes controlling the level of other intracellular signals that interact with the Ca²⁺ signalling system. For example, tetrodotoxin (TTX) (Figure 11.10 below) blocks voltage-gated Na⁺ channels, preventing the generation of an action potential. As a result, voltage-gated Ca²⁺ channels at the nerve terminal are prevented from opening, and the cellular event is stopped. In contrast, experimental addition of an extracellular medium high in K⁺ will depolarise a cell and open voltage-gated Ca²⁺ channels. As a result, a muscle will contract, and a bioluminescent cell will flash. On the other hand, the Ca²⁺ channel blocker amlodipine acts directly on Ca²⁺ signals in smooth muscle in blood vessels, and therefore is used to treat high blood pressure. In contrast, β-adrenergic blockers, such as atenolol, which are also used in some patients to treat hypertension or severe migraine, interact with Ca²⁺ signalling indirectly through cyclic AMP.

Any one of five changes in the Ca²⁺ signalling system can be caused by a pharmacological agent, and as consequence inhibit or activate a cellular event:

1. A change in the concentration of cytosolic free Ca²⁺ in the resting cell.
2. A change in the concentration of free and total Ca²⁺ inside an intracellular organelle.
3. The timing and/or magnitude of the cytosolic free Ca²⁺ signal, produced by the primary stimulus and responsible for triggering the cellular event.

4. The potency of a secondary regulator (see Chapter 2 for definition), which works by affecting the efficacy of a component in the Ca^{2+} signalling system.
5. The activity of the Ca^{2+} target, e.g. calmodulin, so that the timing and magnitude of the cellular event is affected or whether it happens at all.

11.3 Drugs Used Clinically on Intracellular Ca²⁺

Organs targeted clinically by drugs which act on Ca^{2+} signalling include the brain, peripheral nerves, heart and smooth muscle, bone, liver, endocrine pancreas, lung, and the immune system. Clinically used pharmaceuticals can have at least three names: the chemical name, the generic trade name, and the name used on the packet by the drug company (see Chapter 2). For example, amlodipine, a generic Ca^{2+} channel blocker, is used to treat hypertension and angina, and has over 30 names in different countries. There are many drugs used clinically that interact with Ca^{2+} signalling (Box 11.2):

Box 11.2 Some clinical pharmaceuticals that work by affecting Ca²⁺ signalling.

1. Anaesthetics: general and local.
2. Pain relievers that affect the nociceptor pathway.
3. Drugs that reduce blood pressure, e.g. Ca^{2+} channel blockers.
4. Drugs that increase or decrease the heart – its beat, rhythm, oxygen supply and failure, e.g. anti-arrhythmics and cardiac glycosides.
5. Muscle relaxants.
6. Hormone and neurotransmitter receptor blockers, e.g. 5-HT, α- and β-adrenergic effectors.
7. Substances, other than insulin, used to treat diabetes, e.g. sulphonyl ureas.
8. Anti-allergics.
9. Drugs that work in the brain to treat mental problems, e.g. anti-depressives and anti-psychotics.
10. Drugs that work on the Ca^{2+} receptor.
11. Cyclic nucleotide phosphodiesterase inhibitors (e.g. Viagra).

These interact with Ca^{2+} signalling in one of six ways:

1. Ca^{2+} channel blocker.
2. Receptor blocker.
3. Ca^{2+}-binding protein inhibitor.
4. Enzyme inhibitor or activator.
5. Lipid bilayer.
6. Indirect.

11.4 Anaesthetics

General and local anaesthetics (Figure 11.3a, b) interact with Ca^{2+} signalling, but the precise mechanism by which general anaesthetics put a patient to 'sleep' is still not fully understood. Local anaesthetics silence nerves close to the site of injection by blocking Na^+ and/or Ca^{2+} channels, thereby preventing the generation of an action potential. Local anaesthetics have been used experimentally to manipulate Ca^{2+} signalling in a wide range of excitable and non-excitable cells (Table 11.2).

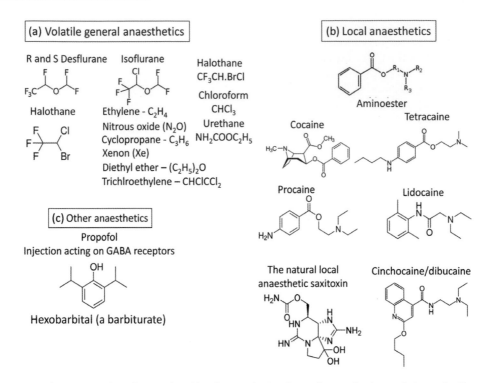

Figure 11.3 Some examples of general and local anaesthetics. General anaesthetics work through effects on the lipid bilayer and membrane receptors. Effects on Ca^{2+} signalling are indirect. Local anaesthetics act particularly by inhibiting sodium influx through voltage-gated Na^+ channels in the neuronal cell membrane, but also affect Ca^{2+} signalling by effects on the phospholipid bilayer.

11.5 Ca^{2+} Channel Effectors

11.5.1 Classes of Ca^{2+} Channel Blocker

Calcium channel blockers are natural or man-made substances that block Ca^{2+} channels (Table 11.3). Their effective concentration range varies from nano- to micro-molar, depending on the compound and cell type. There are three chemical classes of blocker (Figure 11.4):

1. Small organic compounds, e.g. dihydropyridines.
2. Peptides, e.g. conotoxins from cone snails.
3. Inorganic cations, e.g. La^{3+} and Gd^{3+}.

   Ca^{2+} channel blockers interact with either ionotropic or metabotropic receptors. Ionotropic receptors act on ion channels directly. Metabotropic receptors affect ion channels indirectly via an intracellular second messenger, such as cyclic AMP. The term *ionotropic* is distinct from *inotropic*. The latter describes agents that affect muscle contraction, e.g. the heart beat. Voltage-gated Ca^{2+} channels are 'ionotropic' when part of an ionotropic receptor, and consist of four subunits: α_1, $\alpha_2\delta$, β and γ. The pore in the Ca^{2+} channel allows Ca^{2+} to enter when the membrane potential drops, and is formed by the α_1 subunit. Drug binding to the Ca^{2+} channel is usually on the α_1 or $\alpha_2\delta$ subunit.

Table 11.2 Some effects of local anaesthetics on cells involving intracellular Ca^{2+}.

Anaesthetic	Effect
Effects on nerves	
Procaine	Inhibition of Na^+ and K^+ conductance; increase of Ca^{2+} efflux
Tetracaine (amethocaine)	Increases Ca^{2+} efflux
Cocaine	Inhibits uptake of noradrenaline in adrenergic nerves, causing vasoconstriction
Effects on muscle	
Tetracaine (amethocaine)	Increases Ca^{2+} efflux; inhibition of smooth muscle contraction induced by acetylcholine or histamine; inhibition of action potential propagation in striated and smooth muscle
Cinchocaine (dibucaine)	Inhibition of smooth muscle contraction induced by acetylcholine or histamine
Procaine	Inhibition of Na^+ and K^+ conductance and action potential propagation in striated and smooth muscle
Effects on exocytotic secretion	
Cinchocaine (dibucaine)	Inhibition of insulin secretion stimulated by glucose or Ba^{2+}
Tetracaine (amethocaine)	Inhibition of Ca^{2+}-dependent histamine release from mast cells
Various	Inhibition of Ca^{2+}-mediated secretion of adrenaline from the adrenal medulla stimulated by acetylcholine or K^+
Effects on hormone action	
Cinchocaine (dibucaine)	Inhibition of glucose uptake and adrenaline-stimulated lipolysis in adipocytes
Tetracaine (amethocaine)	Inhibition of glucagon/cyclic AMP-stimulated gluconeogenesis and glycogenolysis in liver
Procaine	Increase in glucose 6-phosphate-independent glycogen synthesis in adipocytes
Effects on organelles and vesicles	
Various	Inhibition of mitochondrial Ca^{2+} transport; inhibition of Ca^{2+}-induced increase in Na^+ permeability in phospholipid vesicles

For references, see Table 9.16 in Campbell (1983).

Table 11.3 Examples of Ca^{2+} channel blockers used clinically.

Drug	Class of Ca^{2+} blocker	Use in treatment
Nifedipine	Dihydropyridine	Anti-hypertensive
Amlodipine	Dihydropyridine	Anti-hypertensive
Verapamil	Phenylalkylamine	Anti-arrhythmic (class IV)/anti-angina
Diltiazem	benzothiazepine	Anti-arrhythmic (class IV)/anti-angina
Nimodipine	Dihydropyridine	Alleviates migraine
Zonisamide	Sulphonamide	Reduces epileptic attacks and pain
Gabapentin/pregabalin	Hexanoic acid derivative	Reduces epileptic attacks and pain
Ziconotide	ω-Conotoxin peptide	Reduces pain

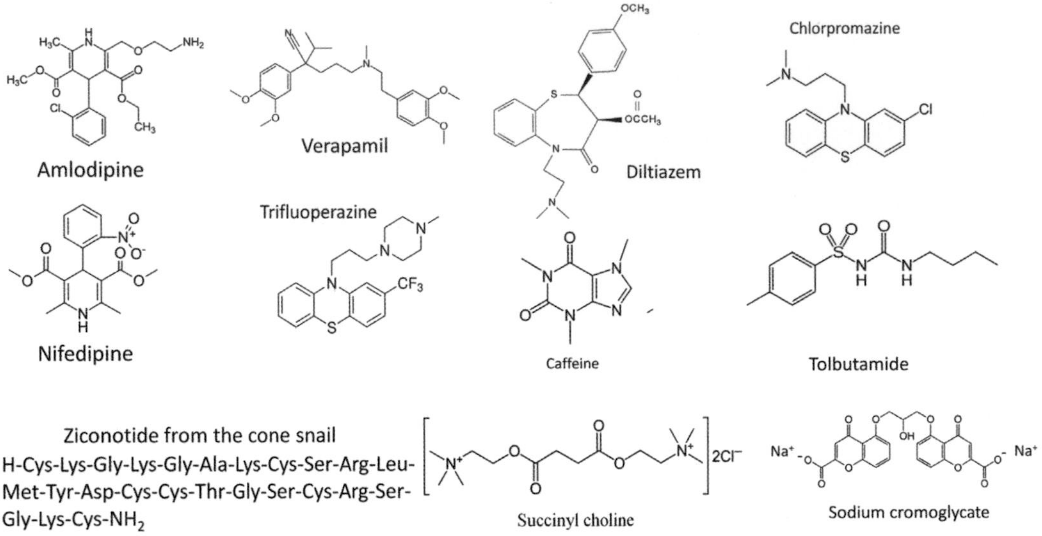

Figure 11.4 Some pharmacological agents that act on Ca^{2+} signalling.

There are two main families of voltage-gated Ca^{2+} channel: HVA and LVA, designated by whether they are high or low voltage activated (see Chapter 5). They are subdivided as:

- L-type HVA Ca^{2+} channels – skeletal and heart muscle, endocrine cells and the retina.
- P/Q- and N-type HVA Ca^{2+} channels – nerve terminals and the dendritic tree of nerves.
- R-type Ca^{2+} channels – the cell body of nerves and the dendrites.
- T-type LVA Ca^{2+} channels – nerve cell bodies, the dendrites, with some in heart myocytes.

Pharmacological agents that block specifically each type of these Ca^{2+} channels are used both clinically, or experimentally to identify a particular Ca^{2+} channel involved in a cell response. Three classes of L-type Ca^{2+} blocker are (Figure 11.4):

1. Dihydropyridines (e.g. nifedipine and amlodipine).
2. Phenylalkylamines (e.g. verapamil).
3. Benzothiazepines (e.g. diltiazem).

Mibefradil blocks both L- and T-type Ca^{2+} channels, but was withdrawn from clinical use as it interfered with drug metabolism. A Ca^{2+} channel can exist in one of three modes:

- Mode 0: the channel does not open when the cell is depolarised.
- Mode 1: the channel probability of opening is low and the opening time is short.
- Mode 2: the channel has a high probability of opening when the cell is depolarised and individual channel opening times are long.

The channel is mostly in either mode 1 or 2, being in mode 1 about 70% of the time and only in mode 0 less than 1% of the time. The channel switches randomly between the modes; mode 0 is favoured when the channel binds dihydropyridine. Each class of Ca^{2+} channel blocker binds to the α_1 subunit of the Ca^{2+} channel, but not to the Ca^{2+} pore itself. These agents affect mainly the heart and smooth muscle, being used therefore to treat heart arrhythmias, angina and hypertension. Better knowledge of the types of Ca^{2+} channels in nerves has increasingly seen particular Ca^{2+} channel blockers being used to treat pain and migraine. There are also experimental Ca^{2+} channel openers, such as BAY K8644, which increase the probability of L-type Ca^{2+}

channels opening. BAY K8644 opening of cardiac Ca^{2+} channels causes increased contractile force in the heart, and blood vessel constriction through its effect on Ca^{2+} channels in smooth muscle. Nifedipine is a competitive inhibitor of BAY K8644.

The most common clinical use of these drugs is for treating high blood pressure (hypertension) and heart problems, particularly angina pectoris. This is because their main cellular targets are smooth muscle in arteries and cardiac muscle in the myocardium. As a result they affect the dilation of blood vessels, the strength and timing of the heart beat, and electrical conduction within the heart. But, the two anti-convulsant drugs gabapentin (Neurontin) and pregabalin (Lyrica) bind the $\alpha_2\delta$-1 and $\alpha_2\delta$-2 subunits of the voltage-gated Ca^{2+} channel, and can be used to treat chronic neuropathic pain.

It might seem, at first, that it would be impossible to use such potent blockers of excitable cells safely. Complete block of Ca^{2+} channels in the heart would be lethal. But, it is possible to use these Ca^{2+} channel blockers over a wide range of plasma concentrations. The classes of Ca^{2+} channel blocker used in a particular patient depends on whether the clinical problem is hypertension or angina, or both, and whether there are other complicating factors such as heart failure or heart arrhythmias. Some of the clinical effects may be via effects on SOCE in smooth muscle. Long-term effects are mediated through changes in the amount of Ca^{2+} stored within the sarcoplasmic reticulum (SR). Adrenaline and noradrenaline, through an increase in cyclic AMP, increase the strength of the heart beat – a positive inotropic effect – by increasing the level of Ca^{2+} within the SR. Thus, β-blockers such as alprenolol, used clinically, reduce the level of Ca^{2+} in the SR. As a result, the concentration of cytosolic free Ca^{2+} in the myocyte is reduced at each action potential. Thus, less troponin C has Ca^{2+} bound to it, so less myofibrils within each cell contract, and the overall contraction of the myocyte cell and the whole heart is reduced. Ca^{2+} channel blockers result in channels opening for a shorter time, and thus let less Ca^{2+} in overall into the cell during each action potential. The end result is the same as with β-blockers, less Ca^{2+} is released from the SR during each contraction, and a smaller cytosolic free Ca^{2+} during each action potential, and so a smaller contraction. The chronotropic effects of Ca^{2+} channel blockers are explained by inhibition of Ca^{2+} entry, slowing the electrical conduction pathway in the heart, and in the myocyte itself. This causes a change in the shape and frequency of the action potential, the plateau of which is entirely due to opening of Ca^{2+} channels. Thus, all three types of calcium channel blockers have a negative inotropic effect, reducing the strength of contraction in cardiac muscle. Many also have a negative chronotropic effect. This effect is used in the treatment of people with atrial fibrillation or 'flutter'.

Effects may be on the cardiac myocyte or the pacemaker cells in the sinoatrial (SA) node.

11.5.2 Dihydropyridines

Dihydropyridines (Figure 11.4) block L-type Ca^{2+} channels in smooth muscle and the heart, relaxing smooth muscle and increasing blood flow to the heart. There are more than a dozen used regularly clinically with an array of trade names, but all scientifically ending in '-dipine', e.g. amlodipine, nifedipine (originally BAY A1140) and nicardipine. They rapidly lower blood pressure. Patients should be warned that, as a result, they may feel dizzy or faint after the first few doses, leading to activation of a fast heart rate (tachycardia). Nifedipine is used sometimes to delay premature labour, to treat Raynaud's phenomenon, and in high-altitude pulmonary oedema. However, nifedipine can increase death in patients with coronary artery disease. Amlodipine is widely used in the treatment of angina and hypertension, the latter with angiotensin receptor inhibitors such as irbesartan. These can be a better alternative to β-adrenergic blockers such as atenolol, which often causes depression and gut problems. Amlodipine has a high bioavailability, and is chiral, normal preparations being a racemic mixture of (R)-$(+)$ and (S)-$(-)$.

A particular advantage is its long half-life of over 40 h. Amlodipine binds to, and dissociates from, its receptor site on the L-type Ca^{2+} channel relatively slowly. This explains the gradual onset of its action. Its effect to lower blood pressure, typically from 180/110 to 140/80, is via direct effects on vascular smooth muscle, and its effect on angina indirect relative to the heart. In severe coronary artery disease, amlodipine can precipitate severe angina or even a heart attack, as well as excessive low blood pressure. The most common side effect is oedema (e.g. in the foot) and tiredness.

11.5.3 Phenylalkylamines

Verapamil (Figure 11.4) is the phenylalkylamine most commonly used clinically to treat hypertension, angina pectoris, and cardiac arrhythmias. It acts on L-type Ca^{2+} channels in smooth muscle and the heart. Phenylalkylamines have also been used to treat migraine, working through relaxation of blood vessels in the brain. There are five classes of anti-arrhythmic drugs:

- Class I: blockers of Na^+ channels (e.g. lignocaine).
- Class II: β-adrenergic blockers (e.g. propranolol).
- Class III: K^+ channel blockers (e.g. amiodorone).
- Class IV: slow Ca^{2+} channel blockers (e.g. verapamil and diltiazem).
- Class V working via other mechanisms (e.g. digoxin).

Thus, verapamil is a class IV anti-arrhythmic, being more effective than the cardiac glycoside digoxin, decreasing both heart rate and output. It should not be used with β-adrenergic blockers. Verapamil reduces the length of the plateau phased of the cardiac action potential, which is dependent on opening of L-type voltage-sensitive Ca^{2+} channels. But verapamil has little or no effect on the fast phase of the cardiac action potential, which is dependent on opening of voltage-sensitive Na^+ channels. Verapamil has a half-life of 5–12 h. Thus, it is necessary to take it three times a day in order to maintain plasma levels. It is metabolised in the liver to a range of low-acting or inactive metabolites. Side-effects of verapamil include headache, flushing face, dizziness, swelling, increased micturition (urination), tiredness, nausea, ecchymosis, light-headedness and constipation, the result of inhibiting Ca^{2+} channels in arteries in the head, gut and elsewhere.

Verapamil's methoxy analogue D-600 (gallopamil) has increased potency and is too dangerous to use clinically. Both verapamil and D-600 have two enantiomers. The (−) isomer is the most active, both clinically and experimentally, but normal preparations are a racemic mixture of both (−) and (+) forms. Verapamil does not significantly affect directly β-adrenergic action of adrenaline on the heart. This means that there is a natural defence against a moderate overdose, since excessive inhibition of cardiac function will trigger the release of β-adrenergic agonists to compensate for this. Verapamil can inhibit Ca^{2+}-dependent events in non-muscle cells, such as glucose-stimulated insulin release from pancreatic β-cells, β-adrenergic activation of adenylate cyclase, and α-adrenergic activation of some cell types.

11.5.4 Benzothiazapines

The most commonly used benzothiazapine clinically is diltiazem (Figure 11.4), to treat high blood pressure, angina pectoris and some types of heart arrhythmias. Diltiazem is a 1,4-thiazipine, and has properties intermediate between verapamil and dihydropyridines. It is a class II anti-anginal drug, and a class IV anti-arrhythmic, and is used as a cutting agent for cocaine and may even be of use in treating addiction, as it has been shown to reduce craving by blocking Ca^{2+} entry in neurones, induced by dopamine or glutamate. The action of diltiazem as a blocker of Ca^{2+} channels in smooth muscle makes it a strong dilator of blood vessels, including peripheral and coronary arteries, increasing blood flow to the heart muscle. The effects of diltiazem on Ca^{2+} channels in the heart make it a modest negative inotropic and chronotropic agent, reducing myocardial oxygen consumption. Diltiazem also increases the time for each heart beat, as it slows conduction through the AV node. Diltiazem can be as effective as verapamil in treating supraventricular tachycardia. But, it has little or no effect on the sympathetic nervous system. Diltiazem can also be useful in the treatment of atrial fibrillation and flutter, and hypertension. It should not be used in patients with major ventricular, SA or AV node problems, as they cannot compensate for the negative inotropic and chronotropic effects of the drug. As a cream, diltiazem is used in the treatment of anal fissures. Several heart and lung disorders preclude its use. Unwanted side-effects are similar to other Ca^{2+} channel blockers: bradycardia, SA and AV block, palpitations, dizziness, too low blood pressure, headache, gut problems, oedema (e.g. in the feet and ankles) and gum hyperplasia. Less common or rare uses are in liver, eye and skin problems or depression.

11.6   Hypertension

Blood pressure is assessed by measuring the systolic and diastolic pressure, being determined by the strength of the heartbeat and the tone of the smooth muscle in the arteries, both of which are controlled by intracellular Ca^{2+}. A typical blood pressure which is safe is a systolic/diastolic of 120/80. In contrast, a blood pressure of 180/111 is not safe, and could result in a stroke, and long-term damage to the heart and aorta, leading to an aortic aneurism – bursting of the aorta. Calcium channel blockers, such as amlodipine, are a major treatment for high blood pressure, acting on heart muscle, electrical conducting pathways in the heart and arterial smooth muscle. Other treatments of high blood pressure are β-adrenergic inhibitors, such as atenolol, or a mix of ACE (angiotensin-converting enzyme) inhibitors, angiotensin receptor antagonists and α-adrenergic antagonists, which also work by affecting intracellular Ca^{2+}. Calcium channel blockers do not usually significantly affect smooth muscle in veins. A problem with many anti-hypertensive therapies is the activation of acid reflux from the stomach, which can be particularly unpleasant at night. Furthermore, what is not well known is that 'silent' reflux leads to defects in the voice, making talking on the phone, lecturing or in a crowd a strain.

11.7   Arrhythmia, Tachycardia and Bradycardia

By the time someone reaches the age of 80, the heart has contracted and relaxed regularly about once a second some 2500 million times. There are three features of a heartbeat that matter:

1. The frequency of the beat = pulse.
2. The strength of the beat.
3. The regularity of the beat.

The strength and frequency of the human heartbeat is regulated by the sympathetic nervous system and through hormones in the blood, particularly adrenaline. Adrenaline and cardiac glycosides are positive inotropes, increasing the strength contraction. Whereas Ca^{2+} channel blockers and β-adrenergic blockers are negative inotropes, decreasing the strength of the heart beat. Chronotropic agents change the frequency of the heart beat, either by affecting the nerves that control the heart or by affecting the rhythm generated by the SA node. Thus, adrenaline and atropine are positive chronotropes, increasing the heart rate, whereas acetylcholine, digoxin and β-adrenergic blockers are negative chronotropes, decreasing the heart rate. Adrenaline works on Ca^{2+} channels and SR Ca^{2+} release in the myocyte via cyclic AMP. In contrast, atropine blocks muscarinic acetylcholine receptors. But all of these work, directly or indirectly, by affecting Ca^{2+} currents through the plasma membrane, or the size of the Ca^{2+} transient through effects on the amount of Ca^{2+} in the cardiac myocyte SR. Cardiac arrhythmia or dysrhythmia is caused by defects in ion channel activity. Many people suffer from heart palpitations – bradycardia less than 60 beats min^{-1}, or tachycardia greater than 110 beats min^{-1}, without any severe problems. But in some people it can be life threatening, requiring treatment with a pacemaker or drugs. Calcium channel blockers are one such treatment in selected patients.

11.8 Angina

Angina pectoris is chest pain caused by ischaemia in the heart, due to poor blood supply and thus lack of sufficient oxygen. Chest pain from angina pectoris is a warning sign that the coronary arteries are blocked with cholesterol-plaque. If this cracks, a blood clot will form, blocking the coronary artery, and a heart attack follows, with potential lethal consequences. Treatment of angina pectoris therefore involves drugs, such as statins, to lower the blood cholesterol, particularly that carried by low-density lipoprotein (LDL). This enables plaques to regress. In addition, treatment with nitroglycerin dilates blood vessels, to ensure proper oxygenation of the heart muscle. Blood pressure is lowered, taking a load off the heart that requires less oxygen. Angina pectoris is an ever-increasing problem, affecting over 6 million people in the United States alone. Other vasodilators that affect the Ca^{2+}-induced contraction of arterial smooth muscle are $I(f)$ current (IF) and ACE inhibitors. The $I(f)$ current is a 'funny' inward current in heart pacemaker cells carried by a mixture of Na^+ and K^+. Thus, there are four types of compounds that have been developed to treat angina pectoris. All interact with Ca^{2+} signalling directly or indirectly:

1. β-Blockers, often the first line of therapy, e.g. propranolol and atenolol.
2. Organic nitrates and muscle relaxants, e.g. nitroprusside, and natriuretic peptides, which work by increasing cyclic GMP through binding of nitric oxide (NO) to the haem group in guanylate cyclase. This changes dephosphorylation of myosin light chain kinase (MLCK), activated by Ca^{2+}-calmodulin in smooth muscle.
3. K^+ channel blockers, e.g. nicorandil.
4. Ca^{2+} channel blockers, e.g. nifedipine and amlodipine.

11.9 Heart Failure

Heart failure occurs when the heart is unable to supply the body with sufficient blood, and thus oxygen, to meet the requirements of all its tissues. Common causes include heart attacks and other forms of cardiac ischaemia, chronic high blood pressure, disease of heart valves, and damage to heart muscle – cardiomyopathy. The result is shortness of breath, worse when lying flat, coughing, chronic venous congestion, swelling of the ankles and difficulties on exercise. It affects some 11% of people over the age of 65, often undiagnosed until too late. Ca^{2+} channel blockers are too potent to treat heart failure. Agents that interfere with Ca^{2+} signalling for treatment, with other drugs, include ACE inhibitors, β-adrenergic blockers and cardiac glycosides.

11.10   Agents that Affect Adrenergic Receptors

A range of substances activate or inhibit adrenergic receptors, activated physiologically by adrenaline or noradrenaline. They work on α or β adrenergic receptors. They affect intracellular Ca^{2+} through effects on Ca^{2+} channels or intracellular Ca^{2+} stores. α-Receptors are coupled to G-proteins that cause a rise in cytosolic free Ca^{2+} directly. β-Receptors are coupled to G-proteins that activate adenylate cyclase, causing a rise in intracellular cyclic AMP. Adrenergic α- and β-agonists and antagonists are widely used both clinically and experimentally. Synthetic agonists and antagonists are usually selective for either α or β. For example, isoproteronol activates both $β_1$- and $β_2$-receptors, whereas dobutamine selects $β_1$-receptors and salbutamol (albuterol) selects $β_2$-receptors. β-Receptor agonists activate adenylate cyclase; β-receptor antagonists inhibit. Cyclic AMP activation of protein kinase A (PKA) affects Ca^{2+} channel opening and the Ca^{2+} store in heart muscle SR, via phospholamban. So $β_1$-agonists increase the speed and magnitude of the contraction of the heart myocyte. Opening of Ca^{2+} channels in the heart treats cardiac shock, acute heart failure and low heart rate arrhythmias. $β_2$-Agonists relax smooth muscle by closing Ca^{2+} channels in the plasma membrane.

On the other hand, α-receptor agonists can inhibit adenylate cyclase, and thus induce constriction of smooth muscle. $α_1$-Agonists increase cytosolic free Ca^{2+} through activation of phospholipase C (PLC), producing IP_3, release of SR or ER Ca^{2+} and opening of SOCE channels. Phenylephrine is often used experimentally to activate α-receptors *in vitro*. $α_2$-Agonists inhibit release of noradrenaline prejunctionally, and work mainly by inhibiting adenylate cyclase, through the G_i subunit of the G-protein-coupled receptor. They are used to treat blood pressure problems and the brain stem, when its vasomotor is too active. They are also used as sedatives, and for drug addicts who are dependent on heroin or alcohol.

There are two types of adrenergic receptor blocker, both interfering with Ca^{2+} signalling inside cells, particularly smooth and heart muscle:

1. α-Blockers (e.g. phentolamine).
2. β-Blockers (e.g. propranolol).

Phentolamine binds competitively to adrenergic receptors, and thus increases the concentration of adrenaline or noradrenaline necessary for the cell to cross the threshold and fire its cytosolic free Ca^{2+} signal. α-Blockers are used clinically to treat hypertension, often with other drugs such as ACE inhibitors and angiotensin receptor blockers. But they are also used to treat other conditions such as Raynaud's disease and scleroderma. The two types of α-receptor, $α_1$ and $α_2$, can be inhibited selectively or non-selectively by several compounds. Selective α-blockers can also be used to treat behavioural problems, such as panic disorder, post-traumatic stress syndrome and generalised anxiety disorders.

β-Blockers, on the other hand, competitively block β-adrenergic receptors. Propranolol, and its widely used analogue atenolol, are based on the structure of adrenaline. They are used clinically to treat high blood pressure, and also migraine, recurrent nausea, and vomiting. They block noradrenaline and adrenaline receptors, particularly those in sympathetic neurones, which cause the so-called 'fight-or-flight' response. Most of the clinically used β-blockers, such as propranolol, are non-selective. Hence there are often problems with side-effects (e.g. depression and gut problems). Some selective $β_1$-inhibitors, such as atenolol, are used clinically to treat hypertension. Selective $β_2$- and $β_3$-blockers, such as butoxamine and SR 59230A, respectively, do not have much clinical application, but are used experimentally to distinguish particular receptor types. β-Blockers reduce the higher level of cytosolic free Ca^{2+} induced by adrenergic agonists, slowing the heart and reducing the strength of the heart beat. They also reduce smooth muscle tone in the arteries, which would not to be expected, since a β_2-blocker ought to increase vascular tone. Both effects lead to a reduction in blood pressure. Both α- and β-adrenergic agonists and antagonists are widely used experimentally to assess the role of α- and β-adrenergic receptors.

11.11 Cardiac Glycosides

Cardiac glycosides (Figure 11.2) are used to treat heart failure and heart arrhythmias. They are found in the leaves and the ripe seeds of several plants, and in some animals. Digoxin and digitoxin are found in foxgloves (*Digitalis purpurea; Digitalis lanata*), and extracts from plants, such as *Strophanthus*, contain ouabain (g-strophanthin). They have been used by Somali tribesman as an arrow poison for centuries. There is enough toxin on a tip to bring down an animal as large as a hippopotamus, causing respiratory and cardiac arrest. Ouabain has also been used as a human poison. In 2003, in the United States, ouabain was used by a serial killer to kill 40 victims. Ouabain has been used as a treatment in France and Germany, and is used commonly experimentally. But, digoxin is the cardiac glycoside most used clinically. It is used with care in the elderly, and can be dangerous without drugs such as verapamil or erythromycin. Side-effects include gut, brain and eye problems. Cardiac glycosides combat the weakness in heart failure by increasing cardiac output, increasing the force of contraction through a prolonging of the plateau phase of the myocyte action potential. This slows ventricular contraction, giving more time for ventricular filling, and increases blood pressure. This also slows the heart rate. They are therefore positive inotropes and negative chronotropes. Cardiac glycosides bind to the α subunit of the sodium pump, the Na^+/K^+-MgATPase, inhibiting it. In cells with an active Na^+/Ca^{2+} (NCX) exchanger, such as the myocyte, the Na^+ then exchanges with extracellular Ca^{2+}, leading to a small increase in cytosolic free Ca^{2+}. This then gets pumped into the SR. When the action potential opens Ca^{2+} channels in the plasma membrane, Ca^{2+} that enters the cell opens the ryanodine receptors in the SR. Since the concentration of cytosolic free Ca^{2+} in the SR is higher, more is released. The higher cytosolic free Ca^{2+} causes a stronger contraction.

11.12 Benzodiazapines

Benzodiazapines (Figure 11.4) were discovered by chance at Hoffman-La Roche as a result of work to find useful tranquillisers. The first one was chlordiazepoxide, which when given to animals, unexpectedly, produced strong sedative, anti-convulsant and muscle-relaxing effects. It was put on the market as Librium in 1960. However, it became clear by the 1980s that benzodiazapines have one major disadvantage – they can be highly addictive. They act on neurones in the brain, through $GABA_A$ receptors. These are G-protein receptors that open chloride channels, causing hyperpolarisation. Benzodiazapines enhance this action of GABA, and thus make it more difficult for other neurotransmitters to depolarise the neurone necessary for the nerve to generate an action potential and fire. Thus, the link between benzodiazapines and intracellular Ca^{2+} is indirect. Without an action potential reaching the nerve terminal, the voltage-sensitive Ca^{2+} channels there cannot open. So no neurotransmitter release occurs.

11.13   Anti-Psychotic Drugs

Anti-psychotics, sometimes called neuroleptics, discovered originally in the 1950s, are used in the treatment of mental disorders, particularly bipolar disorder, schizophrenia, delusional disorders and psychotic depression. Anti-psychotics interact with the Ca^{2+} signalling system in key parts of the brain associated with behaviour. They often have a sedative effect, and thus can be used as tranquillisers. Some also act as muscle relaxants or anti-histamines. They include phenothiazines, such as chlorpromazine, trifluoperazine, thioxanthenes, such as chlorprothixene, dibenzodiazapines, such as clozapine, diphenylbutylpiperazines such as amperozol, and butyrophenone. Most interact with Ca^{2+} signalling indirectly by binding to dopamine receptors in the brain, particularly D_1 and D_2, but may also bind to 5-HT, catecholamine, muscarinic acetylcholine and histamine H_1 receptors. The most common use of anti-psychotic drugs is in the treatment of schizophrenia. But they can have side effects on muscle via effects on Ca^{2+} signalling, particularly in the head, eyes, neck and back, side effects which can be both painful and even life threatening.

11.14   Stimulants and Drugs of Abuse

Stimulants (analeptics) and drugs of abuse affect mood and behaviour by all interacting indirectly with Ca^{2+} signalling in the brain. Stimulants include caffeine, nicotine, amphetamines, such as ecstasy, noradrenaline reuptake inhibitors (NRIs) and noradrenaline–dopamine reuptake inhibitors (NDRIs), modafinil, ampakines, and yohimbine. Caffeine is the most widely used stimulant in the world, being a key ingredient in tea and coffee, and many soft and energy drinks. Its stimulant effect is via neurones in the brain. But caffeine also affects the heart, and other tissues through its ability to activate ryanodine receptors and inhibit cyclic AMP phosphodiesterase, both of which affect Ca^{2+} signalling. Cannabis has been used medically for over 1000 years. Activation of a cannabinoid receptor can inhibit adenylate cyclase, activate K_{ir} channels, affect Ca^{2+} channels, and influence several genes coding for signalling proteins, such as Raf-1, ERK, JNK, p38, c-Fos and c-Jun. Cannabinoids can inhibit gastrointestinal tract activity, lower blood pressure, and affect the heart, attenuate pain, and regulate bone mass through effects on osteoblasts, osteocytes and osteoclasts. Cannabinoid receptors in neurones are concentrated on the surface of the nerve terminal, which releases the transmitter triggered by a rise in cytosolic free Ca^{2+} as a result of the action potential arriving from the axon opening voltage-gated Ca^{2+} channels. As a result, less transmitter (e.g. GABA) is released. Other drugs of abuse, such as the opioid heroin, act on opioid receptors in the brain and elsewhere. Cocaine and LSD also interact directly or indirectly with Ca^{2+} signalling in brain neurones.

11.15   Analgesics

Analgesics are painkillers that reduce or stop pain. Intracellular Ca^{2+} has a crucial role to play at the site of the pain, and in the pathway from this site to the brain. Pain can be induced by physical, chemical or biological agents. Physical inducers include mechanical and temperature, including burns. Chemical inducers include acids and alkalis, stings, and other noxious chemicals, including inflammatory substance causing toothache or joint pain, and substances such as capsaicin, the 'hot' substance in chillies – extremely painful if you happen to rub your eyes when preparing a curry! Biological inducers of pain include stings and toxic bites. The initial pain receptor is the nociceptor, the key molecular sensor being a transient receptor potential

(TRP) channel. TRPs are ion channels, many letting Ca^{2+} into the cell (see Chapter 4). After stimulation of the initial nociceptor neurone (e.g. by heat, a sharp jab or a noxious chemical), the neurone activates a neuronal pathway that leads to the dorsal node in the dorsal root ganglion of the spinal cord. This then leads to neurones in the thalamus and into to brain, which tells us we have pain. All neuronal conducting pathways require Ca^{2+} signals at the nerve terminal to trigger transmitter release, and Ca^{2+} channels are involved in the generation of the next action potential after summation of miniature endplate potentials in the dendrites. But the action potential is carried down the axon via opening of voltage-gated Na^+ channels. The major transmitter along the nociceptor pathway is glutamate, using the NMDA receptor, which opens Ca^{2+} channels. Firing of the neurones along the pathway can be altered by other substances, including substance P, enkephalin, noradrenaline and adenosine. These can involve cyclic AMP and/or changes in cytosolic free Ca^{2+}. Many agents can trigger the nociceptor neurone or alter the threshold required for it to fire. These include K^+, substance P, 5-HT and histamine, released at the pain site as a result of cell damage or activation of non-neuronal cells. These non-neuronal cells are all activated through a rise in cytosolic free Ca^{2+}. K^+ is released by damaged or dying cells. Substance P is released by the nociceptor cell itself, causing a slow depolarisation of postsynaptic cells, which builds as a result of repetitive firing. It enhances the NMDA receptor, leading to Ca^{2+} influx and activation of NO synthase, producing NO. On the other hand, 5-HT is released by activated platelets and histamine by activated mast cells. Both are triggered by a rise in cytosolic free Ca^{2+}. There are two types of axons in nociceptors: myelinated and unmyelinated. Myelinated nerve fibres (Aδ) are fast, and can transmit action potentials at some 20 m s^{-1}, whereas unmyelinated neurones (C), the most common, are at least ten times slower. This is why, when we stub our toe, or burn a finger, it takes a second or so for the brain to sense the pain. Ouch! TRPV1 is an ion channel allowing Ca^{2+} to enter the cell, having threshold at 42 °C. Cold is sensed by TTRM8. Breaks in the surface of skin and mechanical pain involve TRPA1. Similarly, chemical nociceptors have TRP channels that respond to a wide range of substances, including spices used in cooking. TRPV1, for example, detects capsaicin and spider toxins. Several types of ion channel are found in the peripheral terminal of nociceptors, allowing Ca^{2+} into the cell. These act together as a depolarising mechanism and are a means of elevating cytosolic free Ca^{2+}. When an action potential reaches the central terminal, voltage-gated Ca^{2+} channels open, causing a rise in cytosolic free Ca^{2+}, which then triggers neurotransmitter release. A major neurotransmitter released is glutamate, which then opens Ca^{2+} channels in the next nerve leading to the next action potential. The most important Ca^{2+} channel at the terminal is an alternatively spliced $Ca_v2.2$. This is uniquely sensitive to G-protein-coupled receptors, which can regulate the terminal in a voltage-dependent or voltage-independent manner. The terminal also contains other channels that may let Ca^{2+} in, including TRPV1, TRPA1, evoked potentials (EP) and the G-linked bradykinin B2 receptor, which activates PLC, generating IP_3, and thus a rise in cytosolic free Ca^{2+}, and inhibits adenylate cyclase via G_i. Eicosanoids (e.g. prostaglandins) are also released at the site of pain. They do not activate the nociceptor directly, rather they stimulate 5-HT release and bradykinin formation, enhancing the pain response. Aspirin and other salicylates inhibit eicosanoid synthesis, and thus alleviate pain.

There are four categories of analgesics that act on the nociceptor pain pathway in various parts of the neurones involved:

1. Morphine-like, acting on opioid receptors in the central nervous system.
2. Non-steroidal anti-inflammatory drugs (NSAIDs), such as aspirin and paracetamol.
3. Local anaesthetics.
4. Miscellaneous central nervous system acting drugs.

These can influence Ca^{2+} signals directly via effects on Ca^{2+} channels, or indirectly through intracellular signals such as cyclic AMP.

11.16   Anti-Depressants and Manic Depression

Anti-depressants are used clinically to treat mood disorders, such as depression and anxiety. In depression there appears to be a deficiency of certain neurotransmitters in particular parts of the brain, particularly 5-HT (serotonin), noradrenaline and dopamine, with consequent effects on intracellular Ca^{2+} in the dendrites and cell body of neurones. Thus, the strategy to treat depression has been based on developing drugs that maintain better levels of these transmitters, by inhibiting degradation of the transmitter, or by inhibiting reuptake of the transmitter into the neurone. A longer presence of 5-HT or noradrenaline leads to more G-protein receptors being activated for longer in the dendrites of brain neurones, leading to more PLC activation, more intracellular IP_3, and longer lasting effects on cytosolic free Ca^{2+} and cyclic AMP. However, some anti-depressants inhibit 5-HT binding to its 5-HTa receptor. Unusually, anti-depressants usually take at least two weeks, and sometimes several weeks, to take effect. Thus, the mechanism must involve long-term changes in genes coding for neurotransmitter receptors, and neuronal cell number. Side-effects include nausea, diarrhoea, agitation and headache. Sexual dysfunction is common, including loss of sexual drive, failure to reach orgasm and erectile dysfunction. Tricyclic anti-depressants also can cause dry mouth, blurred vision, drowsiness, dizziness, tremors and skin rashes. All involve effects on intracellular Ca^{2+}, directly or indirectly.

11.17   Diabetes

Diabetes mellitus is a loss of the ability to control blood glucose. Acutely, uncontrolled increases in blood glucose can lead to loss of consciousness and death. Long-term hyperglycaemia leads to atheroma, heart problems, blindness and death of cells in peripheral tissues, such as the feet. There are two main types of diabetes mellitus. Type 1 occurs when the β-cells in the endocrine pancreas have been destroyed. As a result, the insulin level in the blood is severely reduced or even absent if all the β-cells have been lost, and is treated by daily doses of insulin. Type 2 diabetes involves a reduced sensitivity to insulin. Blood insulin levels can be higher than in healthy people. Sulphonylureas, such as tolbutamide (Figure 11.4), block potassium channels. So, the number of K_{ATP} channels opened in the β-cell by a particular concentration of glucose is reduced, leading to a higher concentration of glucose being required to produce a big enough depolarisation for a pancreatic β-cell to cross the Rubicon. Only when the membrane potential is low enough, will sufficient voltage-gated Ca^{2+} channels open to induce a big enough rise in cytosolic free Ca^{2+} to provoke fusion of the insulin-containing granules with the plasma membrane.

11.18   Muscle Relaxants

Muscle relaxants (Figure 11.4) reduce the tone of skeletal muscle. There are two types:

1. Neuromuscular blockers – acting at the neuromuscular junction.
2. Spasmolytics – acting in the central nervous system.

Neuromuscular blockers are used to relax muscle during surgery with anaesthesia, whereas spasmolytics are used to treat muscle spasms and pain (e.g. in the back).

Neuromuscular blockers originated from the poison curare. For centuries South American Indians have extracted 'curare' from the plant *Strychnos toxifera* for use on arrow tips, leading to paralysis of the prey, by blocking the generation of the action potential in skeletal muscle, and thus there is no cytosolic free Ca^{2+} signal to trigger muscle contraction. It can therefore also be used as a muscle relaxant in mild dose. Curares have now been replaced by milder muscle relaxants, such as succinylcholine. Succinylcholine (Figure 11.4) also competes with acetylcholine for its nicotinic receptor. But, in contrast to tubocurarine, succinylcholine is an acetylcholine agonist. Initially, just like acetylcholine, succinylcholine opens the Na^+ and K^+ channels at the muscle endplate. This depolarises the membrane, generating an action potential. Ca^{2+} is released from the SR and contraction occurs. However, acetylcholine is rapidly degraded by acetylcholine esterase, so the muscle 'resets' ready for the next action potential. But, succinylcholine is not hydrolysed by acetylcholine esterase. Rather, its degradation occurs via the enzyme butyrylcholine esterase. This is slow, leaving the succinylcholine bound the acetylcholine receptor for many minutes. As a result, the Na^+ and K^+ channels close, Ca^{2+} is pumped back into the SR, and the muscle fibre relaxes. Succinylcholine has some side-effects, the most problematic being malignant hyperthermia (see Chapter 10), an autosomal dominant condition involving mutations in the ryanodine receptor 1.

Spasmolytics acting in the central nervous system include benzodiazapines such as diazepam, balcofen, clonidine and tizanidine. They reduce muscle spasms, and are used in the treatment of gastrointestinal cholic, but have the side-effects of drowsiness, sedation and dependence. Diazepam interacts with $GABA_A$ receptors, whereas balcofen is a $GABA_B$ receptor agonist. The latter leads to hyperpolarisation, due to the opening the K^+ channels in the dendrites, and thus a greater threshold is required for the nerve to fire in order to generate a Ca^{2+} signal at the terminal for neurotransmitter release. On the other hand, clonodine, and its imidazoline derivative tizanidine, act as α_2-adrenergic receptor agonists. One anti-spasmolytic, dantrolene, acts directly on muscle Ca^{2+}. Dantrolene binds to the ryanodine receptor, inhibiting its opening by the dihydropyridine receptor activated through the action potential. Thus it can be used to treat malignant hyperthermia.

11.19 Anti-Allergics and Anti-Immune Compounds

Allergies cause lung and skin problems, as well affecting other organs in the body. The mechanism of allergy is primarily through IgE antibodies. When exposed to the allergen, the IgE–allergen complex binds to receptors on mast cells. This leads to a cytosolic free Ca^{2+} signal, as a result of opening SOCE channels, causing explosive release of histamine (see Chapter 7). The histamine provokes smooth muscle contraction via an intracellular Ca^{2+} signal. An effective treatment uses cromlyn, which inhibits mast cell activation by blocking Ca^{2+} entry, preventing histamine release in the lung, which would otherwise induce smooth muscle contraction. It is often called a mast cell stabiliser, but this confuses its mechanism of action. It was typically used as the sodium salt, disodium cromoglycate (Figure 11.4), but is has now been replaced by leukotriene receptor antagonists when using non-steroidal therapy. Cromoglycate is used to treat several other allergic conditions, such as of the skin, where mast cell activation needs to be blocked. Though its action appears to be primarily through inhibition of the rise in cytosolic free Ca^{2+} in mast cells, cromoglycate also inhibits Cl^- channels. A structurally related compound is quercetin. This inhibits transport MgATPases, and also inhibits histamine release from mast cells induced by antigen, concanavalin A or ATP by blocking Ca^{2+} entry. Quercetin is a flavanol found in many fruits, vegetables, leaves and grains, and is added to many foods and drinks.

11.20   Xanthines

Xanthines, such as caffeine and theophylline, are found in coffee and tea (Figure 11.4). They have two main intracellular targets:

- Cyclic AMP phosphodiesterase.
- The ryanodine receptor.

They may also interact with adenosine receptors. Inhibition of cyclic AMP degradation leads to an increase in cyclic AMP and thus affects on Ca^{2+} signalling. Activation of ryanodine receptors can cause directly a rise in cytosolic free Ca^{2+}. The main cell targets *in situ* are neurones and the heart, the latter leading to heart palpitations.

11.21   Substances Used Experimentally to Interfere with Intracellular Ca^{2+}

The experimental investigation of intracellular Ca^{2+} requires ways of manipulating components of the Ca^{2+} signalling system in live cells. This requires activators and inhibitors, as well as methods for changing gene expression. Pharmacological compounds target four categories of components of the Ca^{2+} signalling system:

1. Ca^{2+} channels, in the plasma membrane and in organelles.
2. Ca^{2+} pumps, in the plasma membrane and in organelles.
3. Ca^{2+}-binding proteins, including enzymes affected by Ca^{2+} directly or indirectly.
4. Enzymes and other proteins that can lead to changes in cytosolic free Ca^{2+} or Ca^{2+} inside organelles.

Thus, there are eight types of targets for substances, used experimentally to investigate the role of a particular protein in cell activation, which interfere directly with intracellular Ca^{2+} signalling:

1. Direct manipulation of cytosolic free Ca^{2+} via Ca^{2+} buffers and ionophores.
2. Ca^{2+} channels – agonists and antagonists.
3. Ca^{2+} pumps and exchangers – activators or inhibitors.
4. ER/SR – IP_3 or ryanodine receptor agonists and antagonists.
5. Mitochondrial Ca^{2+} effectors.
6. Ca^{2+} targets of Ca^{2+} signalling components.
7. Protein expression of Ca^{2+} signalling components.
8. Indirect on cyclic nucleotides and receptors.

Substances that target these can be small organics, peptides, antibodies, proteins, natural toxins or ions. They cause increases or decreases in cytosolic free Ca^{2+}, affect the Ca^{2+} target directly or through gene expression, affect alternative splicing, knock-out or mutate genes, or interfere with gene expression (e.g. siRNA). Ions that interfere with Ca^{2+} signalling include K^+, Cd^{2+}, La^{3+}, Gd^{3+}, Co^{2+}, Ni^{2+}, vanadate, and anions such as isethionate and glutamate.

11.21.1 Ca²⁺ Buffers and Ionophores

Several compounds have been developed to buffer or modify cytosolic free Ca^{2+} in live cells (see Chapter 4 for structures). Citrate was used to stop blood clotting in clinical samples. It was replaced by EDTA. But, EDTA does not select Ca^{2+} over Mg^{2+}. EGTA was then synthesised to bind Ca^{2+} with high affinity in the presence of millimolar Mg^{2+}. EGTA has the disadvantage that its last two carboxyl groups have a pK_a of 8 and 10, respectively. Thus, at physiological pH around 7, EGTA and EGTA-Ca^{2+} buffers are extremely sensitive to changes in pH. BAPTA, on the other hand has all four of its carboxyls fully ionised at pH 7. BAPTA can get into cells as the acetoxymethyl ester, which is then hydrolysed to release BAPTA itself. There are also photoreleasable BAPTA derivatives available, when exposed to a flash of light.

Ionophores, on the other hand, allow charged molecules to cross hydrophobic lipid bilayers. The Ca^{2+} ionophores A23187 and ionomycin (see Chapter 4, Figure 4.1b for structures) have been widely used at micromolar concentrations to raise the cytosolic free Ca^{2+} concentration in many eukaryotic and bacterial cells, in order to activate cell events. Ionomycin is now the compound of choice, because A23187 is light sensitive and fluorescent, and thus can interfere with measurements using fluorescent Ca^{2+} and pH indicators. Furthermore, the specificity of A23187 is not absolutely clear, as it does bind Mg^{2+} and can also lead to changes in intracellular pH because it exchanges H^+ for other cations. Both ionophores can penetrate into intracellular membranes. As a result, in the absence of extracellular Ca^{2+}, they can lead to a rapid loss of Ca^{2+} from the SR/ER and consequent activation of SOCE.

11.21.2 Ca²⁺ Channels

The four main Ca^{2+} channel targets are:

1. Voltage-gated Ca^{2+} channels in the plasma membrane.
2. Receptor-activated Ca^{2+} channels in the plasma membrane.
3. SOCE channels opening as a result of loss of Ca^{2+} from the SR/ER.
4. Ca^{2+} channels in the SR/ER opened either by IP_3 or Ca^{2+} itself acting on ryanodine receptors.

Manipulation of voltage-gated Ca^{2+} channels affects Ca^{2+} signalled events in excitable cells, such as nerves and muscle. On the other hand, SOCE channels are found in both excitable and non-excitable cells. Pharmacological effects on SOCE are usually mediated via effects on IP_3 or ryanodine receptors in the SR/ER, or effects on the SERCA pumps, which take Ca^{2+} into the SR/ER. Inhibition of IP_3 or ryanodine receptors will prevent or reduce Ca^{2+} release from the SR/ER and thus reduce SOCE channels opening. As a result, this will stop the large global rise in cytosolic free Ca^{2+}. In contrast, inhibition of SERCA pumps will inevitably lead to leakage of Ca^{2+} from the SR/ER. The Ca^{2+} depletion then opens SOCE channels via the STIM/Orai mechanism.

11.21.3 Agents which Open or Close Voltage-Gated Ca²⁺ Channels

Naturally occurring and synthetic openers and blockers of voltage-gated Ca^{2+} channels have been developed for both clinical and experimental use. Patch clamping has established that some are selective for particular Ca^{2+} channels, and so can be used to study the effect of activating or blocking them on the whole cell response. These compounds enable the contribution to be assessed of L-, T-, N-, P/Q- and R-type Ca^{2+} channels to the electrical activity of a particular cell type. Dihydropyridines block voltage-gated L-type Ca^{2+} channels: $Ca_v1.1$, 1.2, 1.3 and 1.4. Verapamil and diltiazem also inhibit L-type Ca^{2+} channels (e.g. slow Ca^{2+} channels such as in smooth muscle). By blocking L-type channels, the causes of residual current can be determined. For example, the cone snail toxin ω-conotoxin GIVA identifies a role of N-type Ca^{2+} channels, $Ca_v2.2$, whereas ω-conotoxin IVA blocks P/Q-type Ca^{2+} channels, $Ca_v2.1$. The compound SNX-482 blocks R-type Ca^{2+} channels, $Ca_v2.3$, in neuronal cell bodies and nerve dendrites. There are fewer compounds for blocking specifically T-type Ca^{2+} channels: $Ca_v3.1$, 3.2 and 3.3. τ-Kurtoxin may be such a compound in some cells, as it binds with high affinity to T-type, Ca_v3, voltage-gated Ca^{2+} channels. Kurtoxin is a 63-amino-acid peptide found in the venom of the South African scorpion *Parabuthus transvaalicus*. It works by affecting the gating of the channel.

Ca^{2+} channel openers and blockers affect the electrical excitability of nerves, muscles and other excitable cells. They also lead to a reduction in cytosolic free Ca^{2+} if the Ca^{2+} channel is primarily responsible for the rise in cytosolic free Ca^{2+}, such as nerve terminals. In the case of cardiac myocytes, complete block of the voltage-gated Ca^{2+} channels in the plasma membrane stops the microdomain Ca^{2+} rise necessary for activating the Ca^{2+} activation of ryanodine receptors on the SR. Thus, there is no global rise in cytosolic free Ca^{2+} and the heart stops beating.

11.21.4 IP₃ Receptor Activators and Blockers

Several compounds have been used experimentally to activate or inhibit all three of the known IP_3 receptors (Figure 11.5). Adenophostin A, originally isolated from cultures of *Penicillium brevicompactum*, is a more potent agonist than IP_3 in opening the IP_3 Ca^{2+} channel. Li^+ was often used in initial studies to inhibit IP_3 receptors, but has other effects, including acting as a Na^+ analogue for the Na^+/Ca^{2+} exchanger in the plasma membrane, and thus leading to a rise in cytosolic free Ca^{2+}. Pentosan polysulphate and heparin were also used in early studies to block IP_3 receptors. Pentosan polysulphate is a synthetic semiheparin used to treat painful bladder syndrome (interstitial cystitis). Heparin, on the other hand, was isolated originally from canine liver. It is still widely used as an anti-coagulant in blood samples and clinically. IP_3 receptors are completely blocked by low-molecular-weight heparin, but not by high-molecular-weight preparations. But, since heparin does not penetrate the plasma membrane, it has to be injected into cells or used in permeabilised cells and homogenates. Two compounds for inhibiting IP_3 receptors are now preferred experimentally: 2-aminoethyldiphenylborate (2-APB) and xestospongin (Figure 11.5). 2-APB inhibits a rise in cytosolic free Ca^{2+} provoked by a primary stimulus, but is not entirely specific, and can affect SOCE, TRP channels and connexins. Xestospongins, on the other hand, are cyclic bis-1-oxaquinolizidines found originally in the marine sponge *Xestospongia exigua*. They are membrane permeable and block IP_3 receptors. Thus, they suppresses antigen-induced mast cell degranulation, block hypoxic preconditioning in hippocampal neurones, and protect neurone against death promoted by mutant presenilin, which causes early-onset Alzheimer's disease. Xestospongins are a family of compounds that inhibit IP_3 receptors with an IC_{50} of 35 μM. The inhibition is reversible, but does not involve competition with the IP_3-binding site. But, xestospongins may also inhibit ryanodine receptors at concentrations around 10–50 μM, and may block K^+ and Ca^{2+} channels. The amino-steroid U-73122 (Figure 11.6) has been widely used to inhibit the generation of IP_3 by inhibiting phospholipase C. But, it can also cause Ca^{2+} release from intracellular stores, producing a rise in cytosolic free Ca^{2+} – the opposite effect from inhibition of PLC alone. On the other hand, istaroxime increases ER Ca^{2+} in failing cardiomyocytes, thereby improving heart function.

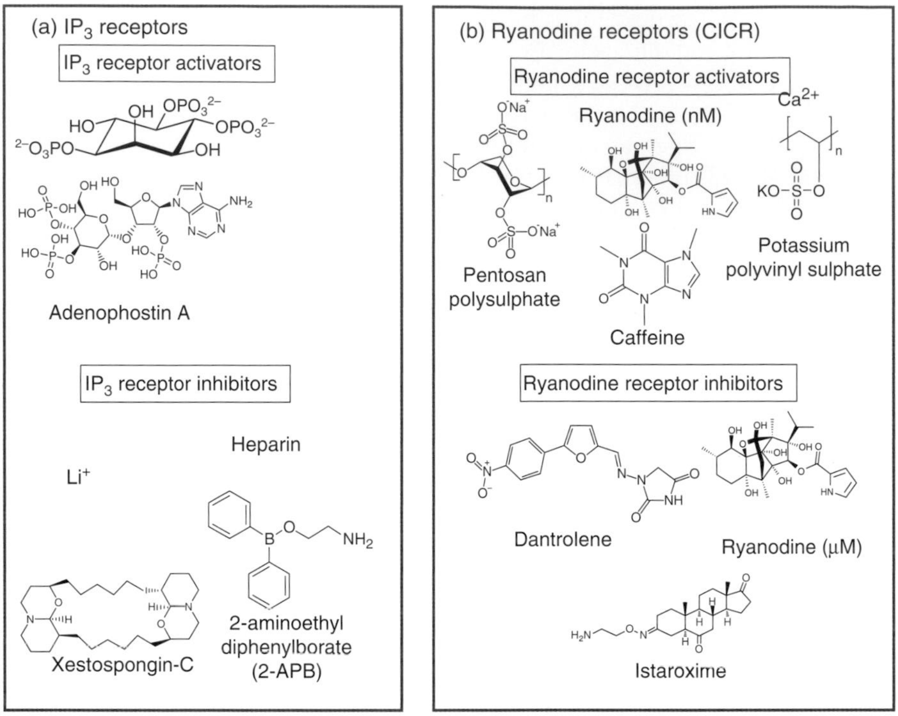

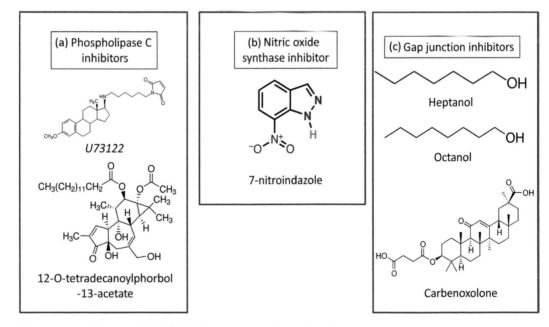

Figure 11.5 (a) IP$_3$ and (b) ryanodine receptor activators and inhibitors. Activators cause a release of Ca^{2+} from the lumen of the ER/SR into the cytosol. Inhibitors block this. Cytosolic and lumenal Ca^{2+} also affects both types of receptor/channel. Adenophostin A, originally isolated from cultures of *Penicillium brevicompactum*, is a more potent activator than IP$_3$ itself, working at nanomolar concentrations compared with micromolar IP$_3$.

Figure 11.6 Inhibitors of (a) PLC, (b) NO synthase and (c) gap junctions.

11.21.5   Ryanodine Receptor Agonists and Antagonists

Ryanodine receptors are responsible for Ca^{2+}-induced Ca^{2+} release from the SR in heart muscle and in several other cell types. All three types of ryanodine receptor are activated by caffeine and other methylxanthines, and by ryanodine at nanomolar concentrations. But at micromolar concentrations ryanodine inhibits the opening of the Ca^{2+} channel. The most commonly used compound to inhibit the ryanodine receptor, both experimentally and clinically, is dantrolene sodium (Figure 11.5). Caffeine increases alertness and improves mental tasks, but also has side-effects, such as increasing the heart rate, dehydration and anxiety. Its effects on ryanodine receptors can explain the action of caffeine to excite neurones and myocytes, but it also inhibits cyclic AMP phosphodiesterase, and can inhibit IP_3 receptors. Compounds related structurally have similar effects, including adenosine, inosine, xanthine and uric acid. Dantrolene can be used to treat ischaemia, as it reduces damage due to reperfusion, which otherwise leads to a catastrophic rise in intracellular Ca^{2+} as a result of the generation of oxygen metabolites.

11.21.6   Plasma Membrane Ca²⁺ Pump and Exchanger Inhibitors

Several substances are used experimentally to inhibit Ca^{2+} pump P-type MgATPases in the plasma membrane or SR/ER. Sodium orthovanadate (Na_3VO_4) inhibits competitively phosphotyrosyl phosphatases, and, at micromolar concentrations, is a potent inhibitor of the Na^+ pump, and Ca^{2+}-MgATPases in the plasma membrane and SR/ER. On the other hand, 7-chloro-4-nitrobenz-2-oxa-1,3-diazole (NBD-Cl) modifies thiol, amino and tyrosine hydroxyl groups in proteins, and also inactivates V-type MgATPases. Bafilomycins (macrolide antibiotics from *Streptomyces griseus*) also inhibit vesicular V-type H^+-MgATPases in exocytotic vesicles and plant vacuoles, but do not inhibit other MgATPases, such as the F_o/F_1-ATPase in mitochondria. Thus, bafilomycin-A1 reduces acidification in vesicles and affects acidocalcisomes (see Chapter 5). The $3Na^+/Ca^{2+}$ exchanger NCX can be inhibited by amiloride. Clinically, amiloride (MK870) is used in the management of hypertension and congestive heart failure, as it is a diuretic working without loss of K^+ (i.e. it is 'potassium sparing'). Its main action *in vivo* is to block the Na^+ channel in the epithelium. In the late distal convoluted tubules of the kidney, this inhibits sodium reabsorption, promoting loss of sodium and water, without losing potassium. In the 1980s amiloride was found to inhibit Na^+/Ca^{2+} exchange. Therefore, it has been used experimentally to investigate the role of Na^+/Ca^{2+} exchange in physiological processes. Amiloride also has other effects on, for example, T-type Ca^{2+} channels, cyclic GMP-gated cation channels and Na^+/H^+ exchange. Another compound which also inhibits Na^+/Ca^{2+} exchange is 3',5'-dichlorobenzamil.

11.21.7   SERCA Pump Inhibitors

Inhibition of SERCA will inevitably lead to a slow depletion of SR/ER Ca^{2+}, followed by SOCE activation. In contrast, inhibition of IP_3 or ryanodine receptors will stop SOCE activation. Two compounds are used routinely to inhibit the MgATPase in the SR/ER (i.e. SERCA1–3): thapsigargin and cyclopiazonic acid, both acting at micromolar concentrations. Thapsigargin is a sesquiterpene lactone found originally in the plant *Thapsia garganica*. Cyclopiazonic acid is a derivative of indole tetramic acid, first isolated from *Penicillium cyclopium*. Both thapsigargin and cyclopiazonic acid cause a slow rise in cytosolic Ca^{2+}, as Ca^{2+} leaks out of the SR/ER (Figure 11.7). As a result, the SR/ER Ca^{2+} store is depleted, and thus STIM is activated to move, and link with Orai in the plasma membrane. Opening of the SOCE channels in the plasma membrane then causes a large rise in cytosolic free Ca^{2+}. Both are, in principle, reversible, but cyclopiazonic

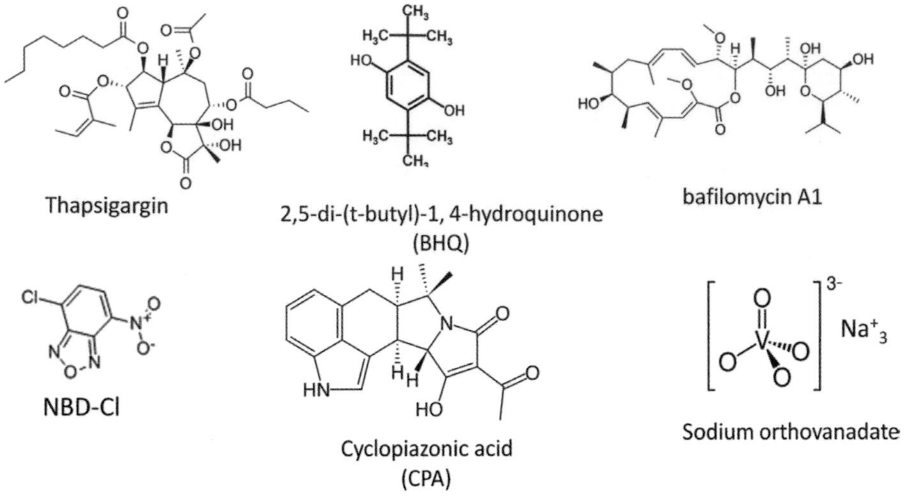

Figure 11.7 Inhibitors of Ca^{2+} pumps in the plasma membrane or ER/SR. These compounds can be relatively specific for either the plasma membrane Ca^{2+}-activated MgATPase or the SERCA pumps.

acid is now the compound of choice, because thapsigargin tends to bind to plastic, and is therefore difficult to remove both rapidly and completely in perfusion systems or microtitre plates. 2,5-Di-*t*-butyl-1,4-benzohydroquinone (BHQ) is also an inhibitor of SERCA pumps, but may also inhibit voltage-gated L-type Ca^{2+} channels, by generating superoxide.

11.21.8   Compounds which Affect SOCE

Fewer compounds have been found that affect SOCE directly. But pharmaceutical companies have found some using high-throughput screening methods. Thus, 1-[2-(4-methoxyphenyl)-2-[3-(4-methoxyphenyl)propoxy]ethyl]-1*H*-imodazole (SKF 96365) and *N*-propylargylnitren-dipine (MRS 1865) can block opening of SOCE by thapsigargin.

11.21.9   Mitochondrial Blockers

A number of compounds have been used to block Ca^{2+} uptake by mitochondria (Figure 11.8). The contribution of mitochondria to cytosolic free Ca^{2+} signals can be assessed by preventing Ca^{2+} uptake using Ruthenium Red and its derivatives. For example, the role of mitochondria in cytosolic free Ca^{2+} oscillations has been established in cells, since these are prevented by Ruthenium Red. Ca^{2+} can also be released from mitochondria by blocking the respiratory chain using CN, azide (N$_3^-$) or antimycin, or by uncoupling the respiratory chain from oxidative phosphorylation using dinitrophenol or FCCP. Respiratory chain blockers and uncouplers inevitably leads to a drop in cytosolic MgATP and a rise in MgADP. This reduces Ca^{2+} efflux by the plasma membrane Ca^{2+} pump and Ca^{2+} influx into the SR/ER via the SERCA pump. Both of these will lead to a gradual rise in cytosolic free Ca^{2+}. Ruthenium Red stains mitochondria and blocks its Ca^{2+} uniporter, preventing Ca^{2+} uptake. However, it is also a potent inhibitor of the ryanodine receptor, with a K_d of around 20 nM, and also inhibits several TRP channels, including TRPM6 and 8, TRPV1–6, and TRPA1, and ion channels in plants. Ruthenium Red has been shown to bind several other proteins, including some Ca^{2+}-MgATPases, myosin light chain phosphatase, tubulin and calmodulin.

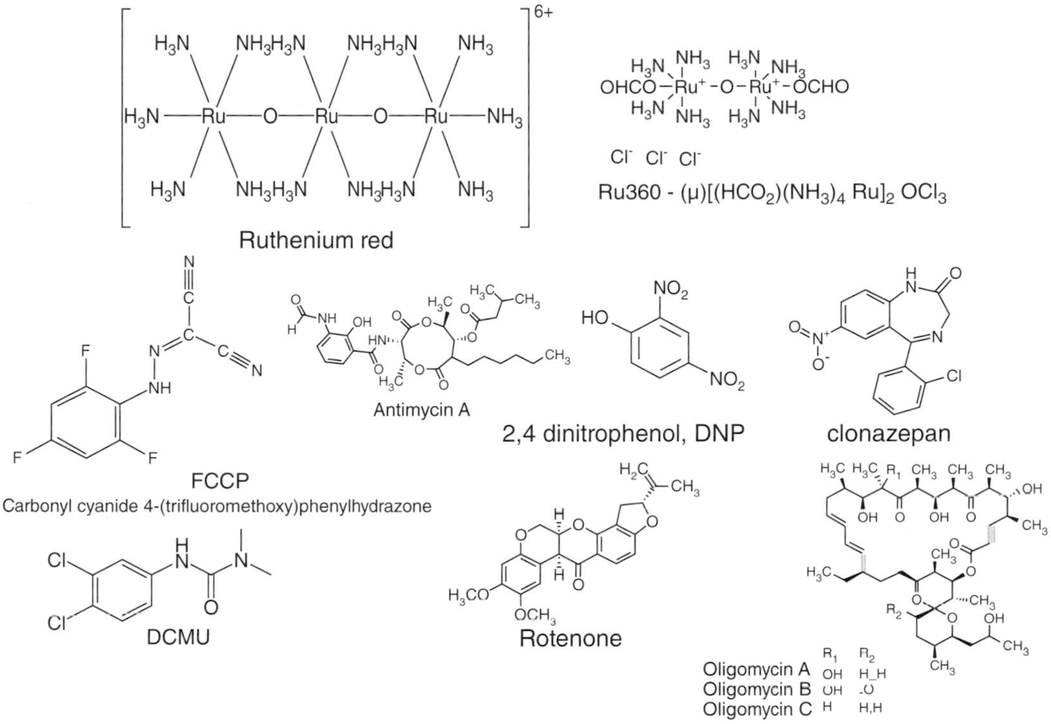

Figure 11.8 Inhibitors of mitochondrial Ca²⁺ uptake. These can act directly on the mitochondrial uptake mechanism, or indirectly through a reduction of the electrochemical potential or drop in MgATP.

11.21.10 Ca²⁺ Target Inhibitors

Several compounds have been developed to inhibit the Ca²⁺ targets inside cells (Figure 11.9). The most widely used are inhibitors of:

- Calmodulin.
- Ca²⁺-calmodulin kinases.
- Ca²⁺-activated phosphatase calcineurin.
- The calpain protease family.

All membrane Ca²⁺-calmodulin inhibitors are membrane permeable, but some penetrate cells better than others. Several are used clinically, but it is not clear whether all their clinical actions can be explained by inhibition of calmodulin. When used experimentally, two structurally different compounds should be used. If they both have the same effects on the cell event, then this provides evidence for a role for calmodulin. The three most widely used experimentally are trifluoperazine (TFP), calmidazolium and W-7 (or its analogue W-13). TFP binds to the hydrophobic domain, preventing calmodulin activating its target. However, the specificity for calmodulin is not fully established. Calmidazolium chloride is also an inhibitor of Ca²⁺-calmodulin activated processes in many tissues. Calmidazolium especially affects the heart, vascular smooth muscle, endothelial and juxtaglomerular cells. But it also has effects not directly related to Ca²⁺-calmodulin. These include blocking voltage-gated Na⁺, K⁺ and L-type Ca²⁺ channels, and release of Ca²⁺ from the SR. Like trifluoperazine, calmidazolium inhibits calmodulin by direct binding, but may also bind to, and inhibit, calmodulin target enzymes.

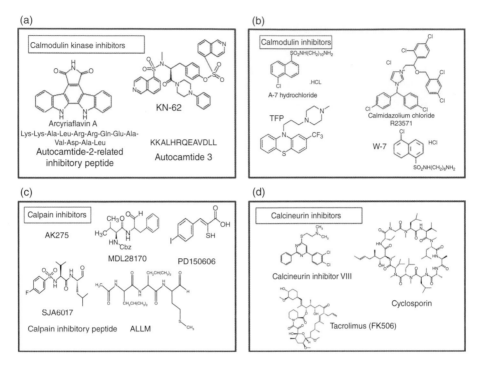

Figure 11.9 Inhibitors of Ca^{2+} target proteins.

Calmidazolium can lead to an increase in cytosolic free Ca^{2+} in some cells, such as slime moulds, as a result of inhibiting SERCA pumps in the ER, causing loss of Ca^{2+} from the ER. As a result, SOCE is activated and Ca^{2+} enters the cell. Calmidazolium also inhibits NO synthase in some cells, yet can lead to an increase in NO in others by raising cyclic GMP. Calmidazolium also inhibits Ca^{2+}-calmodulin *N*-methyltransferase. W-7 and W-13 are naphthalene-sulphonamides. They inhibit Ca^{2+}-calmodulin cyclic nucleotide phosphodiesterase and myosin light chain kinase (MLCK). These also bind to the plasma membrane, affecting electrostatic charge on the membrane and thus the zeta potential. As a result, W-7 and W-13 can affect cell growth and other cellular processes independent of calmodulin. Two other clinically used drugs – chlorpromazine and fendiline – have been used to interfere with Ca^{2+}-calmodulin-activated processes. Chlorpromazine is a potent dopamine receptor inhibitor, but can also inhibit 5-HT, histamine, α-adrenergic and muscarinic receptors, all of which involve intracellular Ca^{2+}. Fendiline, on the other hand, is a voltage-gated Ca^{2+} blocker. So, neither of these compounds is useful experimentally in testing hypotheses regarding the role of calmodulin.

A wide variety of hormones, neurotransmitters, growth factors and other primary cell stimuli work by activating the serine/threonine Ca^{2+}-calmodulin kinases I–IV (CaMKI–IV). These kinases are involved in cell signalling in many cell types, including the brain, memory, the heart and tissue development. Commonly used inhibitors are CaMKII inhibitor 1, arcyriaflavin, autocamtide-2-related peptide, KN-62 and KN-93, and STO-609 (1,8-naphthoylene benzimidazole-3-carboxylic acid). Arcyriaflavin A, originally isolated from the marine ascidian *Eudistoma*, is a potent inhibitor of CaMKII, but also of the CDK-cyclinD complex. As a result of the latter, it can be used as an anti-viral agent. Autocamtide-2-related inhibitory peptide is also a selective CaMKII inhibitor and can be myristoylated for membrane targeting. It is important to use a combination of these inhibitors experimentally to determine the role of Ca^{2+}-calmodulin kinase in a cell event.

Several naturally occurring and synthetic inhibitors of the Ca^{2+}-activated serine/threonine phosphatase calcineurin have been used clinically and experimentally. Three have been widely used clinically in transplantation and for treating skin disease: cyclosporine, pimecrolimus and tacrolimus. These have also been used to treat other conditions where the immune system has been inappropriately activated, such as ulcerative colitis. They inhibit calcineurin by binding to the protein FKBP12 (macrophilin), which then binds to calcineurin. On the other hand, micro-cystins are heptapeptides produced by some cyanobacteria, microcystin LR being the most potent, and inhibit several protein phosphatases, particularly types 1 and 2A, but also calcineurin at much higher concentrations. The fungal metabolite dibefurin, metal-ligating phosphonates, endothal derivatives and PD144795 also inhibit calcineurin, but are not as potent as cyclosporin. Okadaic acid produced by symbiotic dinoflagellates in marine sponges, found first in *Halichondria okadai*, inhibits protein phosphatases. At high concentrations it can inhibit calcineurin.

Calpains are proteases that work in the cytosol of cells at neutral pH. They are C2-type non-selective cysteine proteases activated by Ca^{2+}. They are ubiquitous, and play an important role in cell shape, amoeboid cell movement, the cell cycle and apoptosis. More than a dozen calpain inhibitors have been developed for experimental and potential clinical use, but may not be specific, as many target any cysteine-based protease. They include calpain inhibitors I, II and III, ALLM, ALLN, AK275 and MDL28170 as protectants against ischaemia, as well as PD150606, SJA6017, ABT-705253, BDA-410 and SNJ-1945. BDA-410 appears to be highly specific for calpain over other cysteine proteases. PD150606 is non-competitive and selective for calpain over, for example, cathepsins. Clinically, release of calpain after tissue injury can be a danger, such as in the brain. So calpain inhibitors need to act intra- and extra-cellularly *in situ*. *N*-Acetyl-Leu-Leu-Met is also a good inhibitor. There is also an endogenous inhibitor, the protein calpastatin. This compound targets the Ca^{2+}-binding site and is membrane permeable. There are now potent tripeptide-based macrocyclic calpain inhibitors.

11.22 Natural Toxins and Poisons

11.22.1 Natural Stings, Bites and Other Toxin Events

Many animals, plants, fungi and microbes produce toxins in order to capture prey, compete with other organisms, or to defend themselves against attack. Ca^{2+} signalling is involved directly or indirectly in the action of virtually all natural toxins. Spiders and snakes bite, whereas bees, wasps, scorpions and jellyfish sting. Bacteria produce antibiotics as a natural mechanism for disabling other species of bacteria competing for the same nutrients. Dinoflagellates produce toxins that kill predators, and even cause human deaths when contaminating shellfish are eaten during a meal. Several of these toxins act directly on ion channels, including Ca^{2+} channels. Others interact with the Ca^{2+} signalling system indirectly. The result is either local pain, redness of the skin (erythema) and inflammation or, more seriously, a problem at other sites in the body, such as muscle paralysis and coma.

Pain involves Ca^{2+} signals in the nociceptor pathway. The severity of pain depends on how many receptors have been activated and for how long. The agents involved affect neurotransmitter levels in synapses, hormones and paracrines surrounding non-excitable cells. Venoms contain proteins and peptides that have a kinin-like action, activating the nociception neuronal pathway responsible for pain, affecting both Na^+ and Ca^{2+} ion channels. But venoms also contain small molecules such as histamine, 5-HT and prostaglandins that exacerbate the inflammatory response through intracellular Ca^{2+}. Venoms can also initiate the production of antibodies, with a consequent allergic reaction involving Ca^{2+} signalling in mast cells, after a second sting or bite. Toxins that block Ca^{2+} channels are used experimentally, particularly those from cone snails, spiders and scorpions. Others, such as TTX from pufferfish, block Na^+ channels, and thus affect the ability of cells to generate action potentials, and the Ca^{2+} signals that can arise from this. Examples of toxins that interact directly or indirectly with the Ca^{2+} signalling system are (Figure 11.10):

1. Vertebrates:
   a. Pufferfish – TTX.
   b. Snake venoms – acetylcholine esterase inhibitors.
2. Invertebrates:
   a. Spider (tarantula, funnel web) – agatoxins.
   b. Cone snail – conotoxin.
   c. Scorpion – charybdotoxin and agatoxin.
   d. Jellyfish – peptides.
   e. Dinoflagellates and effects on mitochondria.
   f. Stings from bees (mellinin), wasps and hornets.
   g. Ant bites.
   h. Bitter substances from glow-worms and other beetles.
3. Bacteria:
   a. Antibiotics.
   b. Bacterial toxins – antibiotics, pore formers (cholera, endotoxin).
4. Fungi:
   a. Alfatoxin.
   b. Okadaic acid.
5. Plants:
   a. Ryanodine.
   b. Bean – strychnine.
   c. Foxgloves – digitalis.
   d. Potatoes – atropine.
   e. Nettle stings.

Targets for these toxins are Ca^{2+} and other ion channels, neurotransmitter action at both ends of the neurone, SR/ER Ca^{2+}, mitochondria, and pores in the plasma membrane.

11.22.2 Cone Snail Toxins – Conotoxins

Cone snails are marine gastropod molluscs that live in the Indo-Pacific Ocean, particularly in corals. They produce some of the most potent toxins known, and have caused human fatalities. Conotoxins are peptides with 12–68 amino acids and one or more S–S bonds. There are five main classes: α, μ, δ, κ and ω, which act on neurotransmitter receptors and specific ion channels (Table 11.4). All interact with intracellular Ca^{2+} signalling, the ω-conotoxins (e.g. ω-CTXMVIIA (SNX-111)) (Figure 11.10) acting specifically on N-type Ca^{2+} channels in presynaptic nerve terminals, reducing transmitter release (see Chapters 5 and 7). One synthetic ω-conotoxin is used clinically to treat neuropathic pain. Since each class of conotoxin has a different receptor or ion channel target, and each snail produces several different peptides, it is the combination of these effects that is so dangerous, and effective at killing prey.

Venomous cone snails of the genus *Conus*, of which there are over 300 species, were first described in the mid-nineteenth century by Adams in 1848. The main natural targets of cone snail toxins are fish and worms, for food, though a role in defence against predators cannot be ruled out. *Conus* venom causes muscle paralysis, but can also have cardiovascular effects (e.g. ω-CTXMVIIA). The first clue that conotoxins might involve changes in intracellular Ca^{2+} was the

Table 11.4 Conotoxins: classification and mechanism.

Class	Main target and action
α	Inhibits nicotinic acetylcholine receptor
δ	Inhibits inactivation of voltage-gated Na^+ channels
κ	Inhibits K^+ channels
μ	Inhibits voltage-gated Na^+ channels in skeletal muscle
ω	Inhibits N-type Ca^{2+} channels in presynaptic neurones

The nomenclature is based on: a Greek letter = pharmacological and structural group; name of toxin (i.e. conotoxin or CTX); capital letter based on the species name (e.g. G = *Conus geographus*, T = *Conus tulipa*, P = *Conus purpurascens*, Tx = *Conus textile*, M = *Conus magus*); Roman numeral = order of discovery; letter = variant; for example, ω-CTXMVIIA.

observation that they can cause persistent contraction of rat diaphragm muscle *in vitro* – an effect potentiated by caffeine. Intravenous injection caused spastic paralysis and death. The conotoxins acted on skeletal muscle contraction through an effect on Ca^{2+} permeability in excitable membranes. Cardiac and smooth muscle were essentially insensitive, presumably because, unlike neurones, they do not have significant N-type Ca^{2+} channels. Toxic effects on humans include burning pain, swollen arm, local numbness, followed by spread over the body with cardiac and respiratory distress, loss of coordination, drooping eyelids (ptosis), shallow breath and headache. These are reversible and non-lethal. However, fatalities were reported when the symptoms developed to numbness, stiff lips, blurred vision, paralysis and coma.

CKGKGAKCSRLMYDCCTGSCRSGKC-`NH₂

ω-conotoxinMVIIA (SNX 111)

Disulphide bonds 1C-16C, 8C-20C, 15C-25C

PEFTNVSCTTSKECWSVCQRLHNTSRGKCMNKKCRCTS-OH
Charybdotoxin
Disulphide bonds at C7-C28,
C13-C33, C17-C35

ω-agatoxinIVA

H-Lys-Lys-Lys-Cys-Ile-Ala-Lys-Asp-Tyr-Gly-Arg-
Cys-Lys-Trp-Gly-Gly-Thr-Pro-Cys-Cys-Arg-Gly-
Arg-Gly-Cys-Ile-Cys-Ser-Ile-Met-Gly-Thr-Asn-
Cys-Glu-Cys-Lys-Pro-Arg-Leu-Ile-Met-Glu-Gly-
Leu-Gly-Leu-Ala-OH

Disulphide bonds
C4-C20, C12-C25, C19-C36, C27-C34

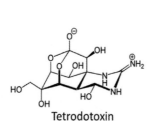

Tetrodotoxin

Figure 11.10 Some naturally occurring toxins that affect Ca^{2+} signalling. The figure shows one example each of a cone snail, spider and scorpion toxin, as there is a range of slightly different compounds produced in the sting or bite of each animal – an example of Darwinian molecular variation in natural pharmacology. The conotoxin and agatoxin shown interact directly with Ca^{2+} signals, whereas charybdotoxin and TTX interact indirectly via inhibition of K^+ and Na^+ channels, respectively. In the agatoxin shown, carbon atoms are coloured grey, nitrogen blue, oxygen red and sulphur atoms yellow. *Source*: Bohog via Wikimedia Commons: https://en.wikipedia.org/wiki/Agatoxin#/media/File:Omega-agatoxin_IVA.png.

Conotoxins are peptides with a very interesting three-dimensional structure, maintained by S–S bonds through cysteine residues, with a two-, three- or four-loop intramolecular structure, leading to some having a 'knot-like' structure. ω-Conotoxins are the only conotoxins that target specifically Ca^{2+} channels. They are relatively specific for N-type Ca^{2+} channels in nerve terminals, though they can also block P- and Q-type. The result is a decrease in neurotransmitter release. Conotoxins can also target presynaptic G-protein receptors. ω-Conotoxins act on N-type voltage-gated Ca^{2+} channels by binding to the $\alpha_2\delta$ subunit, and are so potent that they have been given the name 'King Kong' peptides, some being 100–1000 times more potent as a painkiller than morphine. Ziconotide is a synthetic form of ω-CTXMVII (SNX-111; Prialt). By blocking the N-type Ca^{2+} channel in the spinal cord it is a very powerful painkiller in patients with neuropathic pain, quite different from opioid and local anaesthetic painkillers. It was given FDA and European approval in 2004. However, it has to be given intrathecally (i.e. injected directly into the spinal cord). Conotoxins have a relatively short half-life. So, in an attempt to lengthen this, a circular α-conotoxin was synthesised, α-CTXVc1.1, based on the conotoxin from *Conus victoria*. In this case, the name is α-CTXV as usual with the V = *victoria* and c1.1 = the variant. It is a potent antagonist of nicotinic acetylcholine receptors in muscle, but also inhibits N-type Ca^{2+} channels in the rodent dorsal ganglion, through activation of $GABA_B$ receptors. It is also highly selective for N-type Ca^{2+} currents induced by $GABA_A$-type receptors. This circular conotoxin is absorbed in the gut, so it can be taken orally.

11.22.3 Spider Toxins – Agatoxins

Many spiders produce neurotoxins that block a variety of ion channels. They are usually called agatoxins, after the funnel web spider *Agelenopsis aperta*, to be distinguished from the scorpion toxin agitoxin. Agatoxins are a chemically diverse group of polyamines and peptides (Figure 11.10). Depending on the type, they act as neurotoxins by blocking Na^+ or Ca^{2+} channels gated by voltage or glutamate. For example, ω-agatoxin IVA blocks N-type voltage-gated Ca^{2+} channels. The nomenclature is identical to that used for conotoxins. There are three main subclasses: α, μ and ω. α-Agatoxins are polyamines with an attached aromatic moiety, whereas μ- and ω-agatoxins are peptides. α-Agatoxins block glutamate-activated channels, such as NMDA and AMDA, in the nerve terminals of insects and mammals, thereby modifying the Ca^{2+} signal for neurosecretion, and the amount of neurotransmitter released. This causes irreversible paralysis in insects. On the other hand, μ-agatoxins modify the presynaptic voltage-gated Na^+ channel at the neuromuscular junction of insects, and have little or no effect on other species. This causes a slow, long-lasting paralysis in insects. Thus, α- + μ-agatoxin acting together produce a long-lasting paralysis, which can be terminal. ω-Agatoxins, however, cause muscle spasms leading to progressive paralysis and insect death, and act directly on voltage-gated Ca^{2+} channels. There are four classes of ω-agatoxins. Types IA and IIA block the presynaptic Ca^{2+} channels in insects, resulting in a reduction in neurotransmitter release, whereas type IIIA blocks N, P, Q and R neuronal Ca^{2+} channels, and L-type Ca^{2+} channels in myocardial cells. Type IVA has a high affinity for P/Q Ca^{2+} channels in vertebrates. Thus, each has a preference for different types of Ca^{2+} channel, though, like many toxins, rarely are any absolutely specific. Agatoxins, are useful experimentally for blocking specific neurones, but they are not used clinically.

11.22.4 Scorpion Toxins

A sting from a scorpion toxin can be very painful, sometimes lethal, particularly for children. The venom injected by the sting in their tail contains a mix of substances that affect Ca^{2+} and other ion channels. Scorpions are found world-wide, except Antarctica. Symptoms from a sting are both neurological and non-neurological. Symptoms start immediately with pain. Massive release of neurotransmitters follows causing sweating, nausea and vomiting. Nervous system effects occur through the central nervous system: sympathetic and parasympathetic, somatic, cranial, and peripheral. Multisystem organ failure occurs in extreme cases, leading to death, with neurological and non-neurological symptoms.

Scorpion toxins are usually named after the species from which they were first isolated. For example, noxiustoxin is from the Mexican scorpion *Centruroides noxius* and tityustoxin is from the Brazilian yellow scorpion *Tityus serrulatus*. As with other venoms, scorpion stings contain a wide range of small and macro-molecules, including mucopolysaccharides, hyaluronidase, phospholipase, 5-HT, histamine, enzyme inhibitors (e.g. protease inhibitors) and neurotoxic peptides. Like the other toxins, scorpion neurotoxins act on a variety of ion channels. Like conotoxins and agatoxins, scorpion toxins are peptides held in constrained three-dimensional structures by S–S bridges, subtle changes in the scaffold of scorpion toxins affecting their ion channel selectivity. Scorpion toxins have application as insecticides and clinically in humans. One isolated from *Buthus martensii* is used to treat epilepsy.

The best characterised are scorpion α- and β-toxins, acting on Na⁺ channels, and charybdotoxin acting on K⁺ channels (K_Tx). All three ultimately disturb indirectly the Ca²⁺ signal responsible for the end-response of a nerve, muscle or inflammatory cell. The α- and β-toxins cause Na⁺ channels to stay open, so the nerve cannot repolarise, and is therefore blocked. This stops the Ca²⁺ signal being repeated at the nerve terminal, preventing further neurotransmitter release, and so conduction to the next excitable cell is stopped. Charybdotoxin is a 37-amino-acid peptide in the venom of the deathstalker scorpion *Leiurus quinquestriatus hebraeus* (also known as the Israeli yellow scorpion) helping to paralyse its prey. It is named after a Sicilian whirlpool and a mythical Greek monster, Charybdis – one to avoid! It blocks the pore of the Ca²⁺-activated BK channel (large conductance) in skeletal muscle, and the voltage-gated SK channel. It may also interfere with the SR MgATPase in heart myocytes. The block of K⁺ channels causes hyperexcitability of neurones in the nervous system. The structure and action of charybdotoxin are similar to two other scorpion toxins – margatoxin and iberiotoxin. Charybdotoxin is positively charged and binds negative amino acids in the ion pore with nanomolar affinity. A key amino acid is lysine at position 27. A similar toxin – iberiotoxin – with 37 amino acids, from the Indian red scorpion *Buthus tumulus* also blocks Ca²⁺-activated K⁺ channels.

11.22.5 Pufferfish and Octopus Toxins

Tetrodotoxin (TTX), found in pufferfish, some octopus, and other marine species, is made by symbiotic bacteria within these animals. It causes rapid paralysis. TTX is useful experimentally as a highly potent blocker of voltage-gated Na⁺ channels. The effect of TTX on Ca²⁺ signals is thus indirect, by stopping the generation of Na⁺-dependent action potentials, and thus voltage-gated Ca²⁺ channels and SR Ca²⁺ release in muscle.

11.22.6 Jellyfish Sting Toxins

Jellyfish toxins affect ion channels, particularly Ca²⁺. Box jellyfish found in the sea near North Australia, the Philippines and the coast of Japan during their summer produce a very painful sting. One of the world's deadliest animals to humans is *Chironex yamaguchi* near Japan. Its sting can be fatal within 4 min. A severe jellyfish sting can cause cardiac arrest, preventing unlucky swimmers reaching the shore, so they drown. Treatment involves vinegar, an ice pack, and anti-histamines to reduce local inflammation, the severe cutaneous pain and skin redness (erythema). Jellyfish are cnidarians. All Cnidaria have sting cells – cnidocytes, within which is an amazing organelle called the nematocyst. The nematocyst is a bulb-shaped organelle containing a microscopic syringe coiled up within it. It can discharge within less than 1 μs, with an estimated acceleration of several million times that of gravity. The cnidocyte has a hair-like structure, the cnidocil, which, when touched, fires the nematocyst into the prey or predator. The nematocyst capsule contains a high concentration of Ca^{2+}. There are some 30 types of nematocyst. When triggered, water moves into the nematocyst as result of the osmotic gradient and fires the 'syringe'. Cnidocytes can also fire in response to chemical stimuli in the skin of the animal being attacked, as a result of chemosensor activation. A single nematocyst has enough toxin to paralyse one small crustacean. Isolation of the toxins from three box jellyfish, *Carybdea rastonii*, *Carybdea alata* and *Chiropsalmus quadrigatus*, named CaTXA and B, together with cloning the cDNA coding for them, showed that they were proteins of molecular weight around 45 kDa, with about 463 amino acids. The toxins activate ion channels, Ca^{2+} in particular, as well as non-selective ion channels such as TRPV1. Thus, they cause contraction of arterial smooth muscle – an effect dependent on the presence of extracellular Ca^{2+}. This is accompanied by influx of ^{45}Ca, and blocked by the L-type Ca^{2+} channel blockers verapamil and diltiazem. The effect of the toxin is both pre- and post-synaptic. Presynaptic activation of Ca^{2+} channels releases noradrenaline, which, together with postsynaptic activation of ion channels, causes contraction of smooth muscle.

Using *Xenopus* eggs injected with cDNA coding for the non-selective cation channel TRPV1, it has been shown that toxins from four cnidarians, *Physalia physalis*, *Cyanea capillata*, *Chironex fleckeri* and *Aiptasia pulchella*, can desensitize the TRPV1 receptor in the presence of a channel opener, capsaicin, the hot substance in chillies, the toxin acting at a site distal from the capsaicin-binding site. The effect is to prevent shutting of the channel after repetitive stimulation, thereby maintaining activation of nociceptive neurones and thus provoking pain. Activation of TRPV1 in nociceptive neurones, *in vivo*, causes depolarisation and therefore pain. The more severe effects of jellyfish toxins on the heart are also be explained by effects on Ca^{2+} channels. However, the mechanism by which jellyfish toxins cause haemolysis is not clear. The local anaesthetic lidocaine, which blocks Na^+ and Ca^{2+} channels, may be the best current treatment.

11.22.7   Hymenopteran Stings and Bites

Intracellular Ca^{2+} plays an important role in the mechanism by which the stings and bites of hymenopterans cause pain and their allergic reactions. Hymenoptera, from the Greek *hymen* = membrane and *ptera* = wing, are the largest order of insects, and include bees (Apoidea), wasps, hornets and yellow jackets (Vespoidea), and ants (Formicidae). The venom injected contains a range of proteins and small molecules. In bees, wasps and hornets, the main toxin is a protein. Whereas in ants 95% of the venom is made of alkaloid. In animals, including humans, more than 20 stings at once can be fatal, without inducing an allergic reaction. However, just a second sting, several weeks after the first, can provoke a severe anaphylactic reaction. There are three targets for the toxins in hymenopteran stings:

1. The nociception system, causing pain.
2. Red blood cells, leading to haemolysis.
3. Mast cells, leading to an allergic reaction, often severe and can be fatal.

In nociception and allergy, intracellular Ca^{2+} plays a central role in the response to the toxin. The acute pain of stings and bites is through activation of the kallikrein–kinin nociception neuronal pathway. A key player in this is bradykinin, acting through B2 receptors, involving Ca^{2+} channels. A modified bradykinin, [Thr6]-bradykinin, isolated from the social wasp *Polybia occidentalis* inhibited the nociception pathway in rats, through its effect on the bradykinin B2 receptor.

Apitoxin, the bitter, colourless liquid that makes up a bee sting, like many stings and bites, is acid (pH 4.5–5.5). It is a complex mixture of proteins and other molecules. The latter includes apamin, which stimulates cortisol production by the adrenals and adolapin. This blocks cyclo-oxygenase, leading to effects on prostaglandins. Apamin is a polypeptide (CNCKAPETALCAR-RCQQH), with S–S bonds between C1 and C11 and C3 and C15. It selectively blocks SK-type Ca^{2+}-activated K^+ channels found in the central nervous system and smooth muscle. Apamin therefore has been used experimentally to affect hyperpolarisation of cells after an action potential. The toxic liquid also contains two enzymes: PLA2, which degrades membranes, and hyaluronidase, which dilates capillaries, spreading the inflammation, as well as histamine, exacerbating the allergic response, and dopamine and noradrenaline which increase pulse rate. There are also protease inhibitors that may induce bleeding. But the major toxic agent in apitoxin is a 26-amino-acid peptide: melittin (GIGAVLKVLTTGLPALISWIKRKRQQ). This activates PLA2 in plasma membranes, leading to haemolysis of red blood cells. Melittin also forms 'pores' in cell membranes that allow ions, such as Na^+ and Ca^{2+}, into the cell and are large enough to allow larger molecules, such as ATP, to leak out. This is similar to the pores caused by complement and bacterial protein toxins.

Ant bites are also painful and can be dangerous. A bite from the jack jumper ant *Myrmecia pilosula* in Australia can be life-threatening. Once again, the pain involves the nociception neuronal pathway, the more severe reaction being an allergic response, provoked by protein, and an alkaloid component and formic acid. The allergic mechanism involves the production of IgE antibodies after the first sting or bite. A following bite then leads to proteins in the venom binding to IgE molecules, which then bind to the IgE receptors on mast cells. This causes a large rise in cytosolic free Ca^{2+}, which triggers an explosive fusion of vesicles with the plasma membrane and consequent release of histamine. The histamine then triggers an inflammatory response, including contraction of smooth muscle in the airways, as a result of a rise in cytosolic free Ca^{2+} in these cells.

11.22.8 Snake Venoms

Snake bites frighten people, as in several parts of the world they are fatal. The toxic fluid – the venom – contains neuro- and muscular toxins, cytotoxins, coagulants, and enzymes. Their main interaction with Ca^{2+} signalling is indirect, through inhibition of acetylcholine esterase, causing muscles to stop responding to acetylcholine at the neuromuscular junction. Enzymes in snake venom include phosphodiesterase, PLA2, causing haemolysis, hyaluronidase, increasing the permeability of tissues to enzymes, digestive oxidases and proteases, and MgATPases. It is the cytotoxins acting locally and the neurotoxins that interact with the Ca^{2+} signalling system. Some snake toxins, such as dendrotoxin, bind to voltage-gated K^+ channels.

11.23 Plant Toxins and Intracellular Ca²⁺

Plants produce many toxins that affect the Ca^{2+} signalling system. Some of the effects are mediated through direct effects on Ca^{2+} signalling, while others are indirect. Ryanodine releases Ca^{2+} from the SR in muscle and the ER in non-muscle cells, if the ER has ryanodine receptors. Caffeine can also bind to ryanodine receptors and release Ca^{2+} into the cytosol. But caffeine also has an indirect effect on intracellular Ca^{2+} by inhibiting cyclic nucleotide phosphodiesterases, leading to a rise in cyclic AMP. On the other hand, digitalis from foxgloves has its effects on the heart by inhibiting the Na^+ pump, leading to a rise in intracellular Na^+, which in turn causes a rise in cytosolic free Ca^{2+} through the Na^+/Ca^{2+} exchanger.

Strychnine is a famous plant toxin used by many crime writers, from the seeds of the bean *Strychnos ignatii* and tree *Strychnos nux-vomica*. It is one of the bitterest substances known. Bitter receptors involve 25 types of T2R receptors. Strychnine is used as a pesticide, and, at one time, used clinically as a laxative. Strychnine causes muscle convulsions, leading to death through asphyxia and exhaustion. It acts by blocking the glycine receptor in the spinal cord and brain. Any effect on neurones will affect Ca^{2+} signals directly in the dendrites and Ca^{2+} signals in the nerve terminal. On the other hand, the toxin atropine affects intracellular Ca^{2+} directly through muscarinic acetylcholine receptors. Derived from the Greek myth Atropos (one of three fates), atropine comes from the plant *Atropa belladona*, known as deadly nightshade. It is a muscarinic acetylcholine receptor blocker, and is potentially deadly. As a result, it reduces the parasympathetic nervous system on all muscles and glands regulated by this system. Thus, atropine also has effects on the heart and bronchial secretions. All of these involve changes in cytosolic free Ca^{2+}. Atropine blocks all five types of muscarinic acetylcholine receptor (M1–M5). These are G-protein-coupled receptors. M1, M3 and M5 involve G_q and activation of PLC, and thus production of IP_3, release of ER Ca^{2+} into the cytosol and opening of SOCE channels in the plasma membrane. In contrast, M2 and M4 muscarinic receptors involve G_i, inhibition of adenylate cyclase, an increase in K^+ conductance and consequent hyperpolarisation, which increases the threshold for opening voltage-gated Ca^{2+} channels. Many other plants, such as the stinging nettle, cause pain, and thus interact with intracellular Ca^{2+} through the nociceptive pain pathway.

Thus, there are many toxins found in plants that affect Ca^{2+} signalling directly or indirectly in humans, other mammals and invertebrates, and have wide application experimentally. Furthermore, there are several agricultural chemicals and weed killers that interact with the Ca^{2+} signalling system in plants.

11.24   Drugs and the Ca^{2+} Receptor

Parathyroid cells, C-cells in the thyroid, and kidney cells have a Ca^{2+} receptor within the plasma membrane that responds to very small changes in extracellular free Ca^{2+}. This enables the body to maintain the total Ca^{2+} concentration in blood within very small limits, approximately 1.9–2.5 mM, with a plasma free Ca^{2+} of 1.1–1.3 mM. Drugs that interact with the Ca^{2+} receptor are divided into two groups:

1. Calcimimetics, which mimic or potentiate the action of Ca^{2+} on the Ca^{2+} receptor; type I activating in the absence of Ca^{2+}, type II requiring extracellular Ca^{2+}, acting as allosteric activators.
2. Calcilytics, which are Ca^{2+} receptor antagonists.

Several other metal ions, apart from Ca^{2+}, can bind to the Ca^{2+} receptor: $La^{3+} = Gd^{3+} > Be^{2+} > Ca^{2+} = Ba^{2+} > Sr^{2+} > Mg^{2+}$. Changes in expression of the Ca^{2+} receptor play an important role is several pathological conditions, including parathyroid disease and vitamin D deficiency, and bone disorders such as osteoporosis. In primary hyperparathyroidism there is a decrease in Ca^{2+} receptor expression. A decrease in the Ca^{2+} receptor also occurs in secondary hyperparathyroidism in end-stage chronic renal failure. Primary hyperparathyroidism involves an increase in plasma parathyroid hormone, caused either by an adenoma or hyperplasia of the parathyroid gland. This results in bone loss, kidney damage, muscle fatigue and mental depression. Calcimimetics are used to treat this condition as they cause a decrease in plasma parathyroid hormone and plasma Ca^{2+}.

11.25   Bacteria

Antibiotics are natural substances produced by many bacteria, enabling them to compete with other species for nutrients by killing them or stopping growth. The ones that produce a rise in intracellular Ca^{2+} in bacteria are those that damage cell walls, such as penicillin. Damage to bacterial cell walls also can lead to the opening of mechanosensitive channels. These are non-selective for a particular ion, and thus let Ca^{2+} into the cell, as well as allowing substances out of the cell, in an attempt to prevent osmotic lysis. Several bacteria produce proteins that form pores in eukaryotic cell membranes. These include proteins such as streptolysin O. As with the pores formed by the membrane attack complex of complement, the first event inside a cell after formation of a bacterial toxin pore is a large rise in cytosolic free Ca^{2+}. This will activate calmodulin and other Ca^{2+}-sensitive proteins in a eukaryotic cell. These can activate a protection mechanism, which attempts to remove the pore before the cell explodes and dies.

11.26   Ions and Intracellular Ca²⁺

Manipulation of cations and anions outside cells can cause Ca^{2+} signals to be generated inside the cell, and they can inhibit Ca^{2+} signals or its action (Table 11.5). Monovalent cations that can interfere with Ca^{2+} signalling include H^+, Li^+, Na^+, K^+, Rb^+ and Cs^+. Divalent cations that affect Ca^{2+} signalling include elements from Group II, such as Be^{2+}, Sr^{2+}, Ba^{2+} and Ra^{2+}, and also several transition metals, such as V^{2+}, Cr^{2+}, Ni^{2+}, Co^{2+}, Pb^{2+}, Cu^{2+} and Zn^{2+}. Trivalent cations, from the lanthanides, such as La^{3+} and Gd^{3+}, are potent blockers of Ca^{2+} channels. Even radioactive elements, such as uranium, plutonium and polonium, can be toxic at concentrations well below those where they can be detected through the radioactive decay. As a result, they can cause severe disability or even death as a result of disruption of the Ca^{2+} signalling system. This is poorly understood by workers in nuclear power plants, or those who live near them or waste sites.

Table 11.5 Examples of the effect of cations on intracellular Ca^{2+}.

Ion	Effect	Mechanism
Li^+	Inhibits release of ER Ca^{2+}	Inhibits IP_3 receptor
Na^+	Replacement of extracellular Na^+ raises intracellular free Ca^{2+}	Na^+/Ca^{2+} exchange
K^+	Provokes intracellular Ca^{2+} signal	Depolarisation opens membrane Ca^{2+} channels
Rb^+	Replacement for K^+	K^+ analogue
Mg^{2+}	Inhibits Ca^{2+}	Several proteins
Ca^{2+}	High Ca^{2+} inhibits	Binds to sites on external surface
Sr^{2+}	Mimics Ca^{2+}	Ca^{2+} channel and Ca^{2+}-binding protein
Ba^{2+}	Ca^{2+} channel detector in patch clamp electrode	Ca^{2+} channel
Ni^{2+}	Blocks generation of Ca^{2+} signals	Blocks voltage-sensitive Ca^{2+} channels
Co^{2+}	Blocks generation of Ca^{2+} signals	Blocks voltage-sensitive Ca^{2+} channels
Mn^{2+}	Can enter cell via Ca^{2+} channels (e.g. SOCE)	Used to quench fura signals
La^{3+}	Inhibits Ca^{2+} channels and effects of Ca^{2+}	Binds to Ca^{2+} channels and Ca^{2+}-binding proteins
Gd^{3+}	Blocks Ca^{2+} entry	Blocks mechanoreceptors

Cations and anions interact with Ca^{2+} signalling in one of three ways:

1. By generating a Ca^{2+} signal.
2. By altering the cytosolic free Ca^{2+} signal.
3. By altering the action of Ca^{2+} in the cell.

Ions can alter the Ca^{2+} signal by inhibiting or activating Ca^{2+} channels in the plasma membrane or intracellular organelles, by inhibiting Ca^{2+} pumps and transporters in the plasma membrane, or by affecting the level of free Ca^{2+} within an organelle (e.g. the SR/ER, mitochondria, lysosomes or secretory vesicles). In contrast, ions can affect the action of Ca^{2+} by inhibiting or activating Ca^{2+}-binding proteins, such as calmodulin, either by competitive binding with Ca^{2+} or allosterically.

A routine way of opening voltage-gated Ca^{2+} channels is to use an extracellular medium where the Na^+ has been replaced by K^+. This immediately depolarises the cell and opens the Ca^{2+} channels. This is readily seen in many luminous animals. If you want to look for bioluminescence on the beach at night take a potassium chloride 'gun' (i.e. a squidgy bottle containing 0.5 KCl). Turn over a rock and squirt the underside with KCl. If you are lucky you will see flashing hydroids and polychaete worms. Opening Ca^{2+} channels leads to activation of the Ca^{2+}-activated photoprotein inside the photocytes (see Chapter 7).

Many transition metals and lanthanides block Ca^{2+} channels, both voltage- and non-voltage-gated. For example, La^{3+} is a potent blocker of Ca^{2+} channels in barnacles, and Gd^{3+} is a useful blocker of mechanosensitive channels which let Ca^{2+} into cells. Several divalent cations block voltage-gated Ca^{2+} channels: $Ni^{2+} > Cd^{2+} > Co^{2+} > Mn^{2+}$. SOCE channels are permeable to Mn^{2+}, as well as Ca^{2+}. Thus opening of SOCE channels can be studied using the quenching of fluorescent dyes by Mn^{2+}.

Anions used experimentally when investigating Ca^{2+} signalling include Cl^-, phosphate, glutamate, isethionate, SO_4^{2-}, NO_3^{2-}, CO_3^{2-}, CN^-, OH^- and arsenate. Phosphate, oxalate and sulphate tend to precipitate Ca^{2+}. Vanadate, VO_3^{2-}, is often used experimentally to inhibit the plasma membrane MgATPase Ca^{2+} pump. Glutamate is used in some permeabilised cell preparations, as it is the main intracellular anion. In fact, experiments with platelets permeabilised by electric shock only exhibit Ca^{2+}-dependent exocytosis when glutamate is used. Cl^- does not work. In squid axons, isethionate is the major impermeant anion. CN^- blocks cytochrome oxidase in the mitochondria, and can cause a large rise in cytosolic free Ca^{2+}. Several oxyanions, such as EGTA, BAPTA, EDTA, citrate and oxalate, are used to chelate Ca^{2+}. In order to measure $^{45}Ca^{2+}$ uptake into vesicles isolated from the SR or ER, oxalate is used to bind the Ca^{2+} once it has entered the vesicle. On the other hand, EGTA is widely used to generate Ca^{2+} buffers with a defined free Ca^{2+} or to remove virtually all Ca^{2+} outside a cell to prevent Ca^{2+} entry when channels are opened in the plasma membrane. BAPTA is the preferred Ca^{2+} buffer to use inside a cell to prevent a Ca^{2+} signal, since it can be loaded into a cell from its acetoxymethyl ester. All cells require Mg^{2+}. Mg^{2+} binds to all nucleotides, nucleic acids, and is required as a cofactor for many enzymes. Mg^{2+} enters cells via specific transporters, and interacts with Ca^{2+} binding to many inorganic and organic ligands (see Chapter 6). However, there is little evidence Mg^{2+} affects Ca^{2+} signalling physiologically. Key oxygen-containing high ligands have evolved to select Ca^{2+} at micromolar concentrations when Mg^{2+} is millimolar.

11.27 Antibodies and Intracellular Ca^{2+}

Specific, high affinity antibodies to components of the Ca^{2+} signalling system have been invaluable as tools to locate them in tissues and within the cell. Antibodies can also be injected into live cells, and used to inhibit specific components of the Ca^{2+} signalling system. Location of specific components, and their movement, in live cells is carried out now by tagging the protein with green fluorescent protein (GFP) using genetic engineering. The advantage of using an antibody is that it is possible to be highly accurate about the precise location of a protein, even using gold-labelled antibodies and electron microscopy. This is important when a tissue contains several different cell types. But antibodies to components of Ca^{2+} signalling have no therapeutic application at present.

11.28   Manipulation of mRNA

The expression of components of the Ca^{2+} signalling system can be manipulated in three ways:

- Injection of the protein in a live cell, or over-expression using a plasmid or virus containing the cDNA coding for the protein.
- Injection of the mRNA coding for the protein.
- Addition of siRNA.

The last of these has become a routine technique in many studies. But, as with knock-out animals, the cell often compensates for the loss or over-expression of a protein. So the technique may not be entirely specific.

11.29   Summary

The key message is that, since intracellular Ca^{2+} controls every cell in the body, many drugs used clinically interact in some way with Ca^{2+} signalling. Some have direct effects on components of the Ca^{2+} signalling system, others work indirectly on Ca^{2+} signalling by affecting molecules that interact with Ca^{2+}. A wide range of natural and synthetic small organic compounds and peptides that affect intracellular Ca^{2+} in animals, plants and microbes are used clinically (Box 11.3), and are used experimentally to identify a component in the Ca^{2+} signalling pathway. Direct effects include:

- Modification of the concentration of cytosolic free Ca^{2+}, through activation or inhibition of plasma membrane receptors, Ca^{2+} channels, Ca^{2+} pumps and exchangers, or intracellular Ca^{2+} stores and buffers.
- Effects on the Ca^{2+} target, through activation or inhibition of Ca^{2+}-binding proteins, or covalent modifications mediated through Ca^{2+}, such as kinases, phosphatases, phosphodiesterases and proteases.

Box 11.3 Substance used experimentally to investigate Ca²⁺ signalling.

1. Manipulation of intracellular Ca^{2+} – Ca^{2+} buffers, Ca^{2+} ionophores, caged compounds.
2. Ca^{2+} channel agonists and antagonists.
3. ER IP_3 and ryanodine receptor activators and blockers.
4. Inhibitors of Ca^{2+} pumps and exchangers – plasma membrane and ER/SR.
5. Agents that affect store-operated Ca^{2+} entry (SOCE).
6. Inhibitors of mitochondrial metabolism and Ca^{2+} transport.
7. Inhibitors of intracellular Ca^{2+} targets such as calmodulin, calcineurin, calpain, and Ca^{2+} activated kinases.
8. Various anions and cations such as vanadate and La^{3+}.

A wide range of organic and inorganic substances are used experimentally to investigate the Ca^{2+} signalling system (Box 11.3). Natural agents that work via the intracellular Ca^{2+} signalling system include:

- Cone snail toxins.
- Scorpion toxins.
- Spider toxins.
- Jelly fish stings.
- Bee stings.
- Ant bites.
- Snake bites.
- Agents that affect Ca^{2+} in bacteria.
- Antibodies and intracellular Ca^{2+}.
- Monovalent, divalent and trivalent cations and anions that affect intracellular Ca^{2+}.

A key issue is the specificity of the pharmaceutical agent, together with sensitivity (i.e. how much drug is needed). A drug that acts only at millimolar concentrations would be of little use. The molecular diversity of drugs that target the same component (e.g. a particular voltage-gated Ca^{2+} channel) is a pharmaceutical example of Darwinian molecular variation. Combinatorial chemistry has been a popular approach of the pharmaceutical industry since the 1980s. Bioluminescent and fluorescent Ca^{2+} probes have played an important role in such high-throughput screens. There is still a great future for drug discovery in targeting the Ca^{2+} signalling system. However, there is an urgent need to discover the precise mechanisms responsible for the long-term side effects of drugs, pollutants and food additives through Ca^{2+} signalling.

Recommended Reading

*A must read.

Books

Campbell, A.K. (1983) Intracellular Calcium: Its Universal Role as Regulator (Monographs in Molecular Biophysics and Biochemistry), Chichester: John Wiley & Sons Ltd. Contains historical references to Ca^{2+} and drug action.

*Campbell, A.K. (2015) Intracellular Calcium. Chapter 11, Pharmacology and intracellular Ca^{2+}. Chichester: John Wiley & Sons Ltd, pp 499–562. The role of Ca^{2+} in action of drugs used clinically and experimentally, and toxins, fully referenced.

*Heilbrunn, L.V. (1943) An Outline of General Physiology, 2nd edn. Philadelphia: Saunders. Classic text with many historical references to the role of Ca^{2+} in drug action.

Reviews

Adams, M.E. (2004) Agatoxins: ion channel specific toxins from the American funnel web spider, *Agelenopsis aperta. Toxicon*, **43**, 509–525. How the armoury toxins from the funnel web spider work on particular ion channels.

Bergantin, L.B. & Caricati-Neto, A. (2016) Challenges for the pharmacological treatment of neurological and psychiatric disorders: Implications of the $Ca^{2+}/cAMP$ intracellular signalling interaction. *Eur. J. Pharmacol.*, **788**, 255–260. How drugs that affect intracellular Ca²⁺ can be used to treat mental problems.

Bourinet, E., Francois, A. & Laffray, S. (2016) T-type calcium channels in neuropathic pain. *Pain*, **157** (Suppl 1), S15–22. The role of particular Ca²⁺ channels in a type of pain.

Breitwieser, G.E. (2014) Pharmacoperones and the calcium sensing receptor: Exogenous and endogenous regulators. *Pharmacol. Res.*, **83**, 30–37. How small molecule 'protein chaperones', that allow misfolded mutant proteins to fold and route correctly, can affect the calcium receptor on the outside of cells.

*Kapturczak, M.H., Meier-Kriesche, H.U. & Kaplan, B. (2004) Pharmacology of calcineurin antagonists. *Transplant. Proc.*, **36**, 25–32S. Drugs that act on a key Ca²⁺ activated phosphatase.

Lee, S. (2013) Pharmacological inhibition of voltage-gated Ca²⁺ channels for chronic pain relief. *Curr. Neuropharmacol.*, **11**, 606–620. Use of Ca²⁺channel blockers to treat chronic pain.

*Mata, R., Figueroa, M., Gonzalez-Andrade, M., Alberto Rivera-Chavez, J., Madariaga-Mazon, A. & Del Valle, P. (2015) Calmodulin inhibitors from natural sources: an update. *J. Nat. Prod.*, **78**, 576–586. The search for compounds in nature that inhibit a ubiquitous intracellular Ca²⁺ target.

*Mir, R., Karim, S., Kamal, M.A., Wilson, C.M. & Mirza, Z. (2016) Conotoxins: structure, therapeutic potential and pharmacological applications. *Curr. Pharmaceut. Design*, **22**, 582–589. Toxins that affect Ca²⁺ and other ion channels from cone snails with their lethal harpoon.

Nemeth, E.F. & Goodman, W.G. (2016) Calcimimetic and calcilytic drugs: feats, flops, and futures. *Calcified Tissue Int.*, **98**, 341–358. The efficacy of drugs that either activate or block the Ca²⁺ receptor on the outside surface of many cells.

Ortner, N.J. & Striessnig, J. (2016) L-type calcium channels as drug targets in CNS disorders. *Channels*, **10**, 7–13. Application of blockers of a particular class of Ca²⁺ channel can be used to treat problems in the brain and other parts of the central nervous system.

Pellicena, P. & Schulman, H. (2014) CaMKII inhibitors: from research tools to therapeutic agents. *Frontiers Pharmacol.*, 5. The pathway of kinases activated by a rise in cytosolic free Ca²⁺ from the laboratory to the hospital bed.

Pollesello, P., Papp, Z. & Papp, J.G. (2016) Calcium sensitizers: What have we learned over the last 25 years? *Int. J. Cardiol.*, **203**, 543–548. Drugs that affect Ca²⁺ in the heart.

*Sousa, S.R., Vetter, I. & Lewis, R.J. (2013) Venom peptides as a rich source of Ca(v)2.2 channel blockers. *Toxins*, **5**, 286–314. How substances from the venoms of cone snails, spiders, snakes, assassin bugs, centipedes and scorpions block a particular Ca²⁺ channel, and why side effects limit clinical use.

*Zamponi, G.W. (2016) Targeting voltage-gated calcium channels in neurological and psychiatric diseases. *Nat. Rev. Drug Discovery*, **15**, 19–34. How to target Ca²⁺channels in the nerves, and treatment of mental disease.

Research Papers

*Anekonda, T.S. & Quinn, J.F. (2011) Calcium channel blocking as a therapeutic strategy for Alzheimer's disease: The case for isradipine. *Biochim. Biophys. Acta-Mol. Basis Disease*, **1812**, 1584–1590. The use of a Ca²⁺ blocker to treat a brain disease.

Chen, H., Jiao, W., Jones, M.A., Coxon, J.M., Morton, J.D., Bickerstaffe, R., Pehere, A.D., Zvarec, O. & Abell, A.D. (2012) New tripeptide-based macrocyclic calpain inhibitors formed by N-alkylation of histidine. *Chem. Biodivers.*, **9**, 2473–2484. Development of novel inhibitors of a key Ca²⁺-activated protease.

*De Smet, P., Parys, J.B., Callewaert, G., Weidema, A.F., Hill, E., De Smedt, H., Erneux, C., Sorrentino, V. & Missiaen, L. (1999) Xestospongin C is an equally potent inhibitor of the inositol 1,4,5-trisphosphate receptor and the endoplasmic-reticulum Ca^{2+} pumps. *Cell Calcium*, **26**, 9–13. Evidence for a dual action on Ca^{2+} signalling of a toxin from sponges.

Pal, N., Yamamoto, T., King, G.F., Waine, C. & Bonning, B. (2013) Aphicidal efficacy of scorpion- and spider-derived neurotoxins. *Toxicon*, **70**, 114–122. Evidence that Ca^{2+} and other ion channel blockers from scorpions and spiders can hit plant lice, such as greenflies, blackflies, or whiteflies.

*Yang, F., Chen, Z.-Y. & Wu, Y.-L. (2015) Unique interactions between scorpion toxins and small conductance Ca^{2+}-activated potassium channels. *Shengli Xuebao*, **67**, 255–260. Evidence for scorpion toxins affecting K^+ channels which are activated by a rise in cytosolic free Ca^{2+}.

12

Darwin and 4000 Million Years of Intracellular Ca²⁺

Learning Objectives
1. The relevance of Darwin and Wallace's principle of Natural Selection to intracellular Ca^{2+}.
2. How the Ca^{2+} signalling system has evolved.
3. The importance of Ca^{2+} in the origin of life.
4. How to create a low free Ca^{2+} inside a primeval cell without a Ca^{2+} pump.
5. What Darwin has to tell us about gene knock-outs in biomedical research.

12.1 Darwin and Calcium

For nearly 4000 million years the unique chemistry of calcium has been critical for the evolution of the life on our planet (Box 12.1). Darwin knew nothing about intracellular calcium, but he did use the calcified remains of extinct organisms in fossils as key evidence for the evolution of life, and the appearance of new species through the power of Natural Selection. He also worked out how coral reefs form. Yet the Ca^{2+} signalling system has been central to evolution through Natural Selection first described by Charles Darwin (Figure 12.1a) and Alfred Russel Wallace (1823–1913) (Figure 12.1b).

Box 12.1 Special chemistry of Ca²⁺ selected by evolution.

- Enormous Ca^{2+} gradients across the outer membrane of all living cells pose no osmotic problem.
- Thermodynamic properties and chemistry of Ca^{2+} are just right for it to be selected during evolution as a chemical switch for a wide range of biological processes.
- Ca^{2+} has a unique seven or eight coordination for oxygen, with high affinity and high selectivity over Mg^{2+}.
- Ca^{2+} binds to, and comes off, ligands quickly, in micro- to milli-seconds.
- Ca^{2+} has only one redox state, so no direct production of toxic oxygen species.

Their revolutionary principle was first presented to a meeting of the Linnean Society, in London, on 1 July in 1858. Natural Selection is now *the* unifying concept in biology. It is a

Fundamentals of Intracellular Calcium, First Edition. Anthony K. Campbell.
© 2018 John Wiley & Sons Ltd. Published 2018 by John Wiley & Sons Ltd.
Companion Website: http://www.wiley.com/go/campbell/calcium

Figure 12.1 The discoverers of evolution by Natural Selection. The two nineteenth century naturalists who first put forward the concept of Natural Selection and crucially the evidence for it. (a) Charles Robert Darwin (1809–1882), aged 71. Darwin (1902). (b) Alfred Russel Wallace (1823–1913), aged 79. From the frontispiece of his autobiography *My Life*. Wallace (1905).

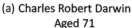

(a) Charles Robert Darwin   (b) Alfred Russel Wallace
        Aged 71   Aged 79

mechanistic *principle* that is applied to the *process* of evolution. There are plenty of existing biological phenomena that provide the evidence it works:

- The beaks of the Darwin finches on the Galapagos surviving an El Niño.
- The beaks of crossbills in Canada only have to vary by 1 mm or so to give the individual a selective advantage over a competitor.
- Guppy colour gives these fish a selective advantage in Colorado rivers and lakes.
- Half an eye is better than none, as shown by the amazing floating gastropod *Nautilus* that has an eye using a pinhole for focus instead of a lens.
- Antibiotic resistance is a natural phenomenon involving Natural Selection.
- Rodents resistant to the poison warfarin have been selected for.
- In our bodies, Natural Selection is crucial to how we make antibodies – the best antibody producers being selected and then killed when not wanted any more.

But, the distinguished astrophysicist Fred Hoyle (1915–2001) argued that Darwin and Wallace were wrong. The numbers just did not add up. Take a Ca^{2+}-binding protein like calmodulin with 148 amino acids. Since there are some 20 amino acids in all proteins, the number of possible combinations for calmodulin is 20^{148} or 3.6×10^{192} – far more than all the stars in the Universe (9×10^{21}). Thus, Hoyle argued that the development of new proteins by random mutation was impossible mathematically. He famously remarked that it was how: 'a tornado sweeping through a junk-yard might assemble a Boeing 747 from the materials therein'. But Hoyle was wrong. Ca^{2+}-binding sites, and the amino acids required at the active centre of most enzymes, only require a handful of amino acids to work. The essential amino acids for an EF-hand Ca^{2+}-binding site are acidic at positions 1, 3, 5, 9 and 12 in the loop, the carbonyl of the peptide bond at position 7, and a glycine at position 6. The number of random possibilities for the seven essential sites would be 20^7 or 1.3×10^9. This is a much more manageable number. Simplistically, if one mutation occurred a day, then it would take some 1300 million years to go through all the possible ones. It took nearly 1000 million years for life to appear after the Earth was formed, plenty of time for random mutations to generate the key Ca^{2+}-binding sites in proteins.

The different molecular and electrical properties of related Ca^{2+} signalling proteins in terms of K_m, K_d, V_{max}, conductance and ion selectivity provide the molecular variations upon which Natural Selection acts. Variation within an individual animal, plant or microbe gives a particular Ca^{2+} signalling protein its selective advantage, by having flexibility in a wide range of physiological, pathological and ecological circumstances. Calmodulin is a highly conserved protein, with high sequence homology, often greater than 80%, between animals, plants and unicellular eukaryotes. Similarly, there is homology between the sequences of other proteins involved in Ca^{2+} signalling, such as Ca^{2+} pumps and exchangers, Ca^{2+} channels, related Ca^{2+}-binding proteins, kinases, and inositol trisphosphate (IP_3) and ryanodine receptors. Yet there are subtle differences in kinetics and biochemical properties between these similar proteins that, when combined in an individual organism, give the individual a selective advantage under particular circumstances.

A good model for the origin of a new enzyme is bioluminescence. But Darwin had a problem with its origins. In Chapter VI of *On the Origin of Species*, 'Difficulties on Theory', he wrote (Darwin, 1859):

> *'The presence of luminous organs in a few insects, belonging to different families and orders, offers a parallel case of difficulty (to the origin of the electric organs of fishes).'*

Darwin could not see how small change by small change could lead, out of the blue, to a completely new phenomenon, through his BIG idea of Natural Selection. Bioluminescence is caused by a chemiluminescent reaction requiring oxygen, so it could only have appeared after oxygen rose in the atmosphere some 2000 million years ago. During the last 1000 million years, several fossil bioluminescent or potentially bioluminescent organisms have been found, which would have been triggered by a rise in cytosolic free Ca^{2+}. These include:

- A 100-million-year-old myctophid (lantern fish) fossil, kept at the Natural History Museum in London, has photocytes clearly visible.
- A 200-million-year old fossil brittle star, almost identical to the extant luminous brittle star *Amphipholis squamata* common in Europe.
- A fossil trilobite *Pricyclopyge dolabra*, nearly 600 million years old, found in South Wales with two secretary organs visible, the evolutionary precursor of the luminous decapod shrimp *Systellaspis* and *Oplophorus*.
- A 500-million-year-old fossil Cambrian ctenophore, *Maotianoascus octonarius*, similar to contemporary luminous sea gooseberry *Pleurobrachia*.

All of the descendants use coelenterazine, oxidised by oxygen to coelenteramide, which emits light. Yet the similarity between the proteins that catalyse this reaction is poor, often less than 10%. Ca^{2+}-activated photoproteins, which have coelenterazine and oxygen tightly bound, cannot have evolved from calmodulin, since the sequence similarity between aequorin and calmodulin is only some 23%, most of which is found in the EF-hand Ca^{2+}-binding sites. All you need for a new enzyme is an appropriate solvent cage. This can be constructed by just a few amino acids. Even albumin, the major protein in mammalian blood, can catalyse coelenterazine chemiluminescence, because it has a binding site for organic substances with three or four amino acids, similar to those in the naturally occurring photoproteins and luciferases.

The key evolutionary questions about intracellular Ca^{2+} are:

1. Why has evolution selected Ca^{2+} and not one of the other 15 cations found in living systems? Why not Na^+, K^+, Zn^{2+}, Fe^{2+} or Cu^{2+}, found in all cells? Why not choose an anion such as Cl^-, SO_4^{2-} or NO_3^- as a signalling molecule?
2. When did Natural Selection kick in as far as intracellular Ca^{2+} is concerned?
3. What was the evolutionary origin of the key proteins required for Ca^{2+} signalling – pumps, channels, buffers, organelle regulation, targets?

4. What was the origin of the key Ca^{2+}-binding motifs (EF-hand, C2, β-sheet structures, acidic amino acid groups), and how did this spread to the diversity of target proteins?
5. How did the role of Ca^{2+} in cell defence, stress and death evolve?
6. What is selective advantage of the molecular biodiversity in the Ca^{2+} signalling system?

Box 12.2 The Darwinian molecular biodiversity in Ca^{2+} signalling.

- Free Ca^{2+} in the cytosol and within organelles in the resting cell or after a stimulus.
- Type of Ca^{2+} signal (timing, oscillation, wave, tide, puff, spark, etc.).
- Location of the Ca^{2+} signal and microdomains.
- Ca^{2+} target protein, its affinity and what it activates or inhibits.
- Isoforms of Ca^{2+} signalling proteins, with different kinetics and turnover.
- Alternative splicing, producing proteins with the same activity, but subtly different biochemical properties.
- Redundancy – the ability of one protein to replace another that has been damaged, knocked-out or down-regulated.
- Regulation of protein level though transcription, translation and degradation.
- Post-translational modification (e.g. phosphorylation, methylation).

Darwinian variations through molecular biodiversity of Ca^{2+} signalling include differences in free Ca^{2+} signals in cells, and a host of differences in protein properties and levels (Box 12.2). These molecular variations mean that the timing and magnitude of the cellular event triggered by Ca^{2+} is slightly different in each cell or individual organism. A leg muscle will contract slightly faster or slower, a heart will respond better or worse to adrenaline, or a pancreatic β-cell will secrete more or less insulin in response to dietary glucose. As a result, the individual organism, thousands of years ago, would be able to sprint faster or run for a longer time when searching for food. Even a 1% improvement in the speed of a cheetah will give it a better chance of catching an antelope, rather than losing it. A fly that sees and recognises the tongue of a lizard has a better chance of escaping. A 1% improvement in the secretion of enzymes into the gut will improve the chance of an animal being able to have sex and reproduce. A 1% improvement in the maturation of an egg, and the ability of a sperm to inject its DNA into it, will give a better chance of producing a new individual of that species. A lymphocyte that has a 1% improvement in how it responds to a pathogen will give the individual a speedier recovery from an infection and survive. All of these are controlled by rises in intracellular Ca^{2+}. Thus, subtle differences in the kinetics and magnitude of Ca^{2+} signalling in muscles, nerves, secretory cells, photoreceptors, defence cells, gametes and cells during growth and development, produce small variations in their timing and magnitude in individual organisms within the population. This produces a population that can respond to changes in circumstances, to different foods, to different dangers and to different environments. This is the power of Natural Selection. Variation, within and between species, together with small change by small change, allows a species to evolve. Even within one individual, Natural Selection allows cells with better Ca^{2+} signalling to survive and not be killed by apoptosis.

Many studies on genome sequences are based on discovering new genes and their *function*. But, Darwin and Wallace taught us that, rather than searching for gene function, we should be asking: What is the *selective advantage* of a gene and its product? Luminous *Aequorea* jellyfish at Friday Harbor, where its Ca^{2+}-activated photoprotein and green fluorescent protein (GFP) were discovered, survive quite happily in the aquarium at Monterey in a non-luminous state, because they do not eat food containing coelenterazine. The question of selective advantage also arises when considering inherited conditions, for example mutations in Ca^{2+} signalling proteins. Inherited 'bad' or 'risk' genes must have a selective advantage in heterozygotes, and even sometimes in homozygotes. Sickle cells protect against malaria. Cystic fibrosis protects against gut infections such as *Salmonella*. Darwin teaches us that there must be a selective advantage of mutations in Ca²⁺ signalling genes, and conditions such as malignity hyperthermia, caused by mutations in the ryanodine receptor. Understanding this selective advantage of variations in the Ca²⁺ signalling system holds the key to understanding how life has evolved on our planet, and what goes wrong in disease. The key to medical genetics is identifying the selective advantage of risk genes.

12.2   Evolution and Ca²⁺

Intracellular Ca²⁺ is a universal regulator in all eukaryotic cells – animal, plant, fungal, protist and microbes, and is likely to play a role in bacteria and archeans. The evolution of this universal role has depended critically on the origin of the huge gradient of free Ca²⁺ that exists across the outer membrane of all living cells. The cytosolic free Ca²⁺ of all cells is in the sub-micro to micro-molar range, producing a gradient of some 10 000 across the outer membrane. The fact that loss of this gradient is lethal is compelling evidence that formation of this gradient was one of the earliest steps in the origin of life. A low intracellular free Ca²⁺ is essential if Ca²⁺ precipitates are to be prevented from forming, and the DNA, RNA and protein machinery is not to be clogged up by binding too much Ca²⁺. Without a low cytosolic free Ca²⁺, it is difficult to see how any of the key biochemical steps in evolution could have occurred.

Evolution selected Ca²⁺ as a universal signal because of the unique chemistry of the Ca²⁺ ion:

- The enormous Ca²⁺ gradients that exist across the outer membrane of all living cells poses no direct osmotic problem, as it would for Na⁺, K⁺ or Cl⁻.
- The thermodynamic properties and chemistry of Ca²⁺ are just right for a chemical switch of a wide range of biological processes.
- Ca²⁺ binds oxygen with high affinity, and importantly Ca²⁺ binds to, and comes off, ligands quickly, in micro- to milli-seconds. In contrast, transition metal cations, such as Zn²⁺, can take minutes to dissociate from a ligand.
- Ca²⁺ has a unique seven or eight coordination for oxygen, which gives Ca²⁺ ligands, such as calmodulin, high selectivity over Mg²⁺. In addition, Ca²⁺ is five times faster than Mg²⁺ at shedding water, needed for rapid entry into Ca²⁺ channels or binding to proteins.
- Ca²⁺ has only one redox state, unlike iron and copper, the latter leading to production of toxic oxygen species.
- Ca²⁺ signalling interacts with oxygen metabolites, but Ca²⁺ alone cannot generate them.
- No anion has the right chemistry to act as an intracellular switch in the same way as Ca²⁺.

Intracellular Ca²⁺ was also crucial to the origin of a new species. A mouse cannot mate with an elephant. The sperm and egg DNA cannot mix, as rates of reaction controlled by Ca²⁺ from each are incompatible, resulting from variations in kinetics of Ca²⁺-activated processes in the male sperm and female egg.

12.3   What is Evolution?

12.3.1   The Word Evolution

The word evolution, from the Latin *volvere* = to roll and '*e*' = out, was first used in 1670 in the *Philosophical Transactions of the Royal Society* to describe the 'natural evolution and growth' of parts of an insect. Similarly, Erasmus Darwin (Figure 12.2), Charles' grandfather, described

Figure 12.2 Erasmus Darwin (1731–1802), aged 61. One of Charles Darwin's grandfathers and the first to describe the process of evolution in modern terms. His great poems *The Botanic Garden* and *The Temple of Nature*, and his medical text *Zoonomia* (Darwin, 1794/1796), also contain the beginnings of the idea of the struggle for existence and Natural Selection. By Joseph Wright of Derby. Date painted: *c.* 1793. *Source:* English Heritage, Down House. communication.team@english-heritage.org.uk.

Erasmus Darwin
1731 – 1802
Aged 61

the 'gradual evolution of the young animal or plant from its egg or seed' in his poem *The Botanic Garden* (Darwin, 1789/1791). It was the French natural philosopher Charles Bonnet (1720–1793) who was the first to use the word 'evolution' in the modern sense, in his 'La Palinénésie philosophique' published in 1790. The last word in *On the Origin* was 'evolve', Charles Darwin using 'transmutation' to describe changes in organisms with time. The geologist Charles Lyell (1797–1875), in 1832, established the word 'evolution' to describe changes in the fossil record.

12.3.2 The Process

Evolution, therefore, is the *process* describing the development of life on our planet. This process was first described in modern terms by Erasmus Darwin, Charles Darwin's, grandfather, in his seminal text *Zoonomia* (Darwin, 1794/1796), and then in his wonderful poem *The Temple of Nature* published in 1803:

> *Organic life beneath the shoreless waves,*
> *Was born and nursed in ocean's pearly caves.*
> *First forms, minute, unseen by spheric glass*
> *Move on the mud and pierce the watery mass.*
> *These, as successive generations bloom,*
> *New forms acquire and larger limbs assume.*
> *And breathing realms of feet, and fin and wing/*
>
> *. . .*
>
> *Arose from rudiments of form and sense*
> *An embryon point or microscopic ens.*

Erasmus' grandson, Charles, with the naturalist Alfred Russel Wallace, proposed the *mechanism* – Natural Selection. The Universe is estimated to be about 13 000 million years old. The "big bang" was followed by the formation by atomic fusion of some 90 naturally occurring elements. Calcium, the fourth most common element in the Universe, was formed by fusion of helium and other smaller elements. The solar system formed into planets about 5000 million years ago. As the Earth cooled, water condensed, providing the essential solvent for life's biochemistry to begin just under 4000 million years ago, and then evolve. It took a further 2000 million years for the life to capture light and make ATP, and a further 1500 million years before large numbers of multicellular animals appeared. We have no convincing models that can cope with the enormity of these timescales. We can only guess! The geological record tells us that the Earth, its atmosphere and oceans were quite different in chemistry and physics from what they are today. A key Rubicon was crossed some 600 million years ago, leading to the Cambrian explosion of multicellular animals and single celled plants, where Ca^{2+} played a major role in hard structures, and, by then, many of the cell events triggered by intracellular Ca^{2+} would have been in place.

During evolution there were at least six holocaust Rubicons. These were mass extinctions, where huge numbers of plant and animal species disappeared. For example, at the Cretaceous–Tertiary (K–T) boundary some 70 million years ago, more than 90% of all organisms became extinct, including the land dinosaurs and marine ammonites. The inability of the Ca^{2+} signalling system to adapt to major environmental changes would have been a central molecular problem in those organisms that became extinct. Organisms disappear from the fossil record either because they adapt or become extinct. Without these there is no evolution.

The process of evolution appears at first glance to be gradual. As Darwin argued in *On the Origin*: '*Natura non facit saltum(s)* [Nature takes no leaps]'. Yet many Rubicons were crossed during the 4000 million years since life began – biochemical processes to form the first dividing cell, photosynthesis, protection against oxygen, shell and bone formation, all the cellular events provoked by a rise in cytosolic free Ca^{2+}, photoreception and developments of the eye, and the emission of light in bioluminescence. A breakthrough in animal evolution, some 500 million years ago, was the ability to deposit calcium phosphate precipitates in the form of first teeth, and then bone, the origin of vertebrates. Before this, controlled formation of calcium carbonate dominated invertebrate evolution. Evidence for the role of intracellular Ca^{2+} in the process of evolution comes from:

1. The geological record, which gives us details of chemical and physical changes in the atmosphere, sea and land mass movement, over 4000 million years.
2. The fossil record, which tells us the sequence of extinct organisms.
3. Phylogeny and cladistics, which tell us relationships between extant and extinct organisms.
4. Genomics and sequence comparisons, within and between extant organisms, giving us a clue to molecular clocks, how features in Ca^{2+} signalling proteins have been conserved through evolution, and the molecular variation that lies at the heart of Natural Selection.
5. Genetic engineering, which identifies key amino acids and structures necessary for function and selective advantage.

12.3.3 Sequence of Earth's Evolution

The major gases in the current atmosphere are nitrogen (78%) and oxygen (21%), the remainder being argon (1%), CO_2 (0.03%) and water vapour (0.1%). But, when the Earth had cooled, so that a crust could form about 4000 million years ago, the atmosphere was quite different. It was weak, containing argon and other gases, such as cyanide, formaldehyde and water vapour. Then, the 'Big Belch' released gases from volcanoes, producing an atmosphere rich in methane, but also with nitrogen, ammonia and water, and hydrogen, the latter escaping from the Earth's gravitational pull. Once the Earth's surface had cooled to less than 100 °C, rainfall condensed to form the oceans. Chemical reactions began, fuelled by energy from the Sun's rays and lightening. This formed the precursors of substances from which life began. The concentration of Ca^{2+} in the oceans has changed over the past 4000 million years, as has the level of oxygen in the atmosphere.

12.3.4 Ca^{2+} and Primeval Life

The control of intracellular Ca^{2+} must have been a crucial feature of cells when life began around 4000 million years ago. The ability to regulate free Ca^{2+} in primeval cells must have been crucial, since Ca^{2+} greater than a few micromolar clogs up cells with precipitates, and messes up DNA and intracellular proteins. Primeval cells chose MgATP to drive Ca^{2+} pumps and other endergonic reactions. Why they made this choice, and not MgGTP or another nucleotide, is a fascinating puzzle? The earliest organisms in the fossil record 3800 million years ago are calciferous microfossils. Thus controlled formation of $CaCO_3$ was an early process in evolution. A Ca^{2+} signalling toolbox developed in early eukaryotes, enabling them to trigger nerves firing, muscle contraction, amoeboid and flagellate movement, vesicular secretion, endocytosis, cell division and differentiation, fertilisation, vision, bioluminescence, defence mechanisms, and programmed cell death. The explosion of invertebrate animals found in the Cambrian fossil record shows that all of these cellular events were in place at least 600 million years ago.

12.3.5 Ca^{2+} and the Origin of the Three Fundamental Cell Types

Three fundamental cell types – Bacteria, Archaea and Eukaryota – formed some 3000 million years ago, either from a primeval cell or independently. Thus, Ca^{2+} signalling molecules in bacteria evolved independently from those in eukaryotes or archaeans. Only a small percentage of genes in Archaea have sequence similarity to genes in Bacteria or Eukaryota. Furthermore, most of the cellular events triggered by intracellular Ca^{2+} in eukaryotes do not occur in Bacteria or Archaea. Thus, these must have evolved after the eukaryotes formed. A key evolutionary step was the formation of organelles in eukaryotes. All of these can move Ca^{2+} in and out through pumps, exchangers and channels. Only a few of these Ca^{2+} transporters occur in Bacteria or Archaea. Most biology textbooks now tell us that organelles such as chloroplasts and mitochondria, both of which have circular DNA, evolved through endosymbiosis. However, the evidence for this is weak. A much more likely origin of organelles, such as the ER, mitochondria, lysosomes, peroxisomes, secretory vesicles, chloroplasts and the vacuole in plants, is membrane invagination. The genomes of mitochondria and chloroplasts are too small to code for the genes necessary for a complete organism. None are involved in Ca^{2+} signalling. Yet, *E. coli* has some 300 essential genes that cannot be knocked-out without killing the cell. Furthermore, there are some 1500 proteins inside a mitochondrion, several of which are involved in mitochondrial Ca^{2+} signalling, but are coded for in the main genome. Mitochondria divide, make proteins, make ATP, and carry out several other biochemical pathways, such as fatty acid oxidation. So if mitochondria originated from an endosymbiont such as *Rickettsia* there are three problems:

1. How did the endocytosed bacterium survive and multiply if its internal environment was oxidising? The cytosol of all cells is reducing, preventing the formation of S–S bonds and damaging oxidative reactions involving reactive oxygen species.
2. Since cells need at least several 100 proteins to survive and replicate, what happened to the proteins essential for nucleotide and nucleic acid, and protein synthesis, and the precursor reactions providing reducing equivalents for the respiratory chain and oxidative phosphorylation?
3. How did the 1500 or so mitochondrial proteins in the main genome become targeted to the mitochondria, if they were lost by the initial endosymbiont?

Ca²⁺ transporters such as MCU and MCU1, and Ca²⁺-sensitive enzymes in pyruvate metabolism, are coded by the main genome, though orthologues occur in some bacteria. Ca²⁺ pumps analogous to those in eukaryotes have been identified in both Gram-negative and Gram-positive bacteria as MgATP Ca²⁺ efflux and P-type ATPases. But, this is misleading. Without measurement of cytosolic free Ca²⁺ in live cells, predictions from genome searches and experiments in vesicles need to be treated with caution, since the three proteins ChaA, YrbG and PitB, identified as potential H⁺/Ca²⁺, Na/Ca²⁺ and Ca²⁺/phosphate symports, respectively, do not regulate cytosolic free Ca²⁺ in live cells. Ion channels have been found in bacteria for K⁺, Na⁺ and Cl⁻. But very few ion channels, equivalent to either ionotropic or metabotropic Ca²⁺ receptor channels in eukaryotic cells, have been found in Bacteria or Archaea. The presence of Ca²⁺ channels has been claimed in some bacteria, analogous to eukaryotic plasma membrane or mitochondrial Ca²⁺ channels, but the evidence they work in live cells is weak. The only credible Ca²⁺ channel identified in bacteria is not a protein – polyhydroxybutyrate, with polyphosphate. This complex is found in many bacteria, and has been found in mitochondria and the plasma membrane of eukaryotes, the latter associated with the Ca²⁺-MgATPase. In addition, although EF-hand proteins have been found through genome searches in some bacteria, there are few sequence similarities to other key Ca²⁺ signalling proteins, such as C2 Ca²⁺ site proteins, or calreticulin and calsequestrin.

Thus, for the endosymbiotic hypothesis to work, the primitive bacterium engulfed by the eukaryote precursor must have lost over 90% of its genes, these being taken up by the nuclear genome. Furthermore, most of the proteins involved in Ca²⁺ signalling must have come from another source. A similar problem exists for the chloroplast genome. No chloroplast genes are involved directly in Ca²⁺ signalling. Chloroplast DNA contains just 60–100 genes, whereas a typical cyanobacterium DNA codes for 1500. There is no direct evidence to support the transfer of genes from the putative endosymbiont to the nuclear genome, followed by targeting signals being engineered on to them. Animals have eaten chlorophyll and the genes coding for it for hundreds of millions of years. Yet there are no green monkeys, analogous to the TV character the green hulk, which was based on the idea that GFP could be expressed in man. Furthermore, mitochondria and chloroplasts do not have cell walls, and there are differences in their genetic code from 'universal' genetic code. Membrane capture of DNA, in the form of a plasmid, is a much more likely origin of the mitochondrion and chloroplast. The ER, a major Ca²⁺ signalling organelle, has no DNA. The evolutionary origin of this and other organelles that process Ca²⁺, such as secretory vesicles, endosomes and lysosomes, is unknown. The endosymbiotic hypothesis just does not work, yet is now embedded in most biology textbooks, as well as school and university curricula. Thus, the arguments I have listed here are controversial. However, in the spirit of Darwin, it is time the evidence for this hypothesis was looked at more critically and alternatives examined. The importance of Ca²⁺ regulation in mitochondria and chloroplasts, together with similarities and differences from bacteria, should be very helpful in this.

12.3.6 Timescale

Cell division can generate large-sized structures remarkably quickly. Imagine a human egg about 120 μm (0.12 mm) in diameter, with a volume about 1 nl (10^{-9} l). Once fertilised by a sperm, if every cell formed divided each day, then within 9 months, the foetus would envelop the whole solar system, being the size of the Milky Way! The same applies to cells in a tumour, if each cell divided every 24 h. Applying this 'baby' calculation to elephants, the Earth would be consumed by elephants within a century or so! What does this tell us? First, cell division must be selective. Once two cells form, each is different, and only one may divide again. Secondly, as well as new cells being formed by division, some cells are dying. So the total number of cells depends on a balance between division and death. Subtle differences in intracellular Ca^{2+} between cells, followed by Natural Selection of these cells, must have been involved from the beginning of multicellular organisms. *E. coli* cells can divide every 20 min, and our cells may take several hours to divide. But 3000 million years ago it would have taken primeval cells months, years or even centuries to divide. If a primeval cell took a year to divide, then, by the end of its first century, it would have formed a living mass the size of the moon. If it took 100 years for a primeval cell to divide, it would take less than 10 000 years to form a living mass greater than that present on the Earth today. This must have been why it took over 3000 million years from the first cell being formed to the Cambrian explosion, when the sea became full of multicellular life, with all the components of the Ca^{2+} signalling system, and a mechanism for keeping the free Ca^{2+} inside cells in the sub-micromolar to micro-molar range.

12.4 Evolution of Ca^{2+} Signalling

12.4.1 Origin of Ca^{2+} Signalling

Given the prevalence of calcium in the Earth's crust, primeval cells will have been surrounded by tens of micromolar or millimolar free Ca^{2+}. At these concentrations, Ca^{2+} competes with Mg^{2+}, binding to ATP and other nucleotides, DNA and RNA, preventing normal function. Micro- to milli-molar Ca^{2+} also precipitate carbonate, phosphate and sulphate. So if a primeval cell was to work, it had to get rid of Ca^{2+}, lowering it to micro-molar levels. There are three ways this can be done across a semipermeable membrane:

1. A membrane potential, positive inside, instead of the negative one in all contemporary cells, with Ca^{2+} at its equilibrium potential.
2. A Ca^{2+} exchanger, a Ca^{2+} anion symport (e.g. phosphate), or a Ca^{2+}/cation antiport, (e.g. Na^+ or H^+).
3. A MgATP Ca^{2+} pump.

But the earliest primeval cells would not have had the latter two mechanisms.

12.4.2 Membrane Potential

All contemporary cells have a plasma membrane potential, typically tens to hundreds of mV, *negative* inside. Without a Ca^{2+} pump, Ca^{2+} would concentrate inside, producing a cytosolic free Ca^{2+} concentration up to 1M, since Ca^{2+} would eventually reach its electrochemical equilibrium. In the absence of such pumps, a primeval a cell could only get rid of Ca^{2+} by generating a Donnan-based membrane potential, *positive* inside (Figure 12.3). Such a Donnan potential can be generated by a gradient of permeant anions, where there is higher concentration inside than out. This will occur if there is a higher concentration of impermeant anions outside the cell than inside. Alternatively, a positive membrane potential inside could be generated by cations, if there is a higher concentration of impermeant cations inside than outside, causing the permeant cations to be higher outside than in.

When a semipermeable membrane has charged ions on one side that are impermeant, then a Donnan potential is set up. Most of the impermeant, large ions in contemporary cells are negatively charged (e.g. DNA, RNA, phospholipids, polysaccharides, proteins), resulting in a Donnan potential, negative inside. This will concentrate Ca^{2+} inside the cell. An example of this is the periplasmic space of *E. coli*, which has impermeant, *negative*, oligosaccharides, setting up

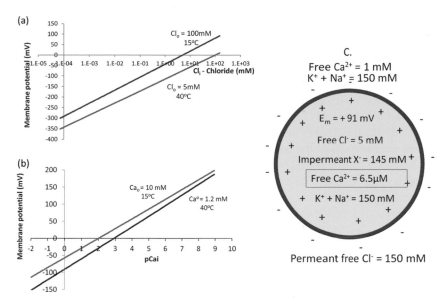

Figure 12.3 Generation of a Donnan potential, positive inside, and its relationship to intracellular Ca²⁺:
Equilibrium potential for Ca²⁺ = $E_m = [RT/(2F\log_{10}e)]\log_{10}(Ca_o/Ca_i)$
Donnan potential for Cl⁻ = $E_m = [RT/(F\log_{10}e)]\log_{10}(Cl_o/Cl_i)$, where $R = 8.3145$ J mol⁻¹ K⁻¹, $F = 96485$ C mol⁻¹ and
$T = 15$ or 40 °C.
Data are calculated for two scenarios: (1) extracellular Cl⁻ = 100 mM and intracellular Cl⁻ = 150 mM to 1.9 nM;
(2) extracellular Cl⁻ = 5 mM and intracellular Cl⁻ = 150 mM to 1.9 nM. The best situation for scenario 1, with a
membrane potential of +10 mV, positive inside, produces a free cytosolic free Ca²⁺ of 44 µM. The best situation
for scenario 2 with a membrane potential of +91 mV positive inside, produces a free cytosolic free Ca²⁺ of 6.5
µM. Both of these would be satisfactory for DNA, RNA and protein synthesis. (a) Plot of equilibrium potential
versus cytosolic free Cl⁻ (Cl_i). (b) Plot of cytosolic free Ca²⁺ against Donnan membrane potential, established,
for example, by a Cl⁻ gradient. (c) A possible primeval cell permeant to Cl⁻, K⁺, Na⁺ and Ca²⁺, but impermeant to
X⁻. With a free Cl⁻ outside of 150 mM, the concentrations of anions would 5 mM for Cl⁻ and 145 mM for X⁻. This
would set up a Donnan membrane potential of +91 mV, positive inside. The Ca²⁺, without any pumps, would
equilibrate to 6.5 µM inside, if the Ca²⁺ concentration outside was 1 mM. This would be low enough to allow
DNA, RNA and proteins to function. Reproduced by permission of Welston Court Science Centre.

a Donnan potential of about 12 mV negative inside the periplasmic space. This concentrates Ca²⁺ to tens of micromolar. On the other hand, a *positive* potential inside a primeval cell could have been derived from *positive* impermeant ions inside. These could have been based on nitrogen, since lysine and arginine -NH₂ groups are positive at acid or neutral pH, in the form of NH₃⁺. There will be less positive free cation, e.g. K⁺, and more free anion, e.g. Cl⁻, inside the cell than outside. These generate a membrane potential, positive inside, predicted by the Nernst equation:

$$E_m = RT/F\left[\log_e\left(K_{out}/K_{in}\right)\right] \tag{12.1}$$
$$E_m = RT/F\left[\log_e\left(Cl_{in}/Cl_{out}\right)\right] \tag{12.2}$$

A membrane potential negative inside the primeval cell would have been disastrous. If the Ca²⁺ outside the primeval cell was just 1 mM, with a membrane potential set at −50 mV inside, the free Ca²⁺ inside would rise rapidly to around 70 mM, which would clog everything up. Even with a potential at −10 mV, the free Ca²⁺ inside would be nearly 3 mM, and over 20 mM if the free Ca²⁺ outside was 10 mM, equivalent to contemporary sea water.

To keep the intracellular free Ca^{2+} to 0.1 μM, in the presence of 1 mM extracellular free Ca^{2+}, equivalent to contemporary mammalian cells, then the equilibrium potential for Ca^{2+} would have to be 123 mV, positive inside. To generate this potential from a chloride gradient would require the extracellular free Cl^- to be 100 times less than that inside. The chloride concentration in the sea is around 550 mM, in human plasma about 110 mM, the intracellular concentration of chloride being as high as 80 mM in a depolarised nerve. But fresh water has only around 0.2% dissolved salts – a typical freshwater lake having only 0.1–0.2 mM Cl^-, just right for Ca^{2+} to be at its equilibrium potential in a primeval cell. Thus, if there was a high concentration of impermeant anions outside the primeval cell, such as silicate clays, a membrane potential, positive inside, could be generated by a permeant anion such as Cl^-. Alternatively, a Donnan potential, positive inside, could be established by an impermeant cation inside the cell, based on the ammonium cation such as polylysine or polyglucosamine. These speculations emphasise the importance of relating the electrophysiology to the biochemistry of Ca^{2+}, if we are to understand how the Ca^{2+} signalling system evolved, and how it works in cells today.

12.4.3 Evolution of Ca^{2+} Signalling Based on Protein Sequences

Software for sequence similarities and alignments are from BLAST and CLUSTAL, or PileUp. The percentage sequence similarities are displayed in a tree-like structure – a dendrogram, e.g. showing the per cent identity of various voltage-gated Ca^{2+} channels (Figure 12.4a). Distances between branch points represent how different one group of related proteins is to another.

For voltage-gate Ca^{2+} channels there are three main clusters for the α_1 subunit, which forms the channel: Ca_v1, Ca_v2 and Ca_v3. $Ca_v1.1$ and $Ca_v1.2$ are 90% sequence-similar, $Ca_v2.1$ and $Ca_v2.2$ are 93% similar, and $Ca_v3.1$ and $Ca_v3.2$ are 88% similar. But, as a group, Ca_v1 and Ca_v2 are only 52% similar, and LVA (low voltage activated) versus HVA (high voltage activated) only 28% similar. Evidence for the evolution of Ca^{2+} pumps and exchangers, Ca^{2+}-binding proteins, IP_3 and ryanodine receptors, the STIM and Orai families, which open store-operated Ca^{2+} entry (SOCE) channels in the plasma membrane, and annexins, is also based on tree-like dendrograms. This approach at least gives us hypotheses about the order of appearance of Ca^{2+} signalling components in evolution, and how they must have changed over tens and hundreds of millions of years. There are two problems with dendrograms. First, it is not easy to correlate directly the distances with absolute time. Secondly, they start with a mythical protein of origin, the presumed start of the family. But we have no idea where this came from.

Alternatively, the evolutionary relationships through sequence similarities can be represented in a tree-like structure based on phylogenetics. Darwin used this in *On the Origin*. For EF-hand proteins these suggest that they evolved by a complex route of gene duplication, splicing and transposition (Figure 12.4b and c). Phylogeny enables Ca^{2+}-dependent processes and mechanisms to be compared, and related to the evolutionary age of organisms. For example, the EF-hand calmodulin is found in all animals and plants which have appeared within the last few 100 000 years, and in protozoa, such as *Paramecium*, that date back hundreds of millions of years. Genome sequencing has revealed thousands of proteins with predicted Ca^{2+}-binding EF-hands in all plants and animals, just a few in bacteria, but very few in Archaeal genomes.

Examination of this phylogenetics of Ca^{2+} signalling suggests that its origins were in cell defence, first to stop Ca^{2+} clogging up the intracellular machinery, and then to enable proteins in the ER and Golgi to fold correctly. This must have started some 1000–2000 million years ago. It was followed by the appearance of cellular events such as movement, secretion and fertilisation that needed to be signalled, followed by excitability. This led to voltage-gated channels and gap junctions, which evolved before nerves. These enabled cells to communicate via nerves and to move using muscles. Jellyfish and other coelenterates have existed at least since

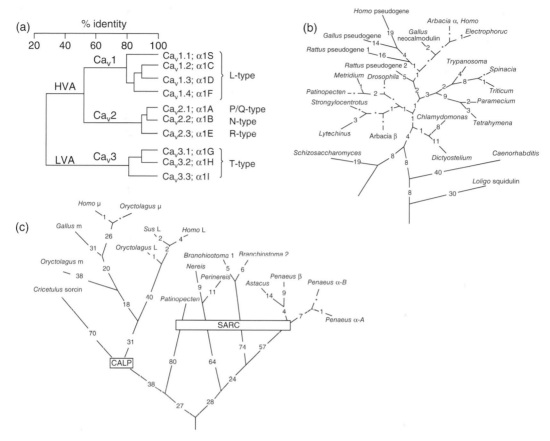

Figure 12.4 Dendrograms for voltage-gated Ca²⁺ channels and EF-hand Ca²⁺-binding proteins. (a) Percent sequence similarities between the three classes of voltage-gate Ca²⁺ channel. Source: Dolphin, A. (2006) A short history of voltage-gated calcium channels. *Br. J. Pharmacol.*, **147**, S56–62. Reproduced with permission from Wiley. (b) Amino acid sequence relationships between members of the calmodulin family. Source: Moncrief, N.D., Kretsinger, R.H. & Goodman, M. (1990) Evolution of EF-hand calcium-modulated proteins. I. Relationships based on amino acid sequences. (1990) *J. Mol. Evol.*, **30**, 522–562. Reproduced with kind permission from Springer Science+Business Media BV. (c) Amino acid sequence relationships between members of the CALP and SARC families. The numbers represent relative differences between proteins from adjoining organisms, proportional to branch length. Source: Moncrief *et al.* (1990). Reproduced with kind permission from Springer Science+Business Media BV.

the Cambrian, 540 million years ago. These lower multicellular animals use Ca²⁺ channels to excite nerves, but not Na⁺ channels. They then transmit electrical signals via gap junctions. Thus, the first ion channels in membranes must have been selective for Ca²⁺. Molecular biodiversity was then introduced via gene mutation and alternative splicing.

An interesting phylogenetic analysis of protein structure in Ca²⁺ signalling comes from examining the system which links loss of Ca²⁺ from the ER via IP₃ or ryanodine receptors to the opening of Ca²⁺ channels in the plasma membrane, SOCE, via the STIM/Orai system. The unusual feature of this system is that it is the dropping off of Ca²⁺ from the Ca²⁺-binding site of STIM facing into the lumen of the ER that provokes it to dimerise, and then relocate to a site close to the plasma membrane. This is followed by linking to Orai in the plasma membrane, and the opening of its Ca²⁺ channel. All of this happens within seconds. This protein bridging is ubiquitous in animals, plants and singled-celled eukaryotes, but is not found in bacteria or archaeans. Furthermore, the role of STIM and Orai in regulating Ca^{2+} is not the same in all organisms.

The *Paramecium* genome suggests that IP$_3$ and ryanodine receptors appeared early in the evolution of eukaryotes, this unicellular eukaryote exhibiting a SOCE mechanism to stimulate exocytosis of the trichocysts via a rise in cytosolic free Ca^{2+}. IP$_3$ receptors have also been found in pathogenic flagellates, such as trypanosomes. The selectivity filters for Ca^{2+} in the IP$_3$ receptor are the amino acid sequences GGVGD and GGVID, and in the ryanodine receptor GGIGD, and have been highly conserved from lower to higher eukaryotes. Both STIM1 and 2 have a single Ca^{2+}-binding site, named EF-SAM, facing the lumen of the ER. STIM1 and 2 must have occurred by gene duplication some 400–500 million years ago, as vertebrates first started to evolve. They both have four core domains: a Ca^{2+}-binding domain (EF-SAM), a single transmembrane domain, a cytosolic coiled-coil spacer (15 nm) and an Orai activation/inactivation domain near the C-terminus. At the C-terminal end in all vertebrates, but not insects, there is a polybasic (PB) domain, which is positively charged, so that it binds phosphatidyl inositol. This strengthens binding to the inner surface of the plasma membrane once it has latched on to Orai, as such basic sequences bind to clusters of inositol phospholipids. STIM and Orai homologues are widely distributed in eukaryotic cells. Interestingly, the STIM/Orai partnership is not always associated with IP$_3$ receptors. Several species, such as the fresh water protozoa *Paramecium* and *Naegleria*, have IP$_3$ receptors but lack STIM and Orai. Conversely, the moss *Physcomitrella* apparently does not have IP$_3$ receptors, but does have Orai homologues in its genome. Similarly Orai-like genes are found in green algae such as *Chlamydomonas* and diatoms such as *Thalassiosira*.

Homologues of plasma membrane Ca^{2+} pumps (Ca^{2+}-MgATPases, PMCAs), and the exchangers NCX and NCKX, are widely distributed in eukaryotic cells. The amount expressed in any cell type depends on the selective advantage of the particular pump. PMCAs have a high affinity for Ca^{2+}, but are relatively slow pumps, compared with the lower Ca^{2+} affinity but fast Na$^+$/Ca^{2+} exchangers NCX or NCKX, the latter also exchanging K$^+$. The ER, thanks to its IP$_3$ or ryanodine receptor Ca^{2+} channels and its SERCA Ca^{2+} pump, releases and takes up Ca^{2+} some 10–100 times faster than Ca^{2+} crossing the plasma membrane. There is also a Ca^{2+} pump, SPCA, taking Ca^{2+} into the Golgi. But strangely, SERCA and SPCA are not found in all cells. The yeast *Saccharomyces cerevisiae* does not have a SERCA homologue, and the plant *Arabidopsis thaliana* does not have an SPCA homologue. Mitochondria use the Na$^+$/Ca^{2+} exchanger NCLX, and the putative Ca^{2+} channels MCU1 with MICU1, and possibly LETM1, to regulate Ca^{2+} inside the inner matrix. Orthologues of these proteins have been found in bacteria, but appear to have been lost in fungi. The subtle differences in these Ca^{2+} pumps and exchangers have selective advantages in a particular organelle, cell type or organism.

In contrast to the cytosol, the Ca^{2+} concentration in the oxidising environment of the ER and Golgi needs to be high. In resting animal cells it may as high as 0.5 mM, though in yeast it is lower, just 10 μM. High Ca^{2+} in these compartments is required to enable proteins there to fold correctly, as Ca^{2+} binds to chaperones such as calreticulin, calnexin and BiP (GRP78). This Ca^{2+}-dependent chaperone system works in all eukaryotes. These proteins are found in the genomes of animals, plants, unicellular protists and algae. If the chaperones fail to compensate for poor protein folding, unfolded proteins build up in the ER, and the cell makes a decision to kill itself by apoptosis. Thus, the Ca^{2+}-dependent chaperone system in the ER and Golgi is likely to be one of the first parts of the Ca^{2+} signalling system to evolve some billion years ago. For this to operate, there had to pumps to allow the ER and Golgi to accumulate Ca^{2+}. These pumps would first allow the cell to defend itself, over short time intervals, to an unwanted burst of Ca^{2+} in the cytosol, e.g. induced by damage to the plasma membrane.

Ion pores are all formed by similar transmembrane domains, either within one protein, such as the voltage-gated Ca^{2+} and Na^+ channels, or from subunits of the same protein, such as the four forming the K^+ channel, six Orai subunits forming the SOCE channel, the formation of gap junctions by connexions, and the four subunits of the IP_3 or ryanodine receptors forming the Ca^{2+} release channel in the sarcoplasmic reticulum (SR)/ER. When formed from one polypeptide, this would have evolved by DNA duplication of the key domains. Plasma membrane Ca^{2+} channels signalled by external stimuli, such as neurotransmitters, hormones and paracrines, would not be needed until a signalling system existed. Similarly, voltage-gated Ca^{2+} channels would have preceded electrical excitability in nerves and muscle in evolution. However, in view of the hypothesis that Ca^{2+} evolved first in the ER and Golgi to aid protein folding, the STIM/Orai system opening Ca^{2+} channels in the plasma membrane to top up ER Ca^{2+} would have been the first plasma membrane channel to appear. Ca^{2+} channels in the plasma membrane include those gated by membrane depolarisation (VGCC), transient receptor potential channels opened by sensory signals (TRP) such as touch, taste, smell, sound and heat, the cyclic nucleotides cyclic AMP and cyclic GMP (CNGCC), and channels opened by glutamate (GLURCC), as well as the plasma membrane channel opened by loss of Ca^{2+} from the ER (SOCE). These channels are widely distributed in eukaryotes. The protozoan *Naegleria gruberi*, which can be signalled to change from amoeboid to flagellate form, has homologues to VGCC, CNGCC and GLURCC Ca^{2+} channels. TRP homologues are found in the yeast *Saccharomyces cerevisiae*, and in the single-celled, biflagellate, green alga found in soil and fresh water, *Chlamydomonas reinhardtii*. ATP-related purinergic (PurRCC) and cholinergic (ChoRCC) Ca^{2+} channels are important in cell–cell communication in higher eukaryotes. Homologues to PurRCC have been found in the genomes of the slime mould *Dictyostelium discoideum* and homologues to ChoRCC have been found in *Chlamydomonas reinhardtii*. This suggests that all of these Ca^{2+} channels, including TRP channels, appeared early in the evolution of eukaryotes.

This analysis therefore suggests an order for the appearance of the components of the Ca^{2+} signalling system (Figure 12.5). All contemporary cells, Bacteria, Eukaryota and Archaea, have high cytosolic K^+ and low Na^+, a very low cytosolic free Ca^{2+}, and medium Mg^{2+}. To reach this state, first, over 3000 million years ago, in order to lower the free Ca^{2+} in the primeval cell, a positive membrane Donnan potential was generated. A few hundred million years later a membrane Ca^{2+} pump appeared, allowing the cell to generate a negative membrane potential, without the cell accumulating large amounts of Ca^{2+}. The appearance of K^+ channels allowed the membrane potential to increase, negative inside. The cell could then develop a high cytosolic K^+, close to its equilibrium potential in most cells. A Ca^{2+}-MgATPase then appeared, targeted to the ER, allowing the cell to protect itself from quick bursts of Ca^{2+} in the cytosol, caused by damage or mechanoreceptor opening in the plasma membrane. As the ER accumulated Ca^{2+}, proteins that helped other proteins to fold (chaperones) appeared and developed Ca^{2+}-binding sites, which aided their binding to unfolded proteins. The appearance of Orai in the plasma membrane allowed the ER to accumulate Ca^{2+}, without large global rises in cytosolic free Ca^{2+}. Phospholipases would already have been in place, some capable of producing IP_3 from phosphatidyl inositol 4,5-bisphosphate (PIP_2). This created the molecular environment for the appearance of IP_3 receptors. But it was not until extracellular primary regulators, such as hormones, began to appear that the environment was there for the appearance of the G-protein ($G_{\alpha q}$) and phospholipase Cβ system to produce IP_3 in a regulated manner. Excitability based on Ca^{2+} channels would then have appeared, followed by excitability based on Na^+ channels. This set the scene for the appearance of nerves and muscle in multicellular organisms. The appearance of muscle provoked the appearance of ryanodine receptors. By the mid-Cambrian, 500 million years ago, all the main components of the Ca^{2+} signalling system would have been

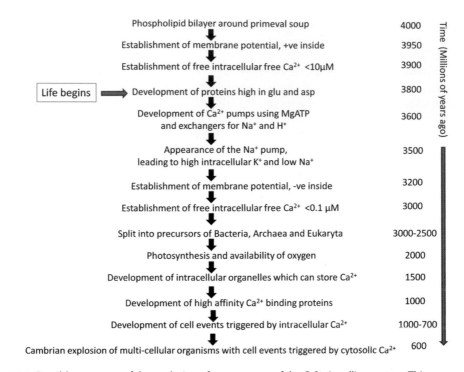

Figure 12.5 Possible sequence of the evolution of components of the Ca²⁺ signalling system. This sequence is speculative, but does highlight the key molecular and electrical events that were necessary between the formation of the first cells 3800 million years ago and the Cambrian explosion of multicellular animals and single celled plants some 600 million years ago. For the Cambrian explosion to occur, virtually all the key cellular events triggered by a rise in cytosolic free Ca²⁺ must have been in place, including muscle contraction, action potentials, secretion, fertilisation and vision. Reproduced by permission of Welston Court Science Centre.

in place. Animals moved using muscle, with nerves firing electrical signals, first conducted via voltage-gated Ca²⁺ channels and gap junctions, then via Na⁺/K⁺ action potentials. These early multicellular animals secreted small and macromolecules, and underwent sexual reproduction, all of which required a Ca²⁺ signalling system to work effectively. All types of Ca²⁺-binding sites were in place by the Cambrian period (541–485 mya). These phylogenetic comparisons do not, however, give us the origin of the first protein in a Ca²⁺ signalling component family – the first Ca²⁺ pump, the first Ca²⁺ channel, the first Ca²⁺-binding protein. Nor do they give us any insights into how the small differences in amino acid sequences, which occurred as a protein evolved in different organisms, fit in with Darwin and Wallace's BIG idea of Natural Selection.

12.4.4 Evolution of Ca²⁺-Binding Sites

A key molecular evolutionary Rubicon was the appearance of a high-affinity Ca²⁺-binding site, selective for Ca²⁺ over Mg²⁺, an essential step for the appearance of Ca²⁺-target proteins. Four types of high-affinity Ca²⁺-binding site have been identified, which satisfy the seven or eight oxygen coordination required for micromolar affinity for Ca²⁺ in the presence of millimolar Mg²⁺:

1. EF-hand (type I), first found in parvalbumin, and its modifications in copines and the S-100 family.
2. C2, first found at the C-terminus of protein kinase C.
3. Types II, III and AB, found in annexin, including those made from the folding of β-sheets.
4. The binding sites in Ca²⁺ pumps.

The oxygen ligand can be from a carboxyl group, the carbonyl in the peptide chain backbone, or water held within the Ca^{2+}-binding site. The strongest binding, and thus highest affinity, comes from negatively charged oxygens from Asp or Glu residues, which dominate EF-hand sites. The weakest, and lowest 'high' affinity, Ca^{2+}-binding sites use more than one H_2O as a ligand, such as in the annexins. Physiologically important low affinity Ca^{2+}-binding sites are found in other Ca^{2+}-binding proteins such as calsequestrin, calreticulin, Excalibur (extracellular Ca^{2+}-binding region), and albumin, using clusters of acidic amino acid residues or pockets formed from the folding of β-pleated sheets, such as the Greek key. Low-affinity Ca^{2+} sites occur in other plasma proteins, such as those involved in blood clotting and the complement cascade, extracellular degradative enzymes, as well as on the surface of cells, including the Ca^{2+} receptor. But, inside cells there are many proteins with low-affinity Ca^{2+} sites. These are non-physiological.

The EF-hand Ca^{2+}-binding site is formed as a loop from the linear sequence, whereas the C2 Ca^{2+}-binding loop forms from the interaction of peptide chains. Similarly, Ca^{2+} sites within transmembrane domains can be formed by the interaction of helices far apart in the linear sequence, such as those in the SERCA pump. The most widely distributed Ca^{2+}-binding sites of this type are found in P-type Ca^{2+}-activated MgATPases.

12.4.5 Origin of the EF-Hand

The most common, and widely distributed, high-affinity Ca^{2+}-binding site, found in all animals, protists, some bacteria and a few archaeans, is the EF-hand, discovered in muscle parvalbumin in 1973. Thousands have now been identified in genomes. The Ca^{2+}-binding site is formed from a 29-amino-acid loop, with at least five acidic residues, giving seven or eight coordination, and an essential glycine at position 6. The full EF-hand has two helices on either end. Many have been identified through database searches using motifs, with no direct evidence for Ca^{2+} binding. In fact, several EF-hand sites thus identified do not bind Ca^{2+} when examined by ^{45}Ca binding or crystallographic structure.

EF-hand Ca^{2+} sites are divided into odd and even, canonical, and pseudo. Canonical (classical) EF-hands are found in proteins like calmodulin and troponin C. But calmodulin has not been found in Bacteria or Archaea. Pseudo EF-hands are found at the N-termini of S-100 and S-100-like proteins, where the Ca^{2+}-binding loop of 14 residues is mainly from the carbonyls at positions 1, 4, 6 and 9. The appearance of the EF-hand was a key step in the evolution of the Ca^{2+} signalling system. The full, canonical EF-hand Ca^{2+}-binding site, found in calmodulin, troponin C and bacterial calerythrin, requires a 29-amino-acid sequence rich in acidic amino acid residues, sandwiched between two α-helices. Ca^{2+} is seven or eight coordinated by oxygen at positions 1, 3, 5, 7, 9 and 12 in the loop. Oxygens come from side-chains at positions 1, 3 and 5, a carbonyl of the peptide bond at position 7, which can be any amino acid, water held at position 9, which is usually Asp (D), Glu (E), Ser (S), Thr (T) or Asn (N), and a bidentate Glu (E) at position 12, the start of the second helix. The third and fifth residues are usually Asp (D), but can be Asn (N). The invariant Glu or Asp at position 12 is essential, and a glycine is usually required at position 6 in the loop for proper folding around Ca^{2+}, as well as an aliphatic hydrophobic residue, Ile (I), Leu (L) or Val (V), at position 8. Only when there is a crystallographic structure can the occurrence also of the EF α-helices on either side of the Ca^{2+}-binding loop be confirmed. A typical database search will look for Dx[DN]x[DN]Gx[ILV][DSTN]xxE (those in brackets being alternatives).

The bacterial Ca^{2+}-binding protein Excalibur, derived from <u>ex</u>tracellular <u>cal</u>cium-<u>b</u>inding <u>r</u>egion, has a Ca^{2+}-binding site of DxDxDGx(2)CE similar to high-affinity EF-hand domains in intracellular Ca^{2+}-binding proteins, where 'x' is any amino acid and numbers in parentheses are used if there is more than one 'x', and is extracellular. The EF-hand like proteins found in bacteria and viruses exhibit a range of structures around the Ca^{2+}-binding loop. Some eukaryote proteins have an odd number of EF-hands. For example, the protease calpain has five EF-hands, which are coupled by dimerization. Whereas STIM1 and 2 have canonical EF-hand motifs which pair with a non-Ca^{2+}-binding EF-hand. Some proteins only have one EF-hand, which acts to attach on to other proteins; e.g. NKD1 and 2 dock on to DVL1, 2 and 3.

The question therefore arises: has the EF-hand moved about within or between genomes during evolution, or did it appear independently on more than one occasion? The latter is possible, because it only takes a few base changes in the DNA coding for a 29-amino-acid sequence to generate an EF-hand. Removing one of the oxygen amino acids can still allow Ca^{2+} binding, but with lower affinity. For example, mutation of Asp119 to Ala in the second EF-hand loop of aequorin reduces its affinity for Ca^{2+} some 50-fold, but Ca^{2+} still triggers the chemiluminescent reaction. So how easy is it to make a Ca^{2+}-binding site?

Comparison of sequences of the same protein from species, whose time of divergence from a common ancestor can be estimated from the fossil record, allows the rate of change of amino acid residues over the past few hundred million years to be estimated. For example, the Ca^{2+}-binding protein parvalbumin took 5 million years to change its amino acid sequence by just 1%. This compares with 3 million years for albumin, and 400 million years for histone H_4. Ca^{2+} likes to bind Asp or Glu residues. Most proteins have an Asp + Glu content greater than Lys + Arg. Calmodulin in bovine brain has nearly 50% of its amino acids acidic (Glu = 26.7%, Asp = 22.7%), compared with 13.4% for basic residues (Lys = 7.2%, Arg 6.2%). The two halves of troponin C are 34% similar, as are the four domains of calmodulin, suggesting these appeared by sequence duplication.

Examination of the triplet code shows that single mutations in the first or second position can cause a neutral amino acid (Ala, Gly, Val), or a basic (Lys), to change to Glu or Asp (Figure 12.6). The mutation of a stop codon can also lead to a Glu, and would extend the protein. Redundancy in the third base of the triplet code is explained by Crick's wobble hypothesis, and shows that Asp and Glu can interchange with each other. There are 64 triplets making up

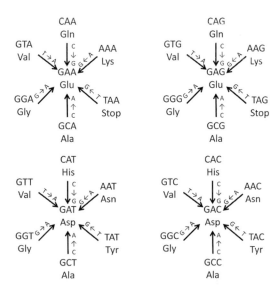

Figure 12.6 Mutation of triplets to form the triplet code for acidic amino acids. Single-base mutations leading to an Asp or Glu in a protein, necessary for the appearance of a Ca^{2+}-binding site. Short arrows represent each mutation. Reproduced by permission of Welston Court Science Centre.

the genetic code. Two, GAT and GAC, code for Asp, and two, GAA and GAG, code for Glu and Asp. A further two, AAT and AAC, code for Asn, and two, CAA and CAG, code for Gln. But a change in just one base of 24 other triplets, two of which are stops codons, will generate an Asp or a Glu (Figure 12.6), making the chance of a single mutation generating an acidic amino acid residue 37.5%, and this is without taking account of mutations to Asn or Gln. Thus, under conditions of high mutation rates, which existed several hundred million years ago when the UV radiation from the Sun, and the emission of radioactive particles from natural rocks, was much higher than it is today, the chance of producing either linear (canonical) or three-dimensional Ca^{2+}-binding sites was high.

Box 12.3 Effects of knocking out some Ca^{2+} signalling genes in mice.

- Calmodulin – lethal.
- Ca^{2+} activated kinases – non-lethal, brain abnormalities.
- Calreticulin – lethal in development.
- Calsequestrin 1 and 2 – non-lethal, death on stress.
- Na^+/Ca^{2+} exchanger NCX – non-lethal.
- α_1 Subunit in most voltage-gated Ca^{2+} channels – non-lethal.
- Ryanodine receptor 1 – non-lethal.
- SERCA1 – null mutants die of respiratory failure.
- SERCA2 – heterozygote KO susceptible to cancer.
- SERCA3 – non-lethal, deafness.

12.5   Darwin and Knock-Outs

Genetic engineering has given us the remarkable ability to knock-out individual genes to provide insights into what they do, and how they are involved in development, function and survival. Important model systems for using knock-outs have been various bacteria, the nematode *Caenorhabditis elegans*, the fruit fly *Drosophila melanogaster*, yeast, the mouse *Mus*, and sheep. But the surprise has been that, when individual genes have been knocked-out, only a few of these were lethal. In many cases, loss of major genes only has subtle effects on development and behaviour (Box 12.3). In the case of Ca^{2+} signalling, the complete loss of calmodulin is lethal. Knocking-out all calmodulin genes in vertebrates kills the embryo, and kills yeast, the mould *Aspergillus*, and the fruit fly *Drosophila*. This explains why calmodulin has been so highly conserved throughout eukaryotes – animals, plants and microbes. It is essential for life, but redundancy of calmodulin genes gives protection against one being damaged or destroyed – the other genes compensate. However, knocking-out individual Ca^{2+}-activated protein kinases is not lethal, but can cause abnormalities in the brain. In addition, the loss of other major proteins involved in Ca^{2+} signalling, including Ca^{2+} channels, Ca^{2+} pumps, Ca^{2+} storage proteins and Ca^{2+} targets other than calmodulin, does not kill the organism, or prevent an embryo surviving to birth. Furthermore, there are several inherited human diseases, where proteins involved in Ca^{2+} signalling have been mutated or lost. Again, these only produce symptoms in specific organs, and not throughout the whole body. For example, in autosomal, dominant, inherited Darier's disease, mutations to the major SR Ca^{2+} pump in heart muscle, SERCA2, loses its function. Yet the only symptoms are skin lesions, and a few psychological problems. Knocking-out SERCA2 causes structural changes in the heart T-tubule, but again is not lethal, Ca^{2+} handling being compensated for by other Ca^{2+} pumps and channels.

Most of the major Ca^{2+} signalling proteins have been knocked-out in mice, *Caenorhabditis elegans*, *Drosophila*, *Neurospora*, and sheep without lethal effects. Amazingly, the only lethal knock-outs have been those removing calmodulin and, in development, calreticulin. All the others, including complete loss of calsequestrin in the SR, Ca^{2+} channels L, T, P, R and N, IP_3 receptors 1, 2 and 3, ryanodine receptors 1, 2 and 3, calmodulin-activated kinases, calpains, annexins, and many others associated with Ca^{2+} signalling are not lethal. Interestingly, the lethal effects of calreticulin can be compensated for by overexpression of the Ca^{2+}-activated phosphatase calcineurin. Non-lethal effects are confirmed by several inherited diseases, where nonsense or other mutations lead to complete loss or major damage of a Ca^{2+} signalling protein. These can lead to death under certain circumstances, but are not lethal *per se*. The mechanisms responsible for compensation are poorly understood.

Similarly, knocking-out plasma membrane receptors, which normally lead to cytosolic free Ca^{2+} signals, is not lethal, but may disturb Ca^{2+} handling by the cells. For example, purinergic receptors are used by macrophages in the immune response. Knocking-out both receptors P2X and Y only affects desensitisation mechanisms in response to UTP, and causes loss of Ca^{2+} signals, which are restored by activating the cells with lipopolysaccharide and interferon-γ. The situation is similar for Ca^{2+} pump P-ATPases – PMCAs, NCXs, SERCAs, SPCAs and PMRs. For example, SPCAs in the Golgi of animal cells are involved in the secretory pathway, similar to the PMRs in yeast, required to get Ca^{2+} into secretory vesicles. Knocking-out individual genes coding for any of these Ca^{2+} pump proteins is not lethal, but can exhibit defects in particular organs. Defects in the *ATP2A1* gene that codes for SERCA1 cause Brody's myopathy that leads to problems relaxing muscle, and stiffness in muscles after exercise, and loss of PMCA1 or SERCA2 can give rise to squamous cancers, and changes in exocytosis normally triggered by a rise in cytosolic free Ca^{2+}. Mice deficient in SERCA3 exhibit deafness, whereas mice with no PMCA2 have problems with balance and those with no PMCA4 show male infertility. Loss of one copy of the SPCA1 in humans causes Hailey–Hailey disease, characterised by skin lesions, whereas loss of its related gene, PMR1, in yeast causes defects in protein glycosylation and protein trafficking in the secretion. In addition, amazingly, mice with the NCX knocked-out survive to adulthood, in spite of the major role this Ca^{2+} pump plays in removing Ca^{2+} that has entered a heart myocyte after the beat.

Calsequestrin is the major Ca^{2+}-binding protein in muscle, where it is found concentrated in the terminal cisternae, so that it can release Ca^{2+} close to troponin C. There are two isoforms: CASQ1, only found in skeletal muscle, and CASQ2, the isoform in adult heart muscle, also forming a quarter of calsequestrin in slow skeletal muscle fibres. Amazingly, down regulation of CASQ2 does not damage Ca^{2+} in the SR or heart contractions. Knocking-out calsequestrin reduces SR Ca^{2+} by less than 15%, casting doubt on its role, entrenched in the literature for decades, as a major a Ca^{2+} store. Knocking-out CASQ1 again is not immediately lethal, though the mice are more susceptible to spontaneous death or stress from high temperature induced by anaesthetics such as halothane. This is similar to malignant hyperthermia in humans, caused by mutations in the ryanodine receptor in skeletal muscle. In contrast, mutations of CASQ2 cause heart arrhythmias, and can cause sudden death, particularly in young people. But again, only under stress are the mutations or knock-outs lethal. Similarly, loss of the SR protein triadin, which binds calsequestrin, can cause sudden death by heart attack in humans. Interestingly, there is a very large loss of triadin and junctin in mice in which CASQ2 has been knocked-out, not caused by transcription. Similarly, calsequestrin and junctin are lost in triadin knock-outs.

Calreticulin is regarded as the major Ca^{2+}-binding protein in the ER of smooth muscle and non-muscle cells. Yet mice embryos with no calreticulin survive initially, but die before birth. But, interestingly, this does not appear to be due to major problems with Ca^{2+} signalling in non-muscle cells. Rather, it is due to calreticulin being required for the development of the heart *in utero*. The heart is one of the first organs to develop in an embryo, as it is essential for further survival. In embryonic cells, calreticulin knock-outs cannot use ER Ca^{2+} stores, and thus cannot activate the serine/threonine Ca^{2+}-activated phosphatase calcineurin required for activation of gene expression through NFATc. Cleverly, by making a double transgenic mouse, with calreticulin knocked-out and calcineurin overexpressed in the heart, viable mature mice were produced.

Knock-outs of the two major classes of voltage-gated Ca^{2+} channel – LVA (T type) and HVA (L, N, P/Q and R types) – have also been generated in mice. Most α_1 knock-out mice survive the embryo and in adulthood. Knock-outs of $Ca_v1.2$ survive for two weeks, when the heart still beats properly, but few animals survive after this. However, by using compensation mechanisms, $Ca_v1.2-/-$ can survive to adulthood, there being a large compensation from $Ca_v1.3$, though they have altered receptors and functions. Mice with the $Ca_v1.2$ gene knocked-out have heart arrhythmias and bradycardia, and are deaf, due to this channel being lost in the ear. But they have no major defects in glucose metabolism or pancreas function. Mice that have lost other Ca^{2+} channels, such as TRPs, have also been produced. These also survive into adults, but can have defects that mimic some human diseases.

The three types of the two key SR/ER receptors, ryanodine and IP_3, which release Ca^{2+} into the cytosol, have also been knocked-out one by one, without being lethal. For example, IP_3 receptor 2 is the main form in atrial myocytes. Loss causes loss of positive inotropic effects, and produces arrhythmias. Loss of IP_3 receptor 3 in lymphocytes has even less of an effect on Ca^{2+} signalling. In contrast, knocking-out both IP_3 receptors 1 and 2 was lethal in the embryo. As with the calreticulin knock-out, death appeared to be due to loss of calcineurin/NFATc signalling in the heart. Knocking-out RyR1 is also non-lethal, and provides evidence that the β subunit of the L-type dihydropyridine receptor interacts with the ryanodine receptor in skeletal muscle, with effects on Ca^{2+} currents measured by patch clamp.

All of this shows that there are feedback mechanisms, where interacting proteins affect each other's level of final expression, and not only through mRNA levels. Changes in expression of compensatory proteins occurs as a result of the initial knock-out. This change in other proteins can be via transcription, translation, or post-transcriptional modifications, particularly protein stability. Compensatory mechanisms have been poorly investigated, but clearly have a selective advantage. There has to be a signalling system, yet to be revealed, which tells the cell to do this. All of this would not be a surprise to Darwin. When a gene is lost or defective, mechanisms come into play that compensate.

12.6   Summary

Calcium, inside and outside cells, has been a key element in evolution since life began some 4000 million years ago. The fossil record as a whole shows that regulation of intracellular Ca^{2+} was essential in all cells, from the earliest 3800 million years ago to the present. The white cliffs of Dover are just one example. These are made of soft white chalk, mainly from the calcified coccoliths formed by coccolithophores, which are single-celled marine algae, still of major importance in the oceans today. They are 70–100 million years old, being formed in the late Cretaceous period. Calcified fossils embedded in them include mollusc bivalves and sea urchins. Without an intracellular Ca^{2+} signalling system Natural Selection cannot act.

The environment in which living cells found themselves has changed enormously over 4000 million years. Changes include the ionic environment, atmospheric composition, humidity, temperature, and exposure to mutagenic UV rays and radioactive particles. Even the length of the day has changed. Some 400 million years ago there were over 400 days in a year, instead of the 365 today. But there was one key constant over all this time. For 3800 million years it has been essential for all cells, whether they were animal, plant or microbe, whether they were Bacteria, Archaea or Eukaryota, to maintain a free Ca^{2+} in the soluble part of the cell, the cytosol, very low, in the micro-molar to sub-micromolar range. Without this low intracellular free Ca^{2+}, damaging inorganic and organic Ca^{2+} precipitates would form, and the essential biochemical processes of energy metabolism, DNA, RNA, protein, lipid and carbohydrate synthesis and degradation could not take place. This low intracellular free Ca^{2+} was likely to be established first by a Donnan potential, before protein Ca^{2+} pumps appeared. Once a 'Ca^{2+} pressure' had been established, the scene was set for Ca^{2+} signalling proteins to appear:

1. Ca^{2+} pumps and exchangers, taking over from the Donnan potential, allowing a membrane potential, negative inside, to develop based on K^+.
2. Intracellular Ca^{2+} stores, in the form of Ca^{2+}-binding proteins, and organelles, such as the ER and mitochondria, with pumps to allow the ER and Golgi to accumulate Ca^{2+}. These would allow the cell to defend itself over short time intervals to an unwanted burst of Ca^{2+} in the cytosol, induced, for example, by damage to the plasma membrane.
3. The Ca^{2+}-dependent chaperone system in the ER and Golgi, the first part of the Ca^{2+} signalling system to have evolved some 1000 million or more years ago.
4. Plasma membrane pumps and exchangers evolving to maintain the cytosolic free Ca^{2+} low, in the presence of high concentrations outside a cell with a negative membrane potential.

These early molecular processes for Ca^{2+} took hundreds of millions of years to develop. It was then possible for processes, capable of being activated by a rise in intracellular Ca^{2+}, to appear and evolve. By the time of Cambrian explosion, some 600 million years ago, the majority of cellular events triggered by a rise in cytosolic free Ca^{2+} were in place:

- Nerve terminal secretion.
- All forms of muscle contraction – heart, skeletal, smooth, catch.
- Vesicular secretion for endo- and exo-cytosis.
- Phagocytosis.
- Invertebrate vision.
- Gamete fusion and egg fertilisation.
- Algal regulation.
- Defence mechanism.
- Cell death by apoptosis.

Some Ca^{2+} processes evolved after this, such as the regulation of calcium phosphate to form teeth, and then bone, around 500 million years ago, to allow the vertebrates to evolve, and have a vision regulated by Ca^{2+} in a different way to invertebrates. But, a key molecular evolutionary change, starting around 600 million years ago, was the development of molecular biodiversity in the Ca^{2+} signalling molecules responsible for triggering each cellular event. This provided a key variation in individuals within the same species, so that Natural Selection could act to select the 'fittest', as the environment, stresses, and competition with other species made it a struggle for existence.

During the billions of years it has taken to produce the complexity and molecular biodiversity of the Ca^{2+} signalling system, many Rubicons have been crossed. In spite of Fred Hoyle's scepticism, there was plenty of time. A new enzyme or protein only needs a few amino acids to generate a new binding site or catalytic centre. Similarly, only a few mutations were required to generate Ca^{2+}-binding sites, such as an EF-hand. Mutation rates would have been far higher billions of years ago compared with today, because the Earth was more radioactive, and there was no ozone layer to protect DNA from mutating UV rays. Furthermore, DNA exchange would have been rife in the first 3000 million years of evolution.

Remarkably, animals in which Ca^{2+} signalling proteins have individually been knocked out survive in the embryo, and can even live to adult-hood. Similarly, inherited diseases, where a Ca^{2+} signalling protein has been lost or has a defective mutation, often show clinical problems restricted to one tissue, or make the individual susceptible to stress.

Darwin and Wallace's BIG idea of evolution by Natural Selection requires variations within and between species, and small change by small change. The evolution of Ca^{2+} signalling is a beautiful example of this at the molecular level. The Ca^{2+} signalling system provides a molecular answer to the question Darwin never fully addressed: What is a species? A mouse cannot mate with an elephant, not because of size, but because the DNA will not mix. The proteins that catalyse reactions in a newly formed zygote, formed from the fusion of a sperm with an egg, have to be compatible with each other. If the proteins arising from the female egg, particularly those activated by Ca^{2+}, occur at a different rate from those coming from the male sperm DNA, then there will chemical chaos. The egg cannot develop and will die. This therefore provides the answer to the old question: Which came first the chicken or the egg? The answer is calcium!

Recommended Reading

*A must read.

Books

*Campbell, A.K. (1994) Rubicon: the fifth dimension of biology. London: Duckworth. Digital events in life over 4000 million years.
*Campbell, A.K. (2015) Intracellular Calcium. Chapter 12, Darwin and 4000 million years of intracellular Ca²⁺. Chichester: John Wiley & Sons Ltd.. The vital role of intracellular Ca²⁺ control in the origin and evolution of life, fully referenced.
*Darwin, C. (1859) On the Origin of Species by Means of Natural Selection, or the Preservation of Favoured Races in the Struggle for Life, 1st edn, 6th Edition 1876. London: John Murray. The two key editions of Darwin's ground breaking work, focussed on the process of evolution and the evidence for the Natural Selection.
*Darwin, E. (1803) The Temple of Nature; or, the Origin of Society. London: J. Johnson. Charles' genius grandfather's wonderful poem of natural history with a clear section on the process of evolution and the origin of life.

Reviews

*Campbell, A.K. (2003). Save those molecules! Molecular biodiversity and life. *J. Appl. Ecol.*, **40**, 193–203. The importance of molecular biodiversity and how this is the key to the origin of a new species.
*Campbell, A.K. (2012) Darwin shines light on bioluminescence. *Luminescence*, **27**, 447–449. How living light provides key evidence for one of the major issues in evolution – the origin of a new enzyme.
Campbell, A.K. & Matthews, S.B. (2015) Darwin diagnosed? *Biol. J. Linnean Soc.*, **116**, 964–984. Comprehensive article on Darwin's 50 year illnes, the evidence he had lactose intolerance, and how this is explained by Ca²⁺ signalling.

Cai, X.J., Wang, X.B., Patel, S. & Capham, D.E. (2015) Insights into the early evolution of animal calcium signaling machinery: A unicellular point of view. *Cell Calcium*, **57**, 166–173. Uses protein comparisons between bacteria and eukaryotes to provide a pathway for the evolutionary origins of the Ca^{2+} signalling system.

Carafoli, E. & Krebs, J. (2016) Why calcium? How calcium became the best communicator. *J. Biol. Chem.*, **291**, 20849–20857. What is special about Ca^{2+} in evolution.

*Case, R.M., Eisner, D., Gurney, A., Jones, O., Muallem, S. & Verkhratsky, A. (2007) Evolution of calcium homeostasis: from birth of the first cell to an omnipresent signalling system. *Cell Calcium*, **42**, 345–350. Important review on how the proteins involved in Ca^{2+} signalling evolved.

Collins, S.R. & Meyer, T. (2012) Evolutionary origins of STIM1 and STIM2 within ancient Ca^{2+} signaling systems. *Trends Cell Biol.*, **21**, 202–211. A key to the origin of store-operated calcium entry (SOCE).

Emery, L., Whelan, S., Hirschi, K.D. & Pittman, J.K. (2012) Protein phylogenetic analysis of Ca^{2+} cation antiporters and insights into their evolution in plants. *Frontiers Plant Sci.*, **3**, Article 1. The role of Ca^{2+} exchangers in plant evolution.

Haynes, L.P., McCue, H.V. & Burgoyne, R.D. (2012) Evolution and functional diversity of the calcium binding proteins (CaBPs). *Front. Mol. Neurosci.*, **5**, article UNSP 9. The molecular biodiversity of Ca^{2+} binding proteins in evolution.

*Plattner, H. & Verkhratsky, A. (2013) Ca^{2+} signalling early in evolution – all but primitive. *J. Cell Sci.*, **126**, 2141–2150. The importance of intracellular Ca^{2+} early in evolution.

Pohorille, A.S.K. & Wilson, M.A. (2005) The origin and early evolution of membrane channels. *Astrobiology*, **5**, 1–17. How channels originated in membranes.

*Williams, R.J. (2006) The evolution of calcium biochemistry. *Biochim. Biophys. Acta*, **1763**, 1139–1146. Comprehensive review focussed on the unique chemistry of Ca^{2+} and how this was exploited in evolution.

Research Papers

Cai, X. (2007). Molecular evolution and functional divergence of the Ca^{2+} sensor protein in store-operated Ca^{2+} entry: Stromal interaction molecule. *PLoS ONE*, **2**, e609. The evolution of STIM as one of the two components of store-operated calcium entry.

Cai, X. (2007). Molecular evolution and structural analysis of the Ca^{2+} release-activated Ca^{2+} channel subunit, Orai. *J. Mol. Biol.*, **368**, 1284–1291. The evolution of Orai as one of the two components of store-operated calcium entry.

Dainese, M., Quarta, M., Lyfenko, A.D., Paolini, C., Canato, M., Reggiani, C., Dirksen, R.T. & Protasi, F. (2009) Anesthetic- and heat-induced sudden death in calsequestrin-1-knockout mice. *FASEB J.*, **23**, 1710–1720. The lethal effects of knocking-out a key Ca^{2+} binding protein in the SR.

Herberger, A.L. & Loretz, C.A. (2012) Vertebrate extracellular calcium-sensing receptor evolution: selection in relation to life history and habitat. *Comp. Biochem. Physiol. D*, **8**, 86–94. Evolution of the Ca^{2+} receptor on the outer surface of many cells.

Pott, C., Philipson, K.D. & Goldhaber, J.I. (2005) Excitation–contraction coupling in Na^+- Ca^{2+} exchanger knockout mice – reduced transsarcolemmal Ca^{2+} flux. *Circ. Res.*, **97**, 1288–1295. Knocking-out a key remover of Ca^{2+} after a heart beat is not lethal.

Prasad, V., Okunade, G.W., Miller, M.L. & Shull, G.E. (2004) Phenotypes of SERCA and PMCA knockout mice. *Biochem. Biophys. Res. Commun.*, **322**, 1192–1203. What happens to mice in which key plasma membrane and SR/ER Ca^{2+} pumps have been knocked out.

Protasi, F., Paolini, C., Canato, M., Reggiani, C. & Quarta, M. (2011) Lessons from calsequestrin-1 ablation in vivo: much more than a Ca^{2+} buffer after all. *J. Muscle Res. Cell Motil.*, **32**, 257–270. How to rescue from death mice in which a key SR Ca^{2+} binding protein has been knocked out.

Sipione, S., Ewen, C., Shostak, I., Michalak, M. & Bleackley, R.C. (2005) Impaired cytolytic activity in calreticulin-deficient CTLs. *J. Immunol.*, **174**, 3212–3219. Loss of a key Ca²⁺ binding protein in the ER is not lethal.

Uchida, K., Aramaki, M., Nakazawa, M., Yamagishi, C., Makino, S., Fukuda, K., Nakamura, T., Takahashi, T., Mikoshiba, K. & Yamagishi, H. (2010) Gene knock-outs of inositol 1,4,5-trisphosphate receptors types 1 and 2 result in perturbation of cardiogenesis. *PLoS ONE*, **5**, e12500. What happens when IP$_3$ receptors are knocked out.

13

They Think It's All Over

13.1 Calcium and the Beauty of Nature

The sixteenth century artist Albrecht Dürer (1471–1528) was one of great figures of the Renaissance in northern Europe. Based in Nuremberg, Germany, not only was he a pioneering painter and a developer of woodcuts, he was also a philosopher and mathematician. He wrote:

> *Be guided by Nature and do not depart from it thinking you can do better yourself. You will be misguided, for truly art is hidden in Nature and he who can draw it out possesses it.*

The story of intracellular calcium is a lovely example of this philosophy. Throughout this book I have tried to show how curiosity about Nature has guided hundreds of imaginative scientists to unravel the molecular intricacies of a universal feature of life – how all cells have a very low free Ca^{2+} inside them, and can exploit the calcium pressure generated as a result to trigger a wide range of cellular events. Intracellular Ca^{2+} plays a central role in:

- Electrical excitation.
- Nerves communicating to other nerves and muscles.
- Muscle contraction and cell movement.
- Secretion by exo- and endo-cytosis.
- The uptake of substances into cells.
- Cell division and differentiation.
- Vision.
- Bioluminescence.
- Photosynthesis, and other processes in plants.
- Defence against stress and attack.
- Cell death.

Fundamentals of Intracellular Calcium, First Edition. Anthony K. Campbell.
© 2018 John Wiley & Sons Ltd. Published 2018 by John Wiley & Sons Ltd.
Companion Website: http://www.wiley.com/go/campbell/calcium

Intracellular Ca^{2+} also plays a role in pathological processes and disease, and is thus a target for many pharmaceuticals. Throughout this book, there have been two central themes concerning what is special about intracellular calcium:

1. Intracellular Ca^{2+} acts as a chemical switch to trigger cell events, and is not itself the energy source. Ca^{2+} is a digital mechanism, but there are analogue processes superimposed on the digital ones.
2. Darwinian molecular variation is the key to understanding how organisms work, and evolved. The evolutionary success of intracellular Ca^{2+} throughout Nature has been a key driving force for Natural Selection.

Charles Darwin did write scientific papers, but most of his important ideas and evidence for Natural Selection are to be found in his books, of which there are over 30, with six editions of *On the Origin of Species*. Therefore, I felt it appropriate to incorporate his BIG idea in this book. In this concluding chapter, I will summarise what we know and what we don't about intracellular Ca^{2+}. In addition, I want to show how intracellular calcium is a lovely example of how curiosity can quite unexpectedly lead to major discoveries and inventions, and even create billion dollar markets. I will end with one of my passions – exciting young people about science, and how crucial engaging with the public is, with Darwin and Wallace as role models through their many books.

13.2 What We Know About Intracellular Ca^{2+}

13.2.1 Principles

A rise in intracellular free Ca^{2+} is the chemical switch that triggers a huge range of cellular events in animal, plant, fungal, protist and bacterial cells (Figure 13.1). We are conceived on a

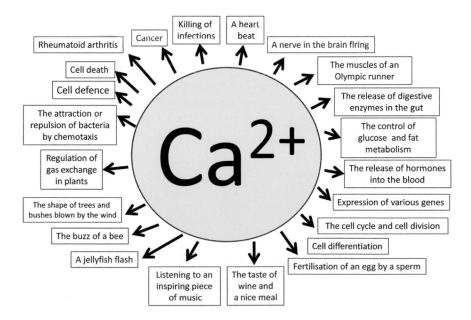

Figure 13.1 Diversity of cellular processes triggered by a rise in intracellular Ca^{2+}. Reproduced by permission of Welston Court Science Centre.

wave of intracellular Ca^{2+}, as our father's sperm gets ready to inject its DNA into our mother's egg. This is followed by a Ca^{2+} wave that moves down the fertilised egg to trigger its first cell division. We are then born on a wave of intracellular Ca^{2+}, as our mother's uterine muscle contracts, and thrusts us out into the world. Throughout our lives we live on waves of intracellular Ca^{2+}, which cause our heart to beat, our nerves to fire, enable us to see, run and move, and allow us to digest and metabolise our food. We then die on a wave of intracellular Ca^{2+}, as Ca^{2+} floods into our dying cells and finishes them off by clogging up proteins and nucleic acids, and wrecking intracellular organelles, such as the mitochondria. This is true in all animals, plants, fungi and many microbes. Quite simply, without intracellular Ca^{2+} signalling life would not exist.

There are six universal truths about intracellular Ca^{2+} (Box 13.1).

Box 13.1 The universal truths about intracellular Ca^{2+}.

- All animal, plant, fungal, protist and microbial cells from the three domains, Eukaryota, Bacteria, Archaea, have a free Ca^{2+} in the cytosol in the micromolar to sub-micromolar range.
- There is a large Ca^{2+} pressure across the outer membrane of all cells.
- Primary stimuli, physical, chemical or biological, cause a large fractional rise in cytosolic free Ca^{2+} up to several micromolar. The nature of the Ca^{2+} signal and its intracellular target determine the magnitude and timing of the cell response.
- The rise in intracellular Ca^{2+} triggers a wide range of cell events through high affinity Ca^{2+} targets inside the cell.
- The Ca^{2+} signalling toolbox is involved in many infectious, inherited and other diseases, and is thus a prime target for drugs used clinically.
- There is a wide molecular biodiversity in the Ca^{2+} signalling system, the Ca^{2+} signals, proteins that control these, and the Ca^{2+} targets.

Molecular diversity occurs through different genes, alternative splicing of the same gene, covalent modifications, and the level of the protein through regulation of transcription, translation and degradation. This is a Darwinian molecular variation that occurs in the same cell type within an organ, in the same cell type between organisms of the same or different species and in different cell types. This molecular biodiversity is the key to understanding how Natural Selection works at the molecular level.

The huge gradient of free Ca^{2+}, some10 000-fold, across the outer membrane of cells is maintained in one of three ways:

- Ca^{2+} pumps driven by the MgATP/MgADP + phosphate reaction.
- The gradient of Na^+ through a Ca^{2+}/Na^+ exchanger.
- By H^+ through a Ca^{2+}/H^+ exchanger.

The first of these dominates in most cells. The Ca^{2+} pressure across the plasma membrane is used by physiological stimuli to raise the concentration of cytosolic free Ca^{2+} up to between 1 and 20 μM. This Ca^{2+} comes from outside and release from intracellular stores (Figure 13.2).

The Ca^{2+} acts as a signal to switch on the cell. A nerve terminal will then fire and secrete its transmitter. A muscle will contract, and a heart will beat. An endocrine or exocrine cell will

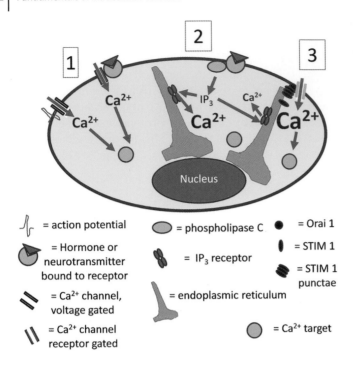

Figure 13.2 Three established scenarios for a rise in cytosolic free Ca^{2+}. The figure shows the main source of the Ca^{2+} rise from outside the cell (1 and 3), or from release of an intracellular store (2).

secrete, and an egg will be fertilised by a sperm. A luminous cell will flash. An eye will perceive light. And a plant will photosynthesise from the Sun's rays. A cell will divide, defend itself or die. The rise in intracellular Ca^{2+} also activates key enzymes involved in intermediary metabolism, in order to increase ATP synthesis, necessary for the cell response. If the level of Ca^{2+} remains too high, the cell will die by necrosis, autophagy or apoptosis. However, Ca^{2+} can also enable a cell to protect itself from stress or damage. The rise in intracellular Ca^{2+} provoked by a primary stimulus is therefore a digital signal, telling the cell to cross the Rubicon and fire. On the other hand, secondary regulators act to modify the effects of the primary signal. They may alter the level of stimulus required to make the cell fire, or they may act in an analogue manner to alter the strength of the cell's response.

Ca^{2+} is also stored within intracellular organelles:

- Endoplasmic reticulum (ER), or sarcoplasmic reticulum in muscle cells.
- Mitochondria.
- Golgi.
- Vesicles – secretory or calcisomes.
- Lysosomes.
- Chloroplasts and the vacuole in plant cells.

Ca^{2+} can be released and taken up from any one of these, to regulate cytosolic free Ca^{2+}, or regulate proteins and structures within the organelle. The major Ca^{2+} store in eukaryotic cells is the endoplasmic reticulum (ER), called sarcoplasmic reticulum in muscle cells. However, plant cells also store substantial amounts of Ca^{2+} in the vacuole, often bound to oxalate. Ca^{2+} is taken up by the ER through a Ca^{2+}-activated MgATPase pump (SERCA), and released through two types of receptor: inositol trisphosphate (IP_3) or ryanodine, the latter being activated by Ca^{2+} itself. There is an intimate relationship between the mitochondria and the ER, involving Ca^{2+} uptake and release between each other. When a cell is activated, Ca^{2+} entry is provoked by opening Ca^{2+} channels in the plasma membrane. These Ca^{2+} channels can be opened in one of four ways:

- Depolarisation of the membrane potential, opening voltage-gated Ca^{2+} channels.
- An agonist binding to cell surface receptor, leading to the opening of plasma membrane Ca^{2+} channels, and release of Ca^{2+} from intracellular stores, particularly the ER.
- Intracellular signals, such as cyclic nucleotides, nitric oxide or oxygen metabolites.
- Store-operated Ca^{2+} entry (SOCE), provoked by loss of Ca^{2+} from the ER.

The SOCE mechanism is wide spread in animal cells. It uses two key proteins that lock into each other to open the Ca^{2+} entry channel:

- STIM in the ER.
- Orai in the plasma membrane.

Ca^{2+} channels are usually inactivated by Ca^{2+}, and can be modulated by phosphorylation and Ca^{2+}-calmodulin. The type of Ca^{2+} signal varies considerably with cell type. It can be a simple transient lasting just 1 s or so, as in a muscle twitch or a heart beat. On the other hand, the Ca^{2+} signal may oscillate, and may change its location within the cell, forming a wave or tide. Small local Ca^{2+} changes can also occur, such as puffs and sparks. The latter may summate, leading to a global Ca^{2+} signal. Ca^{2+} signals can also be transmitted from cell to cell via gap junctions, the action of paracrines and juxtacrines, electrically by action potentials, or through electrotonic spread. Other intracellular signals can cause Ca^{2+} transients or modify them. These include cyclic AMP, cyclic GMP, cyclic ADP ribose, NAADP and sphingosine 1-phosphate.

A variety of Ca^{2+} targets have been identified inside cells. The most important are Ca^{2+}-binding proteins, such as troponin C in muscle and calmodulin. These activate specific proteins and structures that lead to the cell's end-response. These Ca^{2+} targets have Ca^{2+}-binding sites with high affinity, in the micromolar range in the presence of millimolar Mg^{2+}. Two major types of high-affinity Ca^{2+}-binding sites are:

- The EF-hand.
- The C2 domain.

The full EF-hand domain contains a 29-amino-acid loop with 12 amino acids with acidic and oxygen ligands, giving Ca^{2+} the necessary seven or eight coordination for high-affinity binding. Binding of Ca^{2+} exposes a domain enabling the protein to interact with another protein, altering its activity. The major binding site for calmodulin is the IQ domain. Other lower affinity Ca^{2+}-binding sites are formed in proteins by β-sheets in proteins, and clusters of acidic amino acids, e.g. in the SR/ER Ca^{2+} store proteins calsequestrin and calreticulin.

13.2.2 The Pathway of Discovery

The story of intracellular Ca^{2+} began with the experiments of Sydney Ringer at the end of the nineteenth century, when he showed dramatic effects of removing extracellular Ca^{2+} on the physiology and behaviour of the heart and other tissues. But it was not until the 1930s that calcium pioneers such as Lewis Victor Heilbrunn realised that the unique property of Ca^{2+} was to act as a switch *inside* cells, rather than being involved *outside* as Ringer had assumed. Heilbrunn's key experiment was to inject small amounts of ionic solutions into cells such as muscle fibres and eggs. Only calcium was able to trigger a muscle to contract, or an egg to develop and divide. A vital concept was that Ca^{2+} acted as a chemical switch inside cells, and was not the energy source for any of the cell events. This came from the reaction of MgATP= MgADP + phosphate being held far from equilibrium, on the side of MgATP. The application of biochemical techniques on cell homogenates enabled Setsuro Ebashi to discover the sarcoplasmic reticulum (SR) as the major releasable Ca^{2+} store in muscle, and the Ca^{2+} target, troponin C, leading to the discovery of how intracellular Ca^{2+} is released and acts in other cells. A breakthrough occurred when methods for measuring free Ca^{2+} inside living cells were developed, first by Ashley and Ridgway using the Ca^{2+}-activated photoprotein aequorin, extracted from the jellyfish *Aequorea* originally by Osamu Shimomura, and then the inspired development of fluorescent Ca^{2+} indicators by the late Roger Tsien. The key experiment is to measure, and image, free Ca^{2+}, first in the cytosol, and then within specific organelles of live cells. The Ca^{2+} signal can then be correlated with the physiological or pathological event. If intracellular Ca^{2+} is the signal for a cell event, then it must rise before the event occurs. If the Ca^{2+} signal is prevented (e.g. by using a Ca^{2+} chelator), then this should stop the event. The application of molecular biology and live cell imaging, and the study of individual Ca^{2+} channels by patch clamping, has revealed how Ca^{2+} is regulated in cells and how it works. The screening of genomic databases has revealed how universal the components of the Ca^{2+} signalling system are in animal, plant and single-celled eukaryotic cells. Unravelling the intracellular Ca^{2+} story has relied heavily on clever technology:

- Imaging to locate Ca^{2+} signals in live cells.
- Patch clamping to characterise individual Ca^{2+} channels.
- Biochemical methods to purify and study Ca^{2+}-sensitive proteins and organelles.
- DNA sequencing to give us the precise sequences of the proteins involved.
- Genetic engineering to manipulate components of the Ca^{2+} signalling system.
- X-ray crystallography and nuclear magnetic resonance (NMR) to provide their 3D-atomic structures.
- Absorbing, fluorescent and bioluminescent Ca^{2+} indicators, Ca^{2+} chelators, Ca^{2+} ionophores, and pharmacological agents that activate or block specific components of Ca^{2+} signalling.

Natural history has given us a wide range of model systems, and toxins that enable the role of particular proteins and Ca^{2+} channels in a process to be identified.

The story of intracellular Ca^{2+} has also produced many surprises, well illustrated by the role of ER in Ca^{2+} signalling, and the effect of knocking-out specific Ca^{2+} signalling genes. In eukaryotes, the nucleus is usually regarded as the organelle that controls the cell. In fact, it is the ER that is the key decider about the fate of a cell. It is signals that start within the ER, and then communicate to the cytosol, plasma membrane and nucleus, which decide whether the cell fires, survives attack, divides, or dies. Knocking-out Ca^{2+} signalling proteins may cause some problems, but few are actually lethal. The Ca^{2+} signalling system has therefore highlighted a compensatory mechanism that has yet to be discovered.

Box 13.2 What we know about intracellular Ca^{2+}.

1. The free concentration of Ca^{2+} in all cellular compartments.
2. What primary stimuli and secondary regulators do to this Ca^{2+}.
3. How Ca^{2+} goes up, and then returns to rest.
4. What the intracellular Ca^{2+} targets are.
5. What happens to intracellular Ca^{2+} when a cell is under stress.
6. How intracellular Ca^{2+} is involved in many diseases.
7. How intracellular Ca^{2+} is involved in cell death.
8. How various pharmacological agents used clinically and experimentally work through the Ca^{2+} system.

We know a lot about how intracellular Ca^{2+} works in eukaryotic cells – animal, plant, fungal and microbes (Box 13.2):

1. The concentration of free Ca^{2+} in the cytosol and organelles of living cells.
2. The changes in free Ca^{2+} caused by a stimulus, and modified by a secondary regulator.
3. How organelles release and take up Ca^{2+}.
4. How Ca^{2+} gets into cells, and is pumped out.
5. The Ca^{2+} targets which mediate a Ca^{2+}-activated cell event, their three-dimensional structure, and how Ca^{2+}-binding affects their interaction with the machinery responsible for the cell event.
6. How intracellular Ca^{2+} is involved in defence against stress or attack.
7. How intracellular Ca^{2+} is involved in disease.
8. How intracellular Ca^{2+} is involved in cell death.
9. Intracellular Ca^{2+} as a target for pharmaceuticals, both clinical and experimental.

We also know bacteria and archaeans maintain a very low free Ca^{2+} inside. Although there is evidence for Ca^{2+} targets inside them, the role of Ca^{2+} as an intracellular signal is not as widely established as it is in eukaryotes.

Box 13.3 What we still need to know about intracellular Ca^{2+}.

1. The precise details of how a cell generates a Ca^{2+} wave, tide or oscillation.
2. The precise details of how a Ca^{2+} signal in a eukaryotic cell triggers a cell event.
3. How Ca^{2+} works in real systems – intact tissues and whole organisms.
4. Intracellular Ca^{2+} as a regulator in bacteria and archaeans.
5. How intracellular Ca^{2+} mediates events in plants and fungi.
6. How intracellular Ca^{2+} integrates with other signalling networks in whole organisms.
7. The precise role of intracellular Ca^{2+} in many diseases and pathological processes.
8. How the intracellular Ca^{2+} signalling system evolved and what role it played in key steps in the evolution of the three main cell types: Bacteria, Archaea and Eukaryota.
9. The potential for the Ca^{2+} signalling system as a target for new drugs.

13.3 What We Don't Know About Intracellular Ca^{2+}

Although we know a lot about Ca^{2+} as a universal regulator in eukaryotic cells, and some bacteria, there is still much to learn (Box 13.3):

- How the frequency and amplitude of oscillation in free Ca^{2+}, in the cytosol or organelle, may act as an analogue-to-digital converter that tells the cell when it should fire.
- How intracellular Ca^{2+} interacts with other signals, such as cyclic ADP ribose, NAADP and sphingosine 1-phosphate.
- How Ca^{2+} signals are transmitted from cell to cell, through gap junctions.
- How transcription factor pathways are activated by a rise in cytosolic and/or nuclear Ca^{2+}. The calcineurin–NFAT pathway is understood, but interactions of intracellular Ca^{2+} with other gene activation pathways, such as those involving mitogen-activated protein kinase (MAPK), c-*fos*, *jun*, *src* are not so well defined.
- How a cell doubles its contents required for cell division, particularly the many components of the Ca^{2+} signalling system, and its true role in the cell cycle.

- Whether Ca^{2+} is a universal signal in bacteria and archaeans, and, if so, what are the Ca^{2+} channels in the plasma membrane, and intracellular Ca^{2+} targets.
- How the diet of animals, and access to calcium of other organisms, affects Ca^{2+} stores, and the efficiency of the Ca^{2+} signalling system. The recommended daily amount (RDA) of Ca^{2+} is 1–2 g, depending on gender, age and physiology. But no one has investigated the effect of this on Ca^{2+} signalling, or other dietary components such as vitamins.
- How the Ca^{2+} signalling system evolved.

The story in bacteria and archaeans has been confused by the identification of putative Ca^{2+} pumps and other Ca^{2+} signalling components from database searches, without showing that these can really regulate free Ca^{2+} in live cells. A major challenge is to apply the discoveries about intracellular Ca^{2+} in single cell model systems and tissue culture to real organs, whole organisms, and natural physiological processes. For example, the precise role of intracellular Ca^{2+} in activities of the brain, such as memory and sleep, needs to be understood. We also need to understand better the role of intracellular Ca^{2+} in pathology, such as cancer, arthritis, diabetes, Alzheimer's and Parkinson's diseases. A major problem is how the Ca^{2+} signalling system started, and how it evolved. A key question is: what is the selective advantage of the molecular variations in the Ca^{2+} signalling system, produced through:

- Similar proteins, coded for by different genes, with subtly different biochemical and electrical properties.
- Alternative splicing of the same gene, producing proteins with subtly different biochemical and electrical properties.
- Different levels of expression and degradation in individual cells.
- Different types, and levels, of covalent modification.

Intracellular Ca^{2+} has been a key biochemical feature of how life has changed over the 4000 million years since it first began. Without control of intracellular Ca^{2+}, life would never have been able to get going! Lowering the free Ca^{2+} had to be a crucial step in allowing primaeval cells to survive and replicate, even before RNA or DNA synthesis could begin in earnest. There are many evolutionary questions:

- Why has Nature chosen the MgATP = MgADP + phosphate reaction, instead of other nucleotides, to drive synthetic reactions and ion pumps, like Ca^{2+}?
- Why has evolution chosen a particular handedness of molecules – right-handed sugars in polysaccharides and left-handed amino acids in proteins? Is there a mirror image of life out there in space?
- Why are there only 20 amino acids in most proteins?
- What is the origin of the genetic code?
- What really is a species, and how does a new species really appear?

Darwin never really addressed this last question. We now know that a mouse cannot mate with an elephant, not because of size, but because the DNA just will not mix. We need a new concept to understand the pathway of Rubicons that ultimately leads to the appearance of a new species, and how intracellular Ca^{2+} fits into this. The timescale of natural events often bears little relation to the time within which we are able to carry out experiments in the laboratory. Charles Darwin worked out how coral reefs form. But just a 1% increase in reef degradation, as a result of climate change, will lead to complete loss of a reef. It can take months or years for a rampant cancer tragically to kill someone. Just a 1% increase in cell division versus cell death will lead to a large cancer within months. There are currently no methods to detect these small percentage effects directly. Ca^{2+} signals, detected by aequorin in the bacterium *Escherichia coli*, cause changes in gene expression and just a 10% decrease in generation time. Such a difference between competing bacteria in the gut would result in 20 000 more of the bacteria with the slightly faster growth rate.

A fascinating question highlighted in this book is: What role has the molecular variation in the Ca^{2+} signalling system played in Natural Selection? Many evolutionary sequences, and even current research, focus on the selective advantage of structures, such as finch beaks, or behaviour, such as sexual display. In fact, Darwin proposed two types of selection:

1. Natural Selection, which selects the best adapted.
2. Sexual selection, which selects the sexiest.

The array of cellular processes controlled by intracellular Ca^{2+} argues strongly that molecular variations in the Ca^{2+} signalling system must have been crucial to the origin and evolution of all species.

13.4 Intracellular Ca^{2+} at School and University

In view of the importance of Ca^{2+} as a universal intracellular regulator, it is surprising how little of it is involved in the science curriculum in schools, in the United Kingdom at least. For biology pupils, Ca^{2+} features in understanding how the heart works, but there is little else. The clever technologies used to study intracellular Ca^{2+} are missing. What is more, the curriculum is full of major errors. Several school textbooks, and web sites, both in Europe and the United States, still insist on teaching that ATP has an energy-rich bond. This is wrong. As I have pointed out several times in this book, MgATP drives reactions and processes because the cell maintains the MgATP = MgADP + phosphate equilibrium well on the side of MgATP. It is the push towards equilibrium that is the driver for endergonic reactions, including Ca^{2+} pumps and protein phosphorylation. This error is compounded by the statement that ATP is the energy source for bioluminescence. It is not. The energy for this amazing phenomenon comes from oxidation. In physics, students have even been told that potassium is the key cation for a heart to beat! Other errors include penicillin only being able to kill Gram-positive bacteria. This is also wrong – genetic engineers use antibiotics to kill the Gram-negative bacterium *E. coli* every day, in order to select those containing the gene of interest. One revision textbook even claimed that Darwin and Wallace wrote the *On the Origin of Species* together! Serious as these errors are, the real nightmare in the school curriculum is that it lacks the inspiration of curiosity. It also lacks understanding of how discoveries and inventions from science and engineering were actually made. Kids sit in front of screens and fiddle with mobile phones endlessly. But few have any idea what is special about the chemistry and physics of silicon that has revolutionised our society, let alone calcium!

We need to start the teaching of biology with natural history – a curiosity about nature. The story of intracellular calcium shows how relevant this can be to the everyday events we all experience. Even in universities this is still vital. Cancerous cells in an artificial pink fluid, studied in the lab, are not real models for natural processes. Even the timescales in the lab bear little resemblance to the real world. New systems and technology are needed to understand how nature really works, how it goes wrong in disease, and how it evolved. We also need to teach the skills of the naturalist, so well illustrated in Darwin's work and writings – sight, sound, smell, touch and thought. These are relevant to any profession.

Teaching at school has been corrupted by the need for exam ratings, and many university teachers have little knowledge about what their new students have been taught at school and how. We need much better links and collaborations between schools, universities, research institutes, and industry. This is why I set up the Darwin Centre in Wales in 1993 (see Section 13.5). It has been a great success and has inspired thousands of pupils, many of whom have chosen science or engineering as a result. I am sure it is no surprise to learn that many have now heard of intracellular Ca^{2+} and bioluminescence! So how could intracellular Ca^{2+} be better incorporated into the curriculum at school and university? There are four features of the Darwin way:

1. Being curious about nature, natural history, and the mechanisms responsible – natural science.
2. Understanding the pathways of discovery, involving key questions and key experiments, requiring the invention and development of new technology.
3. Showing how curiosity can, quite unexpectedly, lead to major breakthroughs in understanding how the natural world works, what goes wrong in disease, and when the environment is stressed by changes such as global warming.
4. Showing how curiosity can, quite unexpectedly, lead to breakthroughs in medical practice, and even create major markets, with great social and economic benefit.

These are embedded in the DISI model I like to use when I teach: Discovery, Invention, Scholarship and Impact. The pathway of discovery about intracellular Ca^{2+}, for over a century, is a beautiful example of this as a teaching model.

13.5 The Inspiration of Intracellular Ca^{2+}

What is special about Olympic or Paralympic runners, that enables them to participate at all, let alone win a gold medal, while the rest of us would be left on the starting blocks? What is special about a succulent steak or a vegetable curry? What makes us feel inspired by a Rachmaninoff piano concerto, a Beatles oldie, a jazz group, or a popular song? What is going on in our brains when we think, and even have a Eureka moment? What makes a nightingale sing, a bee buzz, a jellyfish flash, or a glow-worm glow? What allows some trees to be blown down by the wind, while an oak can survive gales for hundreds of years? And why is there a diabetic epidemic all over the world, with an increasing number of people suffering dementia and Alzheimer's disease, as they get older? Intracellular Ca^{2+} is central to unravelling the molecular mechanisms responsible for all these events. Wherever we go, what ever we sense, be it a view, a sound, a smell, a taste, a touch or a thought, intracellular calcium is crucial, hidden inside the cells of all living things. Without Ca^{2+} we could not do anything. A bee could not buzz, a bird could not sing, a seed could not be pollinated, and we could not even be born. Intracellular Ca^{2+} is an inspiration for the greatest gift evolution has given us – curiosity.

Our civilisation depends on the continuous generation of new knowledge. This is essential for the growth and maintenance of our culture, our economy, and the health of ourselves, and the planet we guard. For centuries, knowledge generated by science, engineering and mathematics has been the engine that drives our economy, has advanced medicine, and enabled us to understand the ecology of our planet. Science has even been essential for the evolution of the arts. Where would writers be without the ability to reproduce their work in large quantities? Where would artists be without the development of the chemistry for paints? Where would composers or musical performers be without the ability to construct instruments, and record performances? Indeed, where would we all be without the silicon chip revolution? There are ten rules that are required to generate new knowledge in science, well illustrated by the intra-cellular calcium story:

1. *Inspiration* – leading to an idea and key questions.
2. *Logical thinking* – giving us an experimental pathway, leading to a key experiment.
3. *Lateral thinking* – the ability to think 'outside the box', leading to an original idea.
4. *Invention* – the design of a novel technology to answer the key question.
5. *Taking risks* – not by walking into the road with your eyes shut, but rather being prepared to travel intellectually into a domain where no one has travelled before, not knowing for sure what you will find, if anything at all!
6. *Having an open mind, with a positive approach* – discoveries are always made on the basis of positive questions and hypotheses.
7. *Perseverance* – never giving up once you have established in your own mind that you are on the right track.
8. *Hard work* – the pathway to new knowledge is inevitably full of intellectual and physical challenges.
9. *Money* – facilities required for new experiments cost money, sometimes millions of pounds, yet with ingenuity, major discoveries and inventions have been made in the past with little resources.
10. *CURIOSITY* – this is the most important driving force of all, and the starting point for generating new knowledge.

Why does a nerve to fire, a muscle contract or a heart beat? How does a sperm fertilise an egg or a pollen grain fertilise a plant ovum? What goes wrong inside cells when they are attacked or stressed? The biochemist traditionally used the 'grind and find' approach, homogenising animal muscle to isolate components of the Ca^{2+} signalling system in these events. But further progress required lateral thinking. It was essential to measure directly the concentration of free Ca^{2+} in a live cell, intact organs, and whole organisms. Once again, curiosity about nature solved the problem. Learning how a jellyfish flashed gave us the first way of measuring free Ca^{2+} in an intact cell, using the bioluminescent protein aequorin. Then came the ingenuity of chemists and genetic engineers, giving us an array of fluorescent and bioluminescent indicators for Ca^{2+}, pH, other ions, membrane potential and gene expression. This required both logical thinking, and lateral thinking outside the box. And there were huge risks involved. After years of work, it might not even work at all. Lateral thinking was also required to realise that Ca^{2+} is not the energy source for events such as muscle contraction, exocrine secretion or cell division. Rather, Ca^{2+} is a switch – it causes the cell to cross the Rubicon, so that a cell event can occur.

When I started to work on bioluminescence over 40 years ago, many of my medical colleagues were somewhat bemused. Surely I had been brought to Cardiff to do medical research and develop new diagnostic techniques? What on earth was I doing combing the beach at night, and going on 'cruises' to collect deep-sea animals that flashed in the dark? They changed their mind when, with colleagues, we developed a way of replacing radioactive labels in immunoassay and DNA/RNA analysis. This technology is now used in several hundred million clinical tests per year, and brought in some £20 million into my university through patent income, grants and the sale of a spin-out. Yet without curiosity about why a glow-worm glows and a jellyfish flashes, and the development of a super-sensitive detector for photons, none of this would have happened. It even led to other discoveries published in six letters to *Nature*, and to discovering the most common chemistry responsible for bioluminescence in the biggest ecosystem on our planet – the ocean. This is my little story about curiosity. I wonder what yours is, and whether you enjoy enthusing others about your passion for science and experiment. This is the touchstone of most scientists. Yet, it is also important politically. We need to communicate to politicians and the public why we do what we do in universities and research institutes. The weird green fluorescent protein (GFP) has revolutionised biomedical research, and created another billion dollar market.

Scientific discovery and invention are critically dependent on experiments. There are three reasons for doing an experiment. First, having identified a phenomenon and having asked some key questions, data is needed to produce a hypothesis, which provides the mechanism to explain the process concerned. Secondly, you have a hypothesis which needs testing. Thirdly, to answer the key questions, a new technique is required. But, how exactly should one test a hypothesis. In the 1950s, the distinguished philosopher, Karl Popper (1902–1994), argued that, since you could never prove absolutely a scientific hypothesis, you could only disprove it. He wrote this argument in his books, the first of which in 1959 was *The Logic of Scientific Discovery*. Thus, to test the hypothesis 'All swans are white', you should look for a black swan to disprove it. This is his 'falsifiability' principle. Yet this not how real science ever works. Watson and Crick did not consider falsifying their hypothesis that DNA was a double helix. They were too busy working out how DNA replicated and what the genetic code might be. In addition, the 'swan' hypothesis is not even science. Rather it is descriptive natural history. Natural science is about **how** the Universe works, from the Big Bang to how a tiny bacterium has become resistant to antibiotics, and how intracellular Ca^{2+} controls all life.

The success of the intracellular calcium story over more than a century is full of hypotheses, successfully tested with a positive approach, by predicting that things should happen which had not yet been looked for. This led to new discoveries. A good example is the pioneering experiment of Lewis Heilbrunn with his student Floyd Wiercinski published in 1947 (see Chapter 3). His hypothesis was that intracellular Ca^{2+} was the trigger for muscle contraction. So Heilbrunn decided to inject a Ca^{2+} solution into a live muscle fibre from a frog. As predicted, Ca^{2+} triggered the muscle to shorten, but not K^+, Na^+ or Mg^{2+}, nor water. These latter solutions were the controls to show that injection alone was not the cause of the contraction. There have been hundreds of other such positive predictions about Ca^{2+} being the trigger for physiological processes. Another positive prediction was that the cytosolic free Ca^{2+} should rise to micromolar levels before a contraction can be detected. This was proved correct when Ashley and Ridgway injected aequorin as a Ca^{2+} indicator into barnacle muscle. This has been followed by thousands of such demonstrations of free Ca^{2+} in live cells using fluorescent or bioluminescent Ca^{2+} indicators. The hunt for the molecule that released Ca^{2+} from the ER was another nice example of the positive approach.

Michael Berridge, working in Cambridge, United Kingdom, was determined to show that IP_3, produced in the cell through hydrolysis of phosphatidyl inositol 1,4-bisphosphate (PIP_2) was the trigger. He worked with his colleague Robin Irvine – an expert in inositol lipids. Irvine managed to make some IP_3 from a huge volume of red blood cells. Berridge then heard a talk from Streb in the United States, that she had developed a way of permeabilising cells to small molecules. So they sent her some of their IP_3 and Eureka – it caused a large rise in free Ca^{2+}! A breakthrough in intracellular calcium had been made.

I know of no significant discoveries or inventions in science that have ever been made using Popper's negative philosophy. Yet, the 'null' hypothesis has grown like a cancer in our school curriculum. Students are taught to contort positive hypotheses into negative ones. This approach needs to be cut out! It is not the way to inspire young minds. It is frankly Popperian poppycock.

A further issue is money. Many scientific techniques require expensive reagents and equipment. A good confocal microscope for imaging Ca^{2+}-sensitive fluors can cost £200 000 (US$320 000). But, with ingenuity and skill, key experiments can still be done quite cheaply. A famous example involving calcium comes from Sweden. In the early years of the nineteenth century, Magnus Martin af Pontin and Jöns Jacob Berzelius in Stockholm used their kitchen as a lab. They made the crucial discovery that some metals can form an amalgam with mercury. Humphry Davy, with few resources at the Royal Institution in London, was able to use this discovery, with a battery he had made, to electrolyse molten caustic potash and isolate for the first time potassium. This led him to discover the alkali metals and alkaline earths, including calcium.

Thus, the ten rules for generating new knowledge, exemplified by the discoveries of intracellular Ca^{2+} as a universal regulator, hone down to four golden rules:

1. Asking the key question.
2. Doing the key experiment.
3. Always being positive.
4. Always being curious.

Curiosity is the key. This is the force that drives us every day to carry on, never give up, and enjoy our few Eureka moments. There is still much to be curious about intracellular calcium.

13.6 Communicating the Story of Intracellular Ca^{2+} to Others

Like artists, many scientists are happiest when they are shut away from the outside, in a lab doing experiments. This sometimes gives the impression that they are detached from the rest of society, and from a media clambering for more stories to hype up their eager public. Intellectual pursuits like science are difficult, and require the deepest engagement with the inner parts of our brains, if we are to come up with original experiments, ideas and inventions. Yet, I believe that it is also vital we explain why we scientists do what we do. The success of the last 100 years in unravelling the details of how intracellular calcium is regulated, and how it works, is lit up by ingenious inventions, leading to major discoveries. Rightly, this science has been published in peer-reviewed international scientific journals. Unfortunately, these are virtually incomprehensible, and often inaccessible, to the public, and, most importantly, to the next generation of young minds. Of course, PubMed, Web of Science, search engines like Google, and amazing resources such as Wikipedia, are open access to anyone on the click of a mouse. The problem is that the scientific language is comprehensible only to experts in a particular field. Scientific papers on intracellular calcium are full of technical terms, acronyms and assumed knowledge, without which it is barely possible to get beyond the first sentence! It is timely, therefore, that we examine how we can excite the next generation about science, technology, engineering and mathematics (STEM), upon which the successful evolution of our civilisation depends. This requires we engage with the public.

The life-blood of science is experiment, but the conclusion of experiments must be examined and repeated by others if the results are to be generally accepted. It is therefore essential that we publish novel findings and inventions in the international literature. This enables us to assess, through peer review and the reaction of other scientists, what has really been achieved. But, the peer-review system has broken down. It is hopelessly overstretched, meaning that reviewers nowadays only have a short time to go through long manuscripts and grant proposals. In addition, the peer-review system is full of clubs and cliques. Submit a paper outside your clique, and you will have difficulty having a new idea or result accepted. In fact, this happened to me when I submitted my first paper on complement. I had predicted that the first thing that will happen to a cell after the membrane attack complex of complement forms in the plasma membrane would be a rise in intracellular Ca^{2+}, and that this would activate Ca^{2+}-dependent processes and a protection mechanism. I tested this hypothesis, with my friend Paul Luzio, by using the Ca^{2+}- activated photoprotein obelin to show a large rise in cytosolic free Ca^{2+} many minutes before intracellular proteins could be detected in the external medium. However, a distinguished reviewer of the first paper submitted for publication said it was not worth publishing because: 'everyone knows complement always kills cells'. I was not in the complement club at the time. But the reviewer was totally wrong. Yes, the so-called model system of aged sheep erythrocytes is extremely sensitive to complement, and these cells lyse easily. But nucleated cells, real cells, do not lyse so easily. This is because they have protection mechanisms, one of which is activated by intracellular Ca^{2+}. This removes the potentially lethal complexes before the cell bursts. Happily, the paper was published, and others followed to support this novel hypothesis. A further flaw in the current peer-review process is that it is usually secret. This allows some reviewers to get away with prejudice and conflicts of interest. Although the peer-review system has its limitation, it is the best we have come up with, and is vital if good and accurate science is to be made available to everyone. This is the worrying thing about the Internet – much information has not been subject to rigorous peer review.

Important as this is, there is another form of peer review, which I believe is vital in a modern democracy. This is the response of the general public and young budding scientists. This is why it is vital we communicate what we have found out or invented to a wider public, and in particular to use this to inspire the next generation of scientists and engineers. I often ask students: how would you tell your grandmother what your PhD thesis is about? If you cannot explain this in one or two sentences, then you have not found out anything! This enables you to think critically about what has been achieved and what impact your work has had.

In 1993, I founded the Darwin Centre for Biology and Medicine (Canolfan Bywydeg a Meddygaeth Darwin), in Wales, as a vehicle to excite young people about science, to arouse their curiosity about nature, and expose them to cutting edge science. In 1999, the Darwin Centre moved to Pembrokeshire (www.darwincentre.com), a beautiful area in West Wales with surprising links to Darwin. Two families, who are descendants of officers on the *Beagle*, and the descendants of Darwin's cousins, still live there. My dream was to develop a philosophy that would give young, budding scientists the intellectual armoury to discover new knowledge, and then to decide how important this was. It has led to what I like to call 'The Darwin Way', the DISI model:

1. *Discovery*: what have you found out?
2. *Invention*: what have you invented as a result of finding it out?
3. *Scholarship*: what analysis have you carried out that gives us a greater understanding of your field?
4. *Impact*: what impact has your discovery, invention and scholarship had on you, and on the development of science as a whole, biology, medicine and health care, the environment, the economy, culture, education, and the public understanding of science?

Intracellular Ca^{2+} exemplifies the DISI model beautifully. Research into intracellular Ca^{2+} has advanced science by having a huge impact on how we understand a wide range of physiological phenomena work. It has led to new diagnostic and therapeutic substances, and procedures, in medicine. It has created several billion dollar markets, with employment for scientists and others in academic and commercial institutions. It is having a big impact on education and public understanding of science, in my country, Wales and the United Kingdom, as a result of the Darwin Centre, and around the World. In spite of this scientific, medical and economic impact, amazingly, intracellular Ca^{2+} is yet to have much impact on how we manage the environment, and features very little in the school curriculum. This needs to change. We need to bring natural history back into the curriculum, using all the Darwinian skills of a naturalist – sight, sound, smell, taste, touch and thought. The willow warbler and chiffchaff are two birds that I love to find in my garden, for they signal spring. They have returned from their winter vigil in Africa. But, these birds are very difficult to tell apart by looking at them. However, if you listen to their song they are easy to distinguish. How does intracellular Ca^{2+} determine these two songs? The skills of the naturalist are relevant to every profession, even bankers and accountants!

Engagement of science with the public has become a major industry in the United Kingdom and other countries. Many towns and cities have hands-on science centres and museums, which have sections on natural history, science and engineering. Governments, research councils, and charities who fund science tell us that we scientists should be engaging more with the public. But, if we take out difficult words, jargon and maths, then aren't we over simplifying it? And then there is the media, who continue to distort science when it is presented, in the need to hype it up. There are several different agendas involved in public engagement. Some want to make science simple, explaining difficult concepts and dealing with hot topic areas. Hot topics include controversial issues such as genetic engineering and genetically modified (GM) foods, stem cells, designer babies, nuclear power, barrages, green energy, global warming, MMR (measles, mumps and rubella) vaccination, badger culls and so on. Others, like me, want the universities and research institutes to reach out and explain why we do what we do. Some simply want to have a vehicle to link universities, and colleges of further education, to schools, to support programmes such as CREST (http://www.britishscienceassociation.org/crest), and to excite young people about STEM. I have founded an international journal for school students to publish their projects and experiences (www.the youngdarwinian.com). We have clear evidence in Wales that the Darwin Centre has enthused thousands of pupils, many choosing science and engineering at university, and as a career. But some people simply want to use public engagement as a public relations and recruitment exercise. And there are many professions where it is essential to explain scientific concepts in terms understandable by the lay public. This is particularly important in medicine, if patients are to take notice of the advice given them by health care professionals. So how would we explain the intricacies of intracellular Ca^{2+} and their importance to school students or to your grandmother?

The first time I attempted this in a public lecture, I explained and demonstrated how intracellular calcium worked, what it did, and how this was discovered. I even had a demonstration showing how bioluminescence and fluorescence worked. But, at the end, one member of the audience asked: 'What exactly do you mean by intracellular?' I had lost this person in the first few seconds! It is not easy to peal away the jargon and technical terms we use quite happily with scientific colleagues every day. Lectures need buzzy titles to attract people to come and listen. My talks on bioluminescence are often entitled 'Life that sparkles', and on lactose intolerance 'When sugar is not so sweet' or 'Monkeys don't keep cattle'. How about 'The soft side of calcium' for a talk on intracellular calcium? Starting with something familiar helps. Everyone knows we need calcium in our diet for healthy bones and teeth, but if you make a bang and get the audience to stand up, you can explain that tiny amounts of Ca^{2+} have just entered their nerves, their heart cells, and have been released inside their muscle cells to react to this fright. Humour always helps to keep people awake, and hands-on demonstrations keep you in touch with the audience. Some suggestions will be available on the companion web site. I always try to use 'a flash is better than a glow', the natural selector, and the beaks of the finch, when I give talks to schools and the public. Making calcium precipitates by blowing into a solution of calcium chloride, and the scum formed by soap with hard water, gets across some key principles about the chemistry of calcium. Then more sophisticated demonstrations can use microscopes, fluorescence and bioluminescence, to get across how we measure calcium inside cell. Get the students to look at Ca^{2+} in teeth, shells, and bone. And then do something. Move, taste, listen, have an idea, look at pond life. All are regulated by intracellular Ca^{2+}. Get them to feel their pulse when at rest and after running – an increase in heart rate being caused by adrenaline increasing Ca^{2+} signals in the heart. Look at the stars, and see Ca^{2+} as the fifth element. Relate to hot topics, disease mechanisms, and commonly used drugs, such as anti-hypertensives. There is much to be curious about!

It is important to emphasise how model systems have been important in discovering how calcium works inside cells. And, of course, explain what a cell is at the start, and how it maintains its calcium pressure. It will be a surprise to learn that electricity in our heads is carried by ions, and not by electrons as it is in a light bulb. It will also be a surprise to learn that this was first discovered by two scientists from Cambridge University using giant nerves from squid collected at Plymouth in the United Kingdom. Alan Hodgkin and Andrew Huxley won the Nobel Prize for this in 1963. In contrast, Paul Nurse and Tim Hunt won the Nobel Prize in 2001 with Leland H. Hartwell for discovering two protein families, essential for cells to divide, vital in understanding cancer. Paul Nurse used yeast genetics and Tim Hunt sea urchin eggs. On the other hand, Thomas H. Morgan (1866–1945) pioneered genetics by studying mutants of those wretched small flies that will congregate in your fruit bowl – the fruit flies *Drosophila*. And Charles Darwin used an array of animals and plants – barnacles, finches, coral, insectivorous and climbing plants, orchids and earthworms – to develop the evidence for his BIG idea of evolution by Natural Selection. Sidney Brenner and John Sulston, Nobel Laureates in 2002 with H. Robert Horvitz, developed the nematode worm *Caenorhabditis elegans* as a model system, leading to the first animal genomic DNA sequence. I myself have recently been developing the water flea *Daphnia* (Figure 13.3) that I found it in the pond in my garden. *Daphnia* has a heart

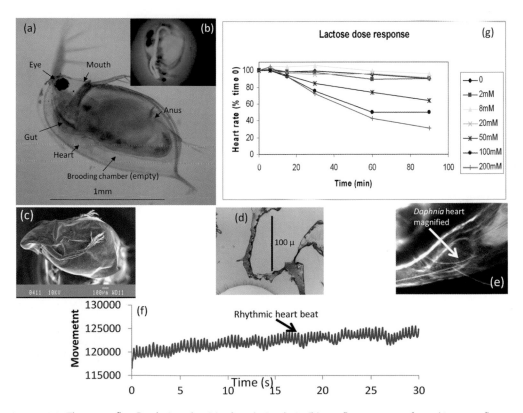

Figure 13.3 The water flea *Daphnia pulex*: (a) a female *Daphnia*; (b) gut fluorescence after taking up a fluor; (c) scanning electron micrograph of a whole *Daphnia* showing the 'coat of mail' which protects it and yet is virtually invisible in water; (d) stained thin section of a *Daphnia* heart; (e) magnified heart beating (f), trace of the heart beat measured using Photek imaging software; (g) effect of lactose to reduce the heart rate and cause arrhythmias, as it does in some people who are lactose intolerant. Reproduced by permission of Welston Court Science Centre.

less than 0.2 mm long, yet it responds to many substances that regulate the human heart and other muscles in the body. I am using it to test our bacteria metabolic toxin hypothesis. The unravelling of intracellular Ca^{2+} as a cell regulator has depended on the development and application of many model systems. Particularly important have been frog heart, muscle from barnacles and crabs, nerves from squid, fruit flies, horseshoe crabs, yeast, tobacco, *E. coli* and mice.

I set up a Public Understanding of Science group (PUSH) at the medical school in Cardiff in 1993. PUS was thought not a good acronym in a medical environment! The PUSH group organise Open Days for Science in Health, now attracting several hundred pupils and teachers for every event. There are tours to experience cutting edge equipment, including imaging systems for seeing intracellular calcium in live cells. There are short talks, a hands-on exhibition and a science theatre – Not the Nine O'Clock Clinic, calcium again featuring in a heart attack. And there are talks about how scientists got to where they are now. The PUSH group involves over 150 members of staff at all levels from several parts of the University. It also runs a monthly series of public lectures and work experience programme for schools.

On the other hand, The Darwin Centre in Pembrokeshire now runs over 150 events each year, involving all the schools there, and the public (Figure 13.4). Pembrokeshire is a popular

Figure 13.4 Public and school events of the Darwin Centre in Wales. The Darwin Centre for Biology and Medicine – bringing cutting edge science and entrepreneurship into everyday life. *Curiosity inspires, discovery reveals*. The figure shows a variety of school events involving the Darwin experience, funded so generously by Dragon LNG, a major importer of liquid natural gas. It also shows a public lecture demonstration given by the author with a glowing flask illustrating Ca^{2+} triggering bioluminescence. The pond dipping and microscope events were at the Welston Court Science Centre which my wife and I set up in Pembrokeshire with my patent income, and supports the Darwin Centre with a lab, library, seminar room and natural history facilities. There is also a science and art event run by my cousin Gwenelin Joy Royston and her husband Bob Royston, showing Darwin finches and other animals Darwin used to develop his BIG idea of evolution by Natural Selection. We also set up a science theatre group, which performed events such as 'Not the Nine O'Clock Clinic'. *Source*: Courtesy of the Darwin Centre.

holiday destination, some visitors having the opportunity to experience the wow factor of a glow-worm hunt. Thousands of school and further education students have been excited by science as a result. We take pupils and teachers out of their classroom into to natural world. The Darwin labs are the rock pools, beaches, cliff tops, river-banks, ponds, moorland, and woods of Pembrokeshire. The model is interesting. The Darwin Centre, the vehicle delivering the public engagement, is a company limited by guarantee and a registered charity. But its success it critically dependent on a partnership with my university at Cardiff, and the Further Education College in Pembrokeshire, as well as being embedded in the local council education programme. Crucially, the funding for staff and a land rover for field work comes from a raft of public and private sources, including the Millennium festival, the European Union, and a major private company, Dragon LNG, who import liquid natural gas and support a community engagement programme. Local and national newspapers, radio and TV give us masses of publicity, showing the success of our events. It is interesting to think that all this energy for public engagement depends on calcium signals in the cells of both the presenters and receivers. And, as you have probably gathered, it really is fun!

13.7 The End of the Beginning

I started this book by pointing out that without intracellular Ca^{2+} we wouldn't enjoy a nice meal with a good bottle of wine, an interesting book, or anything we experience, as intracellular Ca^{2+} is the universal regulator that makes all our organs work. We have learnt an amazing amount about the soft side of Ca^{2+}, the one inside cells, as opposed to the one outside cells in bones, teeth and shells. The fundamental particles of physics and the four main forces – weak and strong nuclear, gravity, and electrical – keep matter together. Yet, this physics has, at present, no direct relevance to biology or to intracellular Ca^{2+}, even though it excites our curiosity. Yet without the laws of physics and chemistry, and the technology arising from them, it would have been impossible to unravel how Ca^{2+} works inside cells. Discovering how Ca^{2+} channels worked depended on intricate electrical equipment, and the laws of electricity. Seeing Ca^{2+} signals in live cells needed clever chemicals to be synthesised, and then observed using cameras originally designed to take pictures of stars thousands of light-years away. Chemistry has given us a way to sequence whole genomes, now in just a few hours. And the silicon chip has given us computers with software that can find Ca^{2+}-binding proteins in genomes, hitherto unknown. We have had decades of developing physics and chemistry *for* biology. A new era will now develop – the physics and chemistry *of* biology.

The story of intracellular Ca^{2+} has been written by thousands of enthusiastic scientists. The Nobel Prize is an inspiration, recognising a supreme achievement in science. But if this is the only thing that matters in science, we might as well all give up. There has been no Nobel Prize for intracellular calcium. Too many have made seminal discoveries, and many are dead anyway. Nowadays, many breakthroughs in science need large teams. The proof that the Higgs boson exists needed many physicists, and large sums of money, to build the Large Hadron Collider. All of this beautiful science has been carried out by hundreds of scientists, engineers and mathematicians. The study of intracellular Ca^{2+} may be an example of the statement made by the late mathematician and philosopher, Bertrand Russell (1872–1970), labelled now through the Internet as one of the world's past great thinkers:

> *In art nothing worth doing can be done without genius; in science even a very moderate capacity can contribute to a supreme achievement.*

This is, I think, a little unkind. Many studies have shown that scientists are amongst the most intelligent people on our planet. I would therefore insist that we replace the words ' very moderate' with 'collaborative' in Russell's statement.

I have had three intellectual passions since I was a small boy – a love of Nature, Natural History, an insatiable curiosity about how it works, Natural Science, and music. I hope that the story of intracellular Ca^{2+} is an example of the music of science. I am now developing software to take intracellular Ca^{2+} into music (Figure 13.5). Curiosity is one of the greatest gifts evolution has given us. Being curious about Nature is the key to science. But curiosity in music, art, history and literature is also inspiring. We need to build bridges between science and the arts. *Curiosity inspires, discovery reveals.* The philosophy of Pablo Picasso (1881–1973) points the way:

I do not seek, I find.

They think it's all over. It isn't quite yet!

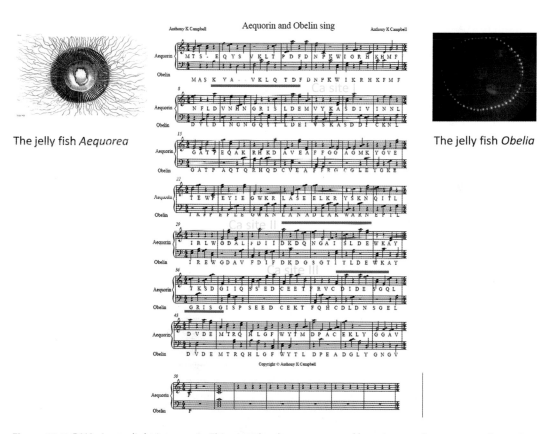

The jelly fish *Aequorea*

The jelly fish *Obelia*

Figure 13.5 DNA sings – light into music. This piece has been composed by using a code to convert the amino acid sequences of aequorin and obelin into notes in scale of C major, typed in through the software Sibelius. The music shows the three predicted Ca^{2+}-binding sites (blue line), heard as high notes when the music is played. The .wav file is available at the companion web site. Drawing of *Aequorea* from Forskål (1776). Score and photograph of Obelia courtesy of Welston Court Science Centre; photographs of Aequorea courtesy of Professor S. Haddock.

Recommended Reading

*A must read.

Books

Campbell, A.K. (1994) Rubicon: the fifth dimension of biology. London: Duckworth. A wide ranging book arguing for the general importance of digital events in biological systems of which intracellular Ca^{2+} is a prime example.
*Campbell, A.K. (2015) Intracellular Calcium. Chapter 13, They think its all over. Chichester: John Wiley & Sons Ltd. Final chapter of the fully referenced book.
Nicholls, D.G. & Ferguson, S.J. (2002) Bioenergetics3. London: Academic Press. Important book on the energy systems in life, with a clear statement of the myth of the ATP energy rich bond, which has confused so many students.

Reviews

*Berridge, M.J. (2012) Calcium signalling remodelling and disease. *Biochem. Soc. Trans.*, **40**, 297–309. Excellent review by a Ca^{2+} pioneer, with ideas about pathology.
*Campbell, A.K. (2012) The production of new knowledge. In: Engwall, L. (Ed.) Scholars in action; past–present–future, the 300th anniversary of the Royal Academy of Sciences. Uppsala: Acta Universitiatis Upsaliensis. Available from the author on request. Article based on a lecture in 2011 with the DISI model for new knowledge.
Carafoli, E. (2013) On beauty and truth in art and science. *Rendiconti Lincei-Scienze Fisiche E Naturali*, **24**, 67–88. The creativity of science and art.
*Russell, B. (1917) Mysticism and logic and other essays. Chapter II, The place of science in a liberal education. Harmondsworth: Penguin. Famous article by a distinguished mathematician and philosopher.
Wang, X., Chen, H.X., Ouyang, Y.Y., Liu, J., Zhao, G., Bao, W. & Yan, M.S. (2014) Dietary calcium intake and mortality risk from cardiovascular disease and all causes: a meta-analysis of prospective cohort studies. *BMC Med.*, **12**, article number: 158. How the dietary intake of Ca^{2+} can affect Ca^{2+} signalling.

Research Papers

Campbell, A.K., Wann, K. & Matthews, S. (2004) Lactose causes heart arrhythmia in the water flea *Daphnia pulex*. *Comp. Biochem. Physiol. B*, **139**, 225–234. Using a water flea as a model system.
Morse, V. & Campbell, A. (2014) Bioluminescence lights up science for schools and the public. *Luminescence*, **29**, 34. Short article on how we have used the wow factor in bioluminescence and Ca^{2+} to excite young people about natural history and science.

Index

Fundamentals of Intracellular Calcium, First Edition. Anthony K. Campbell.
© 2018 John Wiley & Sons Ltd. Published 2018 by John Wiley & Sons Ltd.
Companion Website: http://www.wiley.com/go/campbell/calcium